CONTEMPORARY MATHEMATICS

Titles in this Series

VOLUME 1	**Markov random fields and their applications** Ross Kindermann and J. Laurie Snell
VOLUME 2	**Proceedings of the conference on integration, topology, and geometry in linear spaces** William H. Graves, Editor
VOLUME 3	**The closed graph and P-closed graph properties in general topology** T. R. Hamlett and L. L. Herrington
VOLUME 4	**Problems of elastic stability and vibrations** Vadim Komkov, Editor
VOLUME 5	**Rational constructions of modules for simple Lie algebras** George B. Seligman
VOLUME 6	**Umbral calculus and Hopf algebras** Robert Morris, Editor
VOLUME 7	**Complex contour integral representation of cardinal spline functions** Walter Schempp
VOLUME 8	**Ordered fields and real algebraic geometry** D. W. Dubois and T. Recio, Editors
VOLUME 9	**Papers in algebra, analysis and statistics** R. Lidl, Editor
VOLUME 10	**Operator algebras and K-theory** Ronald G. Douglas and Claude Schochet, Editors
VOLUME 11	**Plane ellipticity and related problems** Robert P. Gilbert, Editor
VOLUME 12	**Symposium on algebraic topology in honor of José Adem** Samuel Gitler, Editor

Titles in this series

VOLUME 13 **Algebraists' homage: Papers in ring theory and related topics** Edited by S. A. Amitsur, D. J. Saltman and G. B. Seligman

VOLUME 14 **Lectures on Nielsen fixed point theory**
Boju Jiang

VOLUME 15 **Advanced analytic number theory. Part I: Ramification theoretic methods**
Carlos J. Moreno

VOLUME 16 **Complex representations of GL(2, K) for finite fields K**
Ilya Piatetski-Shapiro

VOLUME 17 **Nonlinear partial differential equations**
Joel A. Smoller, Editor

VOLUME 18 **Fixed points and nonexpansive mappings**
Robert C. Sine, Editor

VOLUME 19 **Proceedings of the Northwestern homotopy theory conference**
Haynes R. Miller and Stewart B. Priddy, Editors

CONTEMPORARY MATHEMATICS

Volume 19

PROCEEDINGS OF THE

Northwestern Homotopy Theory Conference

Haynes R. Miller and
Stewart B. Priddy, Editors

AMERICAN MATHEMATICAL SOCIETY
Providence • Rhode Island

PROCEEDINGS OF THE CONFERENCE ON HOMOTOPY THEORY

HELD AT NORTHWESTERN UNIVERSITY

MARCH 22–26, 1982

1980 *Mathematics Subject Classifications.* Primary 55-06; Secondary 55P42, 55P45, 55R45, 55T15, 18F25.

Library of Congress Cataloging in Publication Data

Northwestern Homotopy Theory Conference (1982: Northwestern University, Evanston, Ill.)
Proceedings of the Northwestern Homotopy Theory Conference.
(Contemporary mathematics, ISSN 0271-4132; v. 19)
Conference sponsored by Northwestern University, March 22–26, 1982.
Includes bibliographical references.
1. Homotopy theory–Congresses. I. Miller, Haynes R., 1948– . II. Priddy, Stewart, 1940– . III. Northwestern University (Evanston, Ill.) IV. Title. V. Series: Contemporary mathematics (American Mathematical Society); v. 19.
QA612.7.N67 1982 514′.24 83-9941
ISBN 0-8218-5020-2

CONTENTS

Introduction

As part of its emphasis year in topology, Northwestern University sponsored a conference on homotopy theory during the week of March 22-26, 1982. Generous support was provided by the National Science Foundation and the host university. In addition, efforts to bring in several foreign mathematicians were aided by the cooperation and support of the University of Chicago.

The scientific program included 9 plenary lectures, listed below, and numerous seminar talks in parallel sessions. This volume collects papers by participants, corresponding for the most part to lectures given at the conference. All were refereed. It ends with a list of problems presented at a session organized by Bruce Williams.

The editors of this volume would like to express their thanks to the many topologists who served as conscientious referees, to the editors of the Contemporary Mathematics series, and to Karen Walcott, who showed great energy and initiative as secretary both during the conference and afterwards in assembling this book.

Plenary Lectures

John Jones, "The Kervaire Invariant"

Bob MacPherson, "Intersection Homology"

Wu-chung Hsiang, "Some Application of K_{-i}"

Gunnar Carlsson, "Some Remarks on Segal's Burnside Ring Conjecture"

John Morgan "Valuations and the Boundary of Teichmuller Spaces"

Ralph Cohen, "On the Immersion Conjecture"

Ib Madsen, "Odd Order Periodic Homeomorphisms of Spheres"

Bill Dwyer, "Homotopy Methods in Algebraic K-Theory"

Pete Bousfield, "On Stable Homotopy Theory from a Global Viewpoint"

Contemporary Mathematics
Volume 19, 1983

ON WEAKLY ALMOST COMPLEX MANIFOLDS WITH VANISHING DECOMPOSABLE CHERN NUMBERS.

ANDREW BAKER

Abstract:

We describe the subgroup of the complex bordism ring consisting of elements with only the indecomposable Chern monomial giving a non-zero Chern number. In dimensions $4k + 2$ we recover results of Ray, and in dimensions $4k$ we prove a conjecture of Dyer.

In this note we will investigate the subgroup of the complex bordism ring MU_* consisting of classes for which the only non-zero Chern number is that coming from the top dimensional Chern class. In dimensions of form $4k + 2$, we recover results of [Ra]; in dimensions of form $4k$, we prove an old conjecture of E. Dyer [Dy].

Theorem Let $X_n \in MU_{2n}$ be in the subgroup of elements for which all decomposable Chern numbers are zero. Then X_n is a generator if and only if $c_n(X_n)$ takes the value (up to sign)

2, if $n = 1$,

$(2k)!$, if $n = 2k + 1$, $k > 1$,

$d_k(2k - 1)!$, if $n = 2k$, where d_k is the denominator of $B_{2k}/2k$, for B_{2k} the $2k$-th Bernoulli number.

1980 Mathematics Subject Classification. 55R50, 57R77

Our method is to construct certain families of coaction primitives in K_*MU which by the Hattori-Stong Theorem must come from π_*MU. We give a construction for generators in dimensions $4k+2$ (some of which are given in [Ra]), but are unable to do this in the remaining cases.

This work grew out of the author's Ph.D. thesis together with joint work with Nigel Ray and Francis Clarke which will be described in [Ba-Cl-Ra-Sc] - see also [Ba-Ra].

For all notation, basic definitions, etc. see [Ad]. Throughout, E will denote a commutative ring spectrum equipped with a given complex orientation $x^E \in E^2(CP_+^\infty)$ and we assume that E_* is torsion-free. We also have a basis $\beta_0^E = 1, \beta_1^E, \beta_2^E, \cdots$ of $E_*(CP_+^\infty)$ consisting of the duals of powers of x^E. The left coaction (left unit)

$$\psi_E\colon E_*(CP_+^\infty) \longrightarrow (E \wedge E)_*(CP_+^\infty)$$

is described by

$$\psi_E \beta^E(T) = \beta^R(\exp^R(\log^L T)) \tag{1}$$

where $\beta^E(T) = \sum_{0\leqslant j} \beta_j^E T^j$ for a formal indeterminate T, $\exp^E$ and $\log^E$ denote the exponential and logarithmic series for the E-theory formal group law, and R,L denote the images of those under the left and right units.

We can replace T by $\exp^E T$ and get

$$\psi_E \beta^E(\exp^E T) = \beta^R(\exp^R T) \tag{2}$$

- so $\beta^E(\exp^E T)$ is primitive. In fact the coefficients are of form $\frac{1}{n!}(\beta_1^E)^n$. See [Se], [Mi].

Now use the canonical map $CP^\infty \longrightarrow BU$ to embed the above in $E_*(BU_+)$. It is well known that we can usefully identify β_n^E with the elementary symmetric function $\Sigma t_1 \cdots t_n$ - hence

$$\beta^E(T) = \prod_i (1 + t_i T) \tag{3}$$

We also have the Newton functions Σt_1^n which correspond to elements

$s_n^E \in E_{2n}(BU_+)$ (these are of course diagonal primitives). Next we can apply the natural logarithm power series ℓn to give

$$\ell n\, \beta^E(T) = \sum_{1\leq k} \frac{(-1)^{k-1}}{k} s_k^E T^k$$

which on replacing T by $\exp^E T$ becomes

$$\ell n\, \beta^E(\exp^E T) = \sum_{1\leq k}\sum_{1\leq n} \frac{(-1)^{k-1}}{k} \frac{A^E(n,k)}{n!} s_k^E T^n \tag{4}$$

Here we define $A^E(n,k) \in E_{2(n-k)}$ by

$$(\exp^E T)^k = \sum_{k\leq n} \frac{A^E(n,k)}{n!} T^n \tag{5}$$

(actually care needs to be taken if E_* has torsion in which case we use the universality of MU for complex orientations). It turns out that $\frac{A^E(n,k)}{k!} \in E_{2(n-k)}$ (exercise for the reader). For $E = K$, $A^K(n,k) = k!\, S(n,k)u^{n-k}$, where $S(n,k)$ is a Stirling number of the second kind and $K_* = \mathbb{Z}[u,u^{-1}]$ - in this case $\exp^K T = \frac{e^{uT} - 1}{u}$.

Now define $\Sigma_{1}^{E}{}_{n} \in E_{2n}(BU)$ by

$$\Sigma_n^E = \sum_k \frac{(-1)^{k-1}}{k} A^E(n,k) s_k^E \tag{6}$$

From (2) and (4) we have

$$\Psi_E \sum_{1\leq n} \frac{\Sigma_n^E}{n!} T^n = \sum_{1\leq n} \frac{\Sigma_n^R}{n!} T^n \tag{7}$$

- so Σ_n^E is primitive (in fact in the sense of the Hopf algebra structure as well).

We leave the reader to verify that Σ_n^E could also be defined inductively using the Bott homomorphism

$$B_*: E_*(BU) \longrightarrow E_{*+2}(BU).$$

We have

$$\Sigma_{n+1}^E = B_* \Sigma_n^E \tag{8}$$

(9) $\Sigma_n^E = \underline{e}(u^n)$ where $\underline{e}$ denotes the E-theory Hurewicz homomorphism and $u^n \in \pi_{2n}BU = K_{2n}$ is the usual generator. This approach is taken in [Ba] - see [Ad] for details.

We now introduce the E-theory Thom isomorphism

$$\Phi^E: E_*(BU_+) \longrightarrow E_*MU$$

Let $\Phi^E\beta_n^E = b_n^E$, $\Phi^E s_n^E = p_n^E$, $\Phi^E\Sigma_n^E = \overline{\Sigma}_n^E$. Then we can again identify b_n^E with $\Sigma t_1 \cdots t_n$, and p_n^E with Σt_1^n.

The coaction becomes

$$\psi_E b^E(T) = \frac{1}{T}\exp^R(\log^L T) \cdot b^R(\exp^R(\log^L T)) \tag{10}$$

since under the homomorphism induced by the inclusion $MU(1) \longrightarrow \Sigma^2 MU$ we have $\beta_n^E \longmapsto b_{n-1}^E$. Beware - our $b^E(T)$ denotes $\sum_{0\leq j} b_j^E T^j$, not as in [Ad]! Again replace T by $\exp^E T$:

$$\psi_E b^E(\exp^E T) = \left(\frac{\exp^R T}{\exp^L T}\right) \cdot b^R(\exp^R T) \tag{11}$$

Now applying ℓn we obtain

$$\psi_E \sum_{1\leq n} \frac{\overline{\Sigma}_n^E}{n!} T^n = \sum_{1\leq n} \frac{\overline{\Sigma}_n^E}{n!} + \ell n\left(\frac{\exp^R T}{T}\right) - \ell n\left(\frac{\exp^L T}{T}\right) \tag{12}$$

We obtain a primitive series by adding $\ell n \frac{\exp^E T}{T}$ to the $\overline{\Sigma}$ series:

$$\psi_E \sum_{1\leq n} \frac{\overline{\Sigma}_n^E}{n!} T^n + \ell n\left(\frac{\exp^E T}{T}\right) = \sum_{1\leq n} \frac{\overline{\Sigma}_n^R}{n!} T^n + \ell n\left(\frac{\exp^R T}{T}\right). \tag{13}$$

Specialising to $E = MU$, we recall that

$$MU_* \otimes \mathbb{Q} = H_*MU \otimes \mathbb{Q}$$

$$= \mathbb{Q}[b_1^H, b_2^H \cdots]$$

and $$\exp^{MU} T = Tb^H(T).$$ Hence

$$\ell n(\frac{\exp^{MU}T}{T}) = \ell n \prod_i (1 + t_i T).$$

So $\overline{\Sigma}_n^{MU} + (-1)^{n-1}(n-1)!\ p_n^H \in MU_*MU \otimes \mathbb{Q}$ is primitive and we need only determine exactly which multiples are in MU_*MU, hence come from π_*MU via mu. To do this let $E = K$, and note that

$$\frac{\exp^K T}{T} = \frac{e^{uT} - 1}{uT}$$

Differentiating (13) with respect to T tells us that

(14) $$\sum_{1\leq n} \frac{\overline{\Sigma}_n^K}{(n-1))!} T^{n-1} + \frac{1}{T}\left[\frac{1}{2} uT + \sum_{2\leq j} \frac{Bj}{j!} u^j T^j\right]$$

is primitive. So we obtain as indivisible primitives in K_*MU the following (this makes use of the fact that $B_{2k+1} = 0$ if $k > 0$):

(15) $$2\overline{\Sigma}_1^K + u \in K_2MU$$

$$\overline{\Sigma}_{2k+1}^K \in K_{4k+2}MU$$

$$d_k(\overline{\Sigma}_{2k}^K + \frac{B_{2k}}{2k} u^{2k}) \in K_{4k}MU$$

where d_k is as in the statement of the Theorem.

Returning to the case $E = MU$ we now have

(16) $$2\overline{\Sigma}_1^{mu} + 2p_1^H \in MU_2MU$$

$$\overline{\Sigma}_{2k+1}^{mu} + (2k)!\ p_{2k+1}^H \in MU_{4k+1}MU$$

$$d_k\ \overline{\Sigma}_{2k}^{mu} - d_k(2k-1)!\ p_{2k}^H \in MU_{4k}MU$$

as our indivisible primitives. In fact, these must come from elements of π_*MU whose integral homology Hurewicz images are

$$2p_1^H,\ (2k)!\ p_{2k+1}^H,\ - d_k(2k-1)!\ p_{2k}^H.$$

Since $s_n^H = (\Phi^H)^{-1}\ p_n^H$ is dual to c_n^H in the monomial basis for $H^*(BU_+)$, the result about Chern numbers follows.

Hence we have proved the Theorem.

By [Ra], we can construct $(S^{4k+2}) \in MU_{4k+2}$ by suitably choosing a normal U-structure on S^{4k+2} so that

$$\underline{k}(S^2) = 2\bar{\Sigma}^K_1 + u$$

$$\underline{k}(S^{2k+2}) = \bar{\Sigma}^K_{4k+3}$$

$$\underline{k}(S^{8k+2}) = 2\bar{\Sigma}^K_{4k+1}, \quad \text{if} \quad k \geqslant 1.$$

We can realise $\frac{1}{2}(S^{8k+2})$ by taking <u>any</u> $8k+2$ dimensional bounding U-manifold M which is not a Spin manifold and using the existence of a stable factorisation

$$\begin{array}{ccc} M & \xrightarrow{\ pr\ } & S^{8k+2} \\ & \searrow \quad \nearrow_{pr} & \\ & C_\eta & \end{array}$$

where pr denotes projection onto the top cell, and C_η denotes the mapping cone of the Hopf map $\eta \in \pi^S_1$. The details are routine applications of ideas in [Ra-Sw-Ta] and appear in [Ba]. The simplest such manifold is $CP^2 \times S^{8k-2}$.

Unfortunately, we have no general construction in dimension $4k$.

<u>Remarks</u> 1) The above can be used to calculate the e-invariant of an element in the image of the J-homomorphism. The details involve the observation that the mapping cone is a Thom space over a sphere, the fact that

$$\tau_* \bar{\Sigma}^K_n = \frac{B_n}{n}(v^n - u^n) \in K_{2n}K$$

where $\tau\colon MU \to K$ is the canonical Todd orientation, and $v \in K_2K$ is $\eta_R u$. We leave the reader to verify this.

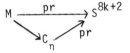

2) R. Stong has pointed out to the author that the above Theorem has appeared in "On the homotopy groups of BPL and PL/O", by G. Brumfiel, Ann. of Math. 88 (1968).

3) The referee has given a variation of the above proof which makes more use of [Mi], and emphasises the role of formal group theory.

Bibliography

[Ad] J. F. Adams, Stable homotopy and generalised homology, University of Chicago Press, Chicago, 1974.

[Ba] A. Baker, Some geometric filtrations on bordism groups, (Ph.D. Thesis) University of Manchester, 1980.

[Ba-Cl-Ra-Sc] A. Baker, F. Clarke, N. Ray, L. Schwartz, "The stable homotopy of BU" in preparation.

[Ba-Ra] A. Baker, N. Ray, "Some infinite families of U-hypersurfaces", to appear in Math. Scand, 50 (1982), 149-66.

[Dy] E. Dyer, "Chern characters of certain complexes", Math. Zeit. 80 (1963) 363-73.

[Mi] H. R. Miller, "Universal Bernoulli numbers and the S^1-transfer", to appear in "Current Trends in Alg. Top." AMS.

[Ra] N. Ray, "Bordism J-homomorphisms", Ill. J. Math. 18, (1974), 290-309.

[Ra-Sw-Ta] N. Ray, R. M. Switzer, L. Taylor, "G-structures, G-bordism, and univeral manifolds", Mem. Amer. Math. Soc., 193 (1977) 1-27.

[Se] D. M. Segal "The cooperation on $MU_*(CP^\infty)$ and $MU_*(HP^\infty)$ and the primitive generators", J. of Pure and App. Alg. 14 (1979) 315-22.

University of Chicago, Chicago, Illinois 60637
June 1982.

Contemporary Mathematics
Volume 19, 1983

THE KERVAIRE INVARIANT PROBLEM

M.G. Barratt*, J.D.S. Jones and M.E. Mahowald*

ABSTRACT. This paper contains an account of a homotopy theoretic approach to the Kervaire invariant problem. Our major object is to explain the relation between the different lines of investigation. We include complete details for one of these lines of investigation - the so-called inductive approach.

§1 INTRODUCTION

In this paper we discuss the results of a homotopy theoretic investigation of the Kervaire invariant problem. This work was begun some time ago by the two senior authors; some of the results were announced in [20]. Indeed, this paper, together with [4], [5] and [6] represent an effort to supply full details for some of the results in [20] and to explore further some of the ideas behind that paper.

It does not seem appropriate here to go into the definition and properties of the Kervaire invariant. Nevertheless we do say something about this invariant, as much as anything to emphasize the geometric roots of the problem. So suppose (M,F) is a 2n-dimensional closed smooth framed manifold, that is F is a stable isomorphism of the stable normal bundle of M with the trivial bundle. We adopt the convention, throughout this paper, that all homology and cohomology groups have mod 2 coefficients unless explicitly stated otherwise.

Then there is a quadratic function, depending on the framing F, $q_F : H^nM \to \mathbf{Z}/2$. Here quadratic means $q_F(x + y) = q_F(x) + q_F(y) + x \cdot y$ where $x \cdot y$ is the mod 2 intersection number of $x, y \in H^nM$. This function was first studied by Kervaire [16], Kervaire and Milnor [17]; it was generalized by Browder [8] and later Brown [9]. An exposition of some of these ideas is given in [13] and [11].

This form q_F has a mod 2 invariant, its Arf invariant, $A(q_F) \in \mathbf{Z}|2$ defined as follows:

1980 Mathematics Subject Classification 55Q45 57R90

* Supported in part by an N.S.F. grant

$$A(q_F) = 1 \Longleftrightarrow |q_F^{-1}(1)| > |q_F^{-1}(0)|$$

and one defines the Kervaire invariant K(M,F) by setting $K(M,F) = A(q_F)$. It turns out that K(M,F) is a framed cobordism invariant of (M,F) and so defines a homomorphism $K:\Omega_{2n}^{fr} \to \mathbb{Z}/2$ where Ω_*^{fr} is the framed cobordism ring. Kervaire and Milnor showed that the following are equivalent:

(i) $K:\Omega_{4k+2}^{fr} \to \mathbb{Z}/2$ is zero.

(ii) The Kervaire exotic sphere Σ^{4k+1} is not diffeomorphic to S^{4k+1}.

(iii) The Kervaire manifold P^{4k+2} has no smooth structure.

(iv) Every framed 4k+2 manifold is framed cobordant to an exotic sphere.

The next major step was taken by Browder. Using the Pontryagin-Thom isomorphism $\Omega_*^{fr} \cong \pi_*^S$, where π_*^S denotes the stable homotopy ring, any problem about framed manifolds is equivalent to a problem in stable homotopy theory, and in [8] Browder proved the following results:

(a) $K:\Omega_{2n}^{fr} \to \mathbb{Z}/2$ is zero if $n \neq 2^k-1$.

(b) If $n = 2^k-1$, then there exists a framed manifold of dimension 2n with non-zero Kervaire invariant if and only if $h_k^2 \in \mathrm{Ext}_A(\mathbb{Z}/2, \mathbb{Z}/2)$ is an infinite cycle in the mod 2 Adams spectral sequence. Here A denotes the mod 2 Steenrod algebra.

In view of Browder's results we concentrate on dimensions of the form $2^{k+1}-2$. One may reformulate the condition in the second of Browder's results more explicitly. Let $\phi_{k,k}$ be the secondary operation defined by the Adem relation

$$Sq^{2^k}\, Sq^{2^k} = \sum_{i=0}^{k-1} Sq^{2^{k}-2^i}\, Sq^{2^i}.$$

The functional $\phi_{k,k}$ operation determines a homomorphism $\Phi_k:\pi^S_{2^{k+1}-2} \to \mathbb{Z}/2$ and Browder shows that using the Pontryagin-Thom isomorphism to identify $\pi^S_{2^{k+1}-2}$ with $\Omega^{fr}_{2^{k+1}-2}$, Φ_k and K may be identified. Then the condition in (b) is simply the condition that Φ_k be non-zero.

We should now introduce some terminology which is both traditional and useful. We write $N = 2^{k+1}-2$ and $K:\pi_N^S \to \mathbb{Z}/2$ for the Kervaire invariant; we usually identify this with Φ_k. Then $\theta_k \in \pi_N^S$ means an element with Kervaire invariant one if such an element exists. One should be careful with this notation; if there is one element of π_N^S with Kervaire invariant one, then precisely half the elements of π_N^S have this property. We will study the following problems.

THE KERVAIRE INVARIANT PROBLEM Does there exist an element $\theta_k \in \pi_N^S$ with $K(\theta_k) = 1$?

THE STRONG KERVAIRE INVARIANT PROBLEM Does there exist an element $\theta_k \in \pi_N^S$ with $2\theta_k = 0$, $K(\theta_k) = 1$?

From the point of view of homotopy theory the second problem seems more natural; certainly the strongest and most useful existence result is that there exists $\theta_k \in \pi_N^S$ with $2\theta_k = 0$, $K(\theta_k) = 1$. Another reason for singling out the strong form of the problem is the following result: There exists an element $\theta_k \in \pi_N^S$ with $2\theta_k = 0$, $K(\theta_k) = 1$ if and only if there is an element $\alpha \in \pi_{2N+1}S^{N+1}$ with $2\alpha = [\iota_{N+1}, \iota_{N+1}]$, see [10]. Here $\iota_n \in \pi_n S^n$ is the class of the identity map and if $x \in \pi_p S^n$, $y \in \pi_q S^n$ then $[x,y] \in \pi_{p+q-1}S^n$ denotes their Whitehead product. This result comes about as follows; if $\alpha \in \pi_{2N+1}S^{N+1}$ is an element such that $2\alpha = [\iota_{N+1}, \iota_{N+1}]$ then, if we continue to write $\alpha \in \pi_N^S$ for the stable element determined by α, $K(\alpha) = 1$, and $2\alpha = 0$. On the other hand if $\theta_k \in \pi_N^S$, $2\theta_k = 0$ and $K(\theta_k) = 1$, then, by the Freudenthal suspension theorem, there exists an element $\alpha \in \pi_{2N+1}S^{N+1}$ such that stably (that is in π_N^S) $\alpha = \theta_k$ and then, for any such α, $2\alpha = [\iota_{N+1}, \iota_{N+1}]$.

The rest of this paper is set out as follows. In §1 we discuss the general approach to the Kervaire invariant problem; all the main results are stated in §2. There are three main points in this approach to the Kervaire invariant problem and in §3 and §4 we give the details for one of these points. The details for the other two will appear in [4] and [5]. The main point, then, of this paper is to give an idea of the general flavour of this work and to explain the relation between various specific points.

§2 THE MAIN RESULTS

There are three major points in this approach to the Kervaire invariant problem.

THE INDUCTIVE APPROACH Here one assumes the existence of $\theta_k \in \pi_N^S$ with $K(\theta_k) = 1$; extra hypotheses on θ_k will give the existence of θ_{k+1}. The cleanest result of this kind is the following well-known theorem.

THEOREM 2.1 *Suppose there is an element* $\theta_k \in \pi_N^S$ *with* $K(\theta_k) = 1$, $2\theta_k = 0$ *and* $\theta_k^2 = 0$. *Then there exists an element* $\theta_{k+1} \in \pi_{2N+2}^S$ *with* $K(\theta_{k+1}) = 1$ *and* $2\theta_{k+1} = 0$.

Actually the method we discuss in §3 for proving this theorem proves rather more.

THEOREM 2.2 *Suppose there exists an element* $\theta_k \in \pi_{2N+1}S^{N+1}$ *such that* $2\theta_k = [\iota_{N+1}, \iota_{N+1}]$. *Then* $\eta\theta_k^2 = 0$ *in* $\pi_{4N+4}S^{2N+3}$ *and*

$$[\iota_{2N+3}, \iota_{2N+3}] \in \langle\theta_k^2, \eta, 2\rangle \subseteq \pi_{4N+5}S^{2N+3}.$$

Here $\eta \in \pi_1^S$ is the class of the Hopf map and we have slipped into the convention of using the same name for a homotopy class and its suspension. So for example $\eta\theta_k^2 \in \pi_{4N+4}S^{2N+3}$ stands for $S^{N+2}(\eta \circ S^N\theta_k \circ \theta_k) = S(\eta \circ (\theta_k \wedge \theta_k))$. Note that the

Toda bracket $\langle\theta_k^2,\eta,2\rangle$ has indeterminacy $2\pi_{4N+5}S^{2N+3}$ and so $[\iota_{2N+3},\iota_{2N+3}] \in 2\pi_{4N+5}S^{2N+3}$ if and only if the Toda bracket $\langle\theta_k^2,\eta,2\rangle$ contains zero. Of course this bracket contains zero if $\theta_k^2 = 0$ and this leads to Theorem 2.1. Clearly everything is tied up in the null-homotopy of $\eta\theta_k^2$ used in forming this bracket. One may extract a stable problem as follows. Write A for the stable complex $S^0 \cup_{\theta_k^2} e^{2N+1}$, then since $\eta\theta_k^2 = 0$ there is a map $f:S^{2N+2} \to A$ such that $p_* f = \eta$, where $p:A \to S^{2N+1}$ is the collapsing map. Refer to such a map as a *null homotopy* of $\eta\theta_k^2$ and call such a map $\phi_{k+1,k+1}$*-carrying* if the functional operation defined using $\phi_{k+1,k+1}$ and f is non-zero.

THEOREM 2.3 *Suppose there exists an element* $\theta_k \in \pi_N^S$ *such that* $K(\theta_k) = 1$, $2\theta_k = 0$. *Then there exists a* $\phi_{k+1,k+1}$ *carrying null homotopy of* $\eta\theta_k^2$; θ_{k+1} *exists if and only if there exists a non-*$\phi_{k+1,k+1}$ *carrying null homotopy of* $\eta\theta_k^2$.

So, the most tangible outcome of the inductive approach is 2.1 but the deep point which emerges is that if there exists an element $\theta_k \in \pi_N^S$ with $2\theta_k = 0$ and $K(\theta_k) = 1$, then to settle whether θ_{k+1} exists or not one really must understand the following question: *why is* $\eta\theta_k^2$ *zero*?

THE UNSTABLE APPROACH This stems from the problem of computing Whitehead products of the form $[\iota_n,\beta]$ where $\beta \in \pi_{j+n}S^n$ is an element in the image of the J-homomorphism. We instantly make the assumption that $j < n-1$ so that $\pi_{j+n}S^n = \pi_j^S$ is the stable group. Next we establish the notation we use for the image of J. Define the numerical function s(n) inductively by the formulas

$$s(0) = 0,\ s(1) = 1,\ s(2) = 3,\ s(3) = 7,\ s(4k+\ell) = 8k + s(\ell).$$

Note that $\pi_{s(k)}SO$ is the k-th non-trivial homotopy group of SO, so write $\beta_k \in \pi_{s(k)}^S$ for the generator of the image of J determined by picking a generator of $\pi_{s(k)}SO$. From now on always assume the stability hypotheses $3s(k) - 2 \le 2n$. A great deal is known about $[\iota_n,\beta_k]$:

$$[\iota_n,\beta_k] = 0 \quad \text{if} \quad \nu_2(n + s(k) + 2) \le k$$

$$[\iota_n,\beta_k] \ne 0 \quad \text{if} \quad \nu_2(n + s(k) + 2) \ge k + 1, \text{ but } n+s(k)+2 \ne 2^{k+1}.$$

Here $\nu_2(a)$ is the exponent of 2 occurring in the prime factorization of the integer a. These results, and many related results are discussed in [19], a complete proof of the first is given in [12],see also [21]. The second is implicit in [19] and [21]; the details are made explicit in [4]. There is an obvious gap in the known results and it should come as no surprise that this is related to the Kervaire invariant problem.

THEOREM 2.4 *Suppose* $n = 2^{k+1}-2-s(k)$ *and let* $\beta \in \pi^S_{s(k)}$ *be any element such that* $e(\beta) = e(\beta_k)$ *where* e *is Adams* e-*invariant* [3]. *Then if* $[\iota_n, \beta] = 0$, *there is an element of* π^S_N *with Kervaire invariant one.*

It is interesting to note that in [19] it is conjectured that, when $n = 2^{k+1}-2-s(k)$, $[\iota_n, \beta_k] = 0$ if and only if $h_k^2 \in \mathrm{Ext}_A(\mathbb{Z}/2, \mathbb{Z}/2)$ is an infinite cycle in the Adams spectral sequence. This conjecture predates Browder's theorem and was one of the reasons for the interest of some homotopy theorists in the question of whether h_k^2 is an infinite cycle before its connection with the Kervaire invariant was known.

We will not say anything about the proof of this theorem, the details are in [4], however we will try to give some idea of why the result might be true. Write $\lambda : \mathbb{R}P^n \to \Omega^{n+1}S^{n+1}$ for the usual map and let $w_n : S^n \to \mathbb{R}P^n$ be the attaching map of the n+1 cell of $\mathbb{R}P^{n+1}$, that is the usual covering map. Then $\lambda_* w_n = [\iota_{n+1}, \iota_{n+1}] \in \pi_n \Omega^{n+1} S^{n+1} = \pi_{2n+1} S^{n+1}$ and so one method of dividing $[\iota_{N+1}, \iota_{N+1}]$ by 2 is to divide w_N by 2. Now according to Adams [2], *stably* w_N fits into the following commutative diagram

$$\begin{array}{ccccc}
S^N & \xrightarrow{w_N} & \mathbb{R}P^N & & \\
 & \searrow^{v_N} & \uparrow & & \\
 & & \mathbb{R}P^{N-s(k+1)} & \xrightarrow{p} & S^{N-s(k+1)}
\end{array}$$

where p is the collapsing map, and $e(p_* v_n) = e(\beta_{k+1})$. Now contemplation of the K-theory of projective space, or e-invariants, or more precisely the J-groups of projective space, as in [21], reveals that if $\alpha \in \pi^S_N \mathbb{R}P^N$ and $2\alpha = w_N$ then, stably, α should fit into the following commutative diagram

$$\begin{array}{ccccc}
S^N & \xrightarrow{\alpha} & \mathbb{R}P^N & & \\
 & \searrow^{\gamma} & \uparrow & & \\
 & & \mathbb{R}P^{N-s(k)} & \longrightarrow & S^{N-s(k)}
\end{array}$$

where $e(p_*\gamma) = e(\beta_k)$. This suggests the following result proved in [4].

THEOREM 2.5 *Suppose* $\alpha \in \pi^S_N \mathbb{R}P^{N-s(k)}$ and $e(p_*\alpha) = e(\beta_k)$. *Then* $\lambda_*(\alpha) \in \pi^S_N$ *has Kervaire invariant one.*

Here $\lambda_* : \pi^S_* \mathbb{R}P^n \to \pi^S_*$ is the homomorphism induced by the stable map $\mathbb{R}P^n \to S^0$ adjoint to the map $\lambda : \mathbb{R}P^n \to \Omega^n S^n \subset \Omega^\infty S^\infty$. Now to deduce Theorem 2.4 we use the following result of [18]. Let $h : \Omega^n S^n \to Q\mathbb{R}P^{n-1}$ be the Snaith map [28], then the following diagram commutes,

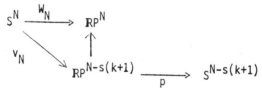

$$\begin{array}{ccccc}
\Omega^n S^n & \longrightarrow & \Omega^{n+1}S^{n+1} & \xrightarrow{H} & \Omega^{n+1}S^{2n+1} \\
{\scriptstyle h}\downarrow & & {\scriptstyle h}\downarrow & & \downarrow \\
Q\mathbb{R}P^{n-1} & \longrightarrow & Q\mathbb{R}P^{n} & \longrightarrow & QS^n
\end{array} .$$

Here the map labeled H is the "James Hopf invariant map" and the unlabeled arrows correspond to the obvious maps. This diagram gives a map of exact sequences

$$\begin{array}{ccccccccc}
\xrightarrow{P_{N-s(k)}} & \pi_{2N-s(k)}S^{N-s(k)} & \xrightarrow{E} & \pi_{2N-s(k)+1}S^{N-s(k)+1} & \xrightarrow{H} & \pi_{2N-s(k)+1}S^{2N-2s(k)+1} & \xrightarrow{P_{N-s(k)}} \\
& {\scriptstyle h_*}\downarrow & & {\scriptstyle h_*}\downarrow & & \| & \\
\longrightarrow & \pi_N^s\mathbb{R}P^{N-s(k)-1} & \xrightarrow[i_*]{} & \pi_N^s\mathbb{R}P^{N-s(k)} & \xrightarrow[p_*]{} & \pi_N^s S^{N-s(k)} & \xrightarrow[\partial]{}
\end{array}$$

where the upper exact sequence is the E.H.P sequence of James and Whitehead; in particular if $\beta \in \pi_{2N-s(k)+1}S^{2N-2s(k)+1}$ then $P_{N-s(k)}(\beta) = [\iota_{N-s(k)},\beta]$. Now suppose there exists an element $\gamma \in \pi_{2N-s(k)+1}S^{N-s(k)+1}$ with $H(\gamma) = \beta$, then from the above diagram of exact sequences, $\partial(\beta) = 0$ and so there exists an element $\alpha = h_*\gamma \in \pi_N^s\mathbb{R}P^{N-s(k)}$ with $p_*\alpha = \beta$. Now suppose that $e(\beta) = e(\beta_k)$, then from 2.5, $K(\lambda_*\alpha) = 1$. Note more; $\lambda_*\alpha = \lambda_*h_*\gamma$ and by the Kahn-Priddy theorem [15], λ_*h_* is the iterated suspension homomorphism, therefore, stably $\gamma \in \pi_N^s$ has Kervaire invariant one.

THE RELATION BETWEEN THE INDUCTIVE AND UNSTABLE APPROACHES

The two approaches are nicely tied together by the following result.

THEOREM 2.6 *Suppose there exists an element* $\theta_{k-1} \in \pi_{2^{k+1}-3-\ell}S^{2^k-1-\ell}$ *with* $2E^\ell\theta_{k-1} = [\iota_{2^k-1},\iota_{2^k-1}]$ *and* $2\ell \geq s(k)$. *Then*

$$E^{2\ell-s(k)}(\eta\theta_{k-1}^2) = [\iota_{N-s(k)},\beta]$$

where $e(\beta) = e(\beta_k)$.

Here E^ℓ means the ℓ-fold suspension homomorphism. If we neglect to omit suspensions then

$$\eta\theta_{k-1}^2 = 0 \text{ stably}$$

$$\eta\theta_{k-1}^2 = [\iota_{N-s(k)},\beta] \text{ on } S^{N-s(k)}$$

where $e(\beta) = e(\beta_k)$, and so if $\eta\theta_{k-1}^2 = 0$ on $S^{N-s(k)}$ then θ_k exists. This result gives us another point of view on null-homotopies of $\eta\theta_{k-1}^2$; since $\eta\theta_{k-1}^2$ is a Whitehead product on $S^{N-s(k)}$ it is canonically trivial on $S^{N+1-s(k)}$. This null-homotopy is the canonical null-homotopy, that is the $\phi_{k,k}$-carrying null homotopy or the null-homotopy used to show that $[\iota_{N+1},\iota_{N+1}] \in \langle\theta_{k-1}^2,\eta,2\rangle$ on S^{N+1}. To decide whether θ_k exists we need to decide whether $\eta\theta_{k-1}^2 = 0$ on $S^{N-s(k)}$.

Finally it is conjectured in [4], see also the problems in this proceedings, that in the notation of the theorem the maximum ℓ is $\ell = s(k-1)$, but $2s(k-1) \geq s(k)$ if and only if $k \geq 4$. It occasionally happens that $k = 1,2,3$ are special cases, but all the results one wants are true in these special cases. So in order to avoid a number of special arguments we implicitly exclude the cases $k = 1,2,3$ where they do not follow the general pattern.

COMPARISON WITH TODA'S WORK ON HOPF INVARIANT ONE There is a revealing and interesting analogy between this work and Toda's approach to the Hopf invariant one problem. We'll write $H:\pi_n^S \to \mathbb{Z}/2$ for the Steenrod-Hopf invariant, and use the symbol $\xi_n \in \pi_{2^n-1}^S$ for an element of Hopf invariant one if such exists. Then, Toda shows that if $\xi_n \in \pi_{2^n-1}^S$ and $H(\xi_n) = 1$ then $2\xi_n^2 = 0$; further there is a $Sq^{2^{n+1}}$ carrying null homotopy of $2\xi_n^2$ so that ξ_{n+1} exists if and only if there is a non $Sq^{2^{n+1}}$ carrying null homotopy of $2\xi_n^2$. Here by a null homotopy of $2\xi_n^2$ we mean a map $S^{2^{n+1}-1} \longrightarrow S^0 \cup_{\xi_n^2} e^{2^{n+1}-1}$ of degree 2 in integral homology. This is the inductive approach.

The analogue of the unstable approach is due to Steenrod rather than Toda; it is the classical observation that there is an element $\xi_n \in \pi_{2^n-1}^S$ with $H(\xi_n) = 1$, if and only if $[\iota_{2^n-1},\iota_{2^n-1}] = 0$. Then the relation between the two approaches is due to Toda; suppose that there is an element $\xi_{n-1} \in \pi_{2^{n-1}-1}^S$ with Hopf invariant one, then on S^{2^n-1}, $2\xi_{n-1}^2 = [\iota_{2^n-1},\iota_{2^n-1}]$.

§3 RECOGNIZING THE WHITEHEAD SQUARE

Suppose n is odd and $n \neq 1,3$ or 7; let A be a space and $f:S^{2n-1} \to A$, $g:A \to S^n$ be two maps, both of which are zero in mod 2 homology, such that $Sgf = 0$ in $\pi_{2n}S^{n+1}$. Then there exists a map $G:SA \cup_{Sf} e^{2n+1} \to S^{n+1}$ extending $Sg:SA \to S^{n+1}$. The functional cohomology operation Sq_G^{n+1} is independent of the choice of extension since $n \neq 1,3$ or 7. Now since the kernel of $S:\pi_{2n-1}S^n \to \pi_{2n}S^{n+1}$ is $\mathbb{Z}/2$ generated by $[\iota_n,\iota_n]$, gf is zero or $gf = [\iota_n,\iota_n]$ and the functional operation Sq_G^{n+1} distinguishes between the two possibilities.

THEOREM 3.1 *With the above notation*

$$gf = [\iota_n, \iota_n] \iff Sq_G^{n+1} \neq 0$$

Proof gf = 0 $\iff$ there is an extension $G: SA \cup_{sf} e^{2n+1} \to S^{n+1}$ which is the suspension of a map $A \cup_f e^{2n} \to S^n$ extending g

$\iff$ there is an extension $G: SA \cup_{Sf} e^{2n+1} \to S^{n+1}$ with zero Hopf invariant

$\iff$ there is an extension $G: SA \cup_{Sf} e^{2n+1} \to S^{n+1}$ with $Sq_G^{n+1} = 0$.

In the second equivalence, as in Boardman and Steer's paper [7], the Hopf invariant is a function $[SX, S^{n+1}] \to [SX, S^{2n+1}]$ which maps suspension elements to zero. In our case, since $G|SA$ is a suspension, it follows that the Hopf invariant of G factors as

$$SA \cup_{Sf} e^{2n+1} \xrightarrow{P} S^{2n+1} \xrightarrow{h} S^{2n+1}$$

where the first map is the collapsing map. Now since there is a map $S^{2n+1} \to S^{n+1}$ whose Hopf invariant has any even degree we may choose our extension G so that the degree of h is either zero or one. However the map p induces an isomorphism in mod 2 homology in dimension 2n+1, and so if H(G) = 0 hp is null-homotopic and the degree of h must be zero. It is now straightforward to check that, in our case, the Hopf invariant of G is zero if and only if G is a suspension. The final equivalence follows from [7] once more using the existence of maps $S^{2n+1} \to S^{n+1}$ with any even Hopf invariant.

COROLLARY 3.2 *There exists an element* $\theta_k \in \pi_N^S$ *with* $K(\theta_k) = 1$ *and* $2\theta_k = 0$ *if and only if* $[\iota_{N+1}, \iota_{N+1}] \in 2\pi_{2N+1}S^{N+1}$.

Proof If k = 1,2 , then $[\iota_{N+1}, \iota_{N+1}] = 0$ and $\theta_1 = \eta^2, \theta_2 = \nu^2$, $2\theta_1 = 2\theta_2 = 0$ where $\eta \in \pi_1^S$, $\nu \in \pi_3^S$ are the Hopf maps. The result is trivially true; now exclude the cases k = 1,2. Then if $[\iota_N, \iota_N] = 2\alpha$ apply 3.1 with n = N+1, $A = S^{2N+1}$, f = 2, g = α. We get a map $G: S^{2N+2} \cup_2 e^{2N+3} \to S^{N+1}$ with $Sq_G^{N+2} \neq 0$. From Adams decomposition of Sq^{N+2}, [1], it must follow that $G|S^{2N+2} = S\alpha$ is detected by the operation $\phi_{k,k}$ so that $K(S\alpha) = 1$. On the other hand if there is an element $\theta_k \in \pi_N^S$ detected by $\phi_{k,k}$, with $2\theta_k = 0$, then by the Freudenthal suspension theorem there is an element $\alpha \in \pi_{2N+1}S^{N+1}$ such that $S\alpha = \theta_k \in \pi_{2N+2}(S^{N+2}) = \pi_N^S$ and therefore $S(2\alpha) = 0$. Now use Adams decomposition of Sq^{N+2} to show that any extension of $S\alpha = \theta_k$ to a map $S^{2N+2} \cup_2 e^{2N+3} \to S^{N+2}$ is detected by Sq^{N+2} and so by 3.1, $2\alpha = [\iota_{N+1}, \iota_{N+1}]$.

Theorem 3.1 is well known, and very useful, but does not seem to be in the literature explicitly. Note that the argument shows how one may recognize an odd multiple of the Whitehead square $[\iota_n, \iota_n]$ when n is even.

§4 THE INDUCTIVE APPROACH

Assume there exists an element $\theta_k \in \pi_N^S$ with $K(\theta_k) = 1$ and $2\theta_k = 0$. The main point is to prove the following result.

THEOREM 4.1 *There exists a stable cell complex*

$$X = S^0 \cup_{\theta_k^2} e^{2N+1} \cup_\eta e^{2N+3} \cup_2 e^{2N+4}$$

with $Sq^{2N+4}: H^0X \to H^{2N+4}X$ *non-zero.*

Adams decomposition of Sq^{2N+4} shows that $\phi_{k+1,k+1}: H^0X \to H^{2N+3}X$ is non-zero; we may summarize the information in a cell diagram

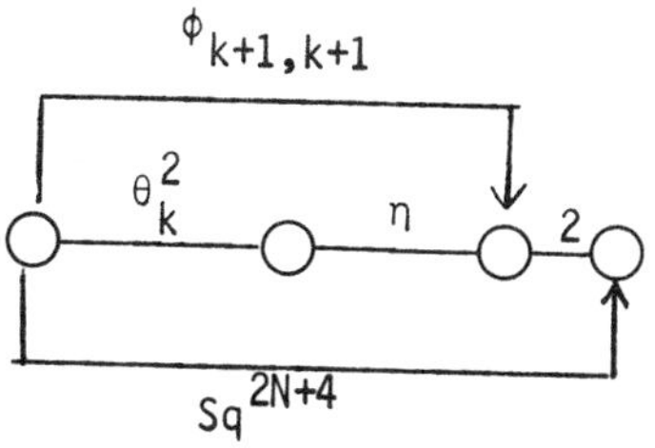

COROLLARY 4.2 (i) $\eta\theta_k^2 = 0$ *in* π_*^S

(ii) $0 \in \langle\theta_k^2, \eta, 2\rangle \subseteq \pi_*^S$

(iii) *There exists a* $\phi_{k+1,k+1}$ *carrying null homotopy of* $\eta\theta_k^2$.

COROLLARY 4.3 *Suppose* $\theta_k^2 = 0$, *then there exists an element* $\theta_{k+1} \in \pi_{2N+2}^S$ *with* $K(\theta_{k+1}) = 1$, $2\theta_{k+1} = 0$.

COROLLARY 4.4 *On* S^{2N+3}, $\eta\theta_k^2 = 0$ *and*

$$[\iota_{2N+3}, \iota_{2N+3}] \in \langle\theta_k^2, \eta, 2\rangle \subseteq \pi_{4N+5}S^{2N+3}.$$

These corollaries cover the results 2.1, 2.2 and 2.3 in the inductive approach. Parts (i) and (ii) are trivial deductions from the existence of the complex X; after all the obstructions to the existence of X are precisely $\eta\theta_k^2$ and $\langle\theta_k^2, \eta, 2\rangle$. The $\phi_{k+1,k+1}$ carrying null homotopy of $\eta\theta_k^2$ is the attaching map of the 2N+4 cell of X. To prove Corollary 4.3, use 4.1 to get a map

$$\phi : S^{2N} \cup_{\eta} e^{2N+2} \cup_{2} e^{2N+3} \to S^{o}$$

such that X is the mapping cone of ϕ, $\phi | S^{2N} = \theta_k^2$, and $Sq_{\phi}^{2N+4} \neq 0$. Now if $\theta_k^2 = 0$, then ϕ factors as

$$\begin{array}{ccc} S^{2N} \cup e^{2N+2} \cup e^{2N+3} & \longrightarrow & S^{2N+2} \cup_2 e^{2N+3} \\ & \phi \searrow \quad \swarrow \psi & \\ & S^{o} & \end{array}$$

and ψ is detected by Sq^{2N+4}. Set $\theta_{k+1} = \psi | S^{2N+2}$, then $2\theta_{k+1} = 0$ and using Adams decomposition of Sq^{2N+4}, $K(\theta_{k+1}) = 1$.

To prove Corollary 4.4 we can find a map

$$\xi : S^{4N+4} \cup_{\eta} e^{4N+6} \cup_{2} e^{4N+7} \longrightarrow S^{2N+4}$$

whose suspension is the stable map ϕ. Let $g : S^{4N+3} \cup_{\eta} e^{4N+5} \to S^{2N+3}$ be any map whose suspension is equal to the restriction of ξ to $S^{4N+4} \cup e^{4N+6}$ and let $f : S^{4N+5} \to S^{4N+3} \cup_{\eta} e^{4N+4}$ be any map whose suspension is the attaching map of the 4N+7 cell of our complex $S^{4N+4} \cup_{\eta} e^{4N+6} \cup_{2} e^{4N+7}$. By construction $S(gf) = 0$ and, if we set $A = S^{4N+3} \cup_{\eta} e^{4N+5}$, there is an extension, ξ , of Sg to a map $SA \cup_{Sf} e^{4N+7} \to S^{2N+4}$ detected by Sq^{2N+4}. Therefore, since we are implicitly excluding the trivial cases $k = 1,2$, we can use 3.1 to see that $gf = [\iota_{2N+3}, \iota_{2N+3}]$. But by the very definition of a Toda bracket

$$gf \in \langle \theta_k^2, \eta, 2 \rangle .$$

We now prove 4.1; there are a number of ways of constructing this complex X, the one we choose uses the quadratic construction, compare [24]. Indeed one of the recurring themes in our presentation of this material is the use of the quadratic construction. Suppose Y is any space, then there is an involution T on the space $S^n \times Y \wedge Y$ defined by $T(w, x \wedge y) = (-w, y \wedge x)$ and by definition $D_2^n Y$ is the space $S^n \times_T Y \wedge Y / S^n \times_T y_0 \wedge y_0$ where y_0 is the base point of Y. We allow $n = \infty$ and then write D_2 rather than D_2^{∞}. The cohomology of $D_2^n Y$ is computed using the cochain complex $W^n \otimes_T H^*Y \otimes H^*Y$ where W^n is the dual of the n-skeleton of the usual resolution of $\mathbf{Z}/2$ by $\mathbf{Z}/2\,[\Sigma_2]$ modules and T acts on $H^*Y \otimes H^*Y$ by setting $T(a \otimes b) = b \otimes a$. The coboundary homomorphism is given by the formula $\delta(e^r \otimes a \otimes b) = e^{r+1} \otimes (a \otimes b + b \otimes a)$ where e^r is the generator of the r-th component of W^n, in particular $e^r = 0$ if $r > n$. We will only use the case $n = \infty$, so from now on we restrict attention to this case. We use the following notation for cohomology classes in H^*D_2Y (note that we use Q_r for a *cohomology operation* rather than the more usual homology operation).

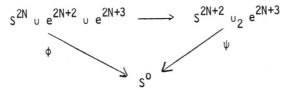

$x \in H^kY \qquad Q_r(x) \in H^{r+2k}D_2Y$ is the class of $e^r \otimes x \otimes x$

$x \in H^kY,\ y \in H^\ell Y, [x,y] \in H^{k+\ell}D_2Y$ is the class of $1 \otimes (x \otimes y + y \otimes x)$

The action of the Steenrod algebra on H^*D_2Y is given by the following formulas:
If $x \in H^kY$, then

$$Sq^tQ_r(x) = \sum_{i\geq 0} \binom{r+k-i}{t-2i} Q_{t+r-2i}(Sq^ix) \quad \text{if} \quad r \geq 1$$

$$Sq^tQ_0(x) = \sum_{i\geq 0} \binom{k-i}{t-2i} Q_{t-2i}(Sq^ix) + \sum_{0\leq i<t/2} [Sq^ix, Sq^{t-i}x].$$

If $x,y \in H^*Y$, then

$$Sq^t[x,y] = \sum_{0\leq i\leq t} [Sq^{t-i}x, Sq^iy]$$

See [26], [27], [25], [22] for proofs.

The quadratic construction D_2 is obviously functorial for maps of spaces; it is also functorial for *stable* maps of spaces [23] (see also [34], [14]). In particular if $f:Y \to S^0$ is a stable map, then we get a stable map $D_2(f):D_2Y \to D_2S^0$. But $D_2(S^0) = \mathbb{R}P^\infty_+$, that is $\mathbb{R}P^\infty$ with a disjoint base point adjoined, and so there is a map $\rho:D_2S^0 \to S^0$ - map the base point to the base point and $\mathbb{R}P^\infty$ to the other point. We write $\gamma(f):D_2Y \to S^0$ for the composite ρD_2f, then γ defines a function (but not a homomorphism in general) $\pi^0Y \to \pi^0D_2Y$ where π^* means stable cohomotopy.

LEMMA 4.5 Suppose $f:Y \to S^0$ is a stable map with the following properties:

(i) $f^*:H^*S^0 \to H^*Y$ is zero

(ii) $Sq^i_f:H^0S^0 \to H^{i-1}Y$ is zero for $i < n$ and $Sq^n_f(a_0) = x$ where $x \in H^{n-1}Y$ and $a_0 \in H^0S^0$ is the non-trivial class.

Then the stable map $\gamma(f):D_2Y \to S^0$ has the following properties.

(iii) $\gamma(f)^*:H^*S^0 \to H^*D_2Y$ is zero.

(iv) $Sq^i_f \quad :H^0S^0 \to H^{i-1}D_2Y$ is zero for $i < 2n$ and $Sq^{2n}_{\gamma(f)}(a_0) = Q_1(x) \in H^{2n-1}D_2Y$.

<u>Proof</u> Condition (iii) is obvious from the fact that the classes $Q_r(x)$ and $[x,y]$ for $x,y \in H^*Y$ generate H^*D_2Y and the obvious naturality formulae $D_2(f)^*Q_r(x) = Q_rf^*x$, and $D_2(f)^*[x,y] = [f^*x,f^*y]$. To prove condition (iv) use the following commutative diagram

$$\begin{array}{ccccc}
S^0 & \longrightarrow & C(\gamma(f)) & \longrightarrow & SD_2Y \\
\uparrow{\scriptstyle\rho} & & \uparrow & & \| \\
D_2S^0 & \longrightarrow & CD_2(f) & \longrightarrow & SD_2Y \\
\| & & \downarrow & & \downarrow{\scriptstyle q} \\
D_2S^0 & \longrightarrow & D_2C(f) & \longrightarrow & D_2SY
\end{array}$$

The top two rows are the cofibration sequences of $\gamma(f)$ and $D_2(f)$; the bottom row is not a cofibration sequence, the maps are those induced by the inclusion $S^0 \to C(f)$ and the projection $C(f) \to SY$. Pick an isomorphism $H^*C(f) \cong H^*S^0 \oplus H^*SY$ and now identify these two groups using this isomorphism. From the formula for $Sq^tQ_0(a_0)$ we find

$$Sq^nQ_0(a_0) = [a_0,\sigma x]$$

$$Sq^{2n}Q_0(a_0) = Q_0(\sigma x)$$

where σ is the suspension isomorphism and x is as in the statement of the lemma. Now we use the fact that

$$q^*Q_0(\sigma x) = \sigma Q_1(x)$$

$$q^*[a_0,\sigma x] = 0$$

(compare [29]). Then a straightforward argument using the above commutative diagram after applying H^* completes the proof.

PROOF OF THEOREM 4.1 Suppose θ_k exists and $2\theta_k = 0$, then there exists a stable map $f:Y = S^N \cup_2 e^{N+1} \to S^0$ such that $f|S^N = \theta_k$; therefore f satisfies conditions (i) and (ii) of 4.5 with $n = N + 2$, $x = a_{N+1}$ (we use the notation a_N, a_{N+1} for the non-zero elements of H^NY and $H^{N+1}Y$); therefore $Sq^{2N+4}_{\gamma(f)}\, a_0 = Q_1(a_{N+1})$. We now analyse the cell structure of D_2Y; the following table gives the cohomology of D_2Y with the action of the Steenrod operations

$$\begin{array}{lll}
2N+3 & Sq^3Q_0(a_N) = Q_1(a_{N+1}) & Sq^2Q_1(a_N) = Q_3(a_N) + Q_1(a_{N+1}) \\
2N+2 & Sq^2Q_0(a_N) = Q_2(a_N) + Q_0(a_{N+1}) & Sq^1Q_1(a_N) = Q_2(a_N) \\
2N+1 & Sq^1Q_0(a_N) = [a_N,a_{N+1}] & Q_1(a_N) \\
2N & Q_0(a_N) &
\end{array}$$

From this one easily deduces the cell structure of the $2N + 3$ skeleton of D_2Y; it is most conveniently described by the following cell diagram:

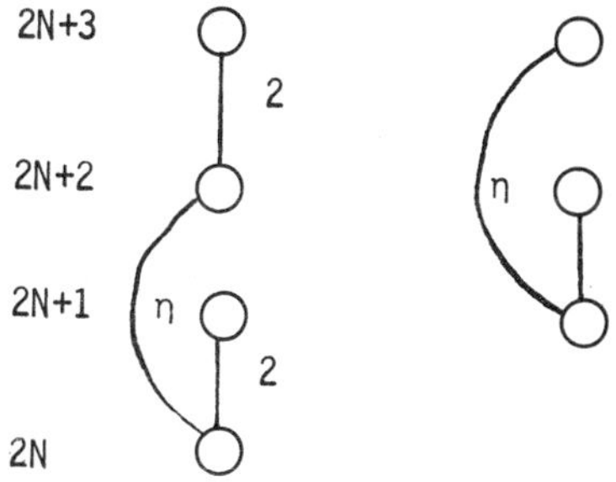

where the cohomology classes dual to the cells in this diagram are given, in the obvious way in the previous table. Now let Z be the subcomplex of D_2Y with cell diagram

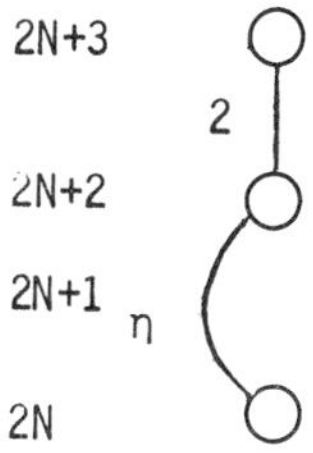

.

Take $g = \gamma(f)|2$, then by construction $Sq_g^{2N+4} \neq 0$ and take X in Theorem 4.1 to be the mapping cone of g.

REFERENCES

[1] Adams, J.F. The non-existence of elements of Hopf invariant one. Ann. Math. 72 (1960) 20-104.

[2] Adams, J.F. Vector fields on spheres. Ann. Math. 75 (1962) 603-632.

[3] Adams, J.F. On the groups J(X) - IV. Topology 5 (1966) 21-71.

[4] Barratt, M.G., Jones, J.D.S., Mahowald, M.E. The Kervaire invariant and the Hopf invariant (to appear).

[5] Barratt, M.G., Jones, J.D.S., Mahowald, M.E. The cup-1-construction and applications to the Kervaire invariant problem. (to appear).

[6] Barratt, M.G., Jones, J.D.S., Mahowald, M.E. The Kervaire invariant in dimension 62. (to appear).

[7] Boardman, J.M., Steer, B.F. On Hopf invariants. Comm. Math. Helv. 42 (1967) 180-221.

[8] Browder, W. The Kervaire invariant of framed manifolds and its generalizations. Ann. Math. 90 (1969) 157-186.

[9] Brown, E.H. Generalizations of the Kervaire invariant. Ann. Math. 95 (1972) 368-383.

[10] Brown, E.H., Peterson, F.P., Whitehead products and cohomology operations. Quart. J. Math., Oxford (2), 15 (1964) 116-120.

[11] Cohen, R.L., Jones, J.D.S., Mahowald, M.E. The Kervaire invariant of immersions. (To appear).

[12] Feder, S., Gitler, S., Lam, K.Y. Composition properties of projective homotopy classes. Pacif. J. Math. 68 (1977) 47-61.

[13] Jones, J.D.S., Rees, E.G. Kervaire's invariant for framed manifolds. Proc. Symp. Pure Math. Vol XXXII A.M.S. (1978).

[14] Jones, J.D.S., Wegmann, S.A. Limits of stable homotopy and cohomotopy groups. (To appear).

[15] Kahn, D.S., Priddy, S.B. Applications of the transfer to stable homotopy theory. Bull. A.M.S. 78 (1972) 981-987.

[16] Kervaire, M. A manifold which does not admit any differentiable structure. Comm. Math. Helv. 34 (1966) 256-270.

[17] Kervaire, M., Milnor, J. Groups of homotopy sphere I. Ann. Math. 77 (1963) 504-537.

[18] Kuhn, N.J. Thesis, University of Chicago, 1980.

[19] Mahowald, M.E. The metastable homotopy of S^n. Mem. A.M.S. No. 72 (1967).

[20] Mahowald, M.E. Some remarks on the Kervaire invariant problem from a homotopy point of view. Proc. Symp. Pure Math. Vol XXII A.M.S. (1971).

[21] Mahowald, M.E. The image of J in the E.H.P sequence. Ann. Math. 116 (1982) 65-112.

[22] May, J.P. A general algebraic approach to Steenrod operations. Lecture Notes in Math. Vol. 168 pp. 153-231, Springer Verlag (1970).

[23] May, J.P. H_∞ ring spectra and their applications. Proc. Symp. Pure. Math. Vol. XXXII A.M.S. (1978).

[24] Milgram, R.J. Symmetries and operations in homotopy theory. Proc. Symp. Pure Math. Vol. XXII A.M.S. (1971).

[25] Milgram, R.J. Unstable homotopy theory from the stable point of view. Lecture notes in Math. Vol. 368 Springer Verlag (1974).

[26] Nishida, G. Cohomology operations in iterated loop spaces. Proc. Japan Acad. 44 (1969) 104-109.

[27] Nishida, G. Extended power operations in homotopy theory. Conference on Algebraic Topology, University of Illinois, Chicago Circle, pp. 253-259 (1968).

[28] Snaith, V.P. A stable decomposition of $\Omega^n S^n X$. J. Lond. Math. Soc. (2) 7 (1974), 577-583.

[29] Steenrod, N.E., Epstein, D.B.A. Cohomology Operations. Ann. Math. Studies No. 50 P.U.P. (1962).

ADDRESSES:

Math. Dept., Northwestern University, Evanston, Illinois 60201, U.S.A.

Maths Institute, University of Warwick, Coventry CV4 7AL, England

Math. Dept. Northwestern University, Evanston, Illinois 60201, U.S.A.

Contemporary Mathematics
Volume 19, 1983

SURGERY THEORY OF IMMERSIONS

J. Scott Carter*

ABSTRACT. The self-intersection sets of immersions provide stable homotopy invariants. A surgery theory is developed which allows stable homotopy classes in $\pi^s_{n+1}(P^\infty)$ and π^s_n to be represented by codimension one immersion in $\mathbb{R}^{n+1}$ without n-tuple points provided two invariants vanish. One invariant is the number of (n+1)-tuple points of a representative immersion counted modulo 2; this number is related to the Kervaire invariant as has been shown by Eccles [5]. The other is an n-tuple point invariant defined and computed herein.

1. BACKGROUND. Two codimension k-immersions $(i_0:M^n_0 \looparrowright \mathbb{R}^{n+k})$ and $(i_1:M^n_1 \looparrowright \mathbb{R}^{n+k})$ are said to be <u>bordant</u> if there is an (n+1)-manifold W^{n+1} with boundary the disjoint union of M_0 and M_1, and if there is an immersion

$$f:(W,\partial W) \looparrowright (\mathbb{R}^{n+k}\times[0,1],\ \mathbb{R}^{n+k}\times(\{0\} \cup \{1\}))$$

which agrees with the disjoint union of the immersions i_0 and i_1 along the boundary

$$f|_{M_j} = i_j:M_j \looparrowright \mathbb{R}^{n+k}\times\{j\}, \qquad \text{for } j = 0, 1.$$

1980 Mathematics Subject Classification primary 57R42, 57R65 (secondary 55Q10, 55N22)

*Research was partially supported by N.S.F. Grant MCS80-02332.

The bordism classes of immersions form an Abelian group Ω_n^{n+k} (imm); if bordisms respect orientations, we denote this group by Ω_n^{n+k} (imm, or). The composition law is disjoint union of immersions, and the inverse of a class is obtained by reflecting a representative immersion through a hyperplane. The stable homotopy interpretation of this group is given by the following:

THEOREM (Wells [13]). There is a Pontryagin-Thom isomorphism

$$\Omega_n^{n+k}\ (\text{imm}) \simeq \pi_{n+k}^s\ (MO(k))$$

where $MO(k)$ is the Thom space of the universal k-plane bundle $\gamma_k \to BO(k)$.

COROLLARY. In the case of codimension one immersions this isomorphism has the form:

$$\Omega_n^{n+1}\ (\text{imm}) \simeq \pi_{n+1}^s\ (P^\infty),$$

and in the oriented theory

$$\Omega_n^{n+1}\ (\text{imm, or}) \simeq \pi_n^s.$$

U. Koschorke and B. Sanderson [12] prove a generalization of the above theorem by using the configuration space model for $\Omega^\infty S^\infty X$, and they identify certain self-intersection invariants with stable James-Hopf invariants.

We will be concerned with the codimension one case where we assume immersions are self-transverse. In this case notice the self-intersection sets of an immersion, $i:M^n \looparrowright \mathbb{R}^{n+1}$,

$$G_k(i) = \{y \in \mathbb{R}^{n+1} \mid \# i^{-1}(y) = k\},$$

form submanifolds of dimension $n+1-k$, for $k = 0,\dots,n+1$. Furthermore

$$\text{Closure } (G_k(i)) = \bigcup_{r\geq k} G_r(i).$$

The self-intersection approach to the subject was innocently introduced by T. Banchoff [1] while discussing immersed surfaces in $\mathbb{R}^3$. He shows: given an immersion

$$i:M^2 \looparrowright \mathbb{R}^3$$

the number of triple points of i is congruent to the Euler characteristic of M modulo 2. In general, the number of $(n+1)$-tuple points of an immersed n-manifold in $\mathbb{R}^{n+1}$ counted modulo 2 defines a homomorphism

$$\psi_{n+1}:\pi_n^s \to \mathbb{Z}/2$$

via Wells' theorem. In 1978, Freedman [6] conjectured that $\psi_{n+1} \neq 0$ if and only if $n = 0, 1, 3$, or 7 and checked this for $n \leq 3$. The work of Koschorke [9, 10] supported Freedman, but the case $n = 7$ was still a mystery. Recently, Eccles [4, 5] has shown:

THEOREM 1. *The homomorphism*

$$\psi_{n+1}:\pi_n^s \to \mathbb{Z}/2$$

is non-trivial if and only if $n = 0, 1$, or 3.

Consider the $(n+1)$-*tuple point invariant in the unoriented theory; this is a homomorphism*

$$\psi_{n+1}:\pi_{n+1}^s(P^\infty) \to \mathbb{Z}/2.$$

2. *If* n *is even,* ψ_{n+1} *is non-trivial if and only if* $n = 0, 2$, *or* 6.

3. <u>If</u> $n \equiv 1 \pmod 4$, <u>then</u> ψ_{n+1} <u>is non-trivial if and only if there is a framed</u> $(n+1)$-<u>manifold</u>, M^{n+1} <u>with Kervaire invariant</u> +1.

Eccles also showed the self-intersection sets provide a complete set of stable homotopy invariants. That is, there are homomorphisms

$$\psi_r : \pi^s_{n+1}(S^1) \to \pi^s_{n+1}(D_r(S^1))$$

such that ψ_r is injective on the p-primary component. This follows from the Kahn-Priddy theorem [8]. The space $D_r(X)$ is obtained from the filtration

$$\Gamma_0(X) \subset \ldots \subset \Gamma_r(X) \subset \ldots \subset \Gamma(X)$$

of the configuration space $\Gamma(X)$ as the quotient

$$D_r(X) = \Gamma_r(X)/\Gamma_{r-1}(X).$$

In the unoriented theory these invariants are homomorphisms of the form

$$\psi_r : \pi^s_{n+1}(P^\infty) \to \pi^s_{n+1}(D_r(P^\infty)).$$

II. RESULTS. Let $\pi(n+1)$ denote either $\pi^s_{n+1}(P^\infty)$ or π^s_n. To what extent are the invariants ψ_r surgery obstructions? That is if $x \in \pi(n+1)$ and

$$\psi_q(x) = 0$$

for all $q \geq r$, is there an immersion (i,M) representing x without r-tuple points? Unfortunately, this question is not always easy to answer. If $r = n+1$, the answer is yes, but even in the case $r = n$, we must change the definition of the invariants to get an affirmative answer. We have the following:

THEOREM A. *There are homomorphisms*

$$C_n : \pi(n+1) \to \Omega_1(BSH_n, \theta) \simeq \begin{Bmatrix} \mathbb{Z}/8 \text{ if } n = 2 \\ \mathbb{Z}/2 \text{ if } n > 2 \end{Bmatrix}$$

such that, for $n = 2$ *or* $n \geq 5$, *if*

$$\psi_{n+1}(x) = 0,$$

and

$$C_n(x) = 0,$$

then x *has a representative immersion without n-tuple points.*

THEOREM B. 1. *If* n *is an odd integer greater than* 1, *then*

$$C_n(x) = 0.$$

2. *If* n *is an even integer greater than* 2, *then*

$$C_n(x) = \psi_{n+1}(x)$$

as integers modulo 2.

3. *In the case* $n = 2$

$$C_2 : \pi_3^s(P^\infty) \to \mathbb{Z}/8$$

is an isomorphism.

This last result is known ([1] and [4]), but our study of n-tuple points was motivated by the nature of this isomorphism.

The group $\Omega_1(BSH_n, \theta)$ consists of bordism classes of triples $(N^1, g, \bar{g})$ where

1. N^1 is a one dimensional manifold,
2. $g:N^1 \to BSH_n$ is a continuous map into the classifying space of the special hyperoctahedral group
$SH_n = \{(a_{jk}) \in SO(n) : (a_{jk})$ is a signed permutation matrix$\}$, and
3. $\bar{g}$ is a vector bundle isomorphism

$$\bar{g}: g^*(\theta) \oplus TN \xrightarrow[\simeq]{} \varepsilon^{n+1} \equiv N \times \mathbb{R}^{n+1}$$

where

$$\theta = ESH_n \times_{SH_n} \mathbb{R}^n$$

is the canonical vector bundle over BSH_n.

PROOF OF A. To prove theorem A we must define the map C_n, compute the group $\Omega_1(BSH_n,\theta)$, and show C_n is a surgery obstruction. Let

$$i: M^n \looparrowright \mathbb{R}^{n+1}$$

be a representative of some class in $\pi(n+1)$.

The construction of C_n is based on Banchoff's [1] analysis of double points of immersed surfaces. Suppose

$$f: S^1 \to \mathbb{R}^{n+1}$$

is one component of the n-tuple point set parametrized by the unit interval $[0,1]$, so that $f(0) = f(1)$. For each $t \in [0,1]$ there are vectors $\ell_1(t)$, $\ell_2(t),\ldots,\ell_n(t)$ which are normal to the immersion (i,M) at the points $p_1(t),\ldots,p_n(t) \in M$ where $p_j(t) \in i^{-1}(f(t))$ for $j = 1,2,\ldots,n$.
It need not be the case that $\ell_j(0) = \ell_j(1)$; in fact, there is a matrix $\sigma = \sigma(i,M) \in SH_n$ such that the initial and final choices of bases are related by the equation:

$$(\ell_1(0),\ldots,\ell_n(0) = \sigma(\ell_1(1),\ldots,\ell_n(1)).$$

Let

$$g_\sigma : S^1 \to BSH_n$$

be the continuous map which wraps the circle around the edge in the 1-skeleton of BSH_n corresponding to σ. Such a map is constructed for each component of the n-tuple manifold $M(n)$. Let

$$g: M(n) \to BSH_n$$

denote the union of these maps. Thus the normal bundle of the n-tuple point manifold is classified by g, for $\ell_1(t), \ell_2(t), \ldots, \ell_n(t)$ form an orthonormal basis for this bundle at the point $f(t)$. Let

$$\bar{g}: g^*(\theta) \oplus TM(n) \to \varepsilon^{n+1}$$

be the canonial vector bundle isomorphism induced from the natural immersion of the n-tuple manifold. The homomorphism C_n is defined by

$$C_n[i,M] = [M(n), g, \bar{g}]$$

where square brackets denote bordism classes.

To compute the bordism group $\Omega_1(BSH_n, \theta)$ one uses the exact sequence 9.3 of Koschorke [11]. In this case the sequence has the form

$$\cdots \to \mathbb{Z}/2 \xrightarrow{\delta} \Omega_1(BSH_n, \theta) \xrightarrow{f} H_1(BSH_n) \to 0.$$

The map f is an Hurewicz homomorphism, and δ is an injection if and only if $n = 2$.

In the case $n = 2$, the group SH_2 is isomorphic to $Z/4$ and this sequence does not split. This can be seen by examining the double point sets of immersed surfaces obtained by twisting the "8" x I below and identifying edges according to the twists. By twisting one end about the indicated axis one half of a turn, one obtains an immersed Klein bottle with double point matrix

$$\sigma(j,K) = \begin{pmatrix} -1 & 0 \\ 0 & -1 \end{pmatrix}.$$

(Banchoff [2] has made a film of this surface.) By twisting one full turn one obtains an immersed torus with trivial double point matrix, but with a framing on the double point curve which does not bord since $\pi_1(SO(3)) \simeq \mathbb{Z}/2$.

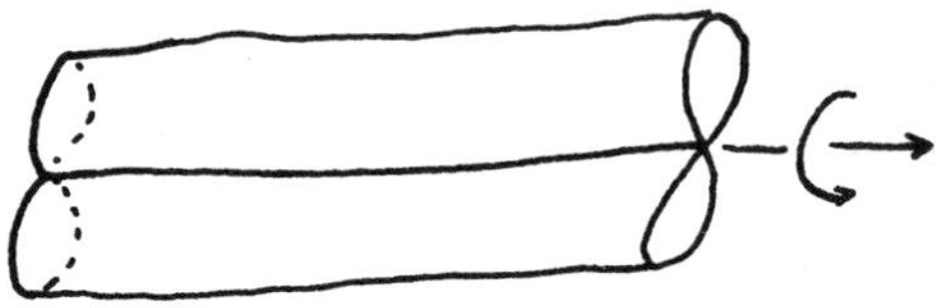

Figure 1

In addition, the isomorphism

$$C_2 : \pi_3^s(P^\infty) \to \mathbb{Z}/8$$

can be seen by examining the double point curve of Boy's surface [7] which represents a generator of the group $\pi_3^s(P^\infty)$.

In the case $n > 2$,

$$SH_n/[SH_n, SH_n] \simeq \mathbb{Z}/2$$

the commutator subgroup being the set of all matrices in the inverse image of the alternating subgroup A_n under the projection

$$SH_n \xrightarrow{\pi} \Sigma_n \to 1$$

onto the symmetry group. This projection takes a matrix (a_{jk}) to the matrix $(|a_{jk}|)$ whose entries are the absolute values of the entries a_{jk}. Recall, the alternating subgroup is not perfect for $n = 3$ and 4. This is one reason that C_n is not the surgery obstruction for these cases; there is 3-torsion detected by the n-tuple points.

Since for $n > 2$

$$\Omega_1(BSH_n,\theta) \simeq H_1(BSH_n) \simeq SH_n/[SH_n,SH_n],$$

we view the map C_n as assigning to a bordism class $[i,M]$ the coset $\sigma(i,M)[SH_n,SH_n]$ of the n-tuple matrix $\sigma = \sigma(i,M)$ associated to the n-tuple curve. Here we are assuming that the n-tuple point manifold is connected; this assumption will be justified in a moment.

To finish the proof we must discuss a method of performing surgery on immersions. For appropriate integers k and r suppose there is an embedded k disk

$$h:D^k \to \mathbb{R}^{n+1}$$

whose interior is in the r-tuple point set and whose boundary is in the (r+1)-tuple point set. The inward pointing normal vectors of $h(\partial D^k)$ define a lift, $\bar{h}$, of $h(\partial D^k)$ to M, for these vectors are all normal to the same sheet of (i,M). Since $h(D^k)$ is in the image of the r-tuple point manifold, there are vectors $z_1(x),\ldots,z_r(x)$ which are normal to (i,M) at $p_j(x)$ where $\{p_j(x)\} = i^{-1}(h(x))$, for each $x \in D^k$. The vectors are combined with a basis for the normal bundle of h in the r-tuple point manifold to form a basis of $\nu(h)_x$. Thus a tubular neighborhood

$$H:D^k \times D^{n+1-k} \to \mathbb{R}^{n+1}$$

is defined with

$$H(x,0) = h(x)$$

which is compatible with the immersion (i,M). In particular, we may remove a neighborhood $N \simeq S^{k-1} \times D^{n-k+1}$ of $\bar{h}(S^{k-1}) \subset M$ and replace the image of this neighborhood in $\mathbb{R}^{n+1}$ with $H(D^k \times S^{n-k})$. This operation is called a (k,r) surgery. Such a surgery will attach an appropriate number of k-handles to the j-tuple point set for $0 < j \leq r+1$; the number of such handles is a binomial coefficient dependent on k and r. A type $(n-r+1,r)$ surgery eliminates that sphere of $(r+1)$-tuple points. These observations follow by choosing the radius of $H(\{x\},D^{n+1-k})$ sufficiently small (see [3] pages 67, 68 for further details).

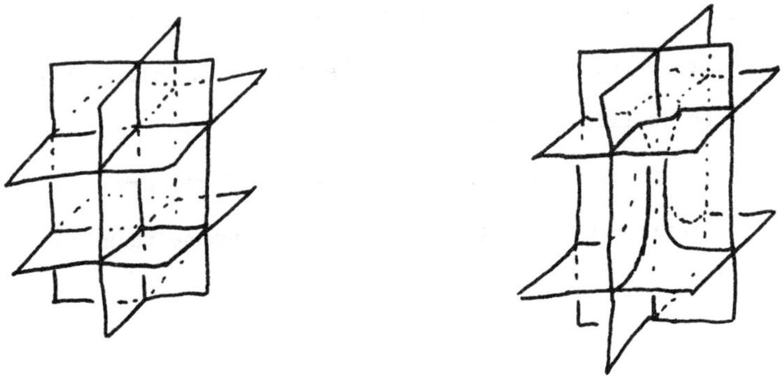

Before (1,2) surgery After surgery

Figure 2

Thus by using surgeries of type $(1,0),\ldots,(1,n)$ we may eliminate pairs of $(n+1)$-tuple points, and by using surgeries of type $(1,0),\ldots,(1,n-1)$ we may assume the n-tuple curve is connected. The first surgeries in these sequences may be necessary in order to inductively define the embedded k disk in the r-tuple manifold. (Technical details may be found on page 71 [3]).

Furthermore, for $n \geq 5$ it is possible to construct immersions which represent the identity of $\pi(n+1)$, but whose n-tuple matrices generate the commutator $[SH_n,SH_n]$. In addition, representatives with n-tuple data in the image of $\delta(\text{in } \Omega_1(BSH,\theta))$ are constructible. These immersions are found on page 75 [3] as examples d,e, and f; they are obtained by performing surgery on the immersion

$$m\colon \bigcup_{j=1}^{n} S^n_j \looparrowright \mathbb{R}^{n+1}$$

which consists of n spheres of dimension n, S_j^n, intersecting in general position. This immersion has a simple closed n-tuple curve. A surgery of type (1,0) is performed along some path with appropriate trivialization. Subsequent surgeries of type $(1,2),\ldots,(1,n-1)$ are performed by following the conventions of lemma 15 [3]. The initial surgery determines the resulting n-tuple matrix.

Given an immersion (i,M) with

$$C_n[i,M] = 0$$

and

$$\Psi_{n+1}[i,M] = 0,$$

we add the above immersions to (i,M) and attach 1-handles in order to assume the n-tuple point data of the result (again denoted (i,M)) consist of a simple closed curve, with trivial n-tuple matrix, and with a framing which extends across a disk. Now push the n-tuple curve, f(t), to the loop

$$\alpha(t) = f(t) + \varepsilon \sum_{j=1}^{n-1} \ell_j(t), \qquad t \in [0,1] \quad (\varepsilon > 0\ ,\ \text{small})$$

where $\ell_j(t)$ are the vectors normal to (i,M) at f(t) as above. This loop is in the non-singular set of (i,M). If it is possible to perform a (2,0) surgery along a disk whose boundary is α, then we may perform subsequent surgeries of type $(2,1),\ldots,(2,n-1)$ in order to eliminate the n-tuple points. Since $n \geq 5$, there is no problem in extending α to an embedded disk, $\bar{\alpha}$. However, this disk may intersect (i,M). By performing appropriate (1,0), (1,1), (1,2), and (2,0) surgeries, we may assume that either a type (2,0) surgery is possible along $\bar{\alpha}$, or that the intersection of $\bar{\alpha}$ and (i,M) is a figure 8.

In the latter case we use a weaker notion of surgery. To perform weak (k,r) surgery we assume the core of the k-handles, $h(D^k)$ intersects the given immersion as follows:

(i) $h(\partial D^k) \subset \{y \in \mathbb{R}^{n+1} \mid \# i^{-1}(y) = r+1\} = G_{r+1}(i)$,

(ii) $h(\text{int } D^k) \subset \{y \in \mathbb{R}^{n+1} \mid r \leq \# i^{-1}(y) \leq r+k\} = G_r^{k+r}(i)$

(iii) $h(\text{int } D^k) \cap G_{r+1}^{k+r}(i)$ is a transverse intersection.

In the case $k = 2$, and $r = 0$, the situation resembles figure 3 where in a neighborhood of the intersection (iii) the immersion, (i,M), appears as the Cartesian product of a figure 8 and a disk D^{n-1}.

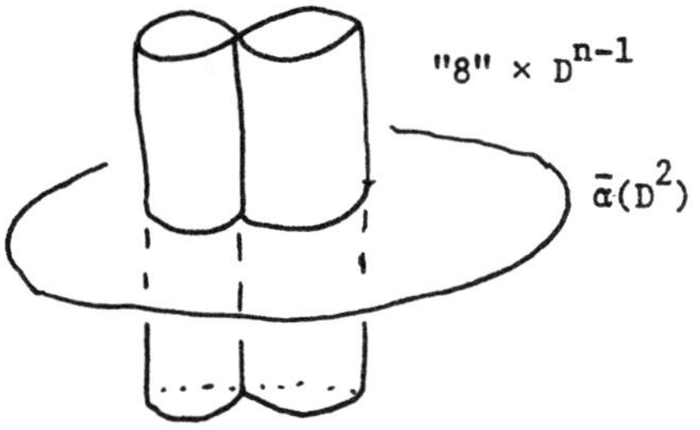

Figure 3

To eliminate the n-tuple points, weak surgeries of type (2,0), (2,1),..., (2,n-1) are performed. The first surgery attaches a 2-handle along the disk $\bar{\alpha}(D^2)$. Subsequent surgeries are performed by inductively defining disks satisfying (i)-(iii) which pass over the previously defined handles. Such surgeries eliminate the given n-tuple curve, but the weak surgeries of type (2,n-3), (2,n-2), and (2,n-1) introduce new (n+1)-tuple and n-tuple points. There are, for example, 2n (n+1)-tuple points which are connected in pairs by arcs of n-tuple points. The n-tuple points which are introduced are a collection of figure 8's and a collection of circles living on 2-spheres of (n-1)-tuple points. The latter n-tuple curves pass three at a time over the (n-1)-tuple point spheres. The (n+1)-tuple points are the intersection of such circles and the "perpendicular" figure 8's.

Although the self-intersection sets of degrees n and $(n+1)$ introduced are ostensibly complicated, there is an advantage in dealing with them. Namely, they may be eliminated by using only surgeries of type $(1,n)$, $(1,n-1)$, and $(2,n-1)$. Since $(n+1)$-tuple points are connected in pairs by arcs of n-tuple points, eliminate them by $(1,n)$ surgeries. For each such surgery there are a collection of 1-handles connecting pairs of 2-spheres in the $(n-1)$-tuple surface. Four arcs of n-tuple points travel "parallel" over each such 1-handle. Since these arcs are parallel, the resulting n-tuple curves form a null homologous cycle (modulo 2) in the $(n-1)$-tuple surface. Moreover, there is some lift of the n-tuple curves which is null homologous in the complement of the other lifts; the other lifts live on different handles. Therefore, we may perform $(1,n-1)$ surgeries until each component of the n-tuple curve bounds a disk in the $(n-1)$-tuple surface. Thus $(2,n-1)$ surgeries are possible, and this completes the proof.

In [3], we worked out the details of the last paragraph in the case $n = 4$, where the surgery obstruction is a homomorphism

$$\bar{C}_4 : \pi(5) \to \Omega_1(B\Sigma_4, \theta) \simeq \mathbb{Z}/3.$$

Proof of B. Let (i,M) be an immersion with connected n-tuple curve

$$f : S^1 \to \mathbb{R}^{n+1}$$

parametrized by $t \in [0,1]$. Push f to an arc

$$\alpha(t) = f(t) + \varepsilon \sum_{j=1}^{n} \ell_j(t)$$

$(\varepsilon > 0$ sufficiently small), and attempt to join the point $\alpha(0)$ to $\alpha(1)$. Then compute the modulo 2 intersection number of this closed curve, $\bar{\alpha}$, with the given immersion. Since the intersection occurs in $\mathbb{R}^{n+1}$, this number must be zero.

If n is odd, any $(n+1)$-tuple points present do not contribute to the intersection number. The intersection number is trivial if and only if $\bar{\alpha}$ passes through (i,M) an even number of times. This happens if and only if the parity of the n-tuple matrix is even. Equivalently the n-tuple matrix is in the commutator, and $C_n[i,M] = 0$.

If n is even, each (n+1)-tuple point contributes to the intersection number. Thus the parity of the n-tuple matrix must agree with that of the number of (n+1)-tuple points; i.e.

$$C_n[i,M] = \psi_{n+1}[i,M].$$

This result is analogous to the computation of the double point invariant on Boy's surface. This completes the proof.

BIBLIOGRAPHY

1. Banchoff, T., "Triple points and surgery on immersed surfaces", Proc. of Amer. Math. Soc., 46, No. 3, (1974), 407-413.

2. ________ and Strauss, "An immersion of the Klein bottle in three space", A short film of this surface, (1982).

3. Carter, J. Scott, Surgery on immersions: a geometric approach to stable homotopy, Dissertation: Yale University, May 1982.

4. Eccles, Peter J., "Multiple points of codimension one immersions of oriented manifolds", Math. Proc. Cambr. Phil. Soc., 87 (1980), 213-220.

5. ________, "Codimension one immersions and the Kervaire invariant one problem", Math. Proc. Cambr. Phil. Soc. 90 (1981), 483-493.

6. Freedman, Michael, "Quadruple points of 3-manifolds in S^4", Comment. Math. Helv., 53 (1978), 385-321.

7. Hilbert and Cohn-Vassen, (Trans. by P. Nemenyi), Geometry and the imagination, Chelsea Publishing (1952), 320-321.

8. Kahn, D.S., and Priddy, S.B., "The transfer and stable homotopy theory", Math. Proc. Cambr. Phil. Soc., 83 (1978), 103-111.

9. Koschorke, Ulrich, "On the (n+1)-tuple points of immersed spheres", Gesamthochschule Siegen Mathematische Forschungsberichte 26, (1979).

10. ________, "Multiple points of immersions, and the Kahn-Priddy theorem", Math. Z., 169 (1979), 223-236.

11. ________, Vector fields and other vector bundle morphisms — a singularity approach, Springer-Verlag LNM847, (1981).

12. ________ and Sanderson, Brian, "Self-intersections and higher Hopf invariants", Topology, 17 (1978), 283-290.

13. Wells, R., "Cobordism groups of immersions", Topology, 5 (1966), 281-294.

DEPARTMENT OF MATHEMATICS
YALE UNIVERSITY
NEW HAVEN, CT 06520

Current Address
Department of Mathematics
University of Texas
Austin, TX 78712

Contemporary Mathematics
Volume 19, 1983

The Homology of Function Spaces

F.R. Cohen and L.R. Taylor[1]

In this note we compute the homology of certain function spaces. Specifically, if X and Y are based spaces, we wish to study $H_*(\mathrm{Map}(X,Y);F_p)$, where Map(X,Y) is the space of based maps in the compact-open topology, and where F_p is the prime field of characteristic p, p = 0 or a prime.

In the 1950's, K. Borsuk [2], S. Mardešić [9], and J.C. Moore [10] studied the case $Y = S^n$. In particular, Moore computed the above homology groups through a range under some mild assumptions on X. We complete Moore's result by computing $H_*(\mathrm{Map}(X,S^m);F_p)$ for all * and for most of Moore's spaces X.

We can also compute $H_*(\mathrm{Map}(X,Y);F_p)$ for somewhat more complicated Y (e.g. Y a wedge of spheres) but nothing like a general result has emerged. In particular we pose

Problem: Compute $H_*(\mathrm{Map}(X,\Sigma^m Y);F_p)$ for m>> dimension X. Is the answer a functor of $H_*(X;F_p)$ and $H_*(Y;F_p)$?

To do our calculations we return to an old idea. If $X_0 \subset X_1 \subset \dots \subset X_n = X$ is a filtration of X by NDR pairs (X_i,X_{i-1}) , we get a family of fibrations

$$\mathrm{Map}(X_i/X_{i-1},Y) \longrightarrow \mathrm{Map}(X_i,Y) \longrightarrow \mathrm{Map}(X_{i-1},Y) \quad .$$

Federer [6] produced his spectral sequence by using the cellular filtration on the CW complex X and applying π_* to the resulting family of fibrations. We originally modified this filtration slightly and then showed that the resulting Serre spectral sequences were orientable and collapsed. The actual answer was then easily worked out.

The method of showing that these fibrations behave so well is to study the map

$$\phi\colon \mathrm{Map}(X,Y) \longrightarrow \mathrm{Map}(X,Q(Y)) \quad .$$

We get results because Q(Y) is an ∞-loop space and this makes some calculations rather easy. Indeed, they become so easy that we never need to formally intro-

1980 Mathematics Subject Classification. 54C35.

[1]Both authors are partially supported by the N.S.F.; the first author was partially supported by the Alfred P. Sloan Foundation.

duce any filtration on X at all. We also use Spanier's function space approach to S-duality, [11], to identify Map(X,Q(Y)) in many cases with Q(Z) for some space Z. In particular, if $Y = S^r$, Z is a Spanier-Whitehead r-dual to X.

We introduce some notation. For any space X define

$$d(X) \le r \quad \text{if} \quad H^*(X;\mathbb{Z}) = 0 \quad \text{for all } * > r$$

and

$$b(X) \le s \quad \text{if} \quad \tilde{H}_*(X;\mathbb{Z}) = 0 \quad \text{for all } * < s \ .$$

Define $d_p(X)$ and $b_p(X)$ similarly but using mod-p homology and cohomology instead of the groups with $\mathbb{Z}$ coefficients. Set

$$\ell(X) = d(X) - b(X) + 1 \quad \text{and} \quad \ell_p(X) = d_p(X) - b_p(X) + 1 \ .$$

Let $\beta_i(X) = \dim_{F_p} H_i(X;F_p)$. We will use $\Omega^i Y$ to denote $\mathrm{Map}(S^i,Y)$ and then $(\Omega^i Y)^{\beta_i(X)}$ will denote the Cartesian product of $\beta_i(X)$ copies of $\Omega^i Y$. We say that a space A has finite type if $H_i(A;\mathbb{Z})$ is finitely-generated for all i.

In the statement below of our main theorem we have fixed p and an integer r. We will require that our space X satisfy

$i)_r$: X is a connected CW complex of finite type with $d(X) \le r$ and with $d_p(X) + \ell_p(X) \le r$.

We will require that Y satisfy

$ii)_r$: Y is an r-connected CW complex of finite type: furthermore we require

a) the map $\Omega^j Y \longrightarrow \Omega^j Q(Y)$ is a mod-p homology monomorphism for $0 < j < r$

and

b) if (B,C) is any pair such that B and C satisfy $i)_r$ and so that the inclusion is a mod-p homology monomorphism, then the map $\mathrm{Map}(B,Q(Y)) \longrightarrow \mathrm{Map}(C,Q(Y))$ is a mod-p homology epimorphism.

We can now state

Main Theorem: Fix p and r. Assume that X satisfies $i)_r$ and that Y satisfies $ii)_r$. Then, as vector spaces,

$$H_*(\mathrm{Map}(X,Y);F_p) \simeq \bigotimes_{i=1}^{r} H_*((\Omega^i Y)^{\beta_i(X)};F_p) \ .$$

With slightly stronger hypotheses we have

Addendum: If X is a p-local co-H space, or if Y is a p-local H space, Map(X,Y) is itself a p-local H space. Assume either that X is a p-local co-H space with $b_p(X) > 1$ or that Y is a p-local H space such that the map $Y \longrightarrow Q(Y)$ is a p-local H map. Then, if X satisfies $i)_r$ and Y satisfies $ii)_r$, $H_*(\mathrm{Map}(X,Y);F_p)$ is an associative, commutative ring. Furthermore, if $H_*(\Omega^i Y;F_p)$ is a free commutative ring for all i, $b_p(X) \le i \le d_p(X)$, then $H_*(\mathrm{Map}(X,Y);F_p)$ is also a free commutative ring. The isomorphism in the main theorem is an isomorphism of rings.

Remark: Moore's results [10] apply to spaces more general than CW complexes. The main theorem can also be applied to more general spaces. The technique is to replace the spaces one has by appropriate CW complexes without changing the homology of the mapping space. We choose not to do this here as the details would lead us too far afield.

Remark: If X is a co-H space with an exponent, Map(X,Y) is an H space with an exponent. One of the techniques in [4] for showing that certain spaces have exponents is to split these spaces into products of "simpler" pieces which are known to have exponents and whose homology is known. The main theorem provides many new examples.

Here are some explicit spaces to which our results apply.

Example 1: Let $Y = S^{2n+1}$. Then Y satisfies $ii)_{2n}$. Assume that X satisfies $i)_{2n}$. If $p \neq 2$, then Y is a p-local H space with $Y \longrightarrow Q(Y)$ a p-local H map. The main theorem applies so

$$H_*(\mathrm{Map}(X,S^{2n+1});F_p) \simeq \bigotimes_{i=1}^{2n} H_*((\Omega^i S^{2n+1})^{\beta_i(X)};F_p)$$

as vector spaces. If $p \neq 2$, then the two sides are isomorphic as algebras.

Example 2: Let $Y = S^{n+1}$ and let p be even (i.e. 0 or 2). Then Y satisfies $ii)_n$. Assume X satisfies $i)_n$. Again the main theorem applies.

Example 3: If $Y = \Omega S^{n+1}$ then Y satisfies $ii)_{n-1}$. Let X satisfy $i)_{n-1}$ and let X be a suspension. The main theorem applies to show

$$H_*(\mathrm{Map}(X,S^{n+1});F_p) \simeq \bigotimes_{i=1}^{n} H_*((\Omega^i S^{n+1})^{\beta_i(X)};F_p)$$

as algebras.

Example 4: If $Y = \Omega(S^{n_1} \vee \ldots \vee S^{n_t})$ then Y satisfies $ii)_r$ for $r+2 = \min\{n_1,\ldots,n_t\}$. Let X be a suspension which satisfies $i)_r$. The main theorem applies so

$$H_*(\mathrm{Map}(X, \bigvee_{j=1}^{t} S^{n_j});F_p) \simeq \bigotimes_{i=1}^{r} H_*((\Omega^i(\bigvee_{j=1}^{t} S^{n_j}))^{\beta_i(X)};F_p)$$

as algebras: $H_*(\Omega^i(\bigvee_{j=1}^{t} S^{n_j});F_p)$ can be computed via the Hilton-Milnor theorem [12].

Example 5: Let $Y = S^{2n+1}\{p^{\ell}\}$ be the fibre of the degree p^{ℓ} map, $p^{\ell}: S^{2n+1} \longrightarrow S^{2n+1}$. Then Y satisfies $ii)_{2n-1}$ if p is odd. If X satisfies $i)_{2n-1}$

$$H_*(\mathrm{Map}(X,S^{2n+1}\{p^{\ell}\});F_p) \simeq \bigotimes_{i=1}^{2n-1} H_*((\Omega^i S^{2n+1}\{p^{\ell}\})^{\beta_i(X)};F_p)$$

as vector spaces.

We conclude this section with a problem. Suppose that $Y = S^{2n+1}$ and $d(X) + \ell(X) \leq 2n$. Then X has a (2n+1)-dual, say Z. The proof of the main theorem together with example 1 will give that the natural map

$$\psi: Q(Z) \longrightarrow \mathrm{Map}(X,QS^{2n+1})$$

is a mod-p homology isomorphism. Thus the structure of the homology of $\mathrm{Map}(X,QS^{2n+1})$ as a Hopf algebra over the Steenrod algebra is reasonably well understood [3]. Furthermore, the map

$$\phi: \mathrm{Map}(X,S^{2n+1}) \longrightarrow \mathrm{Map}(X,QS^{2n+1})$$

is a monomorphism in mod-p homology.

What is the image of ϕ_* in terms of the usual generators for $H_*(Q(Z);F_p)$? We remark that the calculations in this paper would be of more use if one could decide this question.

§2: The proof of the main theorem.

In what follows we have fixed a p, an r, and a CW complex Y satisfying $ii)_r$. With this data fixed, consider the following statement, T(X), for a space X.

$$T(X): \quad H_*(\mathrm{Map}(X,Y);F_p) \simeq \bigotimes_{i=1}^{r} H_*((\Omega^i Y)^{\beta_i(X)};F_p) \quad \text{and the map}$$

$$\phi: \mathrm{Map}(X,Y) \longrightarrow \mathrm{Map}(X,Q(Y)) \text{ is a mod-p homology monomorphism.}$$

Our goal is to prove that for any CW complex X satisfying $i)_r$, T(X) is true. Recall the following.

Lemma 2.1: Let X be a CW complex and fix an integer k. Then we can find a CW complex X_0 and a map $i: X_0 \longrightarrow X$ such that

a) $\pi_* X_0 \longrightarrow \pi_* X$ is onto for $* \leq k$

b) $H_*(X_0;Z) \longrightarrow H_*(X;Z)$ is an isomorphism for $* \leq k$

c) $H_*(X_0;Z) = 0$ for $* > k$

d) the dimension of X_0 as a CW complex is at most k+1.

Proof: Let $B \subset X$ be a k-skeleton. Then $H_k(B;Z)$ is free abelian and $H_k(B;Z) \longrightarrow H_k X;Z)$ is onto. Hence the kernel of this map is free abelian so we can attach some k+1 cells to kill this kernel precisely. The resulting complex, X_0, has a map $i: X_0 \longrightarrow X$ with all the desired properties. //

We will need the following result several times in the sequel.

Lemma 2.2: Let X satisfy $i)_r$ and let Y satisfy $ii)_r$. Then Map(X,Y) is (b(Y) - d(X))-connected and of finite type. Furthermore, if $\tilde{H}_*(X;F_p) = 0$, then $\tilde{H}_*(\mathrm{Map}(X,Y);F_p) = 0$ also.

Proof: The connectivity result is a direct consequence of Federer's spectral sequence [6]. In our case, Map(X,Y) is at least simply-connected. The last two results are mod-C arguments: E_2 of the Federer spectral sequence is of finite type, hence so is $\pi_*(\mathrm{Map}(X,Y))$; E_2 of the Federer spectral sequence is torsion prime to p, hence $\pi_*(\mathrm{Map}(X,Y))$ is torsion prime to p. Each of these homotopy results implies the respective homology result.//

The key lemma we need in our proof is

Lemma 2.3: Let $X_3 \xrightarrow{i} X_2 \xrightarrow{g} X_1$ be a cofibre sequence of CW complexes for which the cofibration, i, is a mod-p homology monomorphism. If X_2 satisfies $i)_r$ and if X_3 satisfies $i)_{r-1}$, then X_1 satisfies $i)_r$. If X_1, X_2, and X_3 satisfy $i)_r$ and if $T(X_1)$ and $T(X_3)$ are true, then $T(X_2)$ is also true.

Proof: It is no trouble to show that X_1 satisfies $i)_r$ so let us consider the map of fibrations

$$\begin{array}{ccc}
\mathrm{Map}(X_1,Y) & \xrightarrow{\phi_1} & \mathrm{Map}(X_1,Q(Y)) \\
\downarrow g^* & & \downarrow g^*_\infty \\
\mathrm{Map}(X_2,Y) & \xrightarrow{\phi_2} & \mathrm{Map}(X_2,Q(Y)) \\
\downarrow i^* & & \downarrow i^*_\infty \\
\mathrm{Map}(X_3,Y) & \xrightarrow{\phi_3} & \mathrm{Map}(X_3,Q(Y))
\end{array}$$

First we note that the Serre spectral sequence in mod-p homology for the right-hand fibration collapses: by $ii)_r$, i^*_∞ is a mod-p homology epimorphism and i^*_∞ is an ∞-loop map, so the Serre spectral sequence is a spectral sequence of algebras.

Now suppose $T(X_1)$ is true. Since both ϕ_1 and g^*_∞ are mod-p homology monomorphisms, so is g^*. Thus the Serre spectral sequence in mod-p cohomology collapses and so it does also in mod-p homology. Since $E^2 = E^\infty$,

$E^2 \simeq \bigotimes_{i=1}^{r} H_*((\Omega^i Y)^{\beta_i(X_2)};F_p)$ by our assumptions on $T(X_1)$ and $T(X_3)$. //

We can combine 2.1 and 2.3 as follows: we can assume, without loss of generality, that X is $(b_p(X) - 1)$-connected. To see this, construct the X_0 in 2.1 for $k = b_p(X) - 1$. Then apply 2.3 to $X_0 \xrightarrow{i} X \to X/X_0$ (after making i into a cofibration). Note $\tilde{H}_*(X_0;F_p) = 0$ for all *, so $\tilde{H}_*(\mathrm{Map}(X_0,Y);F_p) = 0$ for all * by lemma 2.2. Hence $T(X_0)$ is true and X_0 satisfies $i)_{b_p(X)}$. Since $b_p(X) \le r - 1$, we have that T(X) is true if $T(X/X_0)$ is.

We will now prove T(X) by induction on $s(X) = \sum_{i=1}^{\infty} \beta_i(X)$.

If X satisfies $i)_r$ and if $s(X) = 0$, then lemma 2.2 shows $\tilde{H}_*(\mathrm{Map}(X,Y);F_p) = 0$ for all * so T(X) is clearly true.

If $s(X) > 0$ and (as we may assume) X is $(b_p(X)-1)$-connected, we can find a cofibre sequence $S^t \xrightarrow{i} X \longrightarrow X_1$ (with $t=b_p(X)$) for which i is a mod-p homology monomorphism. Since X satisfies $i)_r$, lemma 2.3 applies. Since $s(X_1) = s(X) - 1$ we are done: T(X) is true for any complex satisfying $i)_r$.

We next consider our algebra results. If Y is a p-local H space or if X is a p-local co-H space then Map(X,Y) is a p-local H space. Moreover, the map $\phi: \mathrm{Map}(X,Y) \longrightarrow \mathrm{Map}(X,Q(Y))$ is a p-local H map in both cases of the addendum.

The space Map(X,Q(Y)) always has a homotopy associative, homotopy commutative H space structure coming from its ∞-loop space structure. If X is a p-local co-H space the two p-local H space structures on Map(X,Q(Y)) coincide [12] p.126. Since ϕ is a mod-p homology monomorphism, we see that the Pontrjagin ring, $H_*(\mathrm{Map}(X,Y);F_p)$, is associative and commutative.

We need some notation. If A is an algebra, let V(A) denote the module of indecomposables. (The more usual notation, QA, might be confusing.) If Z is a p-local H space, let V(Z) be $V(H_*(Z;F_p))$: $V(Z)^\beta$ denotes the direct sum of β copies of V(Z).

Now $H_*(\Omega^i Y;F_p)$ is an algebra for $i \geq 1$. If Y is also a p-local H space, recall that the two H space structures on $\Omega^i Y$ are homotopic. Hence $V(\Omega^i Y)$ is unambiguous. Consider the following statement

$$\begin{aligned} A(X)\colon\ & H_*(\mathrm{Map}(X,Y);F_p) \text{ is a free associative, commutative algebra} \\ & \text{with } V(\mathrm{Map}(X,Y)) \cong \bigoplus_{i=1}^{r} V(\Omega^i Y)^{\beta_i(X)} . \end{aligned}$$

To prove that A(X) is true in the relevant cases, we again induct on s(X). There are no difficulties if the H space structure on Map(X,Y) comes from Y. If X is a p-local co-H space there are two points to check before the proof can be completed. First note that we can assume X is $(b_p(X)-1)$-connected: since $X_0 \longrightarrow X \to X \vee X \longrightarrow X/X_0 \vee X/X_0$ is null-homotopic, X/X_0 is again a p-local co-H space. Next note that if $t=b_p(X)$ we can choose $i:S^t \longrightarrow X$ so that the cofibre of i has smaller s and is still a p-local co-H space. (This uses $b_p(X)>1$.)

§3. Some sufficient conditions for $ii)_r$.

We still need some examples of spaces Y satisfying $ii)_r$. We will assume throughout this section that Y has the homotopy type of a CW complex and that the basepoint of Y is non-degenerate. We will also consider p to be fixed.

There are two parts to $ii)_r$ and it is convenient to treat the two parts separately. They are stated below as $M_r(Y)$ and $E_r(Y)$.

$M_r(Y)$: Y is of finite type and r-connected the maps $\Omega^j Y \longrightarrow \Omega^j Q(Y)$ induce monomorphisms in mod-p homology for each j, $0 \le j \le r$.

$E_r(Y)$: Y is of finite type and r-connected if (B,C) is any CW pair such that B and C satisfy $i)_r$ and such that the inclusion is a mod-p homology monomorphism, then the map

$$\mathrm{Map}(B,Q(Y)) \longrightarrow \mathrm{Map}(C,Q(Y))$$

is a mod-p homology epimorphism.

We begin with some remarks.

Lemma 3.1: If $M_r(Y)$ is true, so is $M_{r-1}(\Omega Y)$.

Proof: There is an evaluation map $\Sigma\Omega Y \longrightarrow Y$. Furthermore $Q(Y) = \Omega Q(\Sigma Y)$ and the composite $\Omega^{j+1} Y \longrightarrow \Omega^j Q(\Omega Y) = \Omega^{j+1} Q(\Sigma\Omega Y) \longrightarrow \Omega^{j+1} Q(Y)$ is the $(j+1)^{st}$ loop of the inclusion $Y \longrightarrow Q(Y)$.//

Lemma 3.2: If $E_r(\Sigma Y)$ is true, so is $E_{r-1}(Y)$.

Proof: $\mathrm{Map}(X,Q(Y)) = \mathrm{Map}(X,\Omega Q(\Sigma Y)) = \mathrm{Map}(\Sigma X,Q(\Sigma Y))$.//

Lemma 3.3: Let Y_1 be of finite type and r-connected. Moreover, assume Y_1 is a mod-p retract of Y_2. Then, if $M_r(Y_2)$ is true, so is $M_r(Y_1)$.

Proof: We have $i: Y_1 \longrightarrow Y_2$ and $r: Y_2 \longrightarrow Y_1$ so that $r \circ i$ induces an isomorphism on $H_*(Y_1;F_p)$. But then so does $\Omega^j r \circ \Omega^j i$ so the result is an easy diagram chase.//

Lemma 3.4: Let Y_1 be of finite type and r-connected. Furthermore, assume Y_1 is a stable mod-p retract of Y_2. Then, if $E_r(Y_2)$ is true, so is $E_r(Y_1)$.

Proof: This time we know that $Q(Y_1)$ is a mod-p retract of $Q(Y_2)$. Since $d(X) \le r$, $\mathrm{Map}(X,Q(Y_1))$ is a mod-p retract of $\mathrm{Map}(X,Q(Y_2))$. This can be seen from the Federer spectral sequence. Our result is now a diagram chase.//

The next two results on wedges and products require some restrictions. Thus we pause for

Definition 3.5: Let I be an index set and let $\{Y_\alpha\}$ $\alpha \in I$ be a collection of spaces. We say this collection has strong finite type if $\bigvee_{\alpha \in I} Y_\alpha$ has finite type. We say the collection is r-connected if each Y_α is.

Remarks: If the collection has strong finite type then so does each Y_α but not conversely. For each n >0: all but finitely many of the Y_α must have $b(Y_\alpha) > n$: this condition plus each Y_α being of finite type implies the collection has strong finite type.

Lemma 2.2 shows that if $\{Y_\alpha\}$ has strong finite type and is r-connected, and if X satisfies $i)_r$, then $\{\mathrm{Map}(X,Y_\alpha)\}$ has strong finite type and is at least 1-connected.

Finally, let $\{Y_\alpha\}$ have strong finite type with all but finitely many Y_α simply-connected. Then the map $\underset{\alpha}{\times} Y_\alpha \longrightarrow \underset{\alpha}{\Pi} Y_\alpha$ is a weak equivalence where $\underset{\alpha}{\Pi} Y_\alpha$ denotes the Cartesian product and $\underset{\alpha}{\times} Y_\alpha$ denotes the weak product (the subset of $\underset{\alpha}{\Pi} Y_\alpha$ with all but finitely many coordinates being the respective basepoints). The result is clear since $\pi_*(\underset{\alpha}{\Pi} Y_\alpha)$ = direct product $\pi_*(Y_\alpha)$ and $\pi_*(\underset{\alpha}{\times} Y_\alpha)$ = direct sum $\pi_*(Y_\alpha)$. Now we can prove

Lemma 3.6: Suppose $\{Y_\alpha\}$ is a collection of strong finite type and that each $M_r(Y_\alpha)$ is true. Then $M_r(\underset{\alpha}{\Pi} Y_\alpha)$ and $M_r(\underset{\alpha}{\times} Y_\alpha)$ are true.

Proof: We have a map $Q(\underset{\alpha}{\Pi} Y_\alpha) \longrightarrow \underset{\alpha}{\Pi} Q(Y_\alpha)$ and we know that $\Omega^j \underset{\alpha}{\Pi} Z_\alpha = \underset{\alpha}{\Pi} \Omega^j Z_\alpha$. The following diagram commutes

$$\begin{array}{ccccc} \Omega^j(\underset{\alpha}{\Pi} Y_\alpha) & \longrightarrow & \underset{\alpha}{\Pi}\Omega^j Y_\alpha & \xleftarrow{\;4\;} & \underset{\alpha}{\times}\Omega^j Y_\alpha \\ \downarrow 1 & & \downarrow 2 & & \downarrow 3 \\ \Omega^j Q(\underset{\alpha}{\Pi} Y_\alpha) & \longrightarrow & \underset{\alpha}{\Pi}\Omega^j Q(Y_\alpha) & \xleftarrow{\;5\;} & \underset{\alpha}{\times}\Omega^j Q(Y_\alpha) \end{array}$$

Applying the Künneth formula to $\underset{\alpha}{\times}$ (but not to $\underset{\alpha}{\Pi}$) we see that the map labeled 3 is a mod-p homology monomorphism.

Since $\{Q(Y_\alpha)\}$ has strong finite type if $\{Y_\alpha\}$ does, 4 and 5 are $H_*(\ ;F_p)$ isomorphisms. Hence 1 is an $H_*(\ ;F_p)$ monomorphism.//

Lemma 3.7: Suppose $\{Y_\alpha\}$ is a collection of strong finite type and suppose that each $E_r(Y_\alpha)$ is true. Then $E_r(\underset{\alpha}{\vee} Y_\alpha)$ is also true.

Proof: Since $Q(\underset{\alpha}{\vee} Y_\alpha) = \underset{\alpha}{\Pi} Q(Y_\alpha)$ we have $\mathrm{Map}(X,Q(\underset{\alpha}{\vee} Y_\alpha)) = \underset{\alpha}{\Pi}\mathrm{Map}(X,Q(Y_\alpha))$. Since $\{\mathrm{Map}(X,Q(Y_\alpha))\}$ has strong finite type if X satisfies $i)_r$, $\mathrm{Map}(X,Q(\underset{\alpha}{\vee} Y_\alpha))$ and $\underset{\alpha}{\times}\mathrm{Map}(X,Q(Y_\alpha))$ are weak equivalent. The result follows easily from these remarks.//

Our next result produces lots of examples which satisfy E_r.

Proposition 3.8: Let Y be simply-connected and of finite type. Suppose $b(Y) - \ell(Y) \geq r$. Then $E_r(Y)$ is true.

Proof: Since $\ell(Y) \geq 1$, Y is r-connected. The result will follow easily once we can compute $\mathrm{Map}(X,Q(Y))$, so we begin to describe the answer. The first step is to replace X by a space which more accurately reflects the mod-p properties of X. First note that by lemmas 2.1 and 2.2 we can mod out by a p-acyclic complex, X_0, so that $\mathrm{Map}(X/X_0,Q(Y)) \longrightarrow \mathrm{Map}(X,Q(Y))$ is a p-local weak equivalence. Note X/X_0 is $(b_p(X)-1)$-connected.

We can also find $\hat{X} \subset X/X_0$ so that $d(\hat{X}) = d_p(X/X_0)$ and again $\mathrm{Map}(X/X_0,Q(Y))$ —— $\mathrm{Map}(\hat{X},Q(Y))$ is a p-local weak equivalence. If we have a map $h\colon X_1 \longrightarrow X_2$ we can find $\bar{h}\colon X_1/X_{10} \longrightarrow X_2/X_{20}$ and $\hat{h}\colon\hat{X}_1 \longrightarrow \hat{X}_2$ so that

$$\begin{array}{ccc}
\mathrm{Map}(X_2,Q(Y)) & \xrightarrow{\hat{h}^*} & \mathrm{Map}(X_1,Q(Y)) \\
\uparrow & & \uparrow \\
\mathrm{Map}(X_2/X_{20},Q(Y)) & \longrightarrow & \mathrm{Map}(X_1/X_{10},Q(Y)) \\
\downarrow & & \downarrow \\
\mathrm{Map}(X_2,Q(Y)) & \xrightarrow{h^*} & \mathrm{Map}(X_1,Q(Y))
\end{array}$$

commutes.

Thus, in verifying $E_r(Y)$, it suffices to demonstrate the required epicity under the further assumption that B and C satisfy $\mathfrak{f})_r$, where we say that a space satisfies $\mathfrak{f})_r$ if it satisfies $i)_r$ and if X is $(b(X)-1)$-connected with $d(X) + \ell(X) \leq r$.

Now if X satisfies $\mathfrak{f})_r$ there exists a spectrum called the 0-Spanier-Whitehead dual to X, denoted X^*. It follows from the Freudenthal suspension theorem that $\Sigma^r X^*$ is the suspension spectrum of a CW complex.

The Freudenthal suspension theorem also gives that an r-connected complex of finite type with $b(Y) - \ell(Y) \geq r$ is an r-fold suspension. Hence $X^* \wedge Y$ can be taken to be a CW complex.

From Spanier-Whitehead duality we have a map $X \wedge (X^* \wedge Y) \longrightarrow Q(Y)$ which we can adjoint to get $\psi \colon X^* \wedge Y \longrightarrow \mathrm{Map}(X,Q(Y))$.

We remark that ψ is natural in X and Y. Given $h\colon X_1 \longrightarrow X_2$ we get a stable map $h^\# \colon X_2^* \longrightarrow X_1^*$ and the Freudenthal suspension theorem gives that there is a real map $\Sigma^r X_2^* \longrightarrow \Sigma^r X_1^*$ realizing h . Hence we get a map $h^\# \wedge 1 \colon X_2^* \wedge Y \longrightarrow X_1^* \wedge Y$. If $g\colon Y_2 \longrightarrow Y_1$ is an r-fold suspension it is easy to define $h^\# \wedge g \colon X_2^* \wedge Y \longrightarrow X_1^* \wedge Y$ and

$$\begin{array}{ccc}
X_2^* \wedge Y_2 & \xrightarrow{\psi} & \mathrm{Map}(X_2,Q(Y_2)) \\
\downarrow{\scriptstyle h^\# \wedge g} & & \downarrow{\scriptstyle h^* \circ g^\#} \\
X_1^* \wedge Y_1 & \xrightarrow{\psi} & \mathrm{Map}(X_1,Q(Y_1))
\end{array}$$

commutes up to homotopy. For g to be an r-fold suspension, the Freudenthal suspension theorem requires $b(Y_2) \leq b(Y_1)$.

Consider the statement $T(X,Y)$.

$T(X,Y)$: $\psi \colon X^* \wedge Y$ —— $\mathrm{Map}(X,Q(Y)$ adjoints to an equivalence $\psi_\infty \colon Q(X^* \wedge Y)$—— $\mathrm{Map}(X,Q(Y))$ using the ∞-loop space structure in $\mathrm{Map}(X,Q(Y))$.

Our first goal is to prove that $T(X,Y)$ is true if X satisfies $\mathfrak{f})_r$ and if Y is simply-connected of finite type with $b(Y) - \ell(Y) \geq r$.

The proof resembles the proof of the main theorem so much that we will only give a sketch. It suffices to show that ψ_∞ induces a mod-p homology isomorphism for all p. We will induct on $s(X^* \wedge Y)$.

If $s(X^* \wedge Y) = 1$, X and Y are mod-p spheres. For spheres the result is clear and by lemma 2.2 the result follows for mod-p homology spheres.

If $s(X^* \wedge Y) > 1$ we can find $i\colon S^m \longrightarrow X$ or $j\colon S^n \longrightarrow Y$ so that the induced map in mod-p homology is a monomorphism. If i exists, let X_0 be S^m and let Y_0 be Y: if i does not exist (i.e. $s(X^*)=1$) let X_0 be X and let Y_0 be the cofibre of j. Note that $Y \longrightarrow Y_0$ is always an r-fold suspension. Then

$$\begin{array}{ccc} Q(X^*\wedge Y) & \longrightarrow & Map(X,Q(Y)) \\ \downarrow & & \downarrow \\ Q(X_0^*\wedge Y_0) & \longrightarrow & Map(X_0,Q(Y_0)) \end{array}$$

commutes. Since $s(X_0^*\wedge Y_0) < s(X^*\wedge Y)$ we can assume by induction that the bottom map is a mod-p homology isomorphism. The mod-p homology Serre spectral sequence for the left-hand fibration collapses by direct calculation. Hence it also collapses for the right-hand fibration.

The fibres are $Q(X_1^*\wedge Y_1)$ and $Map(X_1,Q(Y_1))$ respectively for a suitable choice of X_1 and Y_1. The ψ_∞ on the fibres commutes with the ψ_∞ on the total spaces. Since $s(X_1^*\wedge Y_1) < s(X^*\wedge Y)$ we have our result.

Once we know that ψ_∞ is an equivalence the proposition is easy to prove. Since we may assume that B and C satisfy $\text{‡})_r$, it suffices to recall that if $i:B \longrightarrow C$ is a mod-p homology monomorphism, $i:B^* \longrightarrow C^*$ is a mod-p homology epimorphism. This proves 3.8.//

Remark: The reader may well wonder how we have reduced Spanier's function space approach to S-duality to such trivial manipulations. What we use periodically is the work of Dyer-Lashof [5] and Araki-Kudo[1]. Life is a great deal easier when one knows that $H_*(Q(Z);F_p)$ is a functor of $H_*(Z;F_p)$ with the new classes being given by operations which are natural with respect to ∞-loop maps.

There exist spaces, such as $J(Y) = \Omega\Sigma Y$, which do not have finite ℓ but which are stably equivalent to a wedge of spaces which do. Precisely, let us consider spaces Y with the property that there is a collection $\{Y_\alpha\}$ of strong finite type which is r-connected so that $b(Y_\alpha) - \ell(Y_\alpha) \geq r$ for each $\alpha\varepsilon I$ and so that there is an equivalence in the stable category between Y and $\vee_\alpha Y_\alpha$. Any space with this property will be called an I_r-space.

Example 3.9: If Y is an I_r-space so is J(Y).

Proof: $J(Y) \underset{S}{\cong} \vee_{k=1} Y^{[k]}$ where $Y^{[k]}$ denotes the k-fold smash. This is due to James[8]. If Y is an I_r-space, so is $Y^{[k]}$ by direct calculation.//

Example 3.10: Let Y be an I_r-space and suppose A is an I_0-space. Let $i:A \longrightarrow Y$ be a map, and let J(Y,A) denote the fibre of the map $Y/A \longrightarrow \Sigma A$ in the Barratt-Puppe sequence for i. Then J(Y,A) is an I_r-space.

Proof: B. Gray [7] has shown that $J(Y,A) \underset{S}{\cong} \vee_{k=0} Y\wedge A^{[k]}$ (where $A^{[0]} = S^0$). Again the result follows from a short calculation.//

Corollary 3.11: $E_r(Y)$ is true for any I_r-space Y.

Proof: By 3.7 and 3.8 $E_r(\vee_\alpha Y_\alpha)$ is true. By 3.4, $E_r(Y)$ is also true.//

Remark: As concrete examples we have: for the Moore space $P^r(p^k)$, E_{r-2} is true ($P^r = S^{r-1} \cup e^r$); $E_{r-2}(\Omega P^{r+1})$ is true; if $S^r\{p^k\} \longrightarrow S^r \xrightarrow{p^k} S^r$ is a fibration, $E_{r-2}(S^r\{p^k\})$ is true.

Corollary 3.11 can be used to get many examples of spaces for which E_r is true. Some of these spaces have interesting retracts so, by 3.4, E_r is true for them also. The situation with regard to M_r is much less satisfactory. Lemma 3.6 deals satisfactorily with products and 3.1 allows us to loop down, but these are usually not useful procedures. Worse still is our substitute for proposition 3.8. We have

Proposition 3.12: Let Y be an r-connected mod-p homology sphere of dimension m+1. If pm is even, $M_r(Y)$ is true: $M_{r-1}(\Omega Y)$ is always true.

Proof: The first part is a direct calculation from [3]. Note that if pm is odd, $M_r(Y)$ is false.

If pm is even, $M_{r-1}(\Omega Y)$ is true by 3.1. If pm is odd, p is odd and (m+1) is even. Hence, at p, $\Omega Y \simeq S^m \times \Omega S^{2m+1}$: since $M_{m-1}(S^m)$ and $E_{2m}(\Omega S^{2m+1})$ are both true, it follows that $E_{m-1}(\Omega S^{m+1})$ is true and hence so is $M_{r-1}(\Omega Y)$.//

The Hilton-Milnor theorem can be used to push on a bit.

Lemma 3.13: Let $\{Y_\alpha\}$ be an r-connected collection of strong finite type. For any $L \varepsilon I^n$ let $Y_L = Y_{\alpha_1} \wedge \ldots \wedge Y_{\alpha_m}$. Suppose $M_r(J(Y_L))$ is true for all $L \varepsilon I^n$ and all n. Then $M_r(J(\vee_\alpha Y_\alpha))$ is true.

Proof: The Hilton-Milnor theorem [12] says that $J(\vee_\alpha Y_\alpha) = \times J(Y_L)$ as L runs over a specified set of sequences. The result follows easily from 3.6 once one checks that the collection $\{Y_L\}$ one acquires is of strong finite type.//

Remarks: Corollary 3.11 and proposition 3.12 give us examples 1,2 and 3 in the introduction. Example 4 follows from 3.11, 3.12, and 3.13. Example 5 follows from 3.11, 3.12 and

Lemma 3.14: Let $i: A \longrightarrow Y$ be an r-fold suspension and suppose $M_r(J(A))$ and $M_r(Y/A)$ are true. Then, if i induces a monomorphism in mod-p homology, $M_r(J(Y,A))$ is also true.

Proof:We have the fibration $J(A) \longrightarrow J(Y,A) \xrightarrow{\beta} Y/A$ and a map $J(Y,A) \longrightarrow J(Y)$ so that the composite $J(A) \longrightarrow J(Y,A) \longrightarrow J(Y)$ is just $J(i)$. Since i is an r-fold suspension and a mod-p homology monomorphism,

$\Omega^j J(A) \longrightarrow \Omega^j J(Y,A)$ is a mod-p homology monomorphism for $0 \le j \le r$. (This uses that $H_*(\Omega^n \Sigma^n X; F_p)$ is functorially determined by $H_*(X;F_p)$.) Hence the mod-p homology Serre spectral sequences for the fibrations

$\Omega^j J(A) \longrightarrow \Omega^j J(Y,A) \longrightarrow \Omega^j Y/A$ collapse.

Let F denote the homotopy theoretic fibre of the map $Q\beta$. We have a map

$Q(J(Y,A)) \longrightarrow Q(J(Y))$ and, just as above, $Q(J(A)) \longrightarrow F \longrightarrow Q(J(Y,A))$ is a mod-p homology monomorphism. This fails to prove triviality of the Serre spectral sequence since $H_*(F;F_p)$ is much larger than $H_*(Q(J(A));F_p)$. So we turn to the map $Q(J(Y,A)) \longrightarrow Q(Y/A)$. There is a map $Y \longrightarrow J(Y,A)$ so that the composite $Y \longrightarrow J(Y,A) \longrightarrow Y/A$ is the collapse map, which is onto in mod-p homology. Hence $Q(J(Y,A)) \longrightarrow Q(Y/A)$ is onto in mod-p homology. Since $Y \longrightarrow Y/A$ is an r-fold suspension, $\Omega^j Q(J(Y,A)) \longrightarrow \Omega^j Q(Y/A)$ is a mod-p homology epimorphism for $0 \le j \le r$ so the mod-p homology Serre spectral sequences for these fibrations collapse.

If $M_r(J(A))$ and $M_r(Y/A)$ are both true, it is an easy diagram chase to show that $M_r(J(Y,A))$ is true.//

Remark: We can apply example 3.10 and lemma 3.14 to show that the following space satisfies $ii)_{r-2}$. Let $\xi: S^{2k+r} \longrightarrow S^{r+k+1}$ be any map with rp even. Then the fibre of ξ satisfies $ii)_{r-2}$.

BIBLIOGRAPHY

1.S. Araki and T. Kudo, Topology of H_n-spaces and H-squaring operations, Mem. Fac. Sci. Kyusyu Univ. Ser. A 10 (1956), 85-120.

2.K. Borsuk, Concerning the homological structure of the functional space S_m^X, Fund. Math. 39 (1952), 25-37.

3.F.R. Cohen, T.J. Lada, and J.P. May, The homology of iterated loop spaces, Lecture Notes in Math. no. 533, Springer, 1976.

4.F. R. Cohen, J.C. Moore, and J.A. Neisendorfer, The double suspension and exponents of the homotopy groups of spheres, Ann. of Math. (2) 110 (1979), 549-565.

5.E. Dyer and R.K. Lashof, Homology of iterated loop spaces, Amer. J. Math. 84 (1962), 35-88.

6.H. Federer, A study of function spaces by spectral sequences, Trans. Amer. Math. Soc. 82 (1956), 340-361.

7.B. Gray, On the homotopy groups of mapping cones, Proc. Adv. Study Inst. Alg. Top. (1970), 104-142, Aarhus, Denmark.

8.I. M. James, Reduced product spaces, Ann. of Math. (2) 62 (1955), 170-197.

9.S. Mardesic, Sur l'homologie de l'espace fonctionnel S_m^X et la structure homologique de X, C.R. Acad.Sci. Paris 242 (1956), 1112-1114.

10.J.C. Moore, On a theorem of Borsuk, Fund. Math. 43 (1956), 195-201.

11.E.H. Spanier, Function spaces and duality, Ann. of Math. (2) 70 (1959), 338-378.

12.G.W. Whitehead, Elements of homotopy theory, Graduate Texts in Math. no. 61, Springer, 1978.

M.I.T., Department of Mathematics,
Cambridge, MA 02139

University of Notre Dame, Dept. of Math
Notre Dame, IN 46556

Contemporary Mathematics
Volume 19, 1983

SOME NEW IMMERSIONS AND NON-IMMERSIONS OF REAL PROJECTIVE SPACES

Donald M. Davis[1]

ABSTRACT. Using obstruction theory applied to certain modifications of the stable normal bundle, we obtain new immersions and nonimmersions of real projective space RP^n in Euclidean space when n has three, four, or five 1's in its binary expansion. A table of known immersions and nonimmersions for $n \leq 191$ is included.

1. MAIN RESULTS AND TECHNIQUES. Let $\alpha(n)$ denote the number of 1's in the binary expansion of n, and $\nu(n)$ the exponent of 2 in n. Our main results are

THEOREM 1.1. a) If $\alpha(n) = 4$ and $n \equiv 2(8)$, then real projective space RP^n cannot be immersed in $(\not\subseteq) R^{2n-13}$.
b) If $\alpha(n) = 5$ and $n \equiv 4(8)$, then $RP^n \not\subseteq R^{2n-17}$.

THEOREM 1.2. If n satisfies the indicated congruence and conditions and $\alpha(n)$ is as below, then RP^n can be immersed in $(\subseteq) R^{2n-d}$:

	n	α(n)	other conditions	d
a)	4(8)	3		9
b)	6(8)	4		9
c)	1(8)	5		12
d)	0(8)		$\alpha(n-1)=6, n\neq 64$	13
e)	2(8)	5		13
f)	14(16)	5	$n\neq 62$	14
g)	12(16)	≥ 5	$\nu(n+4)<7$	12

These results are all new, beginning with $RP^{28} \subseteq R^{47}$. In Section 2 we tabulate all known immersions and nonimmersions of RP^n for $n \leq 191$. This extends and updates previous tables of [Gi], [J2], and [Be2].

Theorems 1.1 and 1.2 are proved by using the methods of obstruction theory introduced in [DM1], [DM2], and [D]. The only novelty here is that instead of applying these methods to the stable normal bundle η_n of

1980 Mathematics Subject Classification: Primary 57R42, 55S35; Secondary 55S40, 55S45, 55T15.

[1]Supported by N.S.F. Research Grant.

RP^n, we apply them to $\eta_n \otimes \xi_n$ or $\eta_n - k\xi_n$, where ξ_n is the Hopf bundle over RP^n and k is a small positive integer

The obstruction theory for nonimmersions is carried out in Section 4, where we obtain the following lower bounds for geometric dimension (gd):

THEOREM 1.3. a) If $\alpha(\ell) = 3$, then $gd((16\ell-8)\xi_{8\ell+2}) > 8\ell-9$.

b) If $\alpha(\ell) = 4$, then $gd((16\ell-8)\xi_{8\ell+4}) > 8\ell-11$.

Theorem 1.1 follows immediately from 1.3 and Sanderson's observation ([S]) that Hirsch's theorem applied to the twisted normal bundle $\eta_n \otimes \xi_n$ implies

$$P^n \subseteq R^{2n-k} \text{ iff } gd((2n-k+1)\xi_n) \leq n-k . \qquad (1.4)$$

Actually, the nonimmersions of 1.1 use lower bounds 2 smaller than those of 1.3.

In Sections 3 and 5 we prove

THEOREM 1.5. If j is as below, then $gd(m\xi_n) \leq e$:

	$\alpha(j)$	other	m	n	e
a)	2		$16j$	$8j+4$	$8j-5$
b)	2		$16j+4$	$8j+6$	$8j-3$
c)	4		2^L-8j-4	$8j+1$	$8j-13$
d)	3	$j \neq 7$	$16j+4$	$8j+8$	$8j-5$
e)	≥ 4		$16j-8$	$8j+2$	$8j-11$
f)	2	$j \neq 3$	$2^L-16j-16$	$16j+14$	$16j-1$
g)	≥ 3	$\nu(j+1)<3$	$2^L-16j-16$	$16j+12$	$16j-3$.

Parts a), b), d), and e) of Theorem 1.2 are immediate from 1.4 and 1.5. Theorem 1.2c follows from 1.5c and Hirsch's theorem ([H]):

$$gd(\eta_{8j+1}) = gd((2^L-8j-2)\xi_{8j+1}) \leq gd((2^L-8j-4)\xi_{8j+1})+2 \leq 8j-11,$$

and 1.2fg is proved similarly.

2. TABLE OF IMMERSIONS AND NONIMMERSIONS. In this section we tabulate all immersions and nonimmersions of RP^n for $n \leq 191$ known to the author. We point out one change from the compendia of [G1], [J2], and [Be2]: for $i \geq 10$ and $i = 7$ immersions of RP^{2^i-1} which would show that James' nonimmersions are best possible are no longer claimed. In 1979 Crabb and Steer found a gap in the argument of [GM2] which had produced these immersions. For $i \leq 9$, $i \neq 7$, the immersions are known by other methods.

In the table below, references are included, with numbers referring to theorems of this paper. The left side gives families of results, while the right side gives exceptional results. "Dim" refers to $P^n \not\subseteq R^n$ for dimensional reasons.

$P^{2^i+d} \subseteq (\not\subseteq) R^{2^{i+1}+e}$

i≥	d	e(⊆)		e(⊄)	
2	0	-1	W	-2	Mi2
3	1	-1	S	-2	Mi2
3	2	0	S	-1	BB
3	3	0	S	-1	BB
4	4	2	A	1	G_1
4	5	6	S	5	AG
4	6	6	S	5	AG
4	7	6	S	5	AG
4	8	7	R	5	AG
4	9	14	S	13	AG
4	10	14	S	13	AG
4	11	14	S	13	AG
4	12	15	2a	13	AG
4	13	18	R	13	AG
4	14	19	2b	14	AD
5	15	21	DM2	14	AD
5	16	23	R	18	AD
5	17	30	S	29	AG
5	18	30	S	29	AG
5	19	30	S	29	AG
5	20	31	2a	29	AG
5	21	34	R	29	AG
5	22	35	2b	30	AD
5	23	37	DM2	30	AD
5	24	39	R	34	AD
5	25	42	R	34	AD
5	26	45	DM2	39	1a
5	27	45	DM2	39	1a
6	28	46	2f	42	AD
6	29	46	2f	42	AD
6	30	46	2f	42	AD
6	31	48	DM2	44	DM2
6	32	51	2d	44	DM2
6	33	62	S	61	AG
6	34	62	S	61	AG
6	35	62	S	61	AG
6	36	63	2a	61	AG
6	37	66	R	61	AG
6	38	67	2b	62	AD
6	39	69	DM2	62	AD
6	40	71	R	66	AD
6	41	74	R	66	AD
6	42	77	DM2	71	1a
6	43	77	DM2	71	1a
6	44	78	2f	74	AD
6	45	78	2f	74	AD
6	46	78	2f	74	AD
6	47	80	DM2	76	DM2
6	48	83	2d	76	DM2
6	49	90	R	76	DM2
6	50	93	DM2	87	1a
6	51	93	DM2	87	1a
6	52	95	Bel	90	AD
6	53	96	DM2	90	AD

$P^n \subseteq (\not\subseteq) R^m$

n	m(⊆)		m(⊄)	
2	3	W	2	Dim
3	4	Mi1	3	Dim
5	7	H	6	Mi2
6	7	H	6	Mi2
7	8	H	7	Dim
12	18	L	17	G1
13	22	S	21	G1
14	22	S	21	G1
15	22	S	21	J1
31	53	DM2	52	J1
60	111	Bel	106	AD
61	114	R	106	AD
62	115	DM2	106	AD
63	115	DM2	114	J1
124	238	Bel	231	1b
125	238	Bel	231	1b
126	238	Bel	232	AD
127	240	DM2	238	J1

$i \geq$	d	$e(\subseteq)$		$e(\not\subseteq)$	
6	54	96	DM2	90	AD
6	55	96	DM2	92	DM2
6	56	99	2d	92	DM2
6	57	102	2c	92	DM2
6	58	103	2e	96	AD
6	59	108	DM2	96	AD
7	60	108	2g	103	1b
7	61	110	Bel	103	1b
7	62	110	Bel	104	AD
7	63	112	DM2	106	D

3. IMMERSIONS IMPLIED BY MPT's. In this section we prove parts (a), (b), (c), and (d) of 1.5, using modified Postnikov towers (MPT's) as in [DM1] and [DM2; Section 2]. The case j odd ($=2\ell-1$) of 1.5d is the hardest of these, and so we consider it in some detail.

Let
$$\begin{array}{ccccc} & & B^0_{16\ell-13} & \longrightarrow & \widetilde{BSp}_{16\ell-13} \\ & & \downarrow & & \downarrow \\ & & E^0_s & \longrightarrow & E_s \\ & & & \searrow \quad \swarrow & \\ RP^{16\ell} & \xrightarrow{h} & QP^{4\ell} & \xrightarrow{g} & BSp \end{array}$$

denote 16ℓ-MPT's as in [DM2; Section 2], and gh classify $(32\ell-12)\xi$. Relations for the k-invariants are obtained similarly to [GM1] or [R]. In the table below $k^s_j \in H^{16\ell-16+j}(E_s)$. These classes correspond to elements of $Ext_A(H^*P_3)$ in [M; p. 56]. We do not list elements in the top degree (16ℓ) because they will be easily handled by Sq^1-indeterminacy.

k^1_6 : $Sq^2Sq^1w_{16\ell-12}$

k^1_7 : $(Sq^4+w_4)w_{16\ell-12}$

k^1_8 : $Sq^1w_{16\ell-8} + Sq^2Sq^3w_{16\ell-12}$

k^1_9 : $Sq^2w_{16\ell-8} + (Sq^4+w_4)Sq^2w_{16\ell-12}$

k^1_{11}: $(Sq^8+w_8)w_{16\ell-12} + w_4w_{16\ell-8}$

k^1_{15}: $(Sq^8+w_8)w_{16\ell-8}$

k^2_7 : $Sq^2k^1_6$

k^2_8 : $Sq^1k^1_8 + Sq^2Sq^1k^1_6$

k^2_{10}: $Sq^2k^1_9 + Sq^3k^1_8 + (Sq^4+Sq^3Sq^1+w_4)k^1_7$

k^2_{12}: $Sq^5Sq^1k^1_7 + (Sq^7+(Sq^4+w_4)Sq^2Sq^1)k^1_6$

k^2_{13}: $Sq^2Sq^1k^1_{11} + w_4Sq^2k^1_8 + Sq^4Sq^2Sq^1k^1_7 + (Sq^8+w_8+w_4^2+w_4Sq^4)k^1_6$

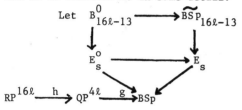

k_{15}^2: $Sq^1k_{15}^1 + Sq^2Sq^3k_{11}^1 + (Sq^7+(Sq^4+w_4)Sq^2Sq^1)k_9^1 + (Sq^8+w_8+Sq^6Sq^2)k_8^1$
$\quad + w_4Sq^2Sq^3k_7^1 + w_4Sq^6k_6^1$

k_8^3 : $Sq^2k_7^2 + Sq^1k_8^2$

k_{12}^3: $Sq^1k_{12}^2 + (Sq^4+w_4)Sq^2k_7^2$

k_{13}^3: $Sq^2k_{12}^2 + (Sq^7+Sq^4Sq^2Sq^1)k_7^2$

k_{15}^3: $Sq^3k_{13}^2 + (Sq^4+Sq^3Sq^1+w_4)k_{12}^2 + Sq^5Sq^1k_{10}^2 + (Sq^9+w_8Sq^1+w_4^2Sq^1)k_7^2$,

k_{12}^4: $Sq^1k_{12}^3 + (w_4+Sq^4)Sq^1k_8^3$

k_{14}^4: $Sq^2k_{13}^3 + Sq^3k_{12}^3 + Sq^4Sq^2Sq^1k_8^3$

k_{15}^4: $Sq^1k_{15}^3 + Sq^2Sq^1k_{13}^3 + (Sq^4+w_4)k_{12}^3 + Sq^7Sq^1k_8^3$

k_{14}^5: $Sq^2Sq^1k_{12}^4$

k_{15}^5: $Sq^1k_{15}^4 + Sq^2k_{14}^4 + (Sq^4+w_4)k_{12}^4$

k_{15}^6: $Sq^2k_{14}^5$

PROPOSITION 3.1. gh lifts to E_3, with only k_8^3 and k_{12}^3 mapping nontrivially.

Proof. $\nu\binom{8\ell-3}{4\ell-\varepsilon} = \begin{cases} \alpha(\ell-1)+1 = \alpha(j) = 3 & \text{if } \varepsilon = 1 \text{ or } 2 \\ \alpha(\ell)-1 = \alpha(j)-\nu(j+1) > 0 & \text{if } \varepsilon = 0 \text{ or } 3 \end{cases}$. Thus by [DM;1.8] $g|QP^{4\ell-2}$ lifts to $B^o_{16\ell-11}$ but not to $B^o_{16\ell-13}$, and hence $g|QP^{4\ell-2}$ lifts to $E_3^o(16\ell-13)$ with $g_3^*(k^3_{16\ell-8}) \neq 0$. [DM1;1.8] also implies $g|QP^{4\ell-1}$ lifts to $B^o_{16\ell-7}$ but not to $B^o_{16\ell-9}$, which combines with the preceding sentence to show $g|QP^{4\ell-1}$ lifts to $E_3^o(16\ell-13)$ with $g_3^*(k^3_{16\ell-4\varepsilon}) \neq 0$ for $\varepsilon = 1$ and 2. Since $\pi_{4*}(E_3^o(16\ell-13),E_3(16\ell-13)) = 0$, there is a similar lifting of $g|QP^{4\ell-1}$ to $E_3(16\ell-13)$.

3.1 follows from the fact that for $s \leq 3$ there is a lifting $\tilde{g}_s$ of gh to E_s such that $\tilde{g}_s|RP^{16\ell-2}$ equals the composite

$$RP^{16\ell-2} \xrightarrow{h} QP^{4\ell-1} \xrightarrow{g_s} E_s,$$

where g_s is the lifting of the preceding paragraph. This fact is true because the only additional k-invariant requiring consideration for $RP^{16\ell}$ is $k^s_{16\ell}$, and this is in Sq^1-indeterminacy for $s \geq 1$, i.e. if $g_s^*(k^s_{16\ell}) \neq 0$, then let g'_s denote the composite

$$RP^{16\ell} \xrightarrow{x^{16\ell-1}\times g_s} K_{16\ell-1}\times E_s \xrightarrow{\mu} E_s$$

where $K_n = K(Z_2,n)$. ∎

Varying the map $\widetilde{g}_3 : RP^{16\ell} \to E_3$ through $K_{16\ell-10}$ changes $\widetilde{g}_3^*(k_{16\ell-\varepsilon})$ only for $\varepsilon = 4$ and 8. This is seen, as in [DM2;p. 365], by using the relations in the MPT above. The new map $\widetilde{g}_3'$ is the composite

$$RP^{16\ell} \xrightarrow{x^{16\ell-10} \times \widetilde{g}_3} K_{16\ell-10} \times E_3 \xrightarrow{\mu} E_3 ,$$

and, for example,

$$\begin{aligned}\widetilde{g}_3'^*(k_{16\ell-4}) &= Sq^4Sq^2x^{16\ell-10} + w_4((32\ell-12)\xi)\cdot Sq^2x^{16\ell-10} + \widetilde{g}_3^*k_{16\ell-4} \\ &= 0 + x^{16\ell-4} + x^{16\ell-4} = 0 .\end{aligned}$$

As in the proof of 3.1, the above analysis ignores the relations for $k^3_{16\ell}$, because it can be changed by varying through $K_{16\ell-1}$, if necessary.

Thus there is a lifting $\widetilde{g}_4 : RP^{16\ell} \to E_4$. The only control which we have over its effect on k-invariants is that the relation listed as k^5_{15} above implies that either both or neither of $k^4_{16\ell-2}$ and $k^4_{16\ell-1}$ map nontrivially. Varying through $K_{16\ell-9}$ changes $k^4_{16\ell-4}$, varying through $K_{16\ell-5}$ changes $k^4_{16\ell-4}$, $k^4_{16\ell-2}$, and $k^4_{16\ell-1}$, and varying through $K_{16\ell-1}$ changes $k^4_{16\ell}$. Thus there is a lifting $\widetilde{g}_4'$ which annihilates all k-invariants, and hence there is a lifting $\widetilde{g}_5$ to E_5. $k^5_{16\ell-1}$ and $k^5_{16\ell}$ are in the indeterminacy, and the relation $Sq^2k^5_{16\ell-2} = 0$ implies $\widetilde{g}_5^*k^5_{16\ell-2} = 0$.

Thus there is a lifting $\widetilde{g}_6$ to E_6. As usual, $k^6_{16\ell}$ is in the indeterminacy. Since $Sq^1k^6_{16\ell} + Sq^2k^6_{16\ell-1} = 0$ (by [M;p.56]), $Sq^1\widetilde{g}_6^*k^6_{16\ell} + Sq^2\widetilde{g}_6^*k^6_{16\ell-1} = 0$. This implies $g_6^*k^6_{16\ell-1} = 0$ once we observe that $\widetilde{g}_6$ extends over $RP^{16\ell+1}$. (E_6 is not part of a $(16\ell+1)$-MPT, but since the map $RP^{16\ell} \to BSp$ extends over larger projective spaces, and there were no k-invariants of degree $> 16\ell$, the liftings all extend.) Thus there is a lifting to E_7, where the only k-invariant is in Sq^1-indeterminacy, and hence a lifting to $E_8 = \widetilde{BSp}_{8\ell-13}$.

The rest of 1.5(a-d) is similar (and easier).

4. PROOF OF NONIMMERSIONS. In this section we use the method of [DM2] to prove

THEOREM 4.1 a) If $\alpha(\ell) = 3$, then

$$RP^{8\ell+2} \xrightarrow{(16\ell-8)\xi} BO\langle 8\rangle \to BO\langle 8\rangle/BO_{8\ell-9}\langle 8\rangle = C_{8\ell-9}$$

is essential.

b) If $\alpha(\ell) = 4$, then

$$RP^{8\ell+4} \xrightarrow{(16\ell-8)\xi} BO\langle 8\rangle \to BO\langle 8\rangle/BO_{8\ell-11}\langle 8\rangle = C_{8\ell-11}$$

is essential.

This immediately implies 1.3, and hence 1.1. Here BO<8> is the classifying space for vector bundles trivial on the 7-skeleton.

We extend slightly the Adams spectral sequence (ASS) charts for $\pi_*(C_N)$ given in [DM2;3.2]. These charts are just $\mathrm{Ext}_{A_2}(\Sigma P_N)$ through degree N+16, and this has been computed completely in [DM3]. Here A_r is the subalgebra of the mod 2 Steenrod algebra A generated by $\{Sq^j : j \le 2^r\}$, $\Sigma P_N = H^*(\Sigma RP_N)$, $RP_N = RP^\infty/RP^{N-1}$, and Z_2 is omitted from the second component of Ext( ). The dot in [DM2;3.2], $N \equiv 7$, $s = 1$, $t-s = N+15$ was a misprint.

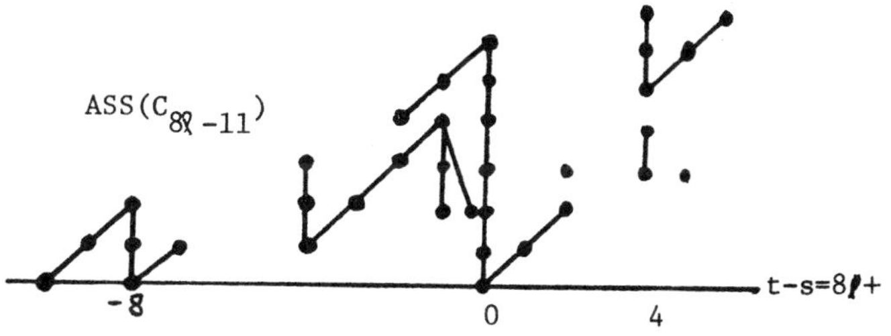

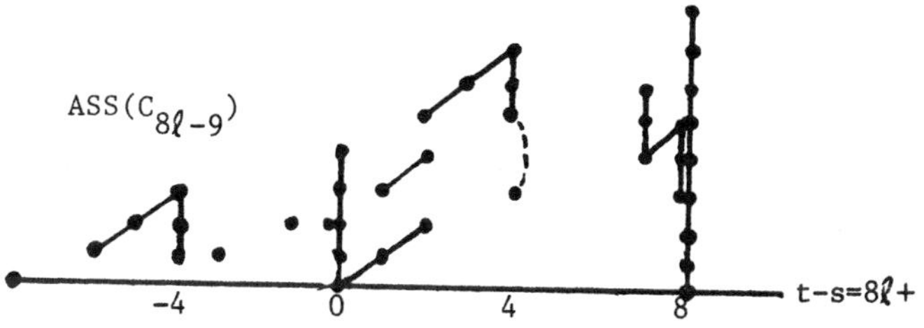

We will also need the following extension of the minimal resolutions of [DM2;Section 6], using the notation of those tables.

H^*P_5 : $16 \le t-s \le 21$

g_{16} : $2\alpha_{15} + 424\alpha_7$

h_{17} : $2g_{16} + 3g_{15} + (46+721)g_8 + 52g_{11} + 561g_6$

k_{17} : $(3+21)h'_{15} + 4h_{14} + 6h_{12} + 7h_{11} + (56+731)h_7$

k_{19} : $3h_{17} + 5h_{15} + 5h'_{15} + 6h_{14} + 463h_7$

k_{20} : $(6+42)h'_{15} + \widehat{11}.21h_7$

ℓ_{19} : $1k_{19} + (3+21)k_{17} + 6k_{14} + (7+421)k_{13} + 27k_{11}$

m_{19} : $51\ell_{14} + (7+421)\ell_{13}$

n_{19} : $1m_{19} + 41m_{15}$

n_{20} : $2m_{19} + (7+421)m_{14}$

p_{19} : $1n_{19} + 23n_{15}$

p_{21} : $2n_{20} + 3n_{19} + (7+421)n_{15}$

H^*P_7 : $20 \le t-s \le 22$

ℓ_{22} : $31k_{19} + (7+421)k_{16} + 462k_{11}$

m_{22} : $1\ell_{22} + (42+51)\ell_{17}$

n_{22} : $1m_{22} + 4m_{19} + 6m_{17}$

We form Adams resolutions of C_N corresponding to the above charts and prove

PROPOSITION 4.2. a) The map of 4.1(a) lifts to a filtration 2 map sending only $k_{8\ell}$ nontrivially.
b) The map of 4.1(b) lifts to a filtration 3 map sending only $\ell_{8\ell}$ nontrivially.

Proof a) We use the pseudo-factorization through $QP^{2\ell}$ as in [DM2;4.1,4.1', 4.1"] and argue as in [DM2;5.1]. We use [DM1;1.8] and

$$\nu\binom{4\ell-2}{2\ell+\varepsilon} = \begin{cases} \alpha(\ell)-1 & \varepsilon = -2 \\ \alpha(\ell)+\nu(\ell) & \varepsilon = -1 \\ \alpha(\ell)-1 & \varepsilon = 0 \\ \alpha(\ell)+\nu(\ell-1) & \varepsilon = 1 \end{cases}$$

to lift $QP^{2\ell}$ to E_2^o by a map sending only $k_{8\ell}$ nontrivially. Comparing ASS's of $\Sigma P_{8\ell-9} \to \Sigma P_{8\ell-9} \wedge bo$, this implies that $QP^{2\ell}$ lifts to E_2 and hence E_2'' sending only $k_{8\ell}$ (and perhaps $k'_{8\ell}$) nontrivially. Thus $RP^{8\ell+2}$ lifts to E_2'', $E_2''\langle 8\rangle$, and E_2', sending only $k_{8\ell}$ (and perhaps $k'_{8\ell}$) nontrivially. But $k'_{8\ell}$ is in primary indeterminacy. Since the MPT of $\xi \to BO\langle 8\rangle$ corresponds to the Adams resolution of $C_{8\ell-9}$, the result follows.
b) A similar method works. One might worry about getting $QP^{2\ell+1}$ past the non bo-primary class in $\pi_{8\ell}(C_{8\ell-9})$, but this is unncecessary because it it not present in $\mathrm{Ext}_A(\Sigma P_{8\ell-9})$, which form the obstructions for lifting $QP^{2\ell+1}$ to E_3. For ASS(P_N) we use the tables of [M]. ∎

4.1 follows from 4.2 and

PROPOSITION 4.3. a) A filtration 2 map $RP^{8\ell+2} \to C_{8\ell-9}$ sending only $k_{8\ell}$ nontrivially is nontrivial.
b) A filtration 3 map $RP^{8\ell+4} \to C_{8\ell-11}$ sending only $k_{8\ell}$ nontrivially is nontrivial

Proof a) We calculate $\mathrm{Ext}_A^{2,2}(H^*C_{8\ell-9}, H^*P^{8\ell+2})$ as in [DM2;3.7] by using the minimal resolution extended above, and find $\hat{k}_{8\ell} \neq 0$. It cannot be hit by a differential in the ASS converging to $[P^{8\ell+2}, C_{8\ell-9}]$ because $\mathrm{Ext}_A^{0,1}(H^*C_{8\ell-9}, H^*P^{8\ell+2}) \approx Z_2$, and the nonzero class survives to give the map $P^{8\ell+2} \to V_{8\ell-9} \to \Omega C_{8\ell-9}$.

b) Similar. We use the minimal resolution to show $\hat{k}_{8\ell} \neq 0 \in \mathrm{Ext}_A^{3,3}(H^*C_{8\ell-11}, H^*P^{8\ell+4})$, and also to show $\mathrm{Ext}_A^{s,s+1}(H^*C_{8\ell-11}, H^*P^{8\ell+4}) = Z_2$ for $s = 0$ or 1, with these classes related by h_o. These latter classes survive to give $P^{8\ell+4} \xrightarrow{i} \Omega C_{8\ell-11}$ and $2i$, so that $\hat{k}_{8\ell}$ cannot be hit by a differential in the ASS converging to $[P^{8\ell+4}, C_{8\ell-11}]$. ∎

5. PROOF OF REMAINING IMMERSIONS. In this section we prove parts e, f, and g of 1.5. We begin by proving the following analogue of [DM2;1.4], which immediate implies 1.5(e).

THEOREM 5.1. If g classifies $(16j-8)\xi$ with $\alpha(j) \geq 4$ and $\mathcal{E} = \mathrm{fibre}(k_o)$, then in

$$\begin{array}{ccccc} & & BO_{8j-11}\langle 8\rangle & & \\ & & \downarrow & & \\ \Omega C\times\mathcal{E} & \xrightarrow{\mu} & \mathcal{E} & \xrightarrow{k_1} & \mathcal{E}/BO_{8j-11}\langle 8\rangle \\ & & \downarrow & & \\ P^{8j+2} & \xrightarrow{g} & BO\langle 8\rangle & \xrightarrow{k_o} & BO\langle 8\rangle/BO_{8j-11}\langle 8\rangle = C, \end{array}$$

i) $k_o g$ is null-homotopic so that there is a lifting ℓ of g,
ii) there is a map $P^{8j+2} \to \Omega C$ such that $k_1\mu(f\times\ell)$ is null-homotopic,
iii) the fibre of k_1 has the same $(8j+2)$-type as $BO_{8j-11}\langle 8\rangle$.

5.1(iii) follows from [DM2;4.3 and following paragraph]. 5.1(i) follows from

PROPOSITION 5.2. i) $k_o g$ lifts to a filtration 3 map annihilating all k-invariants except perhaps k_{8j}.
ii) Any such map in null-homotopic.

Proof The proof of (i) is almost a verbatim copy of that of 4.2. QP^{2j} lifts to E_3^o, sending k_{8j} nontrivially if $\alpha(j) = 4$, and sending all k-invariants trivially if $\alpha(j) > 4$.

To prove (ii) we use the method of [DM2;3.7] to show that $\mathrm{Ext}_{A_2}^{s,t}(H^*\Sigma P_{8j-11}, H^*P^{8j+2})$ is in $0 \leq t-s \leq 1$

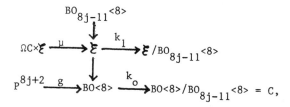

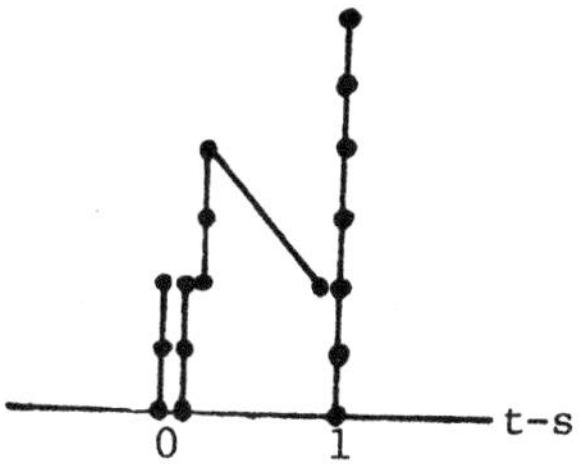

Using the minimal resolution of H^*P_5 of [DM2;Section 6], the element in $\mathrm{Ext}^{4,4}$ is $\hat{\ell}_{14}$, and that of $\mathrm{Ext}^{2,3}$ not in the tower is $\hat{h}_{14} + \hat{h}'_{15}$. The The d_2-differential in the chart for $\mathrm{ASS}(C_5)$ in [DM2;p. 367] implies the one indicated above. The element in $\mathrm{Ext}^{3,3}$ is $\hat{k}_{13} + \hat{k}_{14}$, which is not a map of the type described in (i). Thus a map of (i) has filtration ≥ 4 and hence is 0. ∎

<u>Proof of 5.1(ii)</u> From [DM2;4.3] and the method of [DM2;4.9] $\mathrm{ASS}(\xi/BO_{8j-11}\langle 8\rangle)$ begins

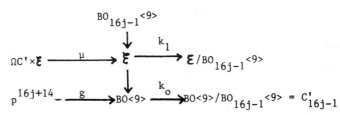

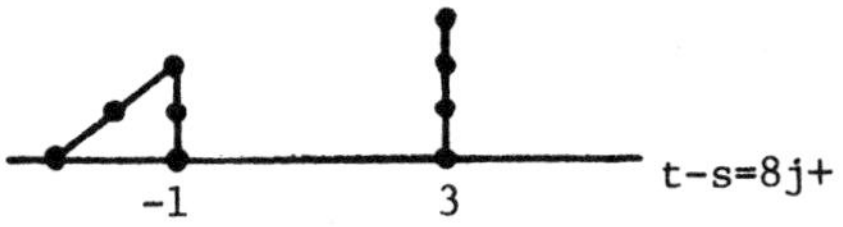

and $[P^{8j+2}, \xi/BO_{8j-11}\langle 8\rangle] \approx Z_8$. Since the bundle is an odd multiple of 8ξ, the method of [DM2;4.11] shows that if f is an appropriate multiple of the map f_o of [DM2;top of p. 375], then the composite of [DM2;4.6] will cancel $\Sigma k_1\ell$ in $[\Sigma P^{8j+2}, \Sigma\xi/BO_{8j-11}\langle 8\rangle]$. ∎

Parts (f) and (g) of 1.5 are immediate consequences of the following result.

THEOREM 5.3. a) If $\alpha(j) \geq 2$ and $\nu(j+1) < 2$, then in

$$\begin{array}{ccccc} & & BO_{16j-1}\langle 9\rangle & & \\ & & \downarrow & & \\ \Omega C'\times\xi & \xrightarrow{\mu} & \xi & \xrightarrow{k_1} & \xi/BO_{16j-1}\langle 9\rangle \\ & & \downarrow & & \\ P^{16j+14} & \xrightarrow{g} & BO\langle 9\rangle & \xrightarrow{k_o} & BO\langle 9\rangle/BO_{16j-1}\langle 9\rangle = C'_{16j-1} \end{array}$$

where ξ = fibre(k_o) and g classifies $(2^L-16j-16)\xi$,

i) k_og is null-homotopic so that there is a lifting ℓ of g,

ii) there is a map $P^{16j+14} \xrightarrow{f} \Omega C'$ such that $k_1\mu(f\times\ell)$ is null-homotopic,

iii) $BO_{16j-1}\langle 9\rangle \to$ fibre(k_1) is a $(16j+14)$-equivalence.

b) The statement obtained from (a) by replacing 2, 1, and 14 by 3, 3, and 12, respectively, is true.

<u>Proof.</u> Part (iii) is easily proved, similarly to 5.1 (iii).

To prove part (a)(i), we argue as in 5.2(i) to see that k_og lifts to a filtration 3 map sending only k_{16j+8} nontrivially if $\alpha(j) = 2$ (and to a filtration 4 map if $\alpha(j) > 2$). The details are similar to [D;4.1], using the ASS chart of [D;2.1] and the coefficients

$$\nu\binom{2^i-4j-4}{4j+\varepsilon} = \alpha(j) + \begin{cases} 0 & \\ 2 & \text{if } \varepsilon = 1 \\ 1 & \qquad\ \ 2 \\ 2 & \qquad\ \ 3 \end{cases}\ \ \begin{matrix} 0 \\ \\ \\ \\ \end{matrix}.$$

This map, $\hat{k}_{16j+8}$, is in primary indeterminacy, and hence $k_o g$ lifts to filtration 4. (a)(i) now follows from

LEMMA 5.4. Any filtration 4 map $P^{16j+14} \to C'_{16j-1}$ is null-homotopic.

<u>Proof</u> The map BO<9> → BO<8> induces $C'_{16j-1} \to C_{16j-1}$ which is an equivalence above filtration 0, and hence it suffices to prove any filtration 4 map $P^{16j+14} \to C_{16j-1}$ is null-homotopic. The ASS converging to $[\Sigma^i P^{16j+14}, C_{16j-1}]$ is

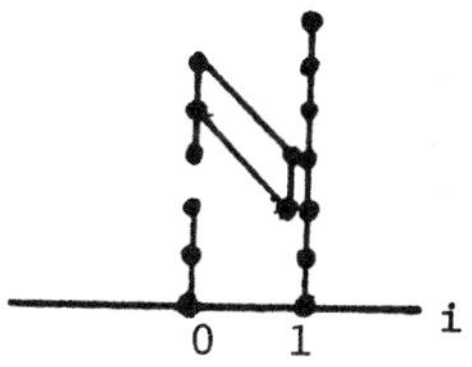

The Ext groups are calculated by the method used in 5.2, using the minimal A_2-resolution of H^*P_7 given in [DM2]. The differential can be proved by naturality from that of [DM2;3.6], or by duality from $[P^{16j+14}, C_{16j-1}] \approx [P^{16j+14}, \Sigma P_{16j-1}^{16j+14} \wedge MO\langle 8\rangle] \approx [P^{16j+16}, \Sigma P_{16j+1}^{16j+16} \wedge MO\langle 8\rangle]$. The differentials in the last ASS follow easily from those in $ASS(P_{16j+1} \wedge MO\langle 8\rangle)$. The above isomorphism utilize results of [DGIM].

The proof of (b)(i) is very similar. The ASS chart for C'_{16j-3} is

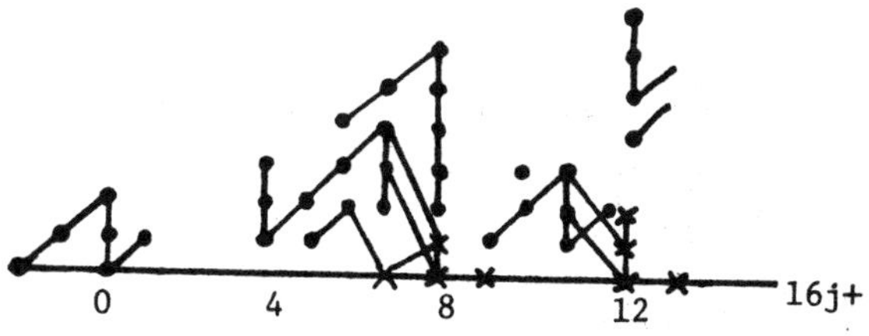

We use towers similar to [DM2;4.1,4.1',4.1"] with BO<9> replacing BO<8>. By the usual comparison with the situation where only bo-primary obstructions are present, we get QP^{4j+3} to lift to E_4, using tables of [M]. (Some care is required to get past the element in [M;p.66,s=2,t-s=k+14]. We use

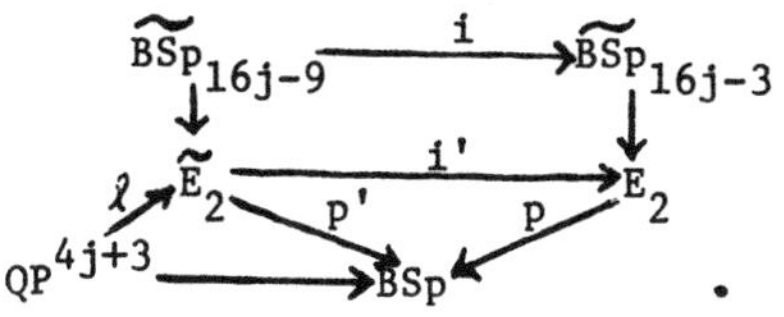

.

ℓ^* annihilates all k-invariants except perhaps the non-bo-primary classes $\tilde{k}_{16j}$ and $\tilde{k}_{16j+12}$. Some calculation with minimal resolutions shows $j'^*(k_{16j+12})$ does not contain $\tilde{k}_{16j+12}$ as a term, and $\ell^*(A(BO)_{12}\cdot k_{16j}) = 0$.) Thus RP^{16j+12} lifts to E_4' sending only $k_{16\ell+8}$ nontrivially. But this is in primary indeterminacy, and maps of higher filtration are in higher order indeterminacy, i.e. killed by differentials in the ASS converging to $[RP^{16j+12}, C'_{16j-3}]_*$, which is seen as in part (a) by comparing with BO<8>.

The proof of a(ii) is in [D;Proof of 4.2]. The details are virtually identical, since the BO-spaces are the same in both cases. The bundle being lifted is different, having $\alpha(j)$ one less in (a)(ii) than in [D], but it is on a projective space of dimension one less. The only significant difference is that we use $P^{n+1} \subseteq R^{2n-14}$ from [DM2;1.1] here rather than $P^n \subseteq R^{2n-14}$ from [DM2;1.2]. The argument here is slightly easier since we don't have the final summand of [D;4.8] to worry about.

We should point out a few minor mistakes in [D;Section 4]. The proof of [D;4.8] refers to a chart, 2.2, which is not present. It is $ASS(\mathcal{E}/BO_{16j-1}\langle 9\rangle)$

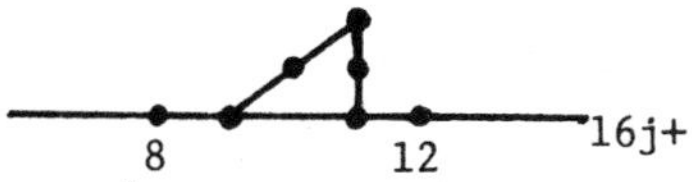

In [D;4.9,4.10, and top of p. 41], G_1 should be replaced by G_2, and G_2 should be replaced by G_4. The kernel of $[P^n, j_3]$ also contains G_1, but it is in primary indeterminacy.

(b)(ii) is similar. $ASS(\mathcal{E}/BO_{16j-3}\langle 9\rangle)$ is

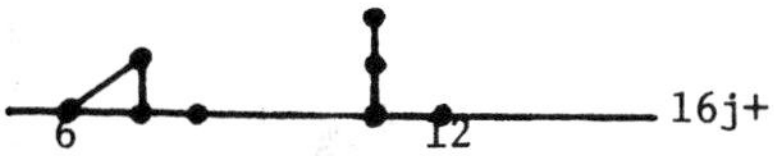

so that $[P^{16j+12}, \mathcal{E}/BO_{16j-3}\langle 9\rangle] \approx Z_8 \oplus Z_2 \oplus Z_2$, with the Z_2's due to filtration 0 Z_2's in $\pi_*(\mathcal{E}/BO_{16j-3}\langle 9\rangle)$ which are in the primary indeterminacy from $\pi_*(\Omega C'_{16j-3})$. We use the fact ([DM2;1.1]) that $(2^L-16j-16)\xi_{16j+15}$ lifts to 16j+1 to deduce that there is $f: P^{16j+12} \to \Omega C'_{16j-3}$ such that $k_1\mu(f\times\ell)$ is 0 or 4G. In the latter case, use [D;4.10] and the hypothesis $\nu(j+1) < 3$ to see that adding to f an appropriate multiple of f_o will cancel the 4G. ∎

BIBLIOGRAPHY

A J. Adem, "Some immersions associated with bilinear maps," Bol. Soc Mat. Mex 13(1968) 95-104.

AG _______ and S. Gitler, "Nonimmersion theorems for real projective spaces," Bol. Soc. Mat Mex 9(1964) 37-50.

AD L. Astey and D.M. Davis, "Nonimmersions of real projective spaces implied by BP, "Bol. Soc. Mat Mex 24(1979) 49-55.

BB P. Baum and W. Browder, "The cohomology of quotients of classical groups," Topology 3(1965) 305-336.

Be1 A.J. Berrick, "The Smale invariants of an immersed projective space," Math. Proc. Camb Phil. Soc. 86(1979) 401-411.

Be2 A.J. Berrick, "Projective space immersions, bilinear maps, and stable homotopy groups of spheres," S.V. Lecture Notes in Math 788(1979) 1-22.

D D.M. Davis, "Connective coverings of BO and immersions of projective spaces," Pac. Jour. Math. 76(1978) 33-42.

DGIM _________, S. Gitler, W. Iberkleid, and M. Mahowald, "The orientability of vector bundles with respect to certain spectra," Bol. Soc. Mat. Mex. 24(1979) 49-55.

DM1 D.M. Davis and M. Mahowald, "The geometric dimension of some vector bundles over projective spaces," Trans. Amer Math Soc 205(1975) 295-316.

DM2 _________, "The immersion conjecture for $RP^{8\ell+7}$ is false," Trans Mex Math Soc 236(1978) 361-383.

DM3 _________, "Ext over the subalgebra A_2 of the Steenrod algebra for stunted real projective space," Amer Math Soc Contemp Math Series, Western Ontario Conference, 1981.

G1 S. Gitler, "Immersion and embedding of manifolds," Proc Symp in Pure Math, Amer Math Soc, 22(1971) 87-96.

G2 _________, "The projective Stiefel manifolds II - Applications," Topology 7(1968) 47-53.

GM1 _________ and M. Mahowald, "The geometric dimension of real stable vector bundles," Bol Soc Mat Mex 11(1966) 85-107.

GM2 _________, "The immersion of manifolds," Bull Amer Math Soc 73(1967) 696-700.

H M.W. Hirsch, "Immersion of manifolds," Trans. Amer. Math. Soc. 93(1959) 242-276.

J1 I.M. James, "On the immersion problem for real projective spaces," Bull Amer Math Soc 69(1963) 231-238.

J2 __________, "Euclidean models of projective spaces," Bull London Math Soc 3(1971) 257-276.

L K.Y. Lam, "Construction of nonsingular bilinear maps," Topology 6(1967) 423-426.

M M. Mahowald, "The metastable homotopy of S^n," Mem. Amer. Math. Soc. 72(1967).

Mi1 J. Milnor, "On the immersion of n-manifolds in (n+1)-space," Comm Math Helv 30(1956) 275-284.

Mi2 J. Milnor, "Lectures on characteristic classes," Princeton Univ, 1957, published in 1975 as Annals Math Sudies 76.

R A.D. Randall, "Some immersion theorems for projective spaces," Trans Amer Math Soc 147(1970) 135-151.

S B.J. Sanderson, "Immersions and embeddings of projective spaces," Proc London Math Soc 14(1964) 137-153.

W H. Whitney, "The singularities of a smooth n-manifold in (2n-1)-space," Ann Math 45(1944) 247-293.

DEPARTMENT OF MATHEMATICS
LEHIGH UNIVERSITY
BETHLEHEM, PA. 18015

Contemporary Mathematics
Volume 19, 1983

Simple proofs of some Borsuk-Ulam results

by Albrecht Dold

The starting point of A.Liulevicius' lecture*) (these proceedings) is the following

THEOREM. *If a map* f: $S^m \to S^n$ *commutes with some free actions of a finite group* G $\neq \{1\}$ *on the spheres* S^m, S^n *then* $n \geq m$. *If* $n = m$ *then* f *is not nullhomotopic*.

This can be proved as follows. If $m \geq n$ then, for any free G-actions on S^m, S^n, there is a G - map

(1) $\quad \gamma: S^n \to S^m$

because S^m is (m-1)-connected, hence m-universal ([St], 19.4). The composite map γf: $S^m \to S^m$ has fixed point index

(2) $\quad I(\gamma f) \equiv 0$ mod order of G

- essentially because the fixed point set of γf consists of full G-orbits, so the "number of fixed points $I(\gamma f)$" is divisible by the length of the orbits. If $n < m$, or if f is nullhomotopic, then γf is nullhomotopic, hence

(3) $\quad I(\gamma f) = I(\text{constant}) = 1$

which contradicts (2).

We now provide the details of such a proof. Clearly, we can assume that G is of prime order p, say

---

*) I wrote this note after listening and talking to him at Aarhus, in August 1982.

(4) $G = \{z \in \mathbb{C} \mid z^p = 1\}$, p prime.

Scalar multiplication defines the standard operation of G on the unit sphere in $\mathbb{C}^r$ (resp. $\mathbb{R}^r$ if $p = 2$). We write $\$^k$ to indicate that the k-sphere is taken with this standard operation.

PROPOSITION 1. *If* $\alpha: \$^k \to \$^k$ *commutes with the (standard) G-action then* $I(\alpha) \equiv 0 \bmod p$.

Proof. Let $\pi: \$^k \to L^k = \$^k/G$ denote the projection onto the orbit (lens-)space. The map α induces a map $\bar{\alpha}: L^k \to L^k$. We can deform $\bar{\alpha}$ into a map $\bar{\beta}$ with finitely many fixed points $b_1, \ldots, b_s$, and, by the covering homotopy, deform α into a G-map $\beta: \$^k \to \$^k$ with $\pi\beta = \bar{\beta}\pi$; then $I(\alpha) = I(\beta)$. But the fixed point set of β consists of some of the orbits $\pi^{-1}(b_j)$, and all points x in one such orbit have the same index because π is homeomorphic in the neighborhood of x. Therefore, the contribution of $\pi^{-1}(b_j)$ to $I(\beta)$ is divisible by p, for all j, hence $p/I(\beta)$. □

PROPOSITION 2. *For any free G-action on the k-sphere* S^k *there are G-maps* (i) $\gamma: S^k \to \$^k$, *and* (ii) $\delta: \$^k \to S^k$.

Proof. For (ii) we can refer to [St] because $\$^k/G$ is a cell complex; the proof in [St], second half of p.102, is simple. For (i) we first construct a G-map $\Gamma: S^k \to \$^{2K-1}$ for some (large) K. We take a finite open covering $\mathcal{V} = \{U_1, U_2, \ldots, U_K\}$ of S^k/G which trivializes the G-bundle $\pi: S^k \to S^k/G$; thus, for each $j = 1, \ldots K$ we have a G-map $q_j: \pi^{-1}(U_j) \to G \subset \mathbb{C}$. Also, we take a partition of unity $\rho_j: S^k/G \to [0,1]$ for $\mathcal{V}$ $(\rho_j^{-1}(0,1] \subset U_j$, $j = 1, \ldots, K)$, and we define

$\Gamma: S^k \to \$^{2K-1} \subset \mathbb{C}^K$ with components $\Gamma_j: S^k \to \mathbb{C}$, $j = 1,\dots,K$, by $\Gamma_j(x) = \sqrt{\rho_j(\pi(x))}\, q_j(x)$; note that $\Gamma_j(x) = 0$ if $\pi(x) \notin U_j$.

The G-map Γ induces a map $\bar{\Gamma} : S^k/G \to \$^{2K-1}/G = L^{2K-1}$. We can deform $\bar{\Gamma}$ into a map $\bar{\gamma}$ whose image lies in L^k (cf. LEMMA below), and, by the covering homotopy, deform Γ into a G-map γ whose image lies in $\$^k$. □

LEMMA. _If Z is a cell complex and Y a (para-)compact space of dimension d < ∞ , then every map_ $\eta: Y \to Z$ _can be deformed into a map whose image lies in the d-skeleton_ Z^d _of Z._

Proof. If Z is simplicial with barycentric coordinates $\lambda_j: Z \to [0,1]$, $j \in J$, then $\{\lambda_j \eta\}_{j \in J}$ is a partition of unity on Y which has a d-dimensional refinement $\{\mu_i: Y \to [0,1]\}_{i \in I}$ (i.e. such that at no point $y \in Y$ more than d+1 of the $\mu_i(Y)$ are non-zero). Thus, η factors, up to homotopy, as follows

$$Y \xrightarrow{\mu} N \xrightarrow{\sigma} Z^d \subset Z ,$$

where N = nerve of $\{\mu_i\}$ and σ is simplicial (sending the vertex i into a vertex j such that $\mu_i^{-1}(0,1] \subset \eta^{-1}\lambda_j^{-1}(0,1]$).

In general, Z may not be simplicial but $D = Z^r - Z^{r-1}$ (or Z^r-neighborhood of Z^{r-1}) is, and can be used as above to deform η away from D for $r > d$. □

The _proof of the theorem_ now works as indicated: If $m \geq n$ then by prop.2 we have G-maps $S^n \xrightarrow{\gamma} \$^m \xrightarrow{\delta} S^m$. The composite $\gamma f \delta: \$^m \to \$^m$ satisfies $I(\gamma f \delta) \equiv 0 \mod p$ by prop.1. If $n < m$, of if f is nullhomotopic then $\gamma f \delta$ is nullhomotopic, hence $I(\gamma f \delta) = 1$ - a contradiction. □

REMARK. Essentially the same proof gives the following result. *If a map* $f: X \to Y$ *commutes with some free actions of a nontrivial finite group on* X *and* Y *then*

$$\text{Dimension}(Y) \geq 1 + \text{Connectivity}(X) .$$

If $\text{Dimension}(Y) = 1 + \text{Connectivity}(X) < \infty$ *then* f *is not nullhomotopic* (assuming Y paracompact). □

We conclude with one more result in the same spirit,

THEOREM'. *If* X *is a finite-dimensional paracompact space and* $f: X \to X$ *is a nullhomotopic map then* f *does not commute with any free* G*-action on* X *for any finite group* $G \neq \{1\}$.

Proof. We can assume G of prime order as above (4). Given any free G-action on X, and $k >$ dimension(X), there is a G-map $\gamma: X \to \$^k$ (same proof as for prop.2), and γ is nullhomotopic. If f commutes with the G-action then there is also a G-map $\varepsilon: \$^k \to X$ (cf. below). The composite $\gamma\varepsilon$ satisfies $I(\gamma\varepsilon) \equiv 0 \bmod p$ by prop.1, and $I(\gamma\varepsilon) = 1$ because $\gamma\varepsilon$ is nullhomotopic - a contradiction.

It remains to construct ε. In fact, we can construct a G-map $\varepsilon: Y \to X$ for every free G-space Y such that Y/G is a finite cell complex. By induction on the number of cells we can assume $Y/G = Z \cup e = Z$ with a cell e attached, and a G-map $\varepsilon_Z: \pi^{-1}(Z) \to X$, where $\pi: Y \to Y/G$ is the projection (covering map). Let $\Phi: B \to Y$ a lift of a characteristic map $B \to Y/G$ for e, where B = standard ball. The restriction of Φ to the boundary $\dot{B}$ composes with $f\varepsilon_Z$ to a nullhomotopic map, i.e. $f\varepsilon_Z(\Phi|\dot{B})$ extends to a map η: $B \to X$, and therefore $f\varepsilon_Z$: $\pi^{-1}(Z) \to X$ extends to a G-map ε: $Y \to X$ (such that $\varepsilon\Phi = \eta$). □

Reference: [St] = N.Steenrod, The topology of fibre bundles. Princeton University Press 1951

Mathematisches Institut der Universität
6900 Heidelberg, Germany Fed.Rep.

Contemporary Mathematics
Volume 19, 1983

Alteration of H-structures

J. Harper and A. Zabrodsky

Introduction Let X be an H-space (not necessarily finite) with given multiplication $\mu : X\times X \to X$. Other multiplications are determined, up to homotopy, by $W\varepsilon[X\wedge X,X]$ using the composition

$$\mu_w = \mu \circ \mu \times w \circ \Delta_{X\times X}$$

It's often useful to know how the Hopf algebra structure of $H^*(X)$ can be altered by changing μ. In general, one has the formula

$$\mu_w^*(x) = \mu^*(x) + w^*(x) + \sum_i \mu^*(x_i')w^*(x_i'')$$

where, as usual, $\mu^*(x) = 1\otimes x + x\otimes 1 + \sum_i x_i' \otimes x_i''$. In this paper we construct some elements w which are useful in an inductive procedure towards making $H^*(X)$ primitively generated. As an application, we shall give a proof for the main result in [H] which is substantially simpler than the original proof.

Next we state a useful consequence of our main result. Let X be an H-space with $H^*(X)$ primitively generated in dimensions $<n$ and suppose a choice of primitive generators has been made (suppressed coefficients are Z/pZ and p is an odd prime). Introduce the subspace $G_t \subset \tilde{H}^*(X)$, $t=1,2,\ldots$ to be the subspace spanned by t-fold products of primitive generators ($G_1 \cong Q\tilde{H}^*(X)$ for $*<n$). We can write

$$(0.1)\quad H^n(X\wedge X) = \bigoplus_{s\geq 2}\ \bigoplus_{a+b=s} G_a \otimes G_b$$

where here and below it is understood that elements not of dimension n are excluded from the right side. Note that if λ is a primitive $(p-1)$-st root of unity and ϕ the λ-th power map of X, then the λ-eigenspace E_λ of $(\phi_\lambda \wedge \phi_\lambda)^*$ on $H^n(X\wedge X)$ is given by

$$(0.2)\quad E_\lambda = \bigoplus_{s\equiv 1\,(p-1)}\ \bigoplus_{a+b=s} G_a \otimes G_b ,$$

Roughly speaking, only a certain subspace of E_λ is relevant to the problem of finding some multiplication such that $H^*(X)$ is

0271-4132/83 $1.00 + $.25 per page

primitively generated in dimensions $< n+1$. In particular, those terms in (0.1) where $a+b \not\equiv 1\ (p-1)$ do not matter. Furthermore, even when $a+b \equiv 1\ (p-1)$, there are restrictions. To describe these, write $s \equiv 1\ (p-1)$ as $s = \ell(p-1)+1$ and introduce

$$u = \left[\frac{\ell}{2} + 1\right](p-1)$$

where [ ] is the greatest integer function.

Theorem (0.3) If in (0.2) $E_\lambda \subset \bigoplus_{\ell \geq 1} \bigoplus_{s-u<r<u} G_r \otimes G_{s-r}$ then X has a multipication ν such that $H^*(X)$ is primitively generated in dimensions $<n+1$.

As a simple illustration of the use of this theorem we have

Proposition. Let X be an H-space with $H^*(X)$ exterior on r generators, $r \leq p-1$. Then $H^*(X)$ can be made primitively generated.

Proof. Let n be the lowest dimension where $H^*(X)$ fails to be primitive. Since there are at most $p-2$ generators of dimension $<n$, $G_s = 0$ for $s \geq p-1$. Thus in (0.2) we need only consider $s=p$. Then $\ell=1$ and $u=p-1$, and we have $E_\lambda \subset \bigoplus_{1<r<p-1} G_r \otimes G_{p-r}$. The proposition follows from the theorem by induction on n.

The paper has the following organization. Section 1 exhibits certain elements $w \in [X \wedge X, X]$ and a method for combining them to produce others. In section 2 we prove our main result. The promised application to the discussion of the "torsion molecule" of [H] is made in section 3.

Section 1 We first introduce notation. The j-fold Cartesian product of X is written X^j, and the j-fold smash product $X^{(j)}$. We shall often write induced maps in cohomology in the form

$$[X^{(j)}, X] \xrightarrow{\bigoplus_k I_k} \bigoplus_k \mathrm{Hom}(H^k(X), H^k(X^{(j)})).$$

We use $+$ to denote either the composition of maps induced by a given multiplication μ on X or the addition of maps in Hom. We say that a map $w \in [X^{(j)}, X]$ is (n-1)-connected if $I_k(w) = 0$ for $k<n$ (this is not standard usage). Note that given maps w_1, w_2 with w_2 (n-1)-connected, then

$$(1.1) \quad I_k(w_1+w_2) = I_k(w_1) \quad \text{for} \quad k<n$$

$$I_n(w_1+w_2) = I_n(w_1) + I_n(w_2).$$

Now we recall a form of difference construction discussed in [Z]. Let $c:X\to X$ satisfy the equation $Id+c=*$ where $*$ is the constant map. Then given $f,g:X^j\to X$ which are homotopic on the fat wedge, we have the fact that $f+cg$ factors through $X^{(j)}$ by a map well defined up to homotopy, which we denote by

$$D_j(f,g):X^{(j)}\to X.$$

If $I_k f = I_k g$ for $k<n$, then

(1.2) $D_j(f,g)$ is $(n-1)$-connected

$$I_n(D_j(f,g)) = I_n(f) - I_n(g).$$

In particular, $\mu\circ 1\times\mu$ and $\mu\circ\mu\times 1$ agree up to homotopy on the fat wedge in X^3, so we have

$$A = D_3(\mu\circ\mu\times 1,\ \mu\circ 1\times\mu).$$

If $H^*(X)$ is primitively generated in dimensions $<n$, then straightforward calculation and (1.2) yield that

(1.3) $I_k A = 0, \quad k<n$

$$I\ A = (\bar{\mu}^*\otimes 1 - 1\otimes\bar{\mu}^*)\bar{\mu}^*$$

where, as usual, $\bar{\mu}^*$ is the reduced coproduct. Note that $\bar{\mu}^*$ is not likely to be induced, even though its summands are.

Next we introduce a gradation of $H^*(X^{(j)})$ using the power maps of X. This gradation extends the scope of the gradation discussed in the introduction. Given (X,μ), let $\lambda:X\to X$ denote the λ-th power map with respect to μ. Note that on the module of indecomposibles $QH^*(X)$, λ induces multiplication by the mod p reduction of the integer λ. We shall therefore regard λ as representing both a mod p integer and a self-map. For spaces with mod p cohomology algebra finitely generated, a splitting

$$\chi:QH^*(X)\to H^*(X)$$

is constructed in [CHZ] with the property that for some iterate λ^{p^m}, we have the commutative diagram

$$\begin{array}{ccc} QH^*(X) & \xrightarrow{\chi} & H^*(X) \\ \lambda \downarrow & & \downarrow (\lambda^{p^m})^* \\ QH^*(X) & \xrightarrow{\chi} & H^*(X). \end{array}$$

If $H^*(X)$ is primitively generated in dimensions $< n$, then $\chi QH^*(X) = PH^*(X)$ for $* < n$. Thus we have a set of algebra generators which are λ-characteristic vectors for some iterate of the self-map λ. We grade $\tilde{H}^*(X)$ as follows

(1.4) $F^1 \tilde{H}^*(X) = \text{im}\,\chi$ (abbreviated F^1)

$F^k \tilde{H}^*(X) = \text{span}\,\{F^1 \cdot F^{k-1}\}$

It will be convenient to collect gradations in the same congruence class mod (p-1) (since $\lambda^{p-1} \equiv 1 \bmod p$) so we write

$$\hat{F}^k = \bigoplus_{\ell \equiv k(p-1)} F^\ell .$$

We also define gradation of $\tilde{H}^*(X^{(j)})$ by

$$F^{k_1,\ldots,k_j} H^*(X^{(j)}) = F^{k_1} \otimes \ldots \otimes F^{k_j}$$

and collect them in

$$\hat{F}^{k_1,\ldots,k_j} = \bigoplus_{\ell_i \equiv k_i (p-1)} F^{\ell_1,\ldots,\ell_j} .$$

We write the map induced by the splitting χ as

$$\chi^\#: \text{Hom}(H^*(X), H^*(X^{(j)})) \to \text{Hom}(QH^*(X), H^*(X^{(j)})) .$$

Convention. Henceforth, λ is a primitive (p-1)-st root of unity. Thus $\lambda^a \equiv \lambda^b \bmod p$ if and only if $a \equiv b \bmod (p-1)$. Thus $\hat{F}^k$ is precisely the subspace of λ^k-characteristic vectors of the self-map λ.

Remark. In the sequel, multiplication will undergo numerous changes and self-maps will play a central role. The ground covered will be from a multiplication μ which makes $H^*(X)$ primitively generated in dimensions $< n$ to ν such that $\mu^* = \nu^*$ in dimensions $< n$, and a precise description in terms of our gradation on $H^n(X \wedge X)$ of the deviation from primitivity of ν^* can be given. We use the initial μ to produce the gradation (1.4) and then keep this gradation and the self-maps with respect to μ fixed during the remainder to the process. Thus, in the sequel, λ will denote both an unchanging self-map and a primitive (p-1)-st root of unity. On the other hand, the + operation in $[X^{(j)}, X]$ will always be with respect to the most recently produced multiplication. For this process, the invariance properties expressed in (1.1) are critical.

(1.5) Proposition. Let (X,μ) be an H-space with $H^*(X)$ primitively generated in dimensions $< n$. There is a multiplication ν such that $\mu^* = \nu^*$ in dimensions $< n$ and

$$\bar{\nu}^* | \chi QH^n(X) \subset \bigoplus_{r+s\equiv 1(p-1)} F^{r,s}.$$

Proof. Let $w_i = D_2(\lambda^i\mu, \mu\circ\lambda^i\times\lambda^i)$, $1\le i\le p-2$ and λ^i is the i-th iterate of the self-map λ. Since $H^*(X)$ is primitively generated in dimensions $< n$, we have $I_k(\lambda^i\mu) = I_k(\mu\circ\lambda^i\times\lambda^i)$ for $k < n$, hence by (1.2) w_i is (n-1)-connected. Given $y \in \chi QH^n$, write

$$\bar{u}^*(y) = \sum_{1\le r\le p-1} v_r \quad \text{where} \quad v_r \in \bigoplus_{q\equiv r(p-1)} \bigoplus_{a+b=q} F^{a,b}.$$

Then $W_i^*(y) = \sum_{r\neq 1(p-1)} (\lambda^i - \lambda^{ir}) v_r$ using (1.1) and the fact $(\lambda^i\times\lambda^i)^* | F^{a,b}$ is multiplication by $\lambda^{(a+b)i}$.
Now, using the Vandermonde determinant, one observes that the $(p-2)\times(p-2)$ matrix $(\lambda^i - \lambda^{ir})$ is invertible. Hence we can write

$$\sum_{r\neq 1(p-1)} v_r = \sum_i \alpha_i w_i^*(y)$$

with α_i independent of y. We define

$$\nu = D_2(\mu, \Sigma\alpha_i w_i)$$

and the verification of its properties follows directly from (1.1) and (1.2).

Now we introduce our main construction. We define an action of the polynomial algebra $Z/p[\lambda_1,\ldots,\lambda_j]$ on Hom $(QH^*(X), H^*(X^{(j)}))$ by means of right-left operations. A key property of this action is that the subspace $\chi\# I_n$ ((n-1)-connected maps) is invariant.

To begin, the indeterminant λ_i corresponds to the self-map of $H^*(X^{(j)})$ induced by $1\wedge\ldots\wedge\lambda\wedge\ldots\wedge 1$ with the self-map λ in the i-th place and the other maps are the identity. For $\alpha \in Z/pZ$, we write ϕ_α for the self-map of X inducing multiplication by α on $QH^*(X)$. Of course ϕ_α is another name for α by our earlier convention. But we use this notation to distinguish the right-left nature of the action by the polynomial algebra. To define the action, write a monomial m

$$m = \alpha\,\lambda_1{}^{n_1}\ldots\lambda_j{}^{n_j}$$

and for $v \in \mathrm{Hom}(QH^*(X), H^*(X^{(j)}))$ define mv by choosing any $v' \in (\chi^\#)^{-1}(v)$ and setting

$$mv = (\lambda_1{}^{n_1} \wedge \ldots \wedge \lambda_j{}^{n_j})^* \circ v' \circ \phi_\alpha{}^* \circ \chi \ .$$

This is well defined since $v' \circ \phi_\alpha \circ \chi = \alpha v$ where α is induced by addition in Hom. We can then write

$$mv = \alpha(\lambda_1{}^{n_1} \wedge \ldots \wedge \lambda_j{}^{n_i})^* \circ (v) \ .$$

We extend the action linearly by requiring $(m_1+m_2)v = m_1v + m_2v$. Furthermore, from the definition, we have $m(v_1+v_2) = mv_1 + mv_2$.

From (1.1) it follows that the action of $Z/p[\lambda_1,\ldots\lambda_j]$ on the image under $\chi^\# I_n$ of the set of (n-1)-connected maps is compatible with $+$ induced by any multiplication on X and this subspace is invariant. Also, if v is an induced map, then by construction mv is induced.

Here is an element in $Z/p[\lambda_1,\ldots\lambda_j]$ which is useful.

__Definition__. Let $P(r_1,\ldots,r_j) = \prod_{i=1}^{j}(1-(\lambda_i-\lambda^{r_i})^{p-1})$. Here λ is a primitive (p-1)-st root of unity in Z/p and λ_i are indeterminants. The r_i are integers.

The element $P(r_1,\ldots,r_j)$ behaves like a projection operator.

__(1.6) Lemma__. For $v \in \mathrm{Hom}(QH^*(X), H^*(X^{(j)}))$, $P(r_1,\ldots,r_j)(v)$ is the composition

$$QH^*(X) \xrightarrow{\chi} H^*(X) \xrightarrow{v'} H^*(X^{(j)}) \xrightarrow{\text{proj.}} \hat{F}^{r_1,\ldots r_j} \ .$$

__Proof__. On elements $y \in QH^*(X)$ write

$$v' \circ \chi(y) = \sum v_{s_1} \otimes \ldots \otimes v_{s_j} \ \in \ \hat{F}^{s_1,\ldots,s_j} \ .$$

Then

$$P(r_1,\ldots,r_j)(v) = \sum \bigotimes_i (v_{s_i} - (\lambda^{s_i}-\lambda^{r_i})^{p-1} v_{s_i}) \ =$$

$$\sum_{\bar{r}_i \equiv r_i (p-1)} v_{\bar{r}} \otimes \cdots \otimes v_{\bar{r}_j}$$

since $\lambda^{s_i} \equiv \lambda^{r_i} \mod p \iff s_i \equiv r_i \mod (p-1)$.

(1.7) Lemma. If $H^*(X)$ is primitively generated in dimensions $< n$ and $k < n$, then

$$\Delta^* P(r,s-r)\bar{\mu}^* | \hat{F}^s H^k(X)$$

is multiplication by $\binom{s}{r} \mod p$.

Proof. $\hat{F}^s = \bigoplus_{s_i \equiv s(p-1)} F^{s_i}$. We write $s_i = s + i(p-1)$ and take $0 \le s \le p-2$. Now $(1+\lambda)^s \equiv (1+\lambda)^{s_i} \mod p$. Expanding and collecting terms, we obtain

$$\sum_{r=0}^{p-2} \binom{s}{r} \lambda^r \equiv \sum_{r=0}^{s_i} \binom{s_i}{r} \lambda^r \equiv \sum_{\bar{r}=0}^{p-2} \left(\sum_j \binom{s_i}{\bar{r}+j(p-1)} \right) \lambda^{\bar{r}}$$

where, on the left, we observe the convention $\binom{s}{r} = 0$ if $r > s$. Since the polynomials in λ are of degree $\le p-2$, it follows that the coefficients are equal term wise, and we have

$$\binom{s}{r} \equiv \sum_j \binom{s_i}{r+j(p-1)} \mod p .$$

But $\Delta^* P(r,s-r)\bar{\mu}^* F^{s_i} = \left(\sum_j \binom{s_i}{r+j(p-1)} \right) F^{s_i}$ because F^{s_i} is spanned by s_i-fold products of primitives. The lemma follows.

Finally we construct the elements of $[X^{(2)}, X]$ which will be used for the main result.

(1.8) Proposition. Let (X,μ) be an H-space with $H^*(X)$ primitively generated in dimensions $< n$. Then there are elements $w_a \in [X^{(2)}, X]$ such that each w_a is $(n-1)$-connected and

$$I_n(w_a) \circ \chi = a\, P(a,p-a)\bar{\mu}^* \circ \chi + E_a$$

where E_a is a composition of the form

$$E_a = E_a' \circ P(1,0)\, \bar{\mu}^* \circ \chi$$

and $$E_a' : \hat{F}^{1,0} \to \hat{F}^{a,p-a} .$$

Furthermore E_a' strictly increases the gradation of the left factors from $\hat{F}^{1,0}$, i.e. if $E_a' : F^{r_1,r_2} \to F^{s_1,s_2}$, then $s_1 > r_1$.

Proof. By (1.3), the element $A \in [X^{(3)}, X]$ is (n-1)-connected. By the invariance of the action of $Z/p[\lambda_1,\lambda_2,\lambda_3]$ on the image of (n-1)-connected maps, there is $w_a' \in [X^{(3)}, X]$ such that

$$I_n(w_a') \circ \chi = P(a,a-1,p-a)\, I_n A \circ \chi \; .$$

The composition

$$X^2 \xrightarrow{\Delta \times 1} X^3 \longrightarrow X^{(3)} \xrightarrow{w_a'} X$$

is null-homotopic on the wedge in X^2, hence induces a map $w_a : X^{(2)} \to X$. By construction, and (1.3) w_a is (n-1)-connected. Using (1.6) we obtain the fact that

$$(\Delta^* \otimes 1)\, P(1,a-1,p-a)\, (\bar{\mu}^* \otimes 1)\, |\hat{F}^{r_1,r_2}$$

is 0 unless $r_1 \equiv a \bmod (p-1)$ and $r_2 \equiv (p-a) \bmod (p-1)$.

By (1.7), the action of this term on $\hat{F}^{a,p-a}$ is multiplication by a. Hence we have the equation

$$(\Delta^* \otimes 1)\, P(1,a-1,p-a)\, (\bar{\mu}^* \otimes 1) \circ \bar{\mu}^* \circ \chi = a\, P(a,p-a)\, \bar{\mu}^* \circ \chi \; .$$

Similarily we have

$$(\Delta^* \otimes 1)\, P(1,a-1,p-a)\, (1 \otimes \bar{\mu}^*)\, |\hat{F}^{r_1,r_2}$$

is 0 unless $r_1 \equiv 1 \bmod (p-1)$ and $r_2 \equiv 0 \bmod (p-1)$. In this case the action is a map $E_a' : \hat{F}^{1,0} \to \hat{F}^{a,p-a}$ in which the left gradation (r_1) is strictly increased by $\Delta^* \otimes 1$. The proof is completed by applying (1.3).

Section 2. Here we prove our main theorem. We first introduce some notation. For any integer ℓ we have the equation

$$\ell = \left[\frac{\ell-1}{2}\right] + \left[\frac{\ell}{2} + 1\right]$$

where [ ] is the greatest integer function. For integers s satisfying $s \equiv 1 \bmod (p-1)$, write $s = \ell(p-1) + 1$ and define

$$a(s) = \left[\frac{\ell-1}{2}\right](p-1)+1$$

$$b(s) = \left[\frac{\ell}{2}+1\right](p-1) .$$

Then $s = a(s)+b(s)$ and using the facts $\left[\frac{\ell-1}{2}\right]+1 = \left[\frac{\ell+1}{2}\right]$ and $2\left[\frac{\ell+1}{2}\right] \geq \ell$, we can describe $a(s)$ as the largest integer congruent to $1 \bmod (p-1)$ such that $2a(s) < s$.

(2.1) Theorem. Let (X,μ) be an H-space with $H^*(X)$ primitively generated in dimensions $< n$. Then there is a multiplication ν such that $\mu^* = \nu^*$ in dimensions $< n$ and in dimension n we have

$$\bar{\nu}^* | \chi Q H^n(X) \subset \bigoplus_{s \equiv 1(p-1)} \ \bigoplus_{r \geq r_0(s)} F^{r,s-r}$$

with the following restrictions,

a) Let $r_0(s)$ be the smallest value of r in which $\bar{\nu}^* | \chi QH^n$ has a non-zero component in $F^{r,s-r}$, then $r_0(s) \equiv 1 \bmod (p-1)$.

b) If $F^{b(s)} H^*(X) = 0$, $* < n$ then $\bar{\nu}^* \chi QH^n$ has no non-zero components in $\bigoplus_r F^{r,s-r}$.

Proof. First we recall that the gradation and splitting χ are achieved as in (1.4) and remain fixed. Using (1.5) we obtain a multiplication ν_1 such that $\mu^* = \nu_1^*$ for $* < n$ and

$$\bar{\nu}_1^* | \chi QH^n \subset \bigoplus_{s \equiv 1(p-1)} F^{r,s-r} .$$

Next we use (1.8) to produce $w \in [X^{(2)},X]$ by the equation

$$w = \sum_{p-1 \geq a > 1} a^{-1} w_a .$$

Define $\nu = D_2(\nu_1,w)$. Then $\nu^* = \mu^*$ for $* < n$ and on χQH^n we have the equations

$$(2.2) \quad \bar{\nu}^* = \bar{\mu}^* - w^* = \bar{\mu}^* - \left(\sum_{p-1 \geq a > 1} P(a,p-a) + a^{-1} E_a' P(1,0) \right) \bar{\mu}^* .$$

Note that terms from $\hat{F}^{a,p-a}$ are removed from $\bar{\mu}^*$ but terms from $\hat{F}^{1,0}$ are moved into $\hat{F}^{a,p-a}$ by E'_a. Since E'_a increases the left gradation from $\hat{F}^{1,0}$, we obtain the conclusion that $r_0(s) \equiv 1 \bmod (p-1)$. Now suppose that $F^{b(s)}H^n(X) = 0$. This fact, together with $r_0(s) \equiv 1 \bmod (p-1)$, imply that

$$r_0(s) \geq a(s) + (p-1) = \left[\frac{\ell+1}{2}\right](p-1) + 1 .$$

Thus, if ℓ is odd, $\left[\frac{\ell+1}{2}\right] = \left[\frac{\ell}{2} + 1\right]$ and we find that $r_0(s) > b(s)$. Hence $\bar{\nu}^*$ has no components in $F^{r,s-r}$.
For ℓ even, write $\ell = 2n$. Then from the inequalities $r_0(s) \leq r \leq b(s)-1$ we obtain (after some arithmetic)

$$n(p-1) \geq s-r \geq n(p-1) - (p-3) .$$

Thus $\bar{\nu}^*$ has no components in $\hat{F}^{0,1}$ coming from $\bigoplus_r F^{r,s-r}$.

Let $T : X^2 \to X^2$ be the map interchanging factors. The multiplication $\nu \circ T$ has no components in $\hat{F}^{1,0}$ and conclusion (b) follows by applying (2.2).

Remark 1. The theorem (0.3) follows directly from part (b) of (2.1).

Remark 2. Our construction apparently depends on having A with $I_n A \neq 0$. But if (1.5) has been applied and we obtain a multiplication ν for which $I_n A = 0$, then we can regard $\bar{\nu}^*(y)$ as defining a class in

$$\mathrm{Ext}^{2,*}_{H_*(X)}(Z/p, Z/p) .$$

Up to addition of decomposibles, $\bar{\nu}^*(y)$ has the form

$$\sum_{j=1}^{p-1} \frac{1}{p}\binom{p}{j} z^j \otimes z^{p-j}$$

which satisfies the conclusions of (2.1).

Remark 3. In general, homotopy associativity is not preserved by the constructions in (2.1). A simple example is provided by the Lie group F_4 localized at 3. The mod 3 cohomology is given

by $H^*(F_4) = \Lambda(x_3,x_7,x_{11},x_{15}) \otimes Z/3[x_8]/(x_8^3)$,
and Steenrod operations are

$$P^1 x_3 = x_7 , \quad P^1 x_{11} = x_{15} , \quad \beta x_7 = x_8 .$$

Thus $H^*(F_4)$ is primitively generated in dimensions ≤ 10 for any multiplication. Now in dimension 11 E_λ $(\lambda = -1)$ satisfies $E_\lambda = \{0\}$. Hence, by (0.3) $H^*(F_4)$ can be made primitively generated in dimensions < 12. From the action of the Steenrod algebra, it follows that $H^*(F_4)$ is primitively generated by any multiplication along this in dimensions ≤ 11. Now recall that the main conclusion in [ZA] asserts that no multiplication making the mod p cohomology (p odd) of a finite H-space primitively generated can be homotopy associative, in the presence of p-torsion.

Section 3. Here we give a proof of tha main result of [H].

(3.1) Theorem. For $p \geq 5$, there is a mod p H-space $K(p)$ with $H^*K(p) = \Lambda(x_3,x_{2p+1}) \otimes Z/p[x_{2p+1}]/(x_{2p+2}^p)$ and

$$P^1 x_3 = x_{2p+1} , \quad \beta x_{2p+1} = x_{2p+1} .$$

Proof. There are two parts to the argument. The first involves constructing the complex $K(p)$ and embedding it in an H-space such that the connectivity of the pair is greater than the dimension of $K(p) = 2p^2 + 2p + 2$. For this step, we follow [H], in particular the details on p.47. We shall summarize those details here. The second step of the proof is to refine the embedding in the first step to a pair $(E, K(p))$ in which E is an H-space and the connectivity of the pair is greater than 2 times the dimension of $K(p)$. The existence of an H-structure on $K(p)$ then follows by simple obstruction theory. In [H], the second step involved a lot of calculation. Here we can make use of (2.1). Now for the details. Begin by considering the fibration of loop spaces and loop maps

$$E_1 \to K(Z,3) \xrightarrow{P^p P^1 \iota_3} K(Z_p, 2p^2 + 1) .$$

Then $H^*(E_1) \cong H^*K(p) \otimes S$, as algebras over the Steemrod algebra $\mathcal{Q}$, where S, is a free algebra with $\mathcal{Q}$-generators e_1 and e_2 of dimensions $2p^2 + 2p - 2$ and $2p^2 + 2p$ respectively and all other $\mathcal{Q}$-generators are above the dimension of $H^*K(p)$. We do not yet assert the existence of the space $K(p)$, but are only specifying the cohomology algebra. Following the argument on p. 47 of [H] we obtain the facts that e_2 is the mod p reduction of an integral class $\tilde{e}_2$ and there exists an H-structure on E_1 (no longer homotopy associative) for which e_1 and $\tilde{e}_2$ are primitive. We then have a fibration of H-spaces and H-maps given by

$$E_2 \to E_1 \xrightarrow{e_1, \tilde{e}_2} K(Z/p, 2p^2+2p-2) \times K(Z_{(p)}, 2p^2+2p) .$$

We define $K(p)$ as the $2p^2+2p+2$ homology approximation of E_2, and this concludes the outline of the first step of the proof. For the second step, we have an inclusion $K(p) \subset E_2$ which is a rational equivalence. We can kill cohomology of E_2 above the dimension of $K(p)$ by using Eilenberg-MacLane spaces of type $K(Z/p,n)$ and after a finite number of steps achieve a pair $(E,K(p))$ of connectivity greater than 2 times the connectivity of $K(p)$. We now apply (2.1) to show that such an E can be obtained which is an H-space. Consider the inductive step,

$$E_{i+1} \to E_i \xrightarrow{k} K(Z/p,n)$$

with $n > 2p^2+2p+2$. In dimensions $< n$, $H^*(E_i) \cong H^*(K(p))$ and is primitively generated. Use the multiplication on E_i to induce a gradation and splitting χ as in (1.4). Since $n > 2p^2+2p+2$, $QH^nE_i \cong H^nE_i$ and we use χ to define the map k. Now the maximum cup length from $H^*K(p)$ is $p+1$. If $p \geq 5$ (used here for the first time) then $p+1 < 2p-2$. Thus the multiplication ν from (2.1) could only have non-zero components of $\bar{\nu}^*(k)$ in

$$\bigoplus_{r=1}^{p-1} F^{r,p-r}, \quad \text{the case } s = p .$$

But $\dim F^{r,p-r} \leq p(2p+2) < n$. Hence k is an H-map with respect to ν, and (3.1) follows by induction on n.

Remark 1. (3.1) is true for $p = 3$ and is proved in [H] using a mod 3 splitting of F_4. To use the theory here would require more explicit calculation of the sort described in the first step of the proof for (3.1), because for $p = 3$ none of the cases $s = p$, $s = 2p-1$, $s = 3p-2$ are excluded by (2.1), and dimensional considerations do not exclude the latter two cases.

This work was supported by grants from the National Science Foundation (USA) and the Binational Science Foundation (Israel-USA).

References

[CHZ] G. Cooke, J. Harper and A. Zabrodsky, "Torsion free mod p H-spaces of low rank", Topology 18 (1979).

[H] J. Harper, "H-spaces with torsion", Mem. Amer. Math. Soc. 22 number 223 (1979).

[Z] A. Zabrodsky, Hopf Spaces North Holland (197)

[ZA] A. Zabrodsky, "Implications in the cohomology of H-spaces", Ill. J. Math. 14 (1970).

DEPARTMENT OF MATHEMATICS
UNIVERSITY OF ROCHESTER
ROCHESTER, NY 14627

and

HEBREW UNIVERSITY
INSTITUTE OF MATHEMATICS
JERUSALEM
ISRAEL

Contemporary Mathematics
Volume 19, 1983

POINCARE SPACES WITH THE HOMOLOGY OF A SPHERE

Jean-Claude HAUSMANN

ABSTRACT. We compute the Poincaré-homology-sphere bordism group $\Omega_n^{PHS}(X)$ of a space X. For n≥6, $\Omega_n^{PHS}(X)$ is shown to be isomorphic to $\pi_n(X^+_{\Gamma(x)}) \oplus L_n^h(\hat{\Gamma}(x))$ where $\Gamma(X)$ is the maximal locally perfect subgroup of $\pi_1(x)$ and $\hat{\Gamma}(X)$ its universal central extension. This enables us to construct Poincaré complexes which are homology spheres but are not homotopy equivalent to closed manifolds.

An (oriented) Poincaré space of formal dimension n is a finite CW-complex K together with a class $[K] \in H_n(K;\mathbb{Z})$ (trivial coefficients) such that the homomorphism $-\cap[K] : H^k(K;B) \longrightarrow H_{n-k}(K;B)$ is an isomorphism for any $\pi_1(K)$-module B. In this paper we are interested in Poincaré spaces which are homology spheres, i.e.

$$H_*(K;\mathbb{Z}) = \begin{cases} \mathbb{Z} & \text{if } * = 0,n \\ 0 & \text{otherwise.} \end{cases}$$

For this, we consider for any space X with base point the Poincaré bordism group $\Omega_n^{PHS}(X)$ of homology spheres. The elements of $\Omega_n^{PHS}(X)$ are equivalence classes of pairs (K,f), where K is an oriented Poincaré complex of formal dimension n which is a homology sphere, K has a base point, and $f : K \longrightarrow X$ is a continous map preserving base points. Two such pairs (K_i,f_i) are equivalent if there exists a Poincaré cobordism (W,K_1,K_2) such that the inclusions $K_i \subset W$ induce integral homology isomorphisms and so that

1980 AMS Subject classification. 57P10, 55N22, 18F25.

the maps f_i's extend to a map $F : W \longrightarrow X$. The cobordism W has a base arc joining the base points of K_1 and K_2 which is mapped by f to the base point of X. The set $\Omega_n^{PHS}(X)$ is an abelian group with the oriented connected sum.

The group $\Omega_n^{HS}(X)$ is defined in the same way with the additional restriction that all the Poincaré spaces under consideration are compact manifolds. The group $\Omega_n^{HS}(X)$ was studied in [Ha] and [HV]. One has a natural homomorphism $\Omega_n^{HS}(X) \longrightarrow \Omega_n^{PHS}(X)$.

Let (K,f) represent a class in $\Omega_n^{PHS}(X)$. The image of $\pi_1(K)$ in $\pi_1(X)$ is a finitely generated perfect group which is therefore contained in the maximal locally perfect subgroup $\Gamma(X)$ of $\pi_1(X)$. (A group G is locally perfect if any finitely generated subgroup of G is contained in a finitely generated perfect subgroup of G.) Performing the Quillen plus consgruction gives us a homotopy commutative diagram :

$$\begin{array}{ccc} K & \longrightarrow & X \\ \downarrow & & \downarrow \\ K^+ \cong S^n & \xrightarrow{f^+} & X^+_\Gamma(X) \end{array}$$

This permits us to associate to (K,f) the class $[f^+] \in \pi_n(X^+_{\Gamma(X)})$. One checks that this correspondance gives rise to a homomorphism $\Omega_n^{PHS}(X) \longrightarrow \pi_n(X^+_{\Gamma(X)})$. Theorem 4.1 of [HV] shows that the composed

$$\Omega_n^{HS}(X) \longrightarrow \Omega_n^{PHS}(X) \longrightarrow \pi_n(X^+_{\Gamma(X)})$$

is an isomorphism for $n \geq 5$. Therefore the homomorphism $\Omega_n^{HS}(X) \longrightarrow \Omega_n^{PHS}(X)$ is split injective and one can write

$$\Omega_n^{PHS}(X) = \Omega_n^{HS}(X) \oplus \mathscr{L}_n(X) = \pi_n(X^+_{\Gamma(X)}) \oplus \mathscr{L}_n(X) \quad (n \geq 5)$$

The aim of this paper is to compute $\mathscr{L}_n(X) = \operatorname{coker}(\Omega_n^{HS}(X) \longrightarrow \Omega_n^{PHS}(X))$.

Let $0 \longrightarrow H_2(\Gamma(X)) \longrightarrow \hat{\Gamma}(X) \longrightarrow \Gamma(X) \longrightarrow 1$ be the universal central extension of $\Gamma(X)$ (see [Ke]). Let us denote by $L_n(G)$ the Wall surgery obstruction group for the group G, as defined algebraically in [Sh] (where this group is called $L_n^h(G)$; the group $L_n^s(G)$ would occur if we were doing our theory with simple Poincaré complexes). The homomorphism $1 \longrightarrow G$ gives rise to a short split exact sequence $0 \longrightarrow L_n(1) \longrightarrow L_n(G) \longrightarrow \tilde{L}_n(G) \longrightarrow 0$.

Theorem. For $n \geq 6$, *there is a natural isomorphism :*

$$\mathscr{L}_n(X) \xrightarrow{\cong} \widetilde{L}_n(\hat{\Gamma}(X))$$

Corollary. $\Omega_n^{PHS}(X) \cong \pi_n(X^+_{\Gamma(X)}) \oplus \widetilde{L}_n(\hat{\Gamma}(X)) \quad (n \geq 6)$

Proof of the theorem. Let (K,f) represent a class in $\Omega_n^{PHS}(X)$. As K is a homology sphere, one has $H_2(\pi_1(K)) = 0$, and therefore $\pi_1(K)$ is its own universal central extension by [Ke, Lemma 2]. The theory of [Ke] then show that there is a unique lifting φ :

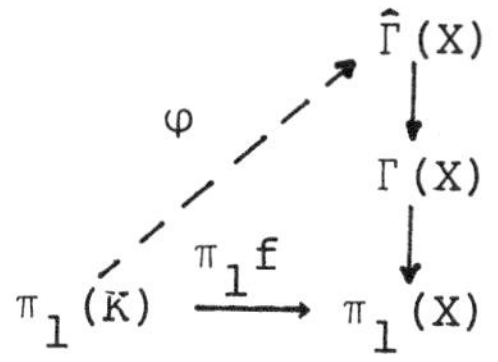

The trivial sphere bundle $\xi = K \times S^p$ is the Spivak bundle for K. Indeed, the Spivak bundle is characterized by the sphericity of the classes of $H_{n+p}(T\xi)$, where $T\xi$ is the Thom space for ξ (see [Br, §1.4]). Since K is a homology sphere, this sphericity comes from the following sequence :

$$\pi_{n+p}(S^{n+p}) = H_{n+p}(S^{n+p}) = H_{n+p}(\Sigma^p K) \twoheadrightarrow H_{n+p}(T\xi)$$

Therefore, the Spivak bundle ξ of K admits a PL-reduction $\tilde{\xi}$ which determines a surgery problem with surgery obstruction $\sigma_{\tilde{\xi}}(K) \in L_n(\pi_1(K))$.

Lemma 1. If (K,f) *and* (K',f') *represent the same class in* $\Omega_n^{PHS}(X)$, *then, for any PL-reduction* $\tilde{\xi}$ *and* $\tilde{\xi}'$ *of the Spivak bundles over* K *and* K', *one has* $\varphi_*(\sigma_{\tilde{\xi}}(K)) = \varphi'_*(\sigma_{\tilde{\xi}'}(K'))$ *in* $\widetilde{L}_n(\hat{\Gamma}(X))$.

Proof : Take a cellular decomposition of K such that $K = K_0 \cup e^n$, where K_0 is a finite complex of dimension $n-1$ [Wa, Lemma 2.9] . The set of Pl-reductions of $\xi | K_0$ contains one element, since it is in bijection with $[K_0, G/PL]$ (see [Wa, Chapter 10]) and K_0 is acyclic. This implies that the surgery problems corresponding to two PL-reductions $\tilde{\xi}$ and $\tilde{\xi}$ of ξ are obtained one from another by connected sum with a surgery problem having target S^n. One deduces that $\sigma_{\tilde{\xi}}(K) = \sigma_{\tilde{\xi}}(K)$ in $\widetilde{L}_n(\pi_1(K))$.

Let (W,K,K') be a Poincaré cobordism such that the inclusion $K \subset W$ and $K' \subset W$ induce an isomorphism on homology and assume that there is a map $f : W \longrightarrow X$ extending f and f'. By obstruction theory, the PL-reduction $\tilde{\xi}$ of ξ extends to a PL-reduction of the Spivak bundle over W and thus induces a Pl-reduction $\tilde{\xi}'$ of ξ'. It follows from the definition of the surgery obstruction (see [Wa, Chapter 9]) that $\sigma_{\tilde{\xi}}(K)$ and $\sigma_{\tilde{\xi}'}(K')$ have same image in $L_n(\pi_1(W))$. By the preceding, one has $\sigma_{\tilde{\xi}'}(K') = \sigma_{\tilde{\xi}'}(K')$ in $\tilde{L}_n(\pi_1(K'))$. Now, $H_1(\pi_1(W)) = H_2(\pi_1(W)) = 0$, hence, like $\pi_1 f$, the homomorphism $\pi_1 F$ admits a unique lifting $\Phi : \pi_1(W) \longrightarrow \hat{\Gamma}(X)$. Thus, one has :

$$\varphi_*(\sigma_{\tilde{\xi}}(K)) = \Phi_*(\sigma_{\tilde{\xi}}(K)) = \varphi'_*(\sigma_{\tilde{\xi}'}(K')) = \varphi'_*(\sigma_{\tilde{\xi}'}(K')) \quad \text{in } \tilde{L}_n(\hat{\Gamma}(X)).$$

Lemma 1 enables us to define a map $\sigma'_n : \Omega_n^{PHS}(X) \longrightarrow \tilde{L}_n(\hat{\Gamma}(X))$. Obviously, σ'_n is a homomorphism and $\Omega_n^{HS}(X)$ $\ker\sigma'_n$. Thus, σ'_n factors through a homomorphism $\sigma_n : \mathscr{L}_n(X) \longrightarrow \tilde{L}_n(\hat{\Gamma}(X))$.

Injectivity of σ_n : This is equivalent to proving that $\ker\sigma'_n \subset \Omega_n^{HS}(X)$. Let (K,f) represent an element of $\ker\sigma'_n$. This means that there is a PL-reduction $\tilde{\xi}$ of the Spivak bundle ξ over K such that the surgery obstruction $\sigma_{\tilde{\xi}}(K) \in L_n(\pi_1(K))$ has image zero in $L_n(\hat{\Gamma}(X))$. Find a finitely presented group G_1 and a commutative diagram

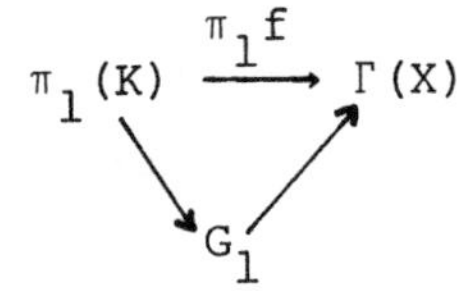

such that $\sigma_{\xi}(K)$ is mapped to zero in $L_n(G_1)$. As $H_2(\hat{\Gamma}(X)) = 0$ and $\hat{\Gamma}(X)$ is locally perfect [Vo, Proposition 5.5] , there exists a finitely presented group G with $H_1(G) = H_2(G) = 0$ and a commutative diagram

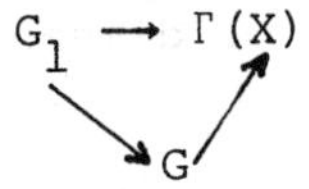

(see [Vo, Lemme 3.5]).

By attaching handles (using Poincaré embeddings) of index 1 and 2 of $K \times I$, one constructs a Poincaré cobordism (W_1,K,K_1) such that $\pi_1(K_1) = \pi_1(W_1) = G$. Thus, $H_2(K_1) = H_2(W_1)$ is free abelian and, as $H_2(G) = 0$, the Hurewicz homomorphism $\pi_2(K_1) \longrightarrow H_2(K_1)$ is surjective. Therefore, one can attach handles of index 3 to W_1 in order to get a Poincaré cobordism (W,K,K') such that $\pi_1(K') = \pi_1(W) = G$ and $H_*(W,K) = H_*(W,K') = 0$. (The handle techniques in Poincaré complexes are known by specialists; a chapter devoted to them will take place in [HV 2].)

By obstruction theory, the PL-reduction $\tilde{\xi}$ extends to W, giving rise to a PL reduction $\tilde{\xi}'$ of the Spivak bundle ξ' over K'. Now, $\sigma_{\tilde{\xi}'}(K') = 0$ in $L_n(\pi_1(K'))$ and therefore there exists a homotopy equivalence $\alpha : M \longrightarrow K'$ where M is a closed manifold. The Poincaré cobordism $W \cup$ (mapping cylinder of α) permits us to show that (K,f) is equivalent to a representative of a class in $\Omega_n^{HS}(X)$.

<u>Surjectivity of σ_n</u> : Let $A = A(X)$ be the theoretical fiber of the map $X \longrightarrow X^+_{\Gamma(X)}$. One has $\pi_1(A) = \hat{\Gamma}(X)$ [HH, Remark 4.4], and thus $\hat{\Gamma}(A) = \hat{\Gamma}(X)$. Considering the commutative diagram :

$$\begin{array}{ccc} \mathcal{L}_n(A) & \longrightarrow & \mathcal{L}_n(X) \\ & \searrow \quad \swarrow & \\ & \tilde{L}_n(\hat{\Gamma}(X)) & \end{array}$$

it suffices to prove the surjectivity of σ_n for the space A.

Let $\sigma_0 \in \tilde{L}_n(\hat{\Gamma}(A)) = \tilde{L}_n(\pi_1(A))$. Let $j : G \longrightarrow \hat{\Gamma}(A)$ be a homomorphism such that G is finitely presented and σ_0 is the image of $\sigma_1 \in L_n(G)$ under the homomorphism $L_n(G) \xrightarrow{j_*} L_n(\hat{\Gamma}(A)) \longrightarrow \tilde{L}_n(\hat{\Gamma}(A))$. For any 2-dimensional finite complex J_1 with $\pi_1(J_1) = G$, there exists a map $J_1 \longrightarrow A$ inducing j on the fundamental groups. As A is acyclic and $\pi_1(A)$ locally perfect, there is a finite 3-dimensional acyclic complex J with a factorisation $J_1 \longrightarrow J \longrightarrow A$ (see [HV], Theorem 3.1 and its proof). By construction, σ_0 is the image of some element $\sigma \in L_n(\pi_1(J))$.

Let T^n be a regular neighbourhood of J in $\mathbb{R}^n$ (J can be embedded in $\mathbb{R}^n$ for $n \geq 6$ by [We, Theorem 1]). Let $\psi : P \longrightarrow \partial T$ be a homotopy equivalence of degree one representing in $\mathcal{S}_{PL}(\partial T)$ the image of σ under the map $L_n(\pi_1(J)) = L_n(\pi_1(\partial T)) \longrightarrow \mathcal{S}_{PL}(\partial T)$

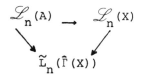

of the surgery exact sequence for ∂T [Wa,Chapter 10] . As T is acyclic, it follows from [HV, Theorem 4.1] that there exists an acyclic manifold T' with $\partial T' = P$ so that ψ extends to a map $\psi' : T' \longrightarrow T$. One may assume that $\pi_1\psi'$ is an isomorphism , since this can be obtained by surgery of index 1 and 2 on T'.

Form the Poincaré complex K by gluing T' to T using $\psi : P' \longrightarrow \partial T$. By construction, one has $\pi_1(J) = \pi_1(K)$ and there is a map $f : K \dashrightarrow A$ extending the map $J \longrightarrow A$ considered above. Also K is a homology sphere. Thus (K,f) represents a class in $\Omega_n^{PHS}(A)$ and we will show that $\sigma_n(K,f) = \sigma_0$ in $\tilde{L}_n(\Gamma(A))$.

Let $g : V \longrightarrow K$ be a normal map of degree one. Call K_1 the image of T and K_2 the image of T' in K. Making g transverse regular to ∂K_1 we obtain a decomposition $V = V_1 \cup V_2$ $(\partial V_1 = \partial V_2)$ and of the map g :

$$g : (V,V_1,V_2,\partial V_i) \xrightarrow{(g,g_1,g_2,\partial g)} (K,K_1,K_2,\partial K_i)$$

As $\pi_1(\partial K_1) = \pi_1(K_i)$, we can apply the $(\pi-\pi)$-theorem [Wa, Theorem 3.3] and find for each $i = 1,2$ a normal cobordism $G_i : (W_i,U_i) \longrightarrow (K_i \times I, \partial K_i \times I)$ from $(g_i,\partial g_i)$ to a homotopy equivalence $g_i' : (V_i',\partial V_i') \longrightarrow (K_i,\partial K_i)$. As K_i is acyclic and $\pi_1(\partial K_i) \cong \pi_1(K_i)$, the surgery exact sequence [Wa, Chapter 10] for K_i shows that $\mathcal{S}_{PL}(K_i) = \{[id]\}$. Thus, we may assume that the map g_i' is a PL-homeomorphism for $i = 1,2$. The surgery obstruction $\sigma(g)$ for the normal map g is equal to the surgery obstruction for the map $G = G_1 \cup G_2 | U_1 \cup U_2 : U_1 \cup U_2 \longrightarrow \partial K_1 \times I$ (relative to the boundary). To have $\partial K_1 \times I$ as the target for G, we must make identifications so that $G|\partial V_1' = g_1'$ and $G|\partial V_2' = \psi \circ g_2'$. As g_1' and g_2' are PL-homeomorphisms, one has the following equations in $\mathcal{S}_{PL}(\partial K_1)$: $\sigma(g)[id] = [\psi \circ g_2'] = [\psi]$. Thus, $\sigma(g)$ and σ both have image $[\psi]$in $\mathcal{S}_{PL}(\partial K_1)$. Hence, the surgery exact sequence tells us that $\sigma - \sigma(g)$ belongs to the image of the map $[\Sigma(\partial K_1), G/PL] \longrightarrow L_n(\pi_1(K_1))$. This map is a homomorphism, since ∂K_1 is a manifold [Wa, Theorem 10.8] . But this image is exactly equal to $L_n(0)$. Indeed, one has the following commutative diagram of groups and homomorphisms :

$$\begin{array}{ccc} \pi_n(G/PL) & \xrightarrow{\cong} & L_n(0) \\ \mu\downarrow & & \downarrow \\ [\Sigma(\partial K_1),G/PL] & \longrightarrow & L_n(\pi_1(K_1)) \end{array}$$

where μ comes from the action $\pi_n(B)\times[Z,B] \longrightarrow [Z,B]$ applied to the constant map. Since $\Sigma(\partial K_1) = S^n$, the homomorphism μ is bijective. Therefore $\sigma-\sigma(g)$ belongs to $L_n(0)$ and thus $f_*(\sigma(g)) = \sigma(K,f) = \sigma_0$ in $\tilde{L}_n(\hat{\Gamma}(A))$.

Remarks :

1) One can define the relative bordism group $\Omega_n^{PHS}(X,Y)$ of Poincaré homology spheres for a pair (X,Y) of spaces. The elements of $\Omega_n^{PHS}(X,Y)$ are represented by maps $f : (A,\partial A) \longrightarrow (X,Y)$, where $(A,\partial A)$ is a finite Poincaré pair and A is acyclic. A homomorphism $\alpha : \Omega_n^{PHS}(X) \longrightarrow \pi_n(X^+_{\Gamma(X)},Y^+_{\Gamma(Y)}) \oplus L_n(\hat{\Gamma}(X),\hat{\Gamma}(Y))$ is defined as in the absolute case. Uning our theorem, the fact that α is an isomorphism for $n\geq 7$ is a direct consequence of the Five Lemma applied to the corresponding long exact sequences.

2) Let Λ be a ring with unit. The case $X = BGL(\Lambda)$ provides the aesthetic formula : $\Omega_n^{PHS}(BGL(\Lambda)) \cong K_n(\Lambda) \oplus L_n(St(\Lambda))$, where $St(\Lambda)$ is the Steinberg group of Λ [Ke, §3] . This generalises [HV, Corollary 4.2].

3) Any non-zero element in $L_n(\hat{\Gamma}(X))$ shows, by our theorem, the existence of Poincaré complexes which are homology spheres but which have not the homotopy type of a manifold. For an example, let us consider an irreducible sufficiently large 3-dimensional homology sphere Σ obtained by gluing the complements of two fibered knots. Let $G = \pi_1(\Sigma) = \hat{\Gamma}(\Sigma)$. By the S. Cappell's version of the Novikov conjecture, one has $L_3(G) \oplus \mathbb{Z}[\frac{1}{2}] \simeq H_3(G;\mathbb{Z}[\frac{1}{2}]) \cong H_3(\Sigma;\mathbb{Z}[\frac{1}{2}]) = \mathbb{Z}[\frac{1}{2}]$. Therefore $\mathscr{L}_n(\Sigma) \otimes \mathbb{Z}[\frac{1}{2}] = \mathbb{Z}[\frac{1}{2}]$ for $n = 4k+3\geq 6$. Observe that $\Sigma^+ \cong S^3$, thus $\Omega_n^{HS}(\Sigma) = \pi_n(S^3)$ for $n\geq 5$ in which case $\pi_n(S^3)$ is finite.

4) Our proofs show that a Poincaré space K^n ($n \geq 6$) with the homology of a sphere is homology cobordant in the Poincaré space category to a Poincaré space obtained by gluing two acyclic manifolds A_1 and A_2 along their boundaries using a homotopy equivalence $\psi : \partial A_1 \longrightarrow \partial A_2$. (Using the techniques of [HV2], we could actually prove that K has the homotopy type of such a Poincaré space). Question : which elements of $\tilde{L}_n(\hat{\Gamma}(X))$ can be represented by (K,f) where K can be obtained by gluing two copies of the same acyclic manifold A using a homotopy equivalence $\psi : \partial A \longrightarrow \partial A$?

BIBLIOGRAPHY

[Br] Browder W. Surgery on simply connected manifolds. Springer-Verlag 1972.

[Ha] Hausmann J-C. Homology sphere bordism and Quillen plus construction, "Algebraic K-theory, Evanston 1976", Springer Lect. Notes 551, 170-181.

[HH] Hausmann J-C.-Husemoller D. Acyclic maps, L'enseignement math. XXV (1979), 53-75.

[HV] Hausmann J-C.-Vogel P. The plus construction and lifting maps from manifolds. Proc. Symp. Pure Math. (AMS) 32 (1978) 67-76.

[HV2] ——————— Geometry in Poincaré spaces. To appear

[Ke] Kervaire M. Multiplicateurs de Schur et K-theorie. Essays on Topology and related topics (in honor of G. De Rahm), Springer 1970.

[Sh] Shaneson J. Wall's surgery obstruction groups for $\mathbb{Z} \times G$. Annals of Math. 90 (1969, 226-234).

[Vo] Vogel P. Un théorème de Hurewicz homologique. Comm. Math. Helv. 52 (1977), 393-413.

[Wa] Wall C.T.C. Surgery on compact manifolds. Academic Press 1970.

[We] Weber Cl. Deux remarques sur les plongements d'un AR dans un espace euclidien. Bull. Ac. Sc. Polonaise XVI, 11 (1968), 851-855.

Université de Genève
Section de Mathématiques
2-4, rue du Lièvre
Case Postale 124

CH-1211 GENEVE 24

Switzerland

Contemporary Mathematics
Volume 19, 1983

RATIONAL ALGEBRAIC K-THEORY OF A PRODUCT OF EILENBERG-MACLANE SPACES

W.-C. Hsiang[1] and R. E. Staffeldt[2]

ABSTRACT. In this paper, we shall explicitly compute the rational algebraic K-theory $A(X)$ of Waldhausen when X is a product of Eilenberg-MacLane spaces. Since Lie groups, Stiefel manifolds, and Grassmannian manifolds are approximated by such products, our results extend to compute at least part of $A(X)$ of these spaces.

1. Introduction and statement of results.

In this paper we continue our effort to compute the rational algebraic K-theory of a simply-connected space X, i.e., the rational homotopy groups of Waldhausen's $A(X)$ [9]. In [6] we computed $\pi_* A(X) \otimes \mathbb{Q}$ in terms of a minimal DGA model K for ΩX as constructed by K. T. Chen in [3], cf. [5]. In that case we took advantage of the fact that K is a tensor algebra on a graded vector space. For a general simply-connected space X, those results are probably the best we can get. However, if X is a product of $K(\pi,n)$'s $(n \geq 2)$, there is another obvious algebraic model K for ΩX. Namely, K may be taken to be a free commutative graded algebra on a graded vector space V with vanishing differential, and one should get even better results for these spaces. In this paper we shall explicitly calculate the rational algebraic K-theory of such products of Eilenberg-MacLane spaces using techniques based on this model. We note also that if $f: X \longrightarrow Y$ is rationally r-connected, then so is $A(f): A(X) \longrightarrow A(Y)$. In particular, Lie groups, Stiefel manifolds, and Grassmannians are all rationally approximable by products of Eilenberg-MacLane spaces, so our results extend to compute at least part of $A(X)$ for such examples. We shall freely use some constructions and lemmas of [4] and [6] and refer the reader to these articles for details.

Our main result is:

---

[1]Partially supported by NSF Grant GP 34324X1
[2]Partially supported by NSF Grant MCS-80-02396

Theorem 1.1. *Let* X *be a simply-connected space rationally equivalent to a product of Eilenberg-MacLane spaces. Suppose that* $H^*(\Omega X;\mathbb{Q}) = S(u_1,\dots,u_n) \otimes \wedge(v_1,\dots,v_m)$, *the tensor product of a symmetric algebra and an exterior algebra. Then the Poincaré series of the rational homotopy groups of* $A(X)$ *is*

$$P.S.(\pi_* A(X)\otimes\mathbb{Q}) = 1 + t^5(1-t^4)^{-1}$$
$$+\ t(1+t)^{-1}\{\prod_i(1+t^{|u_i|+1})(1-t^{|u_i|})^{-1}\prod_j(1+t^{|v_j|})(1-t^{|v_j|+1})^{-1}-1\}$$

For example, one easily obtains the Poincaré series for the rational algebraic K-theory of

A. $\mathbb{C}P^\infty = K(\mathbb{Z},2)$:

$$1 + t^5(1-t^4)^{-1} + t^2(1-t^2)^{-1} \qquad [2],\ [4]$$

B. $\mathbb{H}P^\infty \simeq_{\mathbb{Q}} K(\mathbb{Q},4)$:

$$1 + t^5(1-t^4)^{-1} + t^4(1-t^4)^{-1} \qquad [2],\ [4]$$

C. S^{2n+1} :

$$1 + t^5(1-t^4)^{-1} + t^{2n+1}(1-t^{2n})^{-1} \qquad [2],\ [6]$$

D. $K(\mathbb{Q},2)\times K(\mathbb{Q},4)$, approximating the Grassmannian of two-planes in complex n-space:

$$1 + t^5(1-t^4)^{-1} + [t^2+t^4+t^5-t^6](1-t^2)^{-1}(1-t^4)^{-1}$$

E. $K(\mathbb{Q},3)\times K(\mathbb{Q},5)\times K(\mathbb{Q},7)$ approximating $SU(7)$:

$$1 + t^5(1-t^4)^{-1} + [t^3+t^5+t^8-t^9+t^{10}-t^{11}+t^{12}+t^{13}-t^{14}+t^{15}].$$
$$[(1-t^2)^{-1}(1-t^4)^{-1}(1-t^6)^{-1}]$$

The combinatorial techniques used here are necessarily different from those of [6]. The novelty is to use the Stiefel diagram of $SU(N+1)$ as the catalyzing ingredient.

2. A bookkeeping lemma from [6].

If X is rationally a product of $K(\pi,n)$ spaces $(n \geq 2)$ the rational chains on the Moore loop space of X are weakly equivalent to K, the free graded-commutative algebra on the graded vector space $V = \pi_*(\Omega X) \otimes \mathbb{Q}$. Choose a basis for V and let the set of monomials $\{m_i\}$ in the basis elements be fixed as a basis B for $\overline{K}$, the augmentation ideal of K. We introduce a differential graded vector space $V(\overline{K})$ as follows. $V(\overline{K})$ is the bigraded vector space generated by the symbols

$$(m_1;\ldots;m_k)$$

of bidegree $(k,|m_1|+\ldots+|m_k|)$ where $m_1,\ldots,m_k$ is a finite sequence of monomials, subject to the relations

(A)
$$(m_1;\ldots;m_k) = (-1)^{(|m_k|+|)(\gamma(k-1)+k-1)}(m_k;m_1;\ldots;m_{k-1})$$

where $\gamma(k-1) = |m_1|+\ldots+|m_{k-1}|$.

Define a differential

$$\delta: V(\overline{K})^{(k,*)} \longrightarrow V(\overline{K})^{(k+1,*)}$$

by

$$\delta(m_;\ldots;m_k) = \sum_{i=1}^{k} (-1)^{\gamma(i-1)+(i-1)}\sum(-1)^{|m_i'|+1}(m_1;\ldots;m_i';m_i'';\ldots;m_k)$$

where the second sum is over all the decompositions of the monomial $m_i = m_i'm_i''$.

We find as a special case of Proposition 3.2 of [6]

<u>Lemma 2.1.</u> <u>Let</u> X <u>be a simply connected product of Eilenberg-MacLane spaces and let</u> $V(\overline{K})$ <u>be given as above. Then</u>

$$\pi_q A(X) \otimes \mathbb{Q} \simeq K_q(\mathbb{Z}) \otimes \mathbb{Q} \oplus H^q(V(\overline{K})) .$$

Now in broad outline, the analysis in Section 4 of $H^*(V(\overline{K}))$ consists of two parts, reminiscent of the analysis in [6]. First we observe that $V(\overline{K})$ splits into a sum of subcomplexes and that the homology of a general subcomplex may be derived by symmetrization from the homology of special cases, just as in [6]. Second we tackle the computation of those special cases, meeting difficulties which are overcome by application of the Stiefel diagrams of the Lie groups SU(N+1) . Appearing at the end of Section 4 are the main lemma (Lemma 4.6) and the counting argument which proves the main theorem. Section 3 recalls the Stiefel diagrams with emphasis on the combinatorial properties we need and contains a few further remarks on cohomology and group actions.

3. The Stiefel diagram of SU(N+1).

In this section, we set up a cellular decomposition of a maximal torus $T^N \subset PSU(N+1)$, the projective special unitary group of $(N+1)\times(N+1)$ matrices. T^N is covered by a maximal torus $\tilde{T}$ of SU(N+1) . After a conjugation, we assume that $\tilde{T}$ consists of diagonal matrices, and the Lie algebra $\underline{t}$ of $\tilde{T}$ may be identified with

$$H_0 = \{x = (x^0,\dots,x^N) = \sum x^i e_i \mid \sum x^i = 0\} \subset \mathbb{R}^{N+1} .$$

An isomorphism of the Weyl group $N(\tilde{T})/\tilde{T}$ with S_{N+1} may be chosen so that the action of $\pi \in S_{N+1}$ on H_0 sends e_i to $e_{\pi(i)}$, and this lift of the action on $\tilde{T}$ commutes with the exponential map $\exp:\underline{t} \longrightarrow \tilde{T}$.

To obtain the torus $T \subset PSU(N+1)$ we pass to $\tilde{T}$ modulo the group of translations $\{1,u,\dots,u^N\}$ generated by multiplication by $e^{2\pi i/N+1}I_{N+1}$. This action obviously commutes with the action of $N(\tilde{T})/\tilde{T}$, so S_{N+1} also acts on T and $p:\tilde{T} \longrightarrow T$ is an S_{N+1} equivariant map.

To obtain the cellular subdivisions of T^N , we recall that, in terms of the basis $\{\lambda^i\}$ dual to $\{e_i\}$, the roots of $\underline{t}$ are identified with $\{\lambda^i - \lambda^j \mid i \neq j\}$, and H_0 is cut up into simplices by the hyperplanes

$$\lambda^i - \lambda^j \equiv 0 \quad \mod \mathbb{Z} .$$

The picture of $\underline{t}$ cut up this way is called the Stiefel diagram of SU(N+1) [1], [8].

FIGURE 1

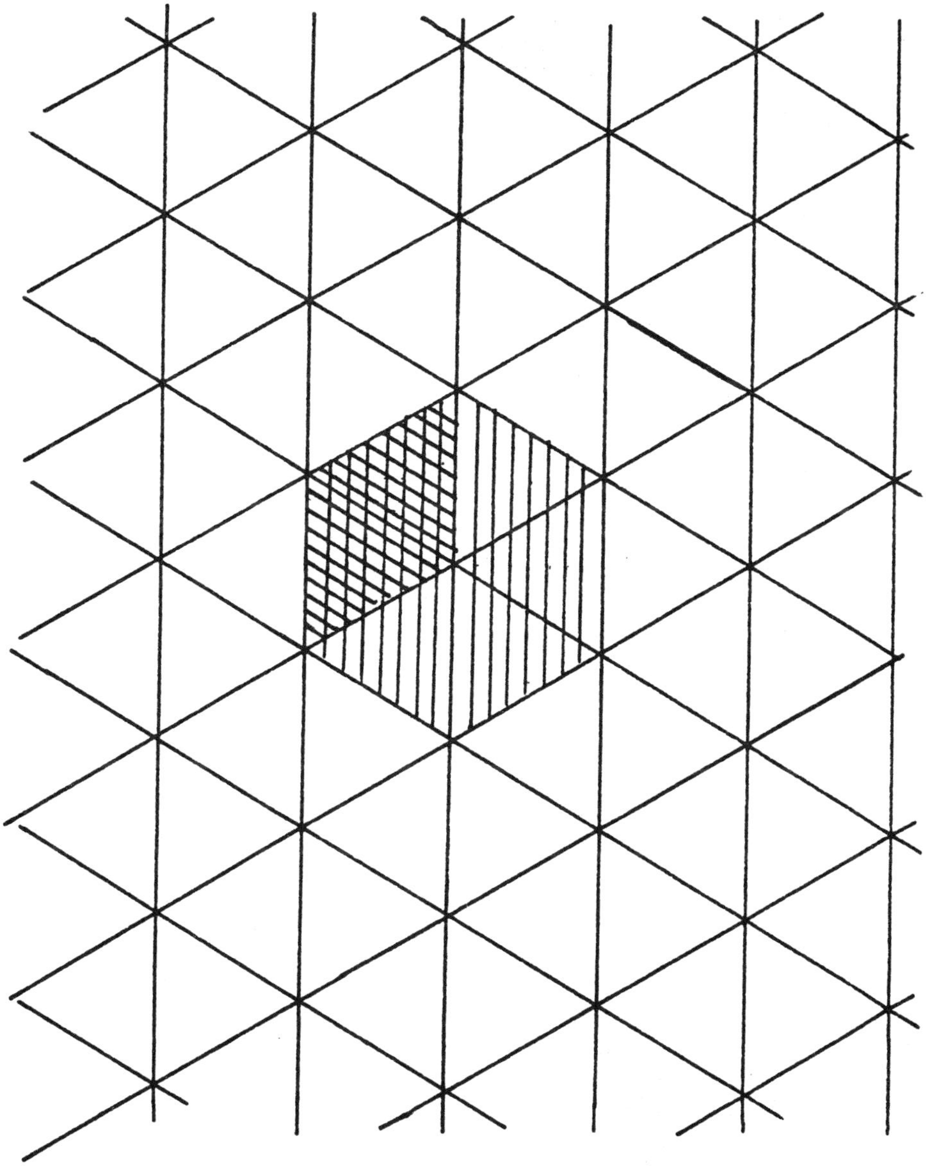

The decomposition of $\underline{t}$ showing fundamental domains for $\tilde{T}^2 \subset SU(3)$ (hexagon) and $T^2 \subset PSU(3)$ (parallelogram).

A fundamental domain for the action of the lattice

$$\tilde{\Lambda} = \mathrm{Ker}(\exp:\underline{t} \longrightarrow \tilde{T})$$

is the union of $(N+1)!$ N-simplices

$$C(\pi) = \{x = \sum x^i e_i \mid x^{\pi(N)} - 1 \leq x^{\pi(0)} \leq \ldots \leq x^{\pi(N)}\} ,$$

where $\pi \in S_{N+1}$. Note that if S_{N+1} acts on H_0 by restricting the conventional action $\pi(e_i) = e_{\pi(i)}$ on $\mathbb{R}^{N+1}$ we have $\pi C(e) = C(\pi)$. Now we may prescribe a cell structure on $\tilde{T}$ by projecting these cells and their faces into $\tilde{T}$ by the exponential map.

For computing cohomology, we must also orient cells by fixing characteristic maps as follows. For the top-dimensional cells define first

$$\sigma(e):\Delta^N \longrightarrow C(e)$$

by setting

$$\sigma(e)(t_0,\ldots,t_N) = t_1 v_1 + \ldots + t_N v_N$$

where $v_0 = 0$,

$$v_1 = \frac{1}{(N+1)}(-Ne_0+e_1+\ldots+e_N),\ldots,v_k = (k-N-1)/(N+1)(e_0+\ldots+e_{k-1})+k/(N+1)(e_k+\ldots+e_N)$$

for $1 \leq k \leq N$. This defines an affine homeomorphism $\Delta^N \longrightarrow C(e)$ from the standard N-simplex $\Delta^N \subseteq \mathbb{R}^{N+1}$. Fix characteristic maps $\sigma(\pi)$ for $C(\pi)$ by taking $\sigma(\pi) = \pi \circ \sigma(e)$. For a proper face C' of $C(\pi)$ the claim is that

$$D\sigma(\pi) = d_{i_1} \ldots d_{i_r} \sigma(\pi)$$

for an appropriate iterated face operator orients the cell. For this we need to observe that we are indeed getting the same characteristic map from (a) all other cells of which C' is a face, or (b) cells of which some translate of C' by an element of $\tilde{\Lambda}$ is a face. Clearly it suffices to see this for faces of $C(e)$. Notice that C' is the image of $d_0 d_{j_2} \ldots d_{j_r}(e)$ where the face operator is in canonical form $0 < j_2 < \ldots < j_r \leq N$. Hence C' is a face of

$$d_0\sigma(e)(\Delta^{N-1}) = d_0\sigma(0N)(\Delta^{N-1}) + (e_N - e_0) ;$$

and, moreover, the maps $d_0\sigma(e)$ and $d_0(\sigma(0N)) + (e_N - e_0)$ from Δ^{N-1} to $\underline{t}$ are the same. Replacing C' and $\sigma(e)$ by $C' + (e_0 - e_N)$ and $\sigma(0N)$ we reduce case (b) to case (a), and then the argument below finishes the claim.

So suppose C' is a face of both $C(e)$ and $C(\pi)$. If a vertex v_i of $C(e)$ is a vertex of both $C(e)$ and $C(\pi)$, then the canonical forms of the face operators obtaining C' from $C(e)$ and $C(\pi)$ are the same and the post-composition by π leaves the map $D\sigma(e)$ unchanged:

$$\pi \circ D\sigma(e) = D\sigma(\pi): \Delta^{N-r} \longrightarrow C' .$$

We next examine the action of the covering translations $\{1, u, \dots, u^N\}$ of the oriented cells of $\tilde{T}$. This we do upstairs in $\underline{t}$ by observing that the translations by $\pi(\lambda_k)$ for all $\pi \in S_{N+1}$ lift the action of u^k to $\underline{t}$, where $\lambda_k = (N+1)^{-1}[k(e_0+\dots+e_{N-k}) + (N-k+1)(e_{N-k+1}+\dots+e_N)]$.

We need the following technical lemma:

Lemma 3.1. (a) *Let* $\zeta \in S_{N+1}$ *denote the cycle* $(N, N-1, \dots, 2, 1, 0)$. *Then the following diagram commutes:*

$$\begin{array}{ccc} \Delta^N & \xrightarrow{\ \sigma(\pi)\ } & \underline{t} \\ \Big\downarrow{\scriptstyle \zeta^{-k}} & & \Big\downarrow{\scriptstyle +\pi(\lambda_k)} \\ \Delta^N & \xrightarrow{\ \sigma((\pi_\zeta)^k\cdot\pi)\ } & \underline{t} \end{array}$$

where $+\pi(\lambda_k)$ *denotes translation by* $\pi(\lambda_k)$, $\pi_\zeta = \pi\zeta\pi^{-1} = (\pi(N), \dots, \pi(0))$, *and* ζ *acts on* Δ^N *sending the* i^{th} *vertex to the* $(i+1)^{th}$ *vertex.*

(b) *The following diagrams also commute, where* δ_p *is the inclusion onto the* p^{th} *face:*

$$\begin{array}{ccc} \Delta^{m-1} & \xrightarrow{\ \delta_j\ } & \Delta^m \\ \Big\downarrow{\scriptstyle (0,\dots,m-1)^{k-1}} & & \Big\downarrow{\scriptstyle (0,\dots,m)^k} \\ \Delta^{m-1} & \xrightarrow{\ \delta_i\ } & \Delta^m \end{array}$$

if $j + k = m + 1 + i \geq m + 1$.

Proof. Both parts are readily proved by induction on k, though the proof of the induction step in (b) requires consideration of several cases. To begin part (a), the equality $v_i + \lambda_1 = \zeta(v_{i+1})$ implies that

$$\begin{array}{ccc} \Delta^N & \xrightarrow{\ \zeta^{-1}\ } & \Delta^N \\ \downarrow & & \downarrow \\ \underline{t} & \xrightarrow{\ +\lambda_1\ } & \underline{t} \end{array}$$

commutes. Post-composition by $\pi \in S_{N+1}$ and substitution of ${}^{\pi}\zeta \cdot \pi = \pi\zeta\pi^{-1} \cdot \pi$ for $\pi\zeta$ establishes the first stage of the induction.

By inductive hypothesis the right and the left squares commute in the diagram

$$\begin{array}{ccccc} \Delta^N & \xrightarrow{\ \zeta\ } & \Delta^N & \xrightarrow{\ \zeta\ } & \Delta^N \\ \Big\downarrow{\scriptstyle \sigma(e)} & & \Big\downarrow{\scriptstyle \sigma(\zeta)} & & \Big\downarrow{\scriptstyle \sigma(\zeta^{k+1})} \\ \underline{t} & \xrightarrow{\ +\lambda_1\ } & \underline{t} & \xrightarrow{\ +\zeta(\lambda_k)\ } & \underline{t} \end{array} .$$

The inductive step is proved by observing that $\zeta(\lambda_k) + \lambda_1 = \lambda_{k+1}$, then post-composing by π.

Part (b) is left to the reader.

<u>Proposition 3.2</u>. (a) $\{1, u, \ldots, u^N\}$ *acts cellularly on* $\tilde{T}$. *Hence* T *may be given a cell structure so that* $p: \tilde{T} \longrightarrow T$ *is a cellular* S_{N+1}*-equivariant map, and, for any coefficient system (over* $\mathbb{Q}$ *)* A *on* T, *the cellular cochains* $C^*(T;A)$ *may be presented as* $C^*(\tilde{T};p^*A)/R^*$ *where* R^* *are the relations induced by the cochain transfer* $\tau^*: C^*(\tilde{T};p^*A) \longrightarrow C^*(T;A)$.

(b) *Let* $D\sigma(\pi) = d_{i_1} \ldots d_{i_p} \sigma(\pi)$ *be a face written in canonical form* $(N \geq i_p > \ldots > i_1 \geq 0)$ *and suppose also that* $i_1 > 0$. *If* $i_p < N$, *the diagram commutes*

$$\begin{array}{ccc} \Delta^{N-p} & \xrightarrow{\ D\sigma(\pi)\ } & \underline{t} \\ \Big\downarrow{\scriptstyle (0\ldots N-p)} & & \Big\downarrow{\scriptstyle +\pi(\lambda_{r+1})} \\ \Delta^{N-p} & \xrightarrow{\ D'\sigma(\pi')\ } & \underline{t} \end{array}$$

where $\pi' = ({}^{\pi}\zeta)^{r+1} \circ \pi$ *and where* D' *has the canonical form determined by* $N > i_{p-r-1} + r + 1 > \ldots > r > r-1 > \ldots > 1 > 0$.

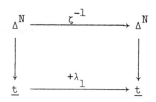

Part (a) of the proposition is obviously a corollary of part (a) of Lemma 3.1, and part (b) of the proposition follows from part (a) and several repetitions of part (b) of Lemma 3.1.

We shall consider the following coefficient system A [7] [10, p. 255] on $T^N \subset PSU(N+1)$. Over each point $x \in T^N$, the stalk of A is (isomorphic to) $\mathbb{Q}$ and each homotopy class of paths from one point to another induces the identity map from $\mathbb{Q}$ to $\mathbb{Q}$. We make A into a non-trivial S_{N+1}-system on T by defining the map $\pi^*A \longrightarrow A$ to be multiplication by $\varepsilon(\pi) = \mathrm{sign}(\pi)$. Of course, we shall also consider the trivial coefficient system on $T^N \subset PSU(N+1)$ which will be denoted by $\mathbb{Q}$.

We now consider the general problem of computing the cohomology of a G-space X with coefficients in a G-system of local coefficients B: i.e., B is a system of local coefficients for which there exist maps of systems $\varepsilon(g):g^*B \longrightarrow B$ satisfying certain obvious conditions of compatibility with the group law. Then G is represented on the cohomology via

$$H^*(X;B) \xrightarrow{g^*} H^*(X;g^*B) \xrightarrow{\varepsilon(g)} H^*(X;B) .$$

In particular, if B is a trivial coefficient system (over X) of vector spaces over a field K, making B into a G-system is equivalent to giving a right representation ρ of G in the vector space B_0 which is the stalk at a point. One can easily prove the following lemma.

Lemma 3.3. *Assume that* B *is a trivial local system over* X *of vector spaces over a field* K. *Then*

(a) $H^q(X;B) \simeq H^*(X;K) \otimes B_0$,

(b) *If* $g^*:H^q(X;K) \longrightarrow H^q(X;K)$ *denotes the usual action of* $g \in G$ *on ordinary cohomology of* K *with coefficients in* K, *the action of* g *on* $H^q(X;B)$ *is given by* $g^* \otimes \rho(g)$.

For the coefficient system A introduced above, $A_0 \simeq \mathbb{Q}$ and the representation ρ is given by the formula $\rho(g)(r) = \mathrm{sign}(g)r$ for $g \in S_{N+1}$ and $r \in A_0$.

4. Technical lemmas and the proof of the main theorem.

Throughout the section we assume that K is a free graded commutative algebra. Recall from Section 2 the differential graded vector spaces $V(\overline{K})$ constructed from a monomial basis $B = \{m_i\}$ for the augmentation ideal $\overline{K}$. We analyse $V(\overline{K})$ in terms of subcomplexes described as follows. Select elements $[x_0,\ldots,x_N]$ from the generating set of K. We do not require selection of distinct elements and $[x_0,\ldots,x_N]$ may be viewed as a set of elements of the generating set, each with a specified multiplicity. In view of the description of the differential δ on $V(\overline{K})$ the subspace $W[x_0,\ldots,x_N]$ of $V(\overline{K})$ spanned by all symbols $(m_1;\ldots;m_k)$, where $\{m_1,\ldots,m_k\}$ runs over all possible ways of forming precisely k-monomials from $[x_0,\ldots,x_N]$ with no element left over and $1 \leq k \leq N+1$, is stable under δ as is its complementary subspace. Our first fact,

$$(V(\overline{K}),\delta) = \oplus(W[x_0,\ldots,x_N],\delta)$$

where the sum is over all selections $[x_0,\ldots,x_N]$ as above, is clear.

We shall call $[x_0,\ldots,x_N]$ the specification of $W[x_0,\ldots,x_N]$ and $N+1$ the rank of the specification. If $\mathbb{Q}S$ denotes the vector space on the set S, we have

$$W[x_0,\ldots,x_N]:0 \longrightarrow \mathbb{Q}\{(x_{(0)}\cdots x_{(N)})\} \xrightarrow{\delta} \cdots$$

$$\xrightarrow{\delta} \mathbb{Q}\{(x_{\sigma(0)};\ldots;x_{\sigma(N)} \mid \sigma \in S_{N+1}\}/(\text{relations}) \longrightarrow 0$$

If $x_0,\ldots,x_N$ are all distinct and of odd degree, we let S_{N+1} act on $W[x_0,\ldots,x_N]$ as follows. Let $\pi \in S_{N+1}$ and let $(m_0;\ldots;m_p)$ be a generator of $W[x_0,\ldots,x_N]$ where $m_0 = x_{i_0}\cdots x_{i_k}, \ldots, m_p = x_{j_1}\cdots x_{j_\ell}$. We set $(m_0;\ldots;m_p)^\pi = ((m_0)^\pi;\ldots;(m_p)^\pi)$ where $(m_0)^\pi = x_{\pi^{-1}(i_0)}\cdots x_{\pi^{-1}(i_k)}, \ldots,$ $(m_p)^\pi = x_{\pi^{-1}(j_1)}\cdots x_{\pi^{-1}(j_\ell)}$. This gives a right action on $W[x_0,\ldots,x_N]$.

Lemma 4.1. *Let $x_0,\ldots,x_N$ be all distinct and of odd degree. Let* A *be the nontrivial S_{N+1}-coefficient system on $T \subset PSU(N+1)$ as described in Section 3. Then there exists an S_{N+1}-equivariant cochain isomorphism of degree* (-1)

$$\Phi: W[x_0,\ldots,x_N] \longrightarrow C^*(T;A) .$$

(In other words, $C^*(T;A)$ is a desuspension of $W[x_0,\ldots,x_N]$.)

Proof. Recall that $\sigma(\pi)$ orients the cell $C(\pi)$ in $\underline{t}$ consisting of vectors $x = \sum x^i e_i$ whose components satisfy the chain of inequalities

$$x^{\pi(N)} - 1 \leq x^{\pi(0)} \leq \ldots \leq x^{\pi(N)} .$$

If we call $x^{\pi(N)} - 1 \leq x^{\pi(0)}$ the zeroth inequality and $x^{\pi(i-1)} \leq x^{\pi(i)}$ the ith inequality the map $D\sigma(\pi) = d_{i_0} \ldots d_{i_p} \sigma(\pi)$ with $0 \leq i_0 < \ldots < i_p \leq N$ orients the face $DC(\pi)$ of $C(\pi)$ obtained by requiring the i_0-th inequalities to be equalities. Now $\pi(\lambda_k)$ carries $C(\pi)$ into the cell $C({}^{\pi}\zeta \cdot \pi) = C(\pi')$ described by the chain

$$x^{\pi(N-k)} - 1 \leq x^{\pi(N-k+1)} \leq \ldots \leq x^{\pi(N-k)} ,$$

which we think of as obtained from the previous chain by bringing the last k inequalities around in front. The face $DC(\pi)$ is carried into the face of $C(\pi')$ described by cyclically permuting the requirements of equality. We note that each cell of the orbit space T has a lift into $\underline{t}$ to a cell $DC(\pi)$ with $i_0 \geq 0$, and that we can also avoid broken chains of equalities that end at the Nth link and "continue" again at the zeroth inequality.

Now we shall define our cochain isomorphism Φ inductively. At the first stage, set

$$\Phi^N((x_{\pi^{-1}(0)};\ldots;x_{\pi^{-1}(N)})) = \text{sign } \pi\{\sigma^*(\pi^{-1})\}$$

where $\sigma^*(\pi^{-1})$ is the cellular cochain which is $+1$ on $\sigma(\pi^{-1})$ and 0 on the other cells. Using the fact that $\xi\sigma(\pi) = \sigma(\xi\pi)$ for $\xi,\pi \in S_{N+1}$, we verify that

$$\Phi^N((x_{\pi^{-1}(0)};\ldots;x_{\pi^{-1}(N)})\xi) = \text{sign}(\xi)(\Phi(x_{\pi^{-1}(0)};\ldots;x_{\pi^{-1}(N)}))^{\xi} .$$

We must also check that this is well-defined by verifying that Φ preserves the relations

$$(x_{\pi^{-1}(0)};\ldots;x_{\pi^{-1}(N)}) = (x_{\pi^{-1}(N)};x_{\pi^{-1}(0)};\ldots;x_{\pi^{-1}(N-1)})$$

which are consequences of relation (A)

$$(m_1;\ldots;m_k) = (-1)^{(|m_k|+1)((k-1)+k-1)}(m_k;m_1;\ldots;m_{k-1})$$

where $\gamma(k-1) = |m_1| + \ldots + |m_{k-1}|$, holding in $V(\overline{K})$. This amounts to checking that

$$\tau^*(\sigma^*(\pi^{-1})) = \mathrm{sign}(\zeta)\tau^*(\sigma^*(\pi^{-1}\zeta^{-1}))$$

where $\zeta = (N,\dots,0)$. But this follows from the definition of cohomology transfer and the fact that in Proposition 3.2 and Lemma 3.1 we have verified that $\mathrm{sign}(\zeta) = (-1)^N$ is exactly the discrepancy in the orientations of $\sigma(\pi^{-1})$ and $\sigma(\pi^{-1}\zeta^{-1}) + \pi^{-1}\zeta^{-1}(\lambda_1)$.

Finally, it is clear that Φ^N is onto in this degree and that both $C^N(T;A)$ and W^{N+1} have dimension $N!$, so Φ^N is also bijective.

Inductively, suppose Φ^N has been extended to a sequence of maps

$$\Phi^r : W^{r+1} \longrightarrow C^r(\tilde{T};p^*A)/(\text{relations})$$

for $N \geq r > p$ which assembles to give a cochain isomorphism in this range.

We define

$$\Phi^* : W^p \longrightarrow C^{p-1}(\tilde{T};p^*A)/R^*$$

by

$$\Phi^p((m_0;\dots;m_p)) = (-1)^{\gamma(i-1)}\chi\{(d_{i+1}D\sigma(\pi))^*\}$$

where $\gamma(i-1) = |m_0| +\dots+ |m_i'|$, and D,π , and $\chi = \pm 1$ are determined by the existing formula

$$\Phi^{p+1}((m_0;\dots;m_i';m_i'';\dots;m_p)) = \chi\{D\sigma(\pi)^*\}$$

and the requirement that the juxtaposition $m_i'm_i'' = m_i$ (i.e., the order of the factors in $m_i'm_i''$ is exactly the order of the factors in m_i .)

We need to check that this is well-defined (independent of the choice of an m_i to aplit and preserving the relations among the generators), that Φ^p is still S_{N+1} equivariant, that Φ^p is an isomorphism of vector spaces, and that the cochain map condition $\delta\Phi^p = \Phi^{p+1}\delta$ is satisfied.

First we compare two definitions of $\Phi^p((m_0;\dots;m_i;\dots;m_j;\dots;m_p))$ in terms of $\Phi^{p+1}((m_0;\dots;m_i';m_i'';\dots;m_p))$ and $\Phi^{p+1}((m_0;\dots;m_j';m_j'';\dots;m_p))$. (We allow $j = i$, in which case we may assume $m_i = m_i'\bar{m}_i n_i''$ and give a similar argument.)

In W^{p+2} there is the element $(m_0;\dots;m_i';m_i'';\dots;n_j';n_j'';\dots;m_p)$ and a definitions

$$\Phi^{p+2}(m_0;\dots;m_i';m_i'';\dots;n_j';n_j'';\dots;m_p) = \{\tilde{\chi}(\tilde{D}\sigma)^*(\pi)\}$$

Applications of the inductive definitions of Φ^{p+1} and Φ^{p} and straightforward computation show that the two signs on $(d_i d_j \tilde{D}\sigma)^*(\pi) = (d_{j-1} d_i \tilde{D}\sigma)^*(\pi)$ are the same multiple of $\tilde{\chi}$ so that Φ^p is well-defined on generators.

Now we must also check that Φ^p carries the relation

$$(m_0;\ldots;m_p) = (-1)^{(|m_p|+1)(\gamma(p-1)+p)}(m_p;m_0;\ldots;m_{p-1})$$

to a relation in R^*. This is where the consequences of Proposition 3.2 for cohomology transfer appear. In most cases we may split $m_i = m_i' m_i''$, $i \neq p$, and use the hypothesis that Φ^{p+1} preserves relations:

$$\Phi^{p+1}((m_0;\ldots;m_i';m_i'';\ldots;m_p)) = (-1)^{\alpha+|k|+1}\Phi^{p+1}((m_p;m_0;\ldots;m_i';m_i'';\ldots;m_{p-1}))$$

where $\alpha = (|m_p|+1)(\gamma(p-1)+p)$, $k = |m_p|$. Evaluating both sides we can say we have

$$(*) \qquad \chi\{D\sigma^*(\pi)\} = (-1)^{\alpha+k+1}\chi'\{D'\sigma(\pi')\}$$

where D',π' are determined by the right half of a commuting diagram

$$\begin{array}{ccccc}
\Delta^p & \xrightarrow{\delta_{i+1}} & \Delta^{p+1} & \xrightarrow{D\sigma(\pi)} & \underline{t} \\
\big\downarrow (0,\ldots,p) & & \big\downarrow (0,\ldots,p+1) & & \big\downarrow +\pi(\lambda_k) \\
\Delta^p & \xrightarrow{\delta_{i+2}} & \Delta^{p+1} & \xrightarrow{D'\sigma(\pi')} & \underline{t}
\end{array}$$

as in Proposition 3.2. Computing $\Phi^p((m_0;\ldots;m_p))$ and $\Phi^p(m_p;m_0;\ldots;m_{p-1})$ from the definitions we first introduce signs $(-1)^{|m_0|+\ldots+|m_i'|}$ on the left and $(-1)^{|m_p|+|m_0|+\ldots+|m_i'|}$ on the right of $(*)$, and then, we introduce d_{i+1} on the left and d_{i+2} on the right, we must also introduce -1 on the right to account for the difference in sign $(0,\ldots,p)$ and sign $(0,\ldots,p+1)$. (Consider the composite diagram.) This produces the correct relation in this case, and the argument in case one has to split $m_p = m_p' m_p''$ is similar, but both m_p' and m_p'' must be brought around in front of the string $m_0,\ldots,m_{p-1}$.

To establish the equivariance property:

$$\Phi^p((m_0;\dots;m_p)^\rho) = \varepsilon(\rho)\Phi^p((m_0;\dots;m_p))^\rho$$

it is easiest to look all the way "upstairs" at the element $(x_{i_0};\dots;x_{i_N})$ which arises from separating each variable factor of each m_i from its neighbor by semicolons. We have arranged that

$$\Phi^N((x_{i_0};\dots;x_{i_N})^\rho) = \varepsilon(\rho)\Phi((x_{i_0};\dots;x_{i_N}))^\rho$$

and we now observe that since $|m_i^\rho| = |m_i|$ we see that the same signs and face operators are added to both sides of the initial equality as we compute $\Phi((m_0;\dots;m_p)^\rho)$ from the left side and $\Phi((m_0;\dots;m_p))^\rho$ from the right. (The requirement to have this relation in the particular case when $m_i^\rho = \pm m_i$ depending on the number of traspositions of odd-degree variables leads to the introduction of S_{N+1}-coefficient systems.)

It is easy to verify that Φ^p is onto $C^p(\tilde{T};p^*A)/R^* \simeq C^p(T;A)$ and to show that $\dim W^{p+1} = \dim C^p(T;A)$ by observing that any scheme for counting a basis for $W^{p+1}[x_0,\dots,x_N]$ consisting of p-tuples $(x_0m_0';\dots;m_p)$ also counts all the p-cells in T by counting paricular representatives in $\underline{t}$ described by chains of inequalities and equalities beginning $x_{i_N} - 1 \leq x_0 \leq \dots$.

Finally, part of the coboundary operator condition has been built into the definition and the rest is taken care of by recalling that $d_{i+1}D\sigma(\pi)$ has incidence number $(-1)^{i+1}$ with $D\sigma(\pi)$. This proves the lemma.

Lemma 4.2. *Let all the variables in the specification* $[x_0,\dots,x_n]$ *be distinct. Then*

$$H^i(W[x_0,\dots,x_n]) \simeq \Lambda^{i-1}H_0^*$$

where H_0^* *is the dual of* $H_0 = \{v \in \mathbb{Q}^{n+1} \mid v^0 + \dots + v^n = 0\}$.

Proof. Let the specification $[x_0,\dots,x_n]$ consist of distinct variables and suppose $|x_i| \equiv 1 \mod 2$ for $0 \leq i \leq n-r-1$ and $|x_i| \equiv 0 \mod 2$ for $n-r \leq i \leq n$. We will compute the cohomology of $W[x_0,\dots,x_n]$ by identifying this complex with the cellular cochains on an n-dimensional subcomplex and subtorus T^n of $T^N \subset PSU(N+1)$, where $N+1 = (n-r)+2(r+1) = n+r+2$. $W[x_0,\dots,x_n]$ supports a group of automorphisms $S_{n-r,r+1} \simeq S_{n-r} \times S_{r+1}$ induced by mod 2-degree-preserving permutations of the variables, and the identification will be equivariant with respect to restriction of the coefficient system A and the S_{N+1} action on T^N to T^n and an $S_{n-r} \times S_{r+1}$ action stabilizing T^n. To develop this, introduce extra variables y_i, $0 \leq i \leq N$, of odd degree; and observe that the assignments

$$x_i \longrightarrow y_i\,, \qquad 0 \leq i \leq n-r-1$$

$$x_{n-r+i} \longrightarrow y_{n-r+2i}y_{n-r+2i+1}\,, \qquad 0 \leq i \leq r\,,$$

extend to an injective map of graded vector spaces $W[x_0,\ldots,x_n] \longrightarrow W[y_0 \ldots, y_N]$. Following this injection by Φ carries $W[x_0,\ldots,x_n]$ to the subspace of $C^*(\tilde{T}^N;p^*A)/R^* \simeq C^*(T^N;A)$ which is fairly clearly identified with the cellular cochains on the torus $T^n \subset PSU(N+1)$ covered by the torus of diagonal matrices $\mathrm{diag}(z_0,\ldots,z_N)$ in $SU(N+1)$ satisfying $z_{n-r+2i} = z_{n-r+2i+1}$ for $0 \leq i \leq r$. The appropriate subgroup of S_{N+1} acting on T^n is the subgroup $S_{n-r,r+1} \simeq S_{n-r} \times S_{r+1}$ defined as the permutations π satisfying $n-r+1 \leq \pi(n-r+2i)+1 = \pi(n-r+2i+1) \leq n+r+1 = N$, and the coefficient system is the restriction of the original system A. As there are no new ideas required to complete the verifications, we leave them to the reader.

Lemma 4.3. *Let $[x_0,\ldots,x_n]$ be a specification of distinct variables where $|x_i| \equiv 1 \bmod 2$ for $0 \leq i \leq n-r-1$, $|x_i| \equiv 0 \bmod 2$ for $n-r \leq i \leq n$. Then on the $\bmod 2$ degree preserving permutation subgroup $S_{n-r,r+1} \simeq S_{n-r} \times S_{r+1}$ we have the character $\varepsilon((\pi,\pi')) = \mathrm{sign}\,\pi$ and as $S_{n-r,r+1}$ spaces*

$$H^i(W[x_0,\ldots,x_n]) \simeq \Lambda^{i-1}H_0^* \otimes \varepsilon$$

for $1 \leq i \leq N+1$, where H_0^* is as in Lemma 4.2 and supports the action $(\lambda^i)^\pi = \lambda^{\pi^{-1}(i)}$ on the basis dual to $\{e_i\}$.

This lemma is a direct consequence of Lemma 3.3 and Lemma 4.2.

The following lemma is basic for determination of invariants in the spaces described in Lemma 4.3.

Lemma 4.4. Let S_k act on $V = \mathbb{Q}^k$ by $e_i \longrightarrow e_{\sigma(i)}$ and let S_k act on V^* by the contragredient representation $\lambda^i \longrightarrow \lambda^{\sigma(i)}$, where λ^i is dual to e_i. Then

(a) there are two alternating functions on V invariant under S_k:

$$1 \in \Lambda^0 V^* \text{ and } \Sigma = \lambda^1 + \ldots + \lambda^k \in \Lambda^1 V^* ;$$

(b) likewise there are two alternating functions which are invariant relative to the character sign ( ) of S_k:

$$1/k(A \wedge \Sigma) = \lambda^1 \wedge \ldots \wedge \lambda^k \in \Lambda^k V^* \text{ and}$$
$$A = (\lambda^1 - \lambda^2) \wedge \ldots \wedge (\lambda^{k-1} - \lambda^k) \in \Lambda^{k-1} V^* .$$

Proof. That these functions have the right invariance properties is clear enough. (It helps in checking this for A to remember S_k is generated by transpositions $(i \;\; i+1)$.)

To show there are no invariants in higher exterior powers of V^* requires mere computation: Let a proposed invariant

$$\Psi = \sum \Psi_I d\lambda^I = \sum_{i_1 < \ldots < i_p} \Psi_{i_1}, \ldots, {}_{i_p} \lambda^{i_1} \wedge \ldots \wedge \lambda^{i_p}$$

be written out with respect to the canonical basis of $\Lambda^p V^*$. Fix attention on a typical p-triple $j_1 < \ldots < j_p$ and let $\sigma = (j_1 j_2)$. Then,

$$\Psi = (\Psi^\sigma) = \Psi_J \lambda^{j_2} \wedge \lambda^{j_1} \wedge \ldots \wedge \lambda^{j_p} + \sum_{I \neq J} \Psi_I \lambda^{\sigma(I)}$$
$$= -\Psi_J \lambda^J + \sum_{I \neq J} \Psi'_I \lambda^I ,$$

which implies $\Psi_J = 0$. Thus $(\Lambda^p V^*)^{S_k} = 0$ for $p \neq 0,1$.

For the rest of the lemma recall the pairing

$$\Lambda^p V^* \otimes \Lambda^{k-p} V^* \longrightarrow \Lambda^k V^*$$

$$\Psi \otimes \Psi' \longrightarrow \Psi \wedge \Psi'$$

which commutes with the diagonal action of S_k on the left and the natural action on the right. Thus, the non-degenerate pairing gives an isomorphism of S_k-spaces

$$\wedge^p V^* \simeq \mathrm{Hom}(\wedge^{k-p} V^*, \wedge^k V^*)$$

Thus the identification of invariant subspace gives

$$(\wedge^p V^*)^{S_k} \simeq \mathrm{Hom}_{S_k}(\wedge^{k-p} V^*, \wedge^k V^*)$$

and the dimension of the right hand space counts the number of invariants relative to the sign character in $\wedge^{k-p}V^*$. Hence, (b) follows from (a) .

In general, a complex $W[z_0,\ldots,z_n]$ with the specifications containing variables some of which appear with high multiplicity maps into a complex $W[x_0,\ldots,x_n]$ where the variables are distinct and the image is an invariant subcomplex split off $W[x_0,\ldots,x_n]$ by an appropriate projection operator as in [6, Lemma 3.1]. For a specification $[x_0,\ldots,x_N]$ with ℓ variables $y_1,\ldots,y_\ell$ with multiplicities $k_1,\ldots,k_\ell$ respectively (i.e., $y_1 = x_i$, $0 \le i \le k_1 - 1$;...; $y_\ell = x_i$, $N - k_\ell + 1 \le i \le N$) and

$$|y_i| \equiv 1 \mod 2 \ , \quad 1 \le i \le m$$

$$|y_i| \equiv 0 \mod 2 \ , \quad m+1 \le i \le \ell$$

the subgroup with respect to which we need invariants is

$S_{k_1} \times\ldots\times S_{k_\ell} \subset S_{k_1+\ldots+k_m, k_{m+1}+\ldots+k_\ell}$ and acting by the restriction of this action.

We extend Lemma 4.4 to the present situation as follows. Let

$$H = \mathbb{Q}^{k_1} \oplus\ldots\oplus \mathbb{Q}^{k_\ell}$$

be acting on by $S_O \times S_E$ in the obvious manner where

$$S_O = S_{k_1} \times\ldots\times S_{k_m}$$

$$S_E = S_{k_{m+1}} \times\ldots\times S_{k_\ell}$$

and let $\varepsilon : S_O \times S_E \longrightarrow \{\pm 1\}$ be the character such that $\varepsilon((\pi,\pi')) = \mathrm{sign}\ \pi$. Set $N(0) = 0$ and $N(i) = k_1 +\ldots+ k_i$ for $i > 0$. The following lemma computes invariants relative to ε in $\wedge^* H^*$.

Lemma 4.5. Write

$$A_{i+1} = \bigwedge_{N(i)+1\leq\nu\leq N(i+1)} (\lambda^{\nu}-\lambda^{\nu+1}) \quad \text{for } m > i \geq 0$$

and

$$\sum_{j+1} = \sum_{N(j)+1\leq\nu\leq N(j+1)} \lambda^{\nu} , \quad \text{for } \ell > j \geq 0 .$$

Then, a basis for the invariants of $S_O \times S_E$ relative to ε in $\wedge^* H^*$ consists of the products

$$A_1 \wedge \ldots \wedge A_m \wedge \sum_{j_1} \wedge \ldots \wedge \sum_{j_r}$$

where $1 \leq j_1 < \ldots < j_r \leq \ell$ runs over all r-tuples for all r .

Proof. One needs to realize first that

$$\wedge^p H^* \simeq \bigotimes_{p_1+\ldots+p_\ell=p} \wedge^{p_1} V_1^* \otimes \ldots \otimes \wedge^{p_\ell} V_\ell^*$$

so $S_O \times S_E$ spaces (where we have written V_i for the ith summand $\mathbb{Q}^{k_i}$ of H .) Second, by multiplying by the character ε , which is itself a tensor product of sign representations of the factors of S_O , the problem is reduced to computation of ordinary invariants of $S_O \times S_E$ in spaces

$$(\varepsilon_1 \otimes \wedge^{p_1} V_1^*) \otimes \ldots \otimes (\varepsilon_m \otimes \wedge^{p_m} V_m^*) \otimes \ldots \otimes \wedge^{p_\ell} V_\ell^* .$$

Third, it is an elementary fact to show that if $G_1 \times G_2$ acts on $W_1 \otimes W_2$ by $(g_1, g_2)(w_1 \otimes w_2) = (g_1 w_1 \otimes g_2 w_2)$ then

$(W_1 \otimes W_2)^{G_1 \otimes G_2} = (W_1 \otimes W_2)^{G_1} \cap (W_1 \otimes W_2)^{G_2} = W_1^G \otimes W_2^G$, so we need to compute

$$(\varepsilon_1 \otimes \wedge^p_1 V_1^*)^{S_{m_1}} \otimes \ldots \otimes (\wedge^{p_\ell} V^*)^{S_\ell} .$$

Applying Lemma 4.4 and pushing the result over to $\wedge^* H^*$ proves the desired result.

Finally, we have the main lemma of this section

Lemma 4.6. *Suppose that the specification* $[x_0,\dots,x_N]$ *is for* ℓ *variables* $y_1,\dots,y_\ell$ *with multiplicities* $k_1,\dots,k_\ell$ *respectively. That is,* $k_1+\dots+k_\ell = N+1$, and $y_1 = x_i$, $0 \le i \le k_1-1$;...; $y_\ell = x_i$, $N-k_\ell+1 \le i \le N$.

(a) *Suppose* $|y_i| \equiv 0 \mod 2$. *Then* $\dim_{\mathbb{Q}} H^j(W[x_0,\dots,x_N]) = \binom{\ell-1}{j-1}$.

(b) *Suppose* $|y_i| \equiv 1 \mod 2$. *Then* $\dim_{\mathbb{Q}} H^j(W[x_0,\dots,x_N]) = \binom{\ell-1}{N+1-j}$.

(c) *Suppose* $|y_i| \equiv 1 \mod 2$ *for* $1 \le i \le m$ *and* $|y_i| \equiv 0 \mod 2$ for $m+1 \le i \le \ell$. *Then* $\dim_{\mathbb{Q}} H^j(W[x_0,\dots,x_N]) = \binom{\ell-1}{j-(N_1-m)-1}$ *where* $N_1 = k_1+\dots+k_m$, *the total number of elements of odd degree*.

Here $\binom{p}{q}$ is a binomial coefficient with the usual conventions for vanishing.)

Proof. This lemma follows from Lemma 4.5 by observing that when restricted to $H_0 = \{V \in \mathbb{Q}^{N+1} \mid v^0+\dots+v^N = 0\}$, $\sum_1+\dots+\sum_\ell = 0$. As $\wedge^* H_0^*$ sits inside $\wedge^* H^*$ with complement $(\sum_1+\dots+\sum_\ell) \wedge (\wedge^* H_0^*)$, it is easy to verify the dimension counts we have given.

At last we are ready to prove the main theorem. In view of Lemma 2.1, it suffices to obtain the formula $P.S.H^*(V(\bar{K}))$ =

$$(1)\quad t(1+t)^{-1}\{\prod_i(1+t^{|u_i|+1})(1-t^{|u_i|})^{-1}\prod_j(1+t^{|v_j|})(1-t^{|v_j|+1})^{-1} -1\}$$

decomposition of $V(\bar{K})$ into $\oplus W[x_0,\dots,x_N]$, the sum being over all specifications and from Lemma 4.6. We organize the counting argument by making use of the fact $H^*(W[x_0,\dots,x_N])$ depends mainly on the number of distinct variables present, an obvious implication of Lemma 4.6. In the right side of formula (1) the factor $\sum(x,t)$ is the contribution to $H^*(V(\bar{K}))$ made by $H^*(W[x_0,\dots,x_N])$ where the specification is for distinct variables of multiplicity one and the other two factors account for the propogation of this contribution as the multiplicities of the variables are increased.

More precisely, Lemma 4.6 says that if the specification $[x_0,\dots,x_N]$ is for y_i with multiplicity $k_i > 0$ for $1 \le i \le \ell$, then the sequence of dimensions in the row $q = \sum|x_i| = \sum k_i|y_i|$ coming from $H^*(W[x_0,\dots,x_N])$ is the same as the sequence of dimensions in the row $q = \sum |y_i|$ coming from $H^*(W[y_1,\dots,y_\ell])$. However there has also been a shift to the right by an amount $N_1 - m = (k_1-1) + \dots + (k_m-1)$ which is just the increase in the multiplicities of the variables of odd degree. Thus we discover two related polynomial summands of the desired Poincaré series: From $H^*(W[y_1;\dots;y_\ell])$ we get

$$t \cdot t^{|y_1|+\dots+|y_\ell|} \cdot (1+t)^{\ell-1} ;$$

from $H^*(W[x_0,\dots,x_N])$ we get

$$t \cdot t^{|y_1|+\dots+|y_\ell|}(1+t)^{\ell-1} \cdot \prod_{1\le j\le m} t^{(k_j-1)(|y_j|+1)} \prod_{j+1\le j\le \ell} t^{(k_j-1)|y_j|} .$$

A moment's reflection shows that the sum of these polynomials over all possible speicifications may indeed be expressed in the form claimed.

References

[1] J. F. Adams, Lectures on Lie Groups, W. A. Benjamin, NY, 1969.

[2] D. Burghelea, Some rational computations of the Waldhausen algebraic K-theory, Comment. Math. Helvetici, 54 (1979), 185-188.

[3] K. T. Chen, Iterated path integrals, BAMS 83 (1977), 831-879.

[4] W. G. Dwyer, W.-C. Hsiang and R. Staffeldt, Pseudo-isotopy and invariant theory, I., Topology 19 (1980), 367-385.

[5] R. M. Hain, Iterated integrals, minimal models, and rational homotopy theory. Thesis, U. ov Illinois at Urbana-Champaign, 1980.

[6] W.-C. Hsiang and R. E. Staffeldt, A model for computing rational algebraic K-theory of simply-connected spaces, Invent. Math. 68 (1982), 227-239.

[7] N. E. Steenrod, Homology with local coefficients, Ann. of Math. (2), 44, (1943), 610-627.

[8] E. Stiefel, Uber eine Beziehung Zwischem geschlossenen Lie'schen gruppen und discontiniuerlichen Bewegungsgruppen euklidischer Raume, Comment. Math. Helv., 14 (1942), 350-380.

[9] F. Waldhausen, Algebraic K-theory of topological spaces, I., Proc. of Symposia in Pure Math., 32 (1978), 35-60.

[10] G. W. Whitehead, Elements of Homotopy Theory. NY: Springer-Verlag (1978).

DEPARTMENT OF MATHEMATICS
PRINCETON UNIVERSITY
PRINCETON, N.J. 08544

DEPARTMENT OF MATHEMATICS
PENNSYLVANIA STATE UNIVERSITY
UNIVERSITY PARK, PA 16802

Contemporary Mathematics
Volume 19, 1983

AN ALGEBRAIC FILTRATION OF H_*BO

Stanley O. Kochman[1]

ABSTRACT. A filtration of $H_*(BO;Z_2)$ by bipolynomial Hopf algebras and A_*-comodules is constructed. Specific polynomial generators are determined. Applications to $H_*(BSO;Z_2)$ are given. Generalizations are made to $H_*(BU;Z_p)$, p an odd prime.

1. INTRODUCTION. In Section 2 we will construct a complete decreasing filtration $B_*(n)$, $n \geq 0$, of H_*BO. (All homology and cohomology with no specified coefficients has Z_2-coefficients.) Each $B_*(n)$ is a sub Hopf algebra and sub A_*-comodule of H_*BO where A_* is the dual of the Steenrod algebra. If $k = 2^{k_1} + \dots + 2^{k_e}$, $0 \leq k_1 < \dots < k_e$, write $e = L(k)$ and $k_1 = M(k)$. Then the $B_*(n)$ are bipolynomial Hopf algebras with one polynomial generator in each degree k with $L(k) + M(k) > n$. We will show that $B_*(n)$, $0 \leq n \leq 3$, can be realized as $B_*(0) = H_*BO$, $B_*(1) = H_*BSO$, $B_*(2) = H_*BSpin$ and $B_*(3) = H_*BO\langle 8\rangle$. However, the $B_*(n)$ for $n \geq 4$ can not be realized as the homology of spaces. However, $B_*(n)$ can be realized as the image of $H_*BO\langle\phi(n)\rangle$ in H_*BO where $BO\langle\phi(n)\rangle$ is the $\phi(n)-1$ connected covering of BO. Section 2 also gives a detailed description of the duals $B^*(n)$ as quotients of H^*BO.

In Section 4 we give two sets of polynomial generators for $B_*(n)$. One set is constructed by the methods of [10] as appropriate elements of the dual basis of monomials in the Stiefel-Whitney classes. The second more interesting set is given by elements of Image τ_{n+1*} where $\tau_{n+1}: BO(2)^{n+1} \to BO(2^{n+1})$ is the iterated tensor product. To this end we

[1]This research was partially supported by a grant from the Natural Sciences and Engineering Research Council of Canada.

compute τ_{n+1*} in Section 3 and prove that Image $\tau_{n+1*} \subset B_*(n)$ in Section 4. The polynomial generators constructed using τ_{n+1*} all lie in $H_*BO(2^{n+1}-1)$ which we prove is the best possible.

In Section 5 we collect some miscellaneous facts about $B_*(1) = H_*BSO$ including a formula for the primitive elements as polynomials in the generators of Section 4. In Section 6 we present the generalization of our results to odd primes p which we phrase in terms of a filtration of the mod p homology of the first space of the Adams splitting [1] of $BU_{(p)}$. The Hopf algebras studied by P. Hoffman [6] can be viewed as the integral generalization of these filtrations.

Write $H_*RP^\infty = H_*BO(1) = Z_2\{A_n \mid n \geq 0\}$ with $A_n \in H_nRP^\infty$. Then $H_*BO(1) \subset H_*BO$ under the canonical map and $H_*BO = Z_2[A_1,\ldots,A_n,\ldots]$ with $\Delta(A_n) = \sum_{i=0}^{n} A_i \otimes A_{n-i}$ and $Sq_*^k(A_n) = (k,n-2k)A_{n-k}$. We will use $(e_1,\ldots,e_t)$ for the multinomial coefficient $(e_1+\ldots+e_t)!/e_1!\ldots e_t!$ even in the case $t = 2$. In cohomology write $H^*BO = Z_2[W_1,\ldots,W_n,\ldots]$ where $W_n \in H^nBO$ is the nth Stiefel-Whitney class. Then $\Delta(W_n) = \sum_{i=0}^{n} W_i \otimes W_{n-i}$ and $Sq^k(W_n) = (k,n-k-1)W_{n+k} + \sum_{i=0}^{k-1} (i,n-k-1)W_{k-i}W_{n+i}$ (Wu Formula). Write PH_nBO or PH^nBO as $\{0,P_n\}$ where

$$P_n = \sum_{e_1+2e_2+\ldots+ne_n=n} [n(e_1,\ldots,e_n)/(e_1+\ldots+e_n)]X_1^{e_1}\ldots X_n^{e_n} \quad \text{(Girard Polynomials)}$$

and X_k is A_k in homology or W_k in cohomology. These facts can be found in [4]. We will let $<\ ,\ >$ denote the Kronecker pairing $H_*X \otimes H^*X \to Z_2$. The diagonal map $X \to X^k$ or $H_*X \to H_*X \otimes \cdots \otimes H_*X$ (k times) will be denoted Δ^k.

This research is motivated by Baker's construction [3] of generators for H_*BSO from a map $g\colon BO(2) \to BSO(3)$. As explained in Corollary 3.5, τ_{2*} includes the information given by g_*. The $B_*(n)$ generalize $H_*BSO = B_*(1)$ and the τ_{n+1} generalize g. Thus the results of Sections 2-5 give new information on the mod two homology and cohomology of BSO, BSpin and BO<8>. Our first set of polynomial generators for H_*BSO are those of Baker [3] while our second set of polynomial generators for H_*BSO, H_*BSpin and $H_*BO\langle 8\rangle$ are similar to those of

[10] which are related to those of Bahri [2]. Other polynomial generators for H_*BSO have been constructed by Papastavridis [11] and in [8], [9]. Our first set of polynomial generators for H_*BSpin and $H_*BO<8>$ are the only known ones with an explicit description as polynomials in the A_n, $n \geq 1$.

2. EXISTENCE OF THE $B_*(n)$. We will construct the $B_*(n)$ in this section and prove that they are sub Hopf algebras and sub A_*-comodules of H_*BO which are bipolynomial. Our method is to construct and study the duals $B^*(n)$ of $B_*(n)$ as quotients of H^*BO. Write $B^*(n) = H^*BO(n)/I_n = B^*(n-1)/J_{n-1}$. We define the $B^*(n)$ by induction on $n \geq 0$ with $B^*(0) = H^*BO$. I_n will be seen to be the ideal in H^*BO generated by one element of the form $\nu_k = W_k$+decomposables for each k with $L(k) + M(k) \leq n$. Then $0 = I_0 \subset I_1 \subset \ldots \subset I_n \subset I_{n+1} \subset \ldots \subset H^*BO$ and the ν_k do not depend on n. There are two equivalent descriptions of J_{n-1}. From one of these it follows that $B^*(n)$ is a quotient Hopf algebra of $B^*(n-1)$ and of H^*BO while from the other it follows that $B^*(n)$ is a quotient A-module of $B^*(n-1)$ and of H^*BO. For technical reasons we keep a record of $PB^*(n)$ as we proceed with this construction in Theorem 2.1. We then derive the properties of the $B^*(n)$ and $B_*(n)$ in several corollaries. In Theorem 2.8 we give a specific description of $PB^*(n)$ which will be used in the second construction of polynomial generators for $B_*(n)$ in Section 4. We conclude by proving that the $B_*(n)$, $n \geq 4$, cannot be realized as the homology of a space but can be realized as the image of $H_*BO\langle\phi(n)\rangle$ in H_*BO.

THEOREM 2.1. There exist quotient Hopf algebras and quotient A-modules $B^*(n) = H^*BO/I_n = B^*(n-1)/J_{n-1}$ with the four properties listed below. Let $\pi_n\colon H^*BO \to B^*(n)$ denote the canonical projection.

(a) $B^*(0) = H^*BO$.

(b) $PB^*(n) = \{P_{n,k} \mid L(k) + M(k) > n\}$ with $\deg P_{n,k} = k$.

(c) J_n is the ideal in $B^*(n)$ spanned by the set of indecomposable primitive elements $\{P_{n,k} \mid L(k) + M(k) = n+1\}$.

(d) J_n is the ideal in $B^*(n)$ spanned by the cyclic A-submodule of $B^*(n)$ generated by $\pi_n(W_{2^n})$.

Proof. We prove this theorem by induction on $n \geq 0$. Let $B^*(0) = H^*BO$. Inductively assume $B^*(n)$ has been constructed and that $B^*(k)$, J_k satisfy (b),(c),(d) for $0 \leq k \leq n-1$. We must check that (b) holds for $PB^*(n)$ and that the ideal J_n' defined in (c) is the same as the ideal J_n'' defined in (d). It will then follow that $B^*(n+1)$ is a quotient Hopf algebra and quotient A-module of $B^*(n)$

and hence a quotient Hopf algebra and quotient A-module of H^*BO. To prove (b) observe that $B_*(n)$ is a sub Hopf algebra of a polynomial Hopf algebra and hence $B_*(n)$ is a polynomial algebra. By rank considerations it follows that

$$PB^k(n) \cong QB_k(n) = \begin{cases} Z_2 & \text{if } L(k) + M(k) > n \\ 0 & \text{otherwise} \end{cases}.$$

Next observe that the $P_{n,k}$ listed in (c) are indecomposable for either k is odd or $k = 2h$ in which case $PB^h(n) = 0$. To prove that $J'_n = J''_n$ we prove that $A\pi_n(W_{2^n})$ equals $P = \{P_{n,k} \mid L(k) + M(k) = n + 1\}$. $B(n)$ is $(2^n - 1)$-connected and hence $\pi_n(W_{2^n})$ is primitive. Thus $A\pi_n(W_{2^n})$ is a set of primitives. By the Wu formula for $0 < k < 2^n$,

$$Sq^k\pi_n(W_{2^n}) = (k,2^n-k-1)\pi_n(W_{2^n+k}) = \begin{cases} \pi_n(W_{2^{n+1}-2^m}) & \text{if } k = 2^n-2^m \\ 0 & \text{otherwise} \end{cases}$$

because for $2^n < j < 2^{n+1}$ $PB^j(n) = 0$ if $j \neq 2^{n+1} - 2^m$. Thus

$A\pi_n(W_{2^n}) = \{\pi_n(W_{2^n})^{2^k} \mid k \geq 0\} \cup \bigcup_{m=0}^{n-1} A\pi_n(W_{2^{n+1}-2^m})$. Using the Wu formula we see by induction on t that

$$(*) \qquad A\pi_n(W_{2^n}) \equiv \bigcup_{\substack{j=2^t+1 \\ L(j)+M(j)=n+1}}^{2^{t+1}-1} A\pi_n(W_j) \text{ modulo } P.$$

Thus $A\pi_n(W_{2^n}) \subset P$. Moreover if $P_{n,j} \in P$ and $2^t < j < 2^{t+1}$ then (*) shows that $P_{n,j} \in A\pi_n(W_{2^n})$. Thus $A\pi_n(W_{2^n}) = P$.

The following corollaries follow easily from Theorem 2.1 and its proof.

COROLLARY 2.2. $B^*(n) = Z_2[\pi_n(W_k) \mid L(k) + M(k) > n]$.

COROLLARY 2.3. $B_*(n)$ is a sub Hopf algebra and sub A_*-comodule of H_*BO.

COROLLARY 2.4. $B_*(n)$ is a $(2^n - 1)$-connected bipolynomial Hopf algebra with one polynomial generator in every degree k with $L(k) + M(k) > n$.

COROLLARY 2.5. $\bigcap_{n=0}^{\infty} B_*(n) = 0$.

COROLLARY 2.6. (a) $B_*(1) = H_*BSO$ and $B^*(1) = H^*BSO$.

(b) $B_*(2) = H_*BSpin$ and $B^*(2) = H^*BSpin$.

(c) $B_*(3) = H_*BO\langle 8\rangle$ and $B^*(3) = H^*BO\langle 8\rangle$.

Proof. (a) $B^*(1) = H^*BO/(W_1)$ which by [4] equals H^*BSO. Thus $B_*(1) = H_*BSO$.

(b) $H^*BSpin = H^*BSO/(AW_2) = B^*(1)/J_1 = B^*(2)$ by [12] and Theorem 2.1(d). Thus $H_*BSpin = B_*(2)$.

(c) $H^*BO\langle 8\rangle = H^*BSpin/(AW_4) = B^*(2)/J_2 = B^*(3)$ by [12] and Theorem 2.1(d). Thus $H_*BO\langle 8\rangle = B_*(3)$.

The following curious result will be used to verify our description of $PB^*(n)$ in Theorem 2.8.

LEMMA 2.7. For $n \geq 1$ the sum of the coefficients of the integral Girard polynomial for P_n is one.

Proof. The assertion of this lemma can be rephrased as

$$\sum_{e_1+2e_2+\ldots+ne_n=n} (-1)^{1+e_1+\ldots+e_n} \frac{n(e_1,\ldots,e_n)}{e_1+\ldots+e_n} = 1.$$

We prove this by induction on $n \geq 1$. We use the notation $H^*(BU;Z) = Z[C_1,\ldots,C_n,\ldots]$ where C_n is the nth Chern class. Then $P_1 = C_1$ so that case $n = 1$ is true. Assume the lemma is true for $1 \leq k < n$. By Newton's formula $P_n = nC_n - \sum_{k=1}^{n-1} C_{n-k}P_k$. Thus the sum of the coefficients of P_n equals $n - (n-1) = 1$.

THEOREM 2.8. Give H_*BO the basis of monomials in the A_k and give H^*BO the dual basis. If $X = \sum \alpha_I A_I \in H_*BO$ with $\alpha_I \in Z_2$ then let $X^* = \sum \alpha_I A_I^*$. In this notation

$$PB^*(n) = \{\pi_n\left(A^*_{k2^{-M(k)}}\right)^{2^{M(k)}} = \pi_n(P_k) \mid L(k) > n\}$$

$$\cup\ \{\pi_n\left(P^{2^{n-L(k)+1}}_{k2^{-M(k)}}\right)^{*2^{L(k)+M(k)-n-1}} \mid L(k) + M(k) > n \geq L(k)\}.$$

Proof. The elements listed in this theorem consist of one element in every degree k such that $L(k) + M(k) > n$. By Theorem 2.1(b), $PB^*(n)$ also has one nonzero element in each of these degrees. It thus suffices to show that the elements listed in this theorem are nonzero primitives. If $L(k) > n$ then $\pi_n(P_k) = \pi_n\left(P_{k2^{-M(k)}}\right)^{2^{M(k)}}$ which is primitive and nonzero because $\pi_n(P_{k2^{-M(k)}})$ is indecomposable in $B^*(n)$. We claim that for $P_{2s+1} \in PH_{2s+1}BO$, and $t = 2^{t_1} + \ldots + 2^{t_u}$, $0 \leq t_1 < \ldots < t_u$, $\left(P^t_{2s+1}\right)^* = \left(P^{2^{t_1}}_{2s+1}\right)^* \ldots \left(P^{2^{t_u}}_{2s+1}\right)^*$. We prove this by induction on t. Note that $X = \left(P^t_{2s+1}\right)^* + \left(P^{2^{t_1}}_{2s+1}\right)^* \ldots \left(P^{2^{t_u}}_{2s+1}\right)^*$ is primitive. Now the coefficient of $W_1^{t(2s+1)}$ in X is $\langle A_{t(2s+1)}, X\rangle$

$$= \langle A_{t(2s+1)}, (P^t_{2s+1})^*\rangle + \left\langle A_{t(2s+1)}, \left(P^{2^{t_1}}_{2s+1}\right)^* \ldots \left(P^{2^{t_u}}_{2s+1}\right)^*\right\rangle$$

$$= \langle P^t_{2s+1}, A^*_{t(2s+1)}\rangle + \left\langle \Delta^u(A_{t(2s+1)}), \left(P^{2^{t_1}}_{2s+1}\right)^* \otimes \ldots \otimes \left(P^{2^{t_u}}_{2s+1}\right)^*\right\rangle$$

$$= \langle P_{2s+1} \otimes \ldots \otimes P_{2s+1}, \Delta^t(A^*_{t(2s+1)})\rangle + \left\langle A_{2^{t_1}(2s+1)}, \left(P^{2^{t_1}}_{2s+1}\right)^*\right\rangle \ldots \left\langle A_{2^{t_u}(2s+1)}, \left(P^{2^{t_u}}_{2s+1}\right)^*\right\rangle$$

$= 0$. Hence $X = 0$ which proves the claim.

Thus if $L(k) + M(k) > n$ and $n \geq L(k)$, $\bar{\Delta}\pi_{n-1}(P^{2^{n-L(k)+1}}_{k2^{-M(k)}})^*$

$$= \pi_{n-1}\left(P^{2^{n-L(k)}}_{k2^{-M(k)}}\right)^* \otimes \pi_{n-1}\left(P^{2^{n-L(k)}}_{k2^{-M(k)}}\right)^*.$$

However $\pi_{n-1}\left(P^{2^{n-L(k)}}_{k2^{-M(k)}}\right)^* = \pi_{n-1}(P^*_{k2^{n-L(k)-M(k)}})$ which is primitive in $B^*(n-1)$ and is thus in $\text{Kernel}[B^*(n-1) \to B^*(n)]$. Hence $\pi_n\left(P^{2^{n-L(k)+1}}_{k2^{-M(k)}}\right)^*$ is primitive in $B^*(n)$. By Lemma 2.7, $\left(P^{2^{n-L(k)+1}}_{k2^{-M(k)}}\right)^* = P^*_{k2^{n-L(k)-M(k)+1}}$ is indecomposable in H^*BO and thus $\pi_n\left(P^{2^{n-L(k)+1}}_{k2^{-M(k)}}\right)^*$ is nonzero in $B^*(n)$. Therefore $\pi_n\left(P^{2^{n-L(k)+1}}_{k2^{-M(k)}}\right)^{*2^{L(k)+M(k)-n-1}}$ is a nonzero primitive in $B^*(n)$.

THEOREM 2.9. For $n \geq 4$ there is no $(2^n - 1)$-connected space X and map $f: X \to BO$ such that in mod two homology f_* is a monomorphism with image $B_*(n)$.

Proof. Assume that $n \geq 4$ and there is a $(2^n - 1)$-connected space X and a map $f: X \to BO$ such that f_* is a monomorphism with image $B_*(n)$. Then $H_{2^n}X = Z_2A$ with $f_*(A) = A_1^{2^n}$. By the Hurewicz Theorem, A is spherical and hence $A_1^{2^n}$ is spherical. However, by [4] the image of the Hurewicz homomorphism in $H_{2^n}(BO;Z)$ is divisible by $(2^{n-1} - 1)!/2$ and thus the mod two Hurewicz homomorphism for BO in degree 2^n is zero, a contradiction. Thus no such $f: X \to BO$ can exist.

THEOREM 2.10. Let $\lambda_k: BO\langle k\rangle \to BO$ denote the canonical map. If $n = 4s + t$, $0 \leq t \leq 3$, then let $\phi(n) = 8s + 2^t$. For $n \geq 0$ there is a canonical map $\gamma_n^*: B^*(n) \to H^*BO\langle\phi(n)\rangle$ such that γ_n^* is one-to-one. γ_n^* identifies $B^*(n)$ with Image $\lambda_{\phi(n)}^*$ as $\mathcal{A}$-modules and Hopf algebras.

Proof. By Theorem 2.1, $I_n = \text{Kernel } \pi_n$ is the ideal in H^*BO spanned by the sub $\mathcal{A}$-module generated by $W_1,\ldots,W_{2^{n-1}}$. By [2a], $W_1,\ldots,W_{2^{n-1}}$ are in Kernel $\lambda_{\phi(n)}^*$. Thus there is a unique map $\gamma_n^*: B^*(n) \to H^*BO\langle\phi(n)\rangle$ such that $\gamma_n^* \circ \pi_n^* = \lambda_{\phi(n)}^*$. By [12], $\lambda_{\phi(n)}^*(W_k)$ is a polynomial generator for $H^*BO\langle\phi(n)\rangle$ if and only if $L(k) + M(k) > n$. By Corollary 2.2 it follows that γ_n^* is one-to-one.

COROLLARY 2.11. $B_*(n) = \text{Image}[\lambda_{\phi(n)*}: H_*BO\langle\phi(n)\rangle \to H_*BO]$.

Proof: Let $\gamma_{n*}: H_*BO\langle\phi(n)\rangle \to B_*(n)$ denote the dual of γ_n^*. Then γ_{n*} is onto and $\lambda_{\phi(n)*}$ equals γ_{n*} followed by the canonical inclusion of $B_*(n)$ in H_*BO. Thus $B_*(n) = \text{Image } \lambda_{\phi(n)*}$.

3. TENSOR PRODUCT IN H_*BO. For $m,n \geq 1$ there is a group homomorphism $\bar{\tau}: O(m) \times O(n) \to O(mn)$ called the tensor product given by $\bar{\tau}(A,B) = C$ where $B = (b_{ij})$ and C is the block matrix

$$C = \begin{pmatrix} Ab_{11} & \cdots & Ab_{1n} \\ \vdots & & \vdots \\ Ab_{n1} & \cdots & Ab_{nn} \end{pmatrix}.$$

Now $\bar{\tau}$ induces $\tau: BO(m) \times BO(n) \to BO(mn)$. Observe that τ is not stable, i.e., neither of the following diagrams commute:

$$\begin{array}{ccc} BO(m) \times BO(n) & \to & BO(m+1) \times BO(n) \\ \downarrow \tau & & \downarrow \tau \\ BO(mn) & \longrightarrow & BO(mn+n) \end{array} \qquad \begin{array}{ccc} BO(m) \times BO(n) & \to & BO(m) \times BO(n+1) \\ \downarrow \tau & & \downarrow \tau \\ BO(mn) & \longrightarrow & BO(mn+m) \end{array}$$

However, it is easy to see that $\bar{\tau}$ and hence τ is associative, i.e., the following diagram commutes:

$$\begin{array}{ccc} BO(m) \times BO(n) \times BO(p) & \xrightarrow{\tau\times 1} & BO(mn) \times BO(p) \\ \downarrow 1\times\tau & & \downarrow \tau \\ BO(m) \times BO(np) & \xrightarrow{\tau} & BO(mnp) \end{array}$$

In Theorem 3.1 we will compute $\tau_*: H_*BO(m) \otimes H_*BO(n) \to H_*BO(mn)$. We then study the iterated tensor product $\tau_k: BO(2)^k \to BO(2^k)$ which will be used in Section 4 to define polynomial generators for the $B_*(n)$.

THEOREM 3.1. The map $\tau: BO(m) \times BO(n) \to BO(mn)$ is given in homology by

$$\tau_*(A_{s_1} \cdots A_{s_m} \otimes A_{t_1} \cdots A_{t_n})$$
$$= \sum_{s_1=s_{11}+\cdots+s_{1n}} \cdots \sum_{s_m=s_{m1}+\cdots+s_{mn}} \; \sum_{t_1=t_{11}+\cdots+t_{1m}} \cdots \sum_{t_n=t_{n1}+\cdots+t_{nm}} \prod_{\gamma=1}^{m} \prod_{\lambda=1}^{n} (s_{\gamma,\lambda}, t_{\lambda,\gamma}) A_{s_{\gamma,\lambda}+t_{\lambda,\gamma}}.$$

Proof. The following diagram commutes because both routes applied to $(X_1,\ldots,X_m,Y_1,\ldots,Y_n)$ give the diagonal matrix whose $km+h$ entry is X_hY_{k+1} for $1 \leq h \leq m$, $0 \leq k < n$.

$$\begin{array}{ccc}
O(1)^{m+n} \xrightarrow{\omega\times\omega} O(m)\times O(n) & \xrightarrow{\bar{\tau}} & O(mn) \\
\downarrow \pi & & \uparrow \omega \\
O(2)^{mn} & \xrightarrow{\det^{mn}} & O(1)^{mn}
\end{array}$$

Here ω is Whitney sum and the km+h entry of $\pi(X_1,\ldots,X_m,Y_1,\ldots,Y_n)$ is $\begin{pmatrix} X_h & 0 \\ 0 & Y_{k+1} \end{pmatrix}$ for $1 \leq h \leq m$, $0 \leq k < n$. We now apply the functor B to the above diagram to obtain the commutative diagram:

$$(*) \qquad \begin{array}{ccc}
BO(1)^{m+n} \xrightarrow{\omega\times\omega} BO(m)\times BO(n) & \xrightarrow{\tau} & BO(mn) \\
\downarrow B\pi & & \uparrow \omega \\
BO(2)^{mn} & \xrightarrow{B\det^{mn}} & BO(1)^{mn}
\end{array}$$

It follows from the following commutative diagram that

$$B\pi_*(A_{s_1} \otimes \ldots \otimes A_{s_m} \otimes A_{t_1} \otimes \ldots \otimes A_{t_n})$$

$$= \sum_{s_1=s_{11}+\ldots+s_{1n}} \cdots \sum_{s_m=s_{m1}+\ldots+s_{mn}} \sum_{t_1=t_{11}+\ldots+t_{1m}} \cdots \sum_{t_n=t_{n1}+\ldots+t_{nm}} \alpha_1 \otimes \ldots \otimes \alpha_{mn}$$

where $\alpha_{km+h} = A_{s_h} A_{t_{k+1}}$ for $1 \leq h \leq m$, $0 \leq k < n$.

$$\begin{array}{ccc}
BO(1)^{m+n} & \xrightarrow{B\pi} & BO(2)^{mn} \\
\downarrow = & & \uparrow \omega^{mn} \\
BO(1)^m \times BO(1)^n \xrightarrow{(\Delta^n)^m \times (\Delta^m)^n} (BO(1)^n)^m \times (BO(1)^m)^n & \xrightarrow{T} & [BO(1)\times BO(1)]^{mn}
\end{array}$$

In this diagram T is the interchange map which makes the diagram commute. It is well known that $B\det_*(A_s A_t) = (s,t)A_{s+t}$. Thus the theorem follows from computing the diagram (*) in homology.

The following definition defines maps which we will see generalize Baker's map [3]:

$$G\colon BO(2) \xrightarrow{\Delta} BO(2) \times BO(2) \xrightarrow{B\det \times 1} BO(1) \times BO(2) \xrightarrow{\omega} BO(3).$$

DEFINITION 3.2. For $k \geq 2$ define $\tau_k\colon BO(2)^k \to BO(2^k)$ as the iterate of the tensor product τ. Let $\tau_1 = 1_{BO(2)}$.

For example $\tau_2 = \tau$, $\tau_3 = \tau(\tau \otimes 1) = \tau(1 \otimes \tau)$ and $\tau_4 = \tau(\tau \otimes \tau)$. The following theorem gives τ_{k*} on those elements which we will use in Section 4. To prove the theorem, however, it is convenient to first compute τ_{k*} in general and then specialize. Let $P(X)$ denote the power set of X.

THEOREM 3.3. For $k \geq 2$, $\tau_{k*}(A_{n_1} \otimes \dots \otimes A_{n_k}) \in H_*BO(2^k - 1)$ and equals

$$\sum_{n_1=n_{F(1,1)}+\dots+n_{F(1,2^{k-1})}} \cdots \sum_{n_k=n_{F(k,1)}+\dots+n_{F(k,2^{k-1})}}$$

$$\prod_{\phi\neq S\in P\{1,\dots,k\}} \binom{n'_{\pi_1(S)},\dots,n'_{\pi_k(S)}}{} A_{n'_{\pi_1(S)}+\dots+n'_{\pi_k(S)}}.$$

Here $P\{1,\dots,\hat{i},\dots,k\} = \{F(i,1),\dots,F(i,2^{k-1})\}$, $\pi_i\colon P\{1,\dots,k\} \to P\{1,\dots,\hat{i},\dots,k\}$ by $\pi_i(S) = S - \{i\}$ and $n'_{\pi_i(S)} = \begin{cases} n_{\pi_i(S)} & \text{if } i \in S \\ 0 & \text{if } i \notin S \end{cases}$.

<u>Proof</u>. It follows from Theorem 3.1 that $\tau_{k*}\left(A_{m_1}A_{m_2} \otimes \dots \otimes A_{m_{2k-1}}A_{m_{2k}}\right)$

$$= \sum_{m_1=m_{11}+\dots+m_{1p}} \cdots \sum_{m_{2k}=m_{2k,1}+\dots+m_{2k,p}}$$

$$\prod_{M=(m_{\varepsilon_1},\dots,m_{\varepsilon_k})\in\{m_1,m_2\}\times\dots\times\{m_{2k-1},m_{2k}\}} \binom{m_{\varepsilon_1,\phi_1(M)},\dots,m_{\varepsilon_k,\phi_k(M)}}{}$$

$$A_{m_{\varepsilon_1,\phi_1(M)}+\dots+m_{\varepsilon_k,\phi_k(M)}}.$$

Here $p = 2^{k-1}$ and ϕ_i is the canonical map

$$\{m_1,m_2\}\times\dots\times\{m_{2k-1},m_{2k}\} \to \{m_1,m_2\}\times\dots\times\widehat{\{m_{2i-1},m_{2i}\}}\times\dots\times\{m_{2k-1},m_{2k}\}$$

followed by a set bijection with $\{1,\dots,2^{k-1}\}$. Now let $m_{2i-1} = n_i$ and $m_{2i} = 0$ for $1 \leq i \leq k$ to obtain the asserted formula for $\tau_{k*}\left(A_{n_1} \otimes \dots \otimes A_{n_k}\right)$.

Observe that the one-to-one correspondence between the $M = \left(m_{\varepsilon_1}, \ldots, m_{\varepsilon_k}\right)$ and $S \in P\{1,\ldots,k\}$ is given by $S = \{(\varepsilon_i + 1)/2 \mid \varepsilon_i \text{ is odd}\}$. Observe that although $\tau_{k*}\left(A_{n_1} \otimes \cdots \otimes A_{n_k}\right)$ appears to be a sum of monomials in the A_t with 2^k factors, the factor indexed by $S = \phi \in P\{1,\ldots,k\}$ is one. Thus $\tau_{k*}\left(A_{n_1} \otimes \cdots \otimes A_{n_k}\right) \in H_*BO(2^k - 1)$.

COROLLARY 3.4. The projection of $\tau_{k*}\left(A_{n_1} \otimes \cdots \otimes A_{n_k}\right) \in H_*BO$ into the indecomposables $QH_*BO = \{A_1,\ldots,A_m,\ldots\}$ is

$$\begin{cases} (n_1,\ldots,n_k)A_{n_1+\cdots+n_k} & \text{if all the } n_i \text{ are nonzero} \\ 0 & \text{if one or more } n_i \text{ are zero} \end{cases}.$$

COROLLARY 3.5. The algebraic relationship between the maps τ_k and Baker's map G is

$$G_*(A_mA_n) = \tau_{2*}(A_m \otimes A_n).$$

COROLLARY 3.6. If $\sigma \in \Sigma_k$ then

$$\tau_{k*}\left(A_{n_{\sigma(1)}} \otimes \cdots \otimes A_{n_{\sigma(k)}}\right) = \tau_{k*}\left(A_{n_1} \otimes \cdots \otimes A_{n_k}\right).$$

COROLLARY 3.7. For $k > h \geq 1$, $\tau_{k*}\left(A_{n_1} \otimes \cdots \otimes A_{n_h} \otimes 1 \otimes \cdots \otimes 1\right)$

$$= \begin{cases} \tau_{h*}\left(A_{m_1} \otimes \cdots \otimes A_{m_h}\right)^{2^{k-h}} & \text{if } n_i = 2^{k-h}m_i \text{ for } 1 \leq i \leq h \\ 0 & \text{if some } n_i \text{ is not divisible by } 2^{k-h} \end{cases}.$$

Proof. Clearly it suffices to prove the case $k = h+1$. In the notation of Theorem 3.3 every summand of $\tau_{h+1*}\left(A_{n_1} \otimes \cdots \otimes A_{n_h} \otimes 1\right)$ appears twice by interchanging the factor for $S \in P\{1,\ldots,h\}$ with the factor for $S \cup \{h+1\}$. The only exception is when these two factors are equal for every $S \in P\{1,\ldots,h\}$. This can only happen if every n_i is even and in this case $\tau_{h+1*}\left(A_{n_1} \otimes \cdots \otimes A_{n_h} \otimes 1\right) = \tau_{h*}\left(A_{n_1/2} \otimes \cdots \otimes A_{n_h/2}\right)^2$.

4. POLYNOMIAL GENERATORS FOR $B_*(n)$. We begin by proving that τ_{n+1*} has image in $B_*(n)$. Our argument in fact shows that τ_2 lifts to BSO and τ_3 lifts to BSpin. In Definition 4.3 we use τ_{n+1*} to define our first set of polynomial generators for $B_*(n)$. We then use the methods of [10] and Theorem 2.8 to define a second set of polynomial generators for $B_*(n)$. We prove in Theorem 4.5 that our first set of polynomial generators which lie in $H_*BO(2^{n+1}-1)$ are optimal in that $B_*(n)$ does not have a set of polynomial generators which lie in $H_*BO(k)$, $k < 2^{n+1}-1$.

THEOREM 4.1. Image $\tau_{n+1*} \subset B_*(n)$.

Proof. Dually it suffices to prove that $\tau^*_{n+1}(I_n) = 0$. By Theorem 2.1(d) it suffices to show that $\tau^*_{n+1}(W_k) = 0$ for $1 \le k \le 2^{n-1}$. We use induction on $n \ge 1$. When $n = 1$, $\tau^*_2(W_1) = 0$ follows from the following commutative diagram.

$$\begin{array}{ccc} BO(2)\times BO(2) & \xrightarrow{\tau_2} & BO(4) \\ \uparrow{\scriptstyle \omega\times\omega} & & \uparrow{\scriptstyle \omega} \\ BO(1)^2\times BO(1)^2 & \xrightarrow{Bt} & BO(1)^4 \end{array}$$

Here $t\colon O(1)^4 \to O(1)^4$ is given by $t(a_1,a_2,a_3,a_4) = (a_1a_3,a_2a_3,a_1a_4,a_2a_4)$. Write $H^*BO(1)^4 = Z_2[X_1,X_2,X_3,X_4]$. Then ω^* is a monomorphism and $(\omega\times\omega)^*\tau^*_2(W_1) = (Bt)^*\omega^*(W_1) = (Bt)^*(X_1+X_2+X_3+X_4) = (X_1+X_3)+(X_2+X_3)+(X_1+X_4)+(X_2+X_4) = 0$. Assume inductively that $\tau^*_n(W_k) = 0$ for $1 \le k \le 2^{n-2}$. By the Wu formula, $\tau^*_n(W_k) = 0$ for $1 \le k \le 2^{n-1}-1$. Consider the following commutative diagram.

$$\begin{array}{ccccc} BO(2)^{n+1} & \xrightarrow{\tau_{n+1}} & BO(2^{n+1}) & \xleftarrow{\quad\omega\quad} & BO(1)^{2^{n+1}} \\ & \searrow{\scriptstyle \tau_n\times 1} & \uparrow{\scriptstyle \tau} & & \uparrow{\scriptstyle Bf} \\ & & BO(2^n)\times BO(2) & \xleftarrow{\omega\times\omega} & BO(1)^{2^n}\times BO(1)^2 \end{array}$$

Here $f\colon O(1)^{2^n}\times O(1)^2 \to O(1)^{2^{n+1}}$ is given by $f(a_1,\ldots,a_{2^n},b_1,b_2) = (a_1b_1,\ldots,a_{2^n}b_1,a_1b_2,\ldots,a_{2^n}b_2)$.

Write $H^*BO(1)^{2^{n+1}} = Z_2[U_1,\ldots,U_{2^{n+1}}]$, $H^*BO(1)^{2^n} = Z_2[X_1,\ldots,X_{2^n}]$ and $H^*BO(1)^2 = Z_2[Y_1,Y_2]$. Now ω^* is a monomorphism and letting σ_k denote the kth elementary symmetric polynomial we have for $1 \le k \le 2^{n-1}$,

$$\begin{aligned}(\omega^* \otimes \omega^*)\tau^*(W_k) &= Bf^*\omega^*(W_k) = Bf^*(\sigma_k(U_1,\ldots,U_{2^{n+1}})) \\ &= \sigma_k(X_1+Y_1,\ldots,X_{2^n}+Y_1,X_1+Y_2,\ldots,X_{2^n}+Y_2) \\ &= \sum_i \alpha_i(X_1,\ldots,X_{2^n})\beta_i(Y_1,Y_2)\end{aligned}$$

where the α_i and β_i are symmetric polynomials. Thus write

$\alpha_i(X_1,\ldots,X_{2^n}) = \omega^*(\overline{\alpha}_i)$ and $\beta_i(Y_1,Y_2) = \omega^*(\overline{\beta}_i)$. Then $\tau^*_{n+1}(W_k) = \sum_i \tau^*_n(\overline{\alpha}_i) \otimes \overline{\beta}_i$.

By the induction hypothesis $\tau^*_n(\overline{\alpha}_i) = 0$ unless $\overline{\alpha}_i = 1$ or $\deg \overline{\alpha}_i = 2^{n-1}$.

If $1 \le k \le 2^{n-1} - 1$ then $\tau^*_{n+1}(W_k) = 1 \otimes \omega^{*-1}(\sigma_k(Y_1,\ldots,Y_1,Y_2,\ldots,Y_2)) = 0$.

If $k = 2^{n-1}$ then $\tau^*_{n+1}(W_k)$

$= 1 \otimes \omega^{*-1}(\sigma_k(Y_1,\ldots,Y_1,Y_2,\ldots,Y_2)) + \tau^*_n\omega^{*-1}(\sigma_k(X_1,\ldots,X_{2^n},X_1,\ldots,X_{2^n})) \otimes 1$

$= 0 + \tau^*_n\omega^{*-1}(\sigma_{2^{n-2}}(X_1^2,\ldots,X_{2^n}^2)) \otimes 1$

$= \tau^*_n(W_{2^{n-2}})^2 \otimes 1 = 0.$

By Corollary 3.7, $\tau_{n+1}\left(A_{k_1} \otimes \ldots \otimes A_{k_{n+1}}\right)^2$

$= \tau_{n+2}\left(A_{k_1} \otimes \ldots \otimes A_{k_{n+1}} \otimes 1\right) \in B_*(n+1)$. It is also well known that if $X \in B_*(0) = H_*BO$ then $X^2 \in B_*(1) = H_*BSO$. The following useful theorem generalizes these two observations.

THEOREM 4.2. If $X \in B_*(n)$ then $X^2 \in B_*(n+1)$.

Proof. J_n is spanned by primitive elements. Thus the composite

$J_n \overset{\psi}{\to} J_n \otimes J_n \overset{\phi}{\to} J_n$ is zero. Now if $X \in B_*(n)$ and $Y \in J_n$ then $\langle X^2,Y\rangle = \langle X \otimes X, \Delta(Y)\rangle = 0$. Thus $X^2 \in B_*(n+1)$.

We now use the preceding two theorems to define elements of $B_*(n)$.

DEFINITION 4.3. (a) If $n \geq 0$ and $L(k) > n$ then define $G(k,n) \in B_k(n)$ by

$$G(k,n) = \tau_{n+1*}\left(A_{2^{k_1}} \otimes \dots \otimes A_{2^{k_n}} \otimes A_{2^{k_n+1}k'}\right) \text{ where } k = 2^{k_1} + \dots + 2^{k_n} + 2^{k_n+1}k',$$

$0 \leq k_1 < \dots < k_n$.

(b) If $n \geq m \geq 0$ and $L(k) > m$ then define $G(k,m,n) \in B_{k2^{n-m}}(n)$ by

$G(k,m,n) = G(k,m)^{2^{n-m}}$.

NOTES: (1) $G(k,0) = A_k$.

(2) The $G(k,1)$ are the polynomial generators of Baker [3].

(3) $G(k,n) \in H_*BO(2^{n+1} - 1)$ and $G(k,m,n) \in H_*BO(2^{n+1} - 2^{n-m})$.

(4) $G(k,m,n) = \tau_{n+1*}\left(A_{2^{k_1+n-m}} \otimes \dots \otimes A_{2^{k_m+n-m}} \otimes A_{2^{k_m+n-m+1}k'} \otimes 1 \otimes \dots \otimes 1\right)$

by Corollary 3.7 where $k = 2^{k_1} + \dots + 2^{k_m} + 2^{k_m+1}k'$, $0 \leq k_1 < \dots < k_m$.

(5) $G(k,n,n) = G(k,n)$.

(6) Theorem 3.3 gives the $G(k,n)$ and $G(k,m,n)$ as explicit polynomials in the A_s.

THEOREM 4.4. $B_*(n) = Z_2[G(k,n),\ G(h2^{L(h)-n-1},\ L(h)-1,\ n) \mid$

$L(k) > n,\ L(h) + M(h) > n \geq L(h)]$.

Proof. By Theorems 4.1 and 4.2 the $G(k,n)$ and $G(h2^{L(h)-n-1}, L(h)-1, n)$ listed above all are in $B_*(n)$. By Corollary 3.4 the $G(k,n)$ are indecomposable in H_*BO and hence are indecomposable in $B_*(n)$. We show the $G(h2^{L(h)-n-1}, L(h)-1, n)$ are indecomposable in $B_*(n)$ by induction on degree. Inductively, for $i < h$ there is no indecomposable element of $B_i(n)$ of the form $A^s_{h2^{L(h)-n-1}}$+monomials of product filtration degree at least s. Thus $G(h,2^{L(h)-n-1}, L(h)-1, n)$ is indecomposable. By Corollary 2.4 our list of indecomposable $G(k,n)$ and $G(h2^{L(h)-n-1}, L(h)-1, n)$ is a complete set of polynomial generators.

NOTE: We have shown that if $k < n$, $L(h) \leq k+1 \leq L(h)+M(h)$ and $X \in B_h(k)$ is a polynomial generator for $B_*(k)$ then $X^{2^{n-k}} \in B_{h2^{n-k}}(n)$ is a polynomial generator for $B_*(n)$.

We now describe the second set of polynomial generators of $B_*(n)$.

THEOREM 4.5. Let $L(k) + M(k) > n$. If $L(k) > n$ let $X(k,n) = \pi_n\left(W^{2^{M(k)}}_{k2^{-M(k)}}\right)^*$ in the dual basis of the basis of $B^*(n)$ of monomials in the $\pi_n(W_t)$ with $L(t) + M(t) > n$. If $n \geq L(k)$ let $X(k,n) = X\left(k2^{L(k)-n-1}, L(k)-1\right)^{2^{n-L(k)+1}}$. Then $B_*(n) = Z_2[X(k,n) \mid L(k) + M(k) > n]$.

Proof. We prove this theorem by induction on $n \geq 0$. The case $n = 0$ asserts that $\left(W^{2^{M(k)}}_{k2^{-M(k)}}\right)^*$ is indecomposable for $k \geq 1$ which is true by Girard's formula for $P_k \in PH^kBO$. Assume that for all $0 \leq m < n$ the theorem is true. By Theorem 4.2, $X(k,n) \in B_*(n)$ for $L(k) + M(k) > n$. If $L(k) > n$ then $X(k,n)$ is indecomposable in H_*BO by Theorem 2.8 and Girard's formula and hence is indecomposable in $B_*(n)$. If $L(k) + M(k) > n$ and $n \geq L(k)$ then

$$\left\langle X(k,n),\ \pi_n\left(P^{2^{n-L(k)+1}}_{k2^{-M(k)}}\right)^{*2^{L(k)+M(k)-n-1}} \right\rangle$$

$$= \left\langle \Delta^{2^{L(k)+M(k)-n-1}}\left(\pi_{L(k)-1}\left(W^{2^{L(k)+M(k)-n-1}}_{k2^{-M(k)}}\right)^{*2^{n-L(k)+1}}\right),\right.$$

$$\left. \pi_{L(k)-1}\left(P^{2^{n-L(k)+1}}_{k2^{-M(k)}}\right)^* \otimes \cdots \otimes \pi_{L(k)-1}\left(P^{2^{n-L(k)+1}}_{k2^{-M(k)}}\right)^* \right\rangle$$

$$= \left\langle \pi_{L(k)-1}\left(W_{k2^{-M(k)}}\right)^{*2^{n-L(k)+1}},\ \pi_{L(k)-1}\left(P^{2^{n-L(k)+1}}_{k2^{-M(k)}}\right)^* \right\rangle$$

$$= \left\langle \pi_{L(k)-1}\left(W_{k2^{-M(k)}}\right)^* \otimes \cdots \otimes \pi_{L(k)-1}\left(W_{k2^{-M(k)}}\right)^*,\right.$$

$$\left. \Delta^{2^{n-L(k)+1}}\ \pi_{L(k)-1}\left(P^{2^{n-L(k)+1}}_{k2^{-M(k)}}\right)^* \right\rangle$$

$$= \left\langle \pi_{L(k)-1}(W_{k2^{-M(k)}})^*, \pi_{L(k)-1}(P^*_{k2^{-M(k)}}) \right\rangle = 1$$

since $\pi_{L(k)-1}(P^*_{k2^{-M(k)}})$ is indecomposable in $B^*(L(k)-1)$.

Thus, by Theorem 2.8, $X(k,n)$ is indecomposable in $B_*(n)$. By Corollary 2.4 $\{X(k,n) \mid L(k) + M(k) > n\}$ is a complete set of polynomial generators for $B_*(n)$.

COROLLARY 4.6. Let $L(k) + M(k) > n$ and let $X \in B_k(n)$ be any polynomial generator. Then X is indecomposable in H_*BO if and only if $L(k) > n$.

Observe that the polynomial generators of $B_*(n)$ given by Theorem 4.4 lie in $H_*BO(2^{n+1} - 1)$ while those of Theorem 4.5 do not lie in any $H_*BO(s)$. The following theorem shows that our first set of generators is optimal in this respect.

THEOREM 4.7. Let $B_*(n) = Z_2[Y_k \mid \deg Y_k = k \text{ and } L(k) + M(k) > n]$. If $Y_k \in H_*BO(s)$ for all k then $s \geq 2^{n+1} - 1$.

<u>Proof</u>. $B_*(n)$ is $(2^n - 1)$-connected. Thus $Y_{2^{n+1}-1}$ must be primitive and $Y_{2^{n+1}-1} = P_{2^{n+1}-1} \in H_*BO(s)$. By Girard's formula $P_{2^{n+1}-1}$ has $A_1^{2^{n+1}-1}$ as a summand. Thus $s \geq 2^{n+1} - 1$.

5. SOME REMARKS ON H_*BSO. We present three results about H_*BSO. In Theorem 5.1 we give properties of the $G(2s+1,1)$ and identify them with B. Gray's odd degree polynomial generators [5]. In Theorem 5.2 we prove that H_*BSO does not have a set of polynomial generators which form an A_*-subcomodule. The remainder of this section studies the primitive elements $PH_*BSO = \{P_2,\dots,P_n,\dots\}$ which in H_*BO are given by the Girard polynomials in the $A_1,\dots,A_n,\dots$. The Girard polynomials are derived from the Newton formula $nA_n = P_n + \sum_{k=1}^{n-1} A_{n-k}P_k$. (See [7, §4].) In Lemma 5.3 we give an analogous formula in H_*BSO and use it in Theorem 5.4 to derive a formula for P_{2n+1}, $n \geq 1$, in H_*BSO as a $Z_2[A_n^2 \mid n \geq 1]$-linear combination of the $\{G(2s+1,1) \mid s \geq 1\}$. Observe that $P_{2^e(2s+1)} = P_{2s+1}^{2^e}$ for $s \geq 1$ and $P_{2^e} = A_1^{2^e}$.

THEOREM 5.1. (a) $\Delta(G(2s+1,1)) = \sum_{k=1}^{s} [G(2k+1,1) \otimes A_{s-k}^2 + A_{s-k}^2 \otimes G(2k+1,1)]$.

(b) $Sq_*^{2k+1}(G(2s+1,1)) = 0.$

$Sq_*^{2k}(G(2s+1,1)) = (k,s-2k)G(2s-2k+1,1).$

(c) $G(2s+1,1) = A_{2s+1} + \sum_{k=1}^{s} A_kA_{2s-k+1} + A_1A_s^2.$

Proof. (a) $\Delta(G(2s+1,1)) = \Delta\tau_{2*}(A_1 \otimes A_{2s})$

$$= (\tau_{2*} \otimes \tau_{2*}) \sum_{i=0}^{2s} \left[(A_1 \otimes A_i) \otimes (1 \otimes A_{2s-i}) + (1 \otimes A_i) \otimes (A_1 \otimes A_{2s-i})\right]$$

$$= (\tau_{2*} \otimes \tau_{2*}) \sum_{j=1}^{s} \left[(A_1 \otimes A_{2j}) \otimes (1 \otimes A_{2s-2j}) + (1 \otimes A_{2s-2j}) \otimes (A_1 \otimes A_{2j})\right]$$

because $\tau_{2*}(1 \otimes A_{2k+1}) = 0$

$$= \sum_{j=1}^{s} \left[G(2j+1,1) \otimes A_{s-j}^2 + A_{s-j}^2 \otimes G(2j+1,1)\right]$$ by Corollary 3.7.

(b) $Sq_*^h(G(2s+1,1)) = Sq_*^h\tau_{2*}(A_1 \otimes A_{2s})$

$= \tau_{2*}(A_1 \otimes Sq_*^h(A_{2s})) = (h,2s-2h)\tau_{2*}(A_1 \otimes A_{2s-h})$

$$= \begin{cases} 0 & \text{if } h = 2r+1 \text{ because } \tau_{2*}(A_1 \otimes A_{2s-2r-1}) = 0 \\ (k,s-2k)G(2s-2k+1,1) & \text{if } h = 2k \end{cases} .$$

(c) This follows easily from the definition of $G(2s+1,1)$ and Theorem 3.3.

THEOREM 5.2. Let $H_*BSO = Z_2[Y_2,\dots,Y_n,\dots]$ with $\deg Y_n = n$. Then

(a) modulo decomposables in H_*BSO: $Sq_*^k(Y_n) \equiv (k,n-2k)Y_{n-k}$ if $n - k \neq 2^t$,

$$Sq_*^k(Y_{k+2^t}) \equiv \begin{cases} Y_{2^t} & \text{if } k = 2^{t-1} \text{ or } k = 2^t, \\ 0 & \text{otherwise;} \end{cases}$$

(b) $\{Y_2,\dots,Y_n,\dots\}$ is not a sub A_*-comodule of H_*BSO.

<u>Proof</u>. (a) If $n \neq 2^t$ then Y_n is indecomposable in H_*BO and hence $Sq_*^k(Y_n) \equiv (k,n-2k)Y_{n-k}$ modulo IH_*BO^2. Thus $Sq_*^k(Y_n) \equiv (k,n-2k)Y_{n-k}$ modulo IH_*BSO^2 if $n - k \neq 2^t$. If $n - k = 2^t$ then $Sq_*^k(Y_n)$ is decomposable if and only if $Sq^k\left(\pi_1(W_2)^{2^{t-1}}\right) = 0$ i.e., if and only if $k \notin \{2^{t-1},2^t\}$. If $n = 2^t$ then by the Cartan formula we only need to compute $Sq_*^k(Y_{2^t})$ modulo IH_*BSO^2 for a specific polynomial generator Y_{2^t}. So take $Y_{2^t} = A^2_{2^{t-1}}$. Then $Sq_*^k\left(A^2_{2^{t-1}}\right) \in IH_*BSO^2$ unless $k = 2^{t-1}$ and $Sq_*^{2^{t-1}}\left(A^2_{2^{t-1}}\right) = A^2_{2^{t-2}}$.

(b) Assume that $\{Y_2,\dots,Y_n,\dots\}$ is a sub A_*-comodule of H_*BSO. Then Sq_*^2 shows that $Y_8 = 0A_8 + A_4^2 + A_5A_3 + \dots$. Now Sq_*^8 shows that

$Y_{16} = 0A_{16} + A_8^2 + A_{10}A_6 + \dots$. Then $Sq_*^1(Y_{16}) = 0A_{15} + A_9A_6 + A_{10}A_5 + \dots$.

This is a contradiction because Y_{15} is indecomposable in H_*BO.

We now derive the Newton formula for H_*BSO.

LEMMA 5.3. In H_*BSO for $s \geq 1$,

$$G(2s+1,1) = P_{2s+1} + \sum_{k=1}^{s-1} A^2_{s-k}P_{2k+1}.$$

Proof. We use induction on $s \geq 1$. $G(3,1) = P_3$ is the unique nonzero element of H_3BSO. Assume that the formula is true for $k < s$. Then

$$\Delta\left(G(2s+1,1) + \sum_{k=1}^{s-1} A_{s-k}^2 P_{2k+1}\right)$$

$$= \sum_{i=1}^{s}\left[G(2i+1,1) \otimes A_{s-i}^2 + A_{s-i}^2 \otimes G(2i+1,1)\right]$$

$$+ \sum_{k=1}^{s-1}\sum_{j=0}^{s-k}\left[A_j^2 P_{2k+1} \otimes A_{s-k-j}^2 + A_j^2 \otimes P_{2k+1}A_{s-k-j}^2\right]$$

$$= \sum_{h=0}^{s}\left[A_h^2 \otimes \left(G(2(s-h)+1,1) + P_{2(s-h)+1} + \sum_{\lambda=1}^{s-h-1} A_{s-h-\lambda}^2 P_{2\lambda+1}\right)\right.$$

$$\left. + \left(G(2(s-h)+1,1) + P_{2(s-h)+1} + \sum_{\lambda=1}^{s-h-1} A_{s-h-\lambda}^2 P_{2\lambda+1}\right) \otimes A_h^2\right]$$

where $P_{2(s-h)+1} \otimes 1 + 1 \otimes P_{2(s-h)+1}$ is omitted if $h = 0$

$$= \left(G(2s+1,1) + \sum_{k=1}^{s-1} A_{s-k}^2 P_{2k+1}\right) \otimes 1 + 1 \otimes \left(G(2s+1,1) + \sum_{k=1}^{s-1} A_{s-k}^2 P_{2k+1}\right).$$

Now $G(2s+1,1) + \sum_{k=1}^{s-1} A_{s-k}^2 P_{2k+1}$ is indecomposable and thus equals the unique nonzero primitive element P_{2s+1} of $H_{2s+1}BSO$.

We now solve the Newton formula of Lemma 5.3 to obtain the Girard formula for H_*BSO.

THEOREM 5.4. In H_*BSO for $s \geq 1$,

$$P_{2s+1} = \sum_{k=0}^{s-1} \sum_{e_1+2e_2+\ldots+ke_k=k} (e_1,\ldots,e_k) A_1^{2e_1} \ldots A_k^{2e_k} G(2s-2k+1,1).$$

Proof. Consider the recursive relation of Lemma 5.3 as simultaneous linear equations in the unknowns P_{2s+1}. The coefficient matrix has (i,j) entry A_{j-i}^2 for $j \geq i$ and is thus upper triangular with ones on the diagonal. The nonhomogeneous constant terms are the $G(2s+1,1)$. By [10, Lemma 2.2] we can solve this system of equations by Cramer's rule to obtain

$$P_{2s+1} = G(2s+1,1) + \sum_{k=1}^{s-1} \sum_{j_1+\dots+j_r=k} A_{j_1}^2 \dots A_{j_r}^2 G(2s-2k+1,1)$$

where the sum $\sum_{j_1+\dots+j_r=k}$ is taken over all sequences $(j_1,\dots,j_r)$ such that $r \geq 1$, all the j_i are positive and $j_1+\dots+j_r=k$. Thus $A_1^{2e_1}\dots A_k^{2e_k}G(2s-2k+1,1)$ appears $(e_1,\dots,e_k)$ times in the above sum.

COROLLARY 5.5. In H_*BSO for $s \geq 1$,

$$P_{2s+1} = \sum_{k=0}^{s-1} \chi(A_k^2)G(2s-2k+1,1).$$

<u>Proof</u>. By [7, Theorem 4.1(v)], $\chi(A_k) = \sum_{e_1+2e_2+\dots+ke_k=k} (e_1,\dots,e_k)A_1^{e_1}\dots A_k^{e_k}$.

Thus this corollary follows from Theorem 5.4.

6. AN ALGEBRAIC FILTRATION OF $H_*(BU_{p,0};Z_p)$, p ODD. Let p be a fixed odd prime. By Adams [1], $BU_{(p)} = \prod_{i=0}^{p-2} BU_{p,i}$ where $BU_{p,0}$ is (2p-3)-connected. In this section we show how the constructions and results of Sections 2-4 generalize from $H_*(BO;Z_2)$ to $H_*(BU_{p,0};Z_p)$. In Theorem 6.2 we construct quotient Hopf algebras and quotient A_p-modules ${}_pB^*(n)$ of $H^*(BU_{p,0};Z_p)$ such that the duals ${}_pB_*(n)$ form a complete decreasing filtration of $H_*(BU_{p,0};Z_p)$ by bipolynomial sub Hopf algebras and sub A_{p*}-comodules. In Theorem 6.5 we study the iterated tensor products τ_n: $BU(p)^{n(p-1)+1} \to BU(p^{n(p-1)+1})$ which we use to construct polynomial generators for ${}_pB_*(n)$ in Theorem 6.8. In Theorem 6.9 we construct a second set of polynomial generators for ${}_pB_*(n)$ using the dual basis of a basis of monomials in the Chern classes.

We will use the following notation in this section. Recall [4] that $H^*(BU;Z) = Z[C_1,\ldots,C_n,\ldots]$ where $C_n \in H^{2n}(BU;Z)$ is the nth Chern class and $\Delta(C_n) = \sum_{i=0}^{n} C_i \otimes C_{n-i}$. Then $H_*(BU;Z) = Z[A_1',\ldots,A_n',\ldots]$ where $A_n' = (C_1^n)^* \in H_{2n}(BU;Z)$ and $\Delta(A_n') = \sum_{i=0}^{n} A_i' \otimes A_{n-i}'$. Let V_s be the image of $C_{s(p-1)}$ in $H^{2s(p-1)}(BU_{p,0};Z_p)$. Then $H^*(BU_{p,0};Z_p) = Z_p[V_1,\ldots,V_s,\ldots]$ with $\Delta(V_s) = \sum_{i=0}^{s} V_i \otimes V_{s-i}$. (Note that if n is not divisible by p-1 then C_n maps to zero in $H^*(BU_{p,0};Z_p)$ because $H^{2n}(BU_{p,0};Z_p) = 0$.) Dually, $H_*(BU_{p,0};Z_p) = Z_p[E_1,\ldots,E_s,\ldots]$ with $E_s \in H_{2s(p-1)}(BU_{p,0};Z_p)$ and $\Delta(E_s) = \sum_{i=0}^{s} E_i \otimes E_{s-i}$. Note that $E_s = (V_1^s)^* = \phi_*(A'_{s(p-1)})$ where ϕ: $BU \to BU_{p,0}$ is the canonical map. [This follows from the observations: (1) $\phi^*(V_s) = (A'^{s}_{p-1})^*$ because neither $\phi^*(V_s)$ nor $(A'^{s}_{p-1})^*$ has $C^{p^e}_{u(p-1)}$ as a summand where $s = up^e$, $p \nmid u$. (2) $(A'^{s}_{p-1})^*$ contains $C_1^{s(p-1)}$ as a summand if and only if s = 1.] We will also need the fact that $P^k(V_s) \equiv (k,s(p-1)-k-1)V_{s+k}$ modulo decomposables. For odd primes we will use the following L and M notation to describe the degrees of the polynomial generators of ${}_pB_*(n)$. Write $s(p-1) = s_1p^{e_1} + \ldots + s_tp^{e_t}$ with $0 \leq e_1 < \ldots < e_t$ and $1 \leq s_i \leq p-1$. Define

$L'(s) = s_1+\dots+s_t$ and $M(s) = e_1$. The following lemma shows that $L'(s)$ is always divisible by p-1. Thus define $L(s) = \frac{L'(s)}{p-1}$.

LEMMA 6.1. The sum of the entries of the p-adic expansion of $s(p-1)$ is divisible by p-1.

Proof. We use induction on $s \geq 1$. The case $s = 1$ is trivial. Assume the lemma is true for s. Write $s(p-1) = s_0p^{e_0}+(p-1)p^{e_0+1}+\dots+(p-1)p^{e_0+u}+s_1p^{e_1}$

$$+\dots+s_kp^{e_k}$$

where $0 \leq u$, $0 \leq e_0$, $e_0+u+1 = e_1 < \dots < e_k$, $1 \leq s_0 \leq p-1$, $0 \leq s_1 \leq p-2$ and $1 \leq s_i \leq p-1$ for $2 \leq i \leq k$. Then $S = s_0 + u(p-1) + s_1 + \dots + s_k$ is divisible by p-1. $(s+1)(p-1) = \begin{cases} (s_0-1)p^{e_0}+(s_1+1)p^{e_1}+s_2p^{e_2}+\dots+s_kp^{e_k} & \text{if } e_0 = 0 \\ (p-1)+s_0p^{e_0}+\dots+s_kp^{e_k} & \text{if } e_0 > 0 \end{cases}$.

Thus the sum of the entries of the p-adic expansion of $(s+1)(p-1)$ is either $S+p-1$ or $S-u(p-1)$ which is divisible by p-1.

We now prove the existence of the ${}_pB_*(n)$ by constructing their duals ${}_pB^*(n)$.

THEOREM 6.2. There exist quotient Hopf algebras and quotient A_p-modules ${}_pB^*(n) = H^*(BU_{p,0};Z_p)/I_n = {}_pB^*(n-1)/J_{n-1}$ with the four properties listed below. Let $\pi_n: H^*(BU_{p,0};Z_p) \to {}_pB^*(n)$ denote the canonical projection.

(a) ${}_pB^*(0) = H^*(BU_{p,0};Z_p)$.

(b) $P_pB^*(n) = \{P_{n,k} \mid L(k) + M(k) > n\}$ with degree $P_{n,k} = 2k(p-1)$.

(c) J_n is the ideal in ${}_pB^*(n)$ spanned by the set of indecomposable primitive elements $\{P_{n,k} \mid L(k) + M(k) = n+1\}$.

(d) J_n is the ideal in ${}_pB^*(n)$ spanned by the cyclic A_p-submodule of ${}_pB^*(n)$ generated by $\pi_n(V_{p^n})$.

Proof. The proof of Theorem 2.1 generalizes to prove this theorem from the following fact:

(*) $A_p\pi_n(V_{p^n}) \equiv \{\pi_n(V_k) \mid L(k) + M(k) = n + 1\}$ modulo decomposables.

To prove (*) it suffices to show:

(1) If $L(h) + M(h) = n + 1$ and $P^{p^t}(V_h) \equiv V_k$ modulo decomposables then $L(k) + M(k) \leq n + 1$.

(2) If $L(k) + M(k) = n+1$ and $k \neq p^n$ there there is some $t \geq 0$ such that $P^{p^t}(V_{k-p^t}) \equiv V_k$ modulo decomposables.

To prove (1) note that we are given that $(p^t, h(p-1)-p^t-1) \neq 0$. If $M(h) < t$ then $M(h+p^t) = M(h)$ and $L(h+p^t) \leq L(h)$. If $M(h) > t$ then $M(h+p^t) = t < M(h)$ and $L(h+p^t) = L(h)+1$. If $M(h) = t$ then $M(h+p^t) = M(h)$ and $L(h+p^t) \leq L(h)$. Thus in all three cases $L(h+p^t) + M(h+p^t) \leq L(h) + M(h) = n+1$.

To prove (2) write $k(p-1) = k_0p^0+\ldots+k_rp^r+(p-1)p^{r+1}+\ldots+(p-1)p^{r+s}$ where $0 \leq k_i \leq p-1$, $k_r \neq p-1$, $-1 \leq r$ and $0 \leq s$. If $s \geq 1$ and $k_0p^0+\ldots+k_rp^r \neq 0$ then take $t = r$. If $s \geq 2$ and $k_0p^0+\ldots+k_rp^r = 0$ then take $t = r+1$. If $s = 0$ and $k_r \neq 0$ then take $t = r-1$ since $1 \leq k_{r-1} \leq p-2$. In all three cases $(p^t, (k-p^t)(p-1)-p^t-1) \neq 0$ and hence $P^{p^t}(V_{k-p^t}) \equiv V_k$ modulo decomposables.

COROLLARY 6.3. There exist $2p^n(p-1)-1$ connected sub Hopf algebras and sub A_{p*}-comodules ${}_pB_*(n)$ of $H_*(BU_{p,0};Z_p)$ which are bipolynomial Hopf algebras with one polynomial generator in each degree k such that $L(k) + M(k) > n$. The ${}_pB_*(n)$ form a complete decreasing filtration of $H_*(BU_{p,0};Z_p)$.

THEOREM 6.4. For all $n \geq 1$ there is no $2p^n(p-1)-1$ connected space X and map $f: X \to BU_{p,0}$ such that in mod p homology f_* is a monomorphism with image ${}_pB_*(n)$.

Proof. From [4] we have that the image of the Hurewicz homomorphism in $H_{2p^n(p-1)}(BU;Z)$ is divisible by $[p^n(p-1)-1]!$. Hence the mod p Hurewicz homomorphism for $BU_{p,0}$ in degree $2p^n(p-1)$ is zero and thus the proof of Theorem 2.9 generalizes.

The following iterated tensor product maps will be used to construct polynomial generators for the ${}_pB_*(n)$.

THEOREM 6.5. For $n \geq 0$ the iterated tensor product map $\tau_n\colon BU(p)^{n(p-1)+1} \to BU(p^{n(p-1)+1})$ has the following properties:

(a) In mod p homology Image $\phi_*\tau_{n*} \subset {}_pB_*(n)$.

(b) $\tau_{n*}\left(A'_{k_1} \otimes \dots \otimes A'_{k_{n(p-1)+1}}\right) \in H_*(BU(p^{n(p-1)+1} - (p-1)^{n(p-1)+1});Z_p)$ and in the notation of Theorem 3.3

$$\phi_*\tau_{n*}\left(A'_{k_1} \otimes \dots \otimes A'_{k_{n(p-1)+1}}\right)$$

$$= \sum_{k_1=k_{F(1,1)}+\dots+k_{F(1,2^{n(p-1)})}} \cdots \sum_{k_{n(p-1)+1}=k_{F(n(p-1)+1,1)}+\dots+k_{F(n(p-1)+1,2^{n(p-1)})}} \prod_{\phi\neq S\in \mathcal{P}\{1,\dots,n(p-1)+1\}}$$

$$\left(k'_{\pi_1(S)},\dots,k'_{\pi_{n(p-1)+1}(S)}\right) E^{\,n(p-1)-\mathrm{card}\, S+1}_{(p-1)\,(k'_{\pi_1(S)}+\dots+k'_{\pi_{n(p-1)+1}(S)})/(p-1)}$$

where $E_k = 0$ if k is not an integer.

(c) The projection of $\phi_*\tau_{n*}\left(A'_{k_1} \otimes \dots \otimes A'_{k_{n(p-1)+1}}\right)$ in $QH_*(BU_{p,0};Z_p)$ is

$$\begin{cases} (k_1,\dots,k_{n(p-1)+1})E_{(k_1+\dots+k_{n(p-1)+1})/(p-1)} & \text{if all the } k_i \text{ are nonzero} \\ 0 & \text{otherwise} \end{cases}.$$

(d) If $\sigma \in \Sigma_{n(p-1)+1}$ then

$$\tau_{n*}\left(A'_{k_{\sigma(1)}} \otimes \dots \otimes A'_{k_{\sigma(n(p-1)+1)}}\right) = \tau_{n*}\left(A'_{k_1} \otimes \dots \otimes A'_{k_{n(p-1)+1}}\right).$$

(e) If $n > m \geq 0$ then $\tau_{n*}\left(A'_{k_1} \otimes \dots \otimes A'_{k_{m(p-1)+1}} \otimes 1 \otimes \dots \otimes 1\right)$

$$= \begin{cases} \left(\tau_{m*}\left(A'_{h_1} \otimes \dots \otimes A'_{h_{m(p-1)+1}}\right)\right)^{p^{n-m}} & \text{if } k_i = p^{n-m}h_i \text{ for all } i \\ 0 & \text{if some } k_i \text{ is not divisible by } p^{n-m} \end{cases}.$$

Proof. (a) Let $\tau_0 = 1_{BU(p)}$. We prove (a) by induction on $n \geq 1$. To prove the case $n = 1$ we use induction on $p \geq k \geq 1$ to see that the iterated tensor product τ'_k: $BU(p)^k \to BU(p^k)$ has the property that in mod p cohomology $\tau'^*_k(C_i) = 0$ for $1 \leq i \leq k-1$. The argument is analogous to the proof of Theorem 4.1 where we factor τ'_k as $\tau \circ (\tau'_{k-1} \times 1)$. The induction step in the proof of (a) is also analogous to the proof of Theorem 4.1. It uses the factorization $\tau_n = \tau \circ (\tau_{n-1} \times \tau'_{p-1})$ and the properties $\tau^*_{n-1}(C_i) = 0$, $1 \leq i \leq p^{n-2}(p-1)$, and $\tau'_{p-1}(C_j) = 0$, $1 \leq j \leq p-2$, to show $\tau^*_n(C_k) = 0$, $1 \leq k \leq p^{n-1}(p-1)$.

(b)-(d) The arguments of Section 3 apply in this context. Note that $\tau_{n*}\left(A'_{k_1} \otimes \dots \otimes A'_{k_{n(p-1)+1}}\right) \in H_*(BU(p^{n(p-1)+1} - (p-1)^{n(p-1)+1});Z_p)$ because

$$\sum_{\phi \neq S \in \mathcal{P}\{1,\dots,n(p-1)+1\}} (p-1)^{n(p-1)+1-\mathrm{card}\, S}$$

$$= \sum_{i=1}^{n(p-1)+1} \sum_{\substack{S \in \mathcal{P}\{1,\dots,n(p-1)+1\} \\ \mathrm{card}\, S = i}} (p-1)^{n(p-1)+1-i}$$

$$= \sum_{i=1}^{n(p-1)+1} (p-1)^{n(p-1)+1-i}(i, n(p-1)+1-i)$$

$$= [1 + (p-1)]^{n(p-1)+1} - (p-1)^{n(p-1)+1}$$

$$= p^{n(p-1)+1} - (p-1)^{n(p-1)+1}.$$

THEOREM 6.6. If $X \in {}_pB_*(n)$ then $X^p \in {}_pB_*(n+1)$.

Proof. ${}_pB^*(n+1) = {}_pB^*(n)/J_n$ and J_n is spanned by primitive elements. Thus the proof of Theorem 4.2 applies.

We now use Theorems 6.5 and 6.6 to define elements of ${}_pB_*(n)$.

DEFINITION 6.7. (a) If $n \geq 0$ and $L(k) > n$ then define $G(k,n) \in {}_pB_k(n)$ by $G(k,n) = \phi_*\tau_{n*}\left(A'_{p^{r_1}} \otimes \ldots \otimes A'_{p^{r_{n(p-1)}}} \otimes A'_{k'}\right)$ where $0 \leq r_1 \leq \ldots \leq r_{n(p-1)}$, $p^{r_{n(p-1)}}$ divides k', $k(p-1) = p^{r_1}+\ldots+p^{r_{n(p-1)}}+k'$, $k' = \alpha p^{r_{n(p-1)}}+k''p^{r_{n(p-1)}+1}$,

and for each j, $\operatorname{card}\{i \mid r_i = r_j\} \leq \begin{cases} p-1 & \text{if } 1 \leq j < n(p-1) \\ p-\alpha-1 & \text{if } j = n(p-1) \end{cases}$.

(b) If $n \geq m \geq 0$ and $L(k) > m$ then define $G(k,m,n) \in {}_pB_{kp^{n-m}}(n)$ by $G(k,m,n) = G(k,m)^{p^{n-m}}$.

NOTES: (1) $G(k,0) = E_k$.

(2) $G(k,n) \in H_*(BU(p^{n(p-1)+1} - (p-1)^{n(p-1)+1});Z_p)$ and

$G(k,m,n) \in H_*(BU(p^{mp+n-2m+1} - (p-1)^{m(p-1)+1}p^{n-m});Z_p)$.

(3) $G(k,m,n) = \phi_*\tau_{n*}\left(A'_{p^{r_1+n-m}} \otimes \cdots \otimes A'_{p^{r_{m(p-1)}+n-m}} \otimes A'_{k'p^{n-m}} \otimes 1 \otimes \cdots \otimes 1\right)$

in the above notation.

(4) $G(k,n,n) = G(k,n)$.

(5) Theorem 6.5(b) gives the $G(k,n)$ and $G(k,m,n)$ as explicit polynomials in the E_s.

THEOREM 6.8. ${}_pB_*(n) = Z_p[G(k,n),\ G(hp^{L(h)-n-1},\ L(h)-1,\ n) \mid$
$L(k) > n \text{ and } L(h) + M(h) > n \geq L(h)]$.

Proof. The proof of this theorem is analogous to the proof of Theorem 4.4.

We conclude by constructing a second set of polynomial generators for the ${}_pB_*(n)$.

THEOREM 6.9. Let $L(k) + M(k) > n$. If $L(k) > n$, let $X(k,n) = \pi_n\left(V^{p^{M(k)}}_{kp^{-M(k)}}\right)^*$ in the dual basis of the basis of ${}_pB^*(n)$ of monomials in the $\pi_n(V_t)$ with $L(t) + M(t) > n$. If $n \geq L(k)$, let $X(k,n) = X(kp^{L(k)-n-1}, L(k)-1)^{p^{n-L(k)+1}}$. Then ${}_pB_*(n) = Z_p[X(k,n) \mid L(k) + M(k) > n]$.

Proof. The proof is analogous to the proof of Theorem 4.5. It requires the following explicit description of $PB^*(n)$:

$$PB^*(n) = \{\pi_n(E^*_{kp^{-M(k)}})^{p^{M(k)}} \mid L(k) > n\}$$

$$\cup \left\{\pi_n\left(P^{p^{n-L(k)+1}}_{kp^{-M(k)}}\right)^{*p^{L(k)+M(k)-n-1}} \;\middle|\; L(k) + M(k) > n \geq L(k)\right\}.$$

The proof of this fact is analogous to the proof of Theorem 2.8.

BIBLIOGRAPHY

1. J. F. Adams, Lectures on generalised cohomology, Category Theory, Homology Theory and Their Applications III (Battelle Institute Conference, Seattle, Wash., 1968) vol. 3, Lecture Notes in Math., no. 99, Springer-Verlag, Berlin, 1969, pp. 1-138.

2. A. Bahri, Polynomial generators for $H_*(BSO;Z_2)$, $H_*(BSpin;Z_2)$ and $H_*(BO<8>;Z_2)$, Current Trends in Algebraic Topology, Can. Math. Soc. Conference Proceedings (to appear).

2a. A. Bahri and M. Mahowald, Stiefel-Whitney classes in $H^*BO<\phi(r)>$, Proc. Amer. Math. Soc. 83(1981), 653-655.

3. A. Baker, More homology generators for BSU and BSO, Current Trends in Algebraic Topology, Can. Math. Soc. Conference Proceedings (to appear).

4. Séminaire Henri Cartan 12ième année: 1959/60. Periodicité des groupes d'homotopie stables des groupes classiques d'après Bott. Deux fasc., 2ième éd., Ecole Normale Supérieure, Sécretariat mathématiques, Paris, 1961.

5. B. Gray, Products in the Atiyah-Hirzebruch spectral sequence and the calculation of MSO_*, Trans. Amer. Math. Soc. 260(1980), 475-483.

6. P. Hoffman, Hopf algebras from branching rules (preprint).

7. S. Kochman, Primitive generators for algebras, Can. J. of Math. 34 (1982), 454-465.

8. __________, Polynomial generators for $H_*(BSU)$ and $H_*(BSO;Z_2)$, Proc. Amer. Math. Soc. 84(1982), 149-154.

9. __________, Integral polynomial generators for the homology of BSU, Proc. Amer. Math. Soc. (to appear).

10. __________, Change of basis in H_*BU, Amer. J. of Math. (to appear).

11. S. Papastavridis, The image of $H_*(BSO;Z_2)$ in $H_*(BO;Z_2)$, (preprint).

12. R. Stong, Determination of $H^*BO(k,\dots,\infty)$ and $H^*BU(k,\dots,\infty)$, Trans. Amer. Math. Soc. 107(1963), 526-544.

Department of Mathematics
The University of Western Ontario
London, Ontario, Canada N6A 5B7

Contemporary Mathematics
Volume 19, 1983

The Cohomology of the Dyer Lashof Algebra

David Kraines and Thomas Lada

ABSTRACT. The Dyer Lashof algebra R is a noncommutative algebra over Z/p similar in form to the mod p Steenrod algebra A_p. In this paper we show that the cohomology of the algebra R_p is isomorphic as A_p modules to $A_p(L)$, the Steenrod algebra for restricted Lie algebras. Many of the results of this paper are implicit in work of S. Priddy and of H. Miller for $p = 2$.

1. Let X be an infinite loop space and let p be a prime. Then X has a product $m: X \times X \to X$ which enjoys strong homotopy commutativity properties. Kudo and Araki [KA] and Browder [B2] exploited these properties to construct mod 2 homology operations similar in type to the Steenrod reduced squares. Dyer and Lashof [DL] generalized this construction to odd primes and, in addition, showed that the operations $Q^k: H_n(X; Z/p) \to H_{n+2k(p-1)}(X; Z/p)$ satisfy excess and Adem relations quite analogous to those satisfied by the Steenrod operations. See [B1], [CLM], [M2], [M3] and [M4] for more details.

DEFINITION 1.1. The algebraic Dyer Lashof algebra R_p is the graded Z/p algebra generated by symbols Q^k for $k \geq 0$ and βQ^k for $k > 0$ of degree $2k(p-1)$ and $2k(p-1)-1$ respectively. The relations are generated by the DL Adem relations

$$\beta^{\varepsilon}Q^rQ^s = \sum_j (-1)^{r+j}\binom{(p-1)(j-s)-1}{pj-r}\beta^{\varepsilon}Q^{r+s-j}Q^j \tag{1.2}$$

for $r > ps$

and

$$\beta^{\varepsilon}Q^r\beta Q^s = \sum_j (-1)^{r+j}\binom{(p-1)(j-s)}{pj-r}\beta^{\varepsilon}\beta Q^{r+s-j}Q^j - \sum_j (-1)^{r+j}\binom{(p-1)(j-s)-1}{pj-r-1}\beta^{\varepsilon}Q^{r+s-j}\beta Q^j$$

for $r \geq ps$ where $\varepsilon = 0$ or 1 and $\beta\beta = 0$.

Let $I = (\varepsilon_1, a_1, \ldots, \varepsilon_k, a_k)$ be a sequence of integers such that $\varepsilon_i = 0$ or 1 and $a_i \geq \varepsilon_i$. The DL excess of I, and of the monomial $Q^I = \beta^{\varepsilon_1}Q^{a_1}\ldots\beta^{\varepsilon_k}Q^{a_k}$, is defined to be

0271-4132/83 $1.00 + $.25 per page

$$e(I) = 2a_1 - \varepsilon_1 - \sum_{j=2}^{k} [2a_j(p-1) - \varepsilon_j].$$

DEFINITION 1.3. The (geometric) Dyer-Lashof algebra, $R_p(0)$, is the quotient of the algebraic Dyer-Lashof algebra R_p by the 2-sided ideal $E(0)$ generated by monomials Q^I of negative excess.

REMARK 1.4. The algebraic Dyer-Lashof algebra is denoted by $R_p(-\infty)$ in [M5]. See also [CLM]. The opposite algebra R_p^{op} is isomorphic to the Λ algebra [M5],[P2], as can be seen from the representation of Λ in [C]. If $X = \Omega^\infty S^\infty$, then $R_p(0) \approx Q\, H_*(X; Z/p)$, the quotient module of indecomposable homology classes [DL],[M4].

A $\mathbb{Z}/p$ basis for R_p consists of the DL admissible monomials Q^I where $I = (\varepsilon_1, a_1, \ldots, \varepsilon_k, a_k)$ satisfies $pa_j - \varepsilon_j \geq a_{j-1}$ for $2 \leq j \leq k$. A $\mathbb{Z}/p$ basis for $R_p(0)$ consists of those DL admissible monomials with non negative excess. By convention $1 = Q^\phi$ is admissible.

REMARK 1.5. If $p = 2$ the Dyer Lashof algebra is usually defined as consisting of operations Q^k for $k \geq 0$ of degree k subject to the Adem relations above that do not contain β. It is easy to check that the mod 2 operation $\beta^\varepsilon Q^k$ in Definition 1.1 of degree $2k-\varepsilon$ satisfies the usual mod 2 relations for $Q^{2k-\varepsilon}$. Thus we need not make special cases in our theorems for p odd and even. In some examples below we will, however, use the standard mod 2 notation for Dyer Lashof operations.

Let A be a graded augmented Z/p algebra with augmentation ideal $\bar{A}$. The cohomology $H^{k,*}(A) = \mathrm{Ext}_A^{k,*}(Z/p, Z/p)$ is defined to be the cohomology of a projective resolution of the A module Z/p. For example, let $\bar{B}_{k,*}A = \bar{A} \otimes \ldots \otimes \bar{A}$ (k times) be the reduced bar construction on A. There are maps

$$d_j : \bar{B}_{k,*}A \to \bar{B}_{k-1,*}A \text{ for } j = 1, \ldots, k-1$$

given by $d_j(a_1 \otimes \ldots \otimes a_k) = a_1 \otimes \ldots \otimes a_j a_{j+1} \otimes \ldots \otimes a_k$.
It is well known that, with $d = \Sigma(-1)^j d_j : \bar{B}_{k,*}A \to \bar{B}_{k-1,*}A$, $\bar{B}_{*,*}A$ is a complex whose cohomology, i.e. the cohomology of $\mathrm{Hom}(\bar{B}_{*,*}A, Z/p)$, is $\mathrm{Ext}_A^{**}(Z/p, Z/p)$ [M1].

The large size of $\bar{B}_{*,*}(A)$ makes direct computation of $\mathrm{Ext}_A^{*,*}(Z/p, Z/p)$ very difficult. If A is a polynomial or exterior algebra, then Koszul constructed a small chain equivalent subcomplex of $\bar{B}_{*,*}(A)$ making computations trivial. Priddy has generalized this construction to the class of Poincare Birkhoff Witt (PBW) algebras which we will briefly discuss. Assume that A has algebra generators $\{a_i : i \in I\}$ and a generating set of relations of the form $\Sigma c_{i,j} a_i a_j = 0$. Note that the length of a monomial in the a_i's is unambiguously defined. We call such algebras homogeneous. A PBW basis for the homogeneous algebra A is an ordered Z/p basis of monomials

$\mathcal{B} = \{1, a_1, a_2, \ldots, a_I, \}$, called admissible monomials, which satisfy the conditions:

i) If $a_{i_1} \cdots a_{i_n} \in \mathcal{B}$ and $1 < k < n$ then $a_{i_1} \cdots a_{i_k}$ and $a_{i_k} \cdots a_{i_n}$ are in $\mathcal{B}$.

ii) If a_I and a_J are in $\mathcal{B}$, then either $a_I a_J \in \mathcal{B}$ or else $a_I a_J = \Sigma c_\alpha a_{k_\alpha}$ with $a_{k_\alpha} \in \mathcal{B}$ and $K_\alpha > (I,J)$ in the lexicographic order (Compare [P1]). In particular for each r and $s \in I$ for which $a_r a_s$ is not admissible, there is a relation

$$a_r a_s = \Sigma c(i,j,r,s) a_i a_j \tag{1.6}$$

where the sum is over admissible length 2 monomials $a_i a_j \in \mathcal{B}$. Indeed the generators $\{a_i\}$ and the relations (1.6) characterize A. If A has a PBW basis we say that A is a PBW algebra.

The Steenrod algebra A_p is not homogeneous since, for example $Sq^2 Sq^3 = Sq^5 + Sq^4 Sq^1$. The algebra $A_p(L)$ of Steenrod operations for restricted Lie algebras is the same as A_p except that the relation $P^0 = 1$ is replaced by $P^0 = 0$. See [M3] and [P2] for more details. Each of these algebras A_p and $A_p(L)$ have a Z/p basis of (Steenrod) admissible monomials. The relation $P^0 = 0$ makes $A_p(L)$ into a homogeneous algebra and the $\mathbb{Z}/p$ basis of admissibles is easily seen to be a PBW basis. See [P1] and [P2]. In the next section we will distinguish between the elements $P^I \in A_p$ and $P^I_L \in A_p(L)$.

The DL Adem relations for R_p in (1.1) are homogeneous. It is not hard to check that the set of DL admissible monomials forms a PBW basis for R_p with the opposite (right to left) lexicographic ordering.

DEFINITION 1.7. Let A be a PBW algebra with generators $\{a_i\}$. Then the Koszul complex $\bar{K}_{**}(A)$ is the submodule of $\bar{B}_{**}(A)$ generated by sums of the form $\omega = \Sigma c_{\alpha,k} a_{i_1,\alpha} \otimes \cdots \otimes a_{i_k,\alpha}$ which satisfy the conditions $d_j \omega = 0$ for $0 < i < k$.

THEOREM 1.8. The homology of $\bar{B}_{*,*}(A)$ is isomorphic to $\bar{K}_{*,*}(A)$.

Proof: See Theorem 3.8 [P1].

EXAMPLE 1.9. If $A = R_p$, then $\bar{K}_{1,*}(A)$ has basis $\{[\beta^\varepsilon Q^r] : \varepsilon + i > 0$ and $\varepsilon = 0$ or $1\}$. $\bar{K}_{2,*}(A)$ is generated by the "Adem relations"

$$[\beta^\varepsilon Q^r \otimes \beta^\delta Q^s - \Sigma\, c_t \beta^\eta Q^{r+s-t} \otimes \beta^\xi Q^t].$$

For any k and r, the element $[\beta Q^{p^k r} \otimes \cdots \otimes \beta Q^r]$ is in $\bar{K}_{k,*}(A)$.

Although Theorem 1.8 theoretically determines the homology and thus the cohomology of R_p, the form of the answer is too complicated. Priddy observed that the dual is far simpler.

DEFINITION 1.10. Let A be a PBW algebra with generators $\{a_i\}$ of degree d_i and relations (1.6). Then the coKoszul complex is the homogeneous Koszul algebra with generators $\{\alpha_i\}$ of degree $d_i + 1$ and generating relations

$$(-1)^{\nu_{i,j}}\alpha_i\alpha_j = -\Sigma\,(-1)^{\nu_{r,s}}c(i,j,r,s)\alpha_r\alpha_s$$

where $\nu_{u,v} = d_u + d_u d_v + 1$.

PROPOSITION 1.11. With the above notation $\mathrm{Ext}_A^{*,*}(Z/p,\ Z/p) \approx \bar{K}^{*,*}(A)$.

Proof: Theorem 2.5 of [P1].

THEOREM 1.12. $\mathrm{Ext}_{R_p}^{*,*}(Z/p,\ Z/p) \approx A_p^{*,*}(L)$.

Proof: The generators of the PBW algebra R_p are the elements $\{\beta^\varepsilon Q^i\colon \varepsilon = 0$ or 1 and $\varepsilon + i > 0\}$. Thus the generators of its cohomology are corresponding classes $\sigma^{\varepsilon,i}$ of degree $2i(p-1) - \varepsilon + 1$. Note that the degree is the same as that of $\beta^\delta P^i \in A_p(L)$ where $\varepsilon + \delta = 1$.

By the equations in (1.2) and (1.10) there are four sets of relations involving $\sigma^{\varepsilon,i}\,\sigma^{\delta,j}$ depending on the values of ε and δ. For example if $\varepsilon = \delta = 1$, then

$$-\sigma^{1,i}\,\sigma^{1,j} = -\Sigma\,(-1)^{r+j}\binom{(p-1)(j-s)-1)}{pj-r-1}\sigma^{1,r}\,\sigma^{1,s}$$

where the sum is taken over pairs (r,s) with $i+j = r+s$ and $r > ps$. In particular there is no relation for $\sigma^{1,i}\,\sigma^{1,j}$ if $i \geq pj$ since there are no such terms on the right side of equation (1.2). If we identify $\sigma^{1,i}$ with P^i and use the equality $\binom{u+v}{u} = \binom{u+v}{v}$ we obtain the Adem relation

$$P^iP^j = \Sigma\,(-1)^{i+s}\binom{(p-1)(j-s)-1}{i-ps}P^{i+j-s}P^s \quad \text{for } i < pj.$$

Similarly if $\varepsilon = 1$ and $\delta = 0$, then the relation becomes

$$\sigma^{1,i}\sigma^{0,j} = -\Sigma\,(-1)^{r+j}\binom{(p-1)(j-s)-1}{pj-r}\sigma^{1,r}\sigma^{0,s}$$
$$+\Sigma\,(-1)^{r+j}\binom{p-1)(j-s)}{pj-r}\sigma^{0,r}\sigma^{1,s}$$

corresponding to the Adem relation

$$P^i\beta P^j = -\Sigma\,(-1)^{i+s}\binom{(p-1)(j-s)-1}{i-ps-1}P^{i+j-s}\beta P^s$$
$$+\Sigma\,(-1)^{i+s}\binom{(p-1)(j-s)}{}\beta P^{i+j-s}P^s\ .$$

Theorem 1.12 also follows from Priddy's computation of $\mathrm{Ext}_{A_p(L)}(Z/p,Z/p)$ and his duality theorem in [P1].

REMARK 1.13. These results imply that there is a nonsingular pairing

$$A_p^{**}(L) \otimes \bar{K}_{**}(R_p) \to Z/p$$

determined by

$$<\beta^\varepsilon P^a,\ \beta^\delta Q^b> = \begin{cases} 1 & \text{if } \delta + \varepsilon = 1 \text{ and } a = b \\ 0 & \text{otherwise} \end{cases}$$

and

$$<\beta^{\varepsilon}P^{a}P^{I},\ \Sigma\ \beta^{\delta}Q^{b} \otimes \omega > = \Sigma <\beta^{\varepsilon}P^{a},\ \beta^{\delta}Q^{b}><P^{I},\ \omega>.$$

In particular, $<\beta P^{I},\omega> = <P^{I},\beta\omega>$.

If p = 2 and we use standard notation, then the pairing is determined by

$$< Sq^{a+1},Q^{b}> = \begin{cases} 1 & \text{if } a = b \\ 0 & \text{otherwise} . \end{cases}$$

For example (with standard notation) $Q^8Q^2 + Q^6Q^4 + Q^5Q^5 = 0$. Thus $\omega = Q^8 \otimes Q^2 + Q^6 \otimes Q^4 + Q^5 \otimes Q^5 \in \bar{K}_{2,10}(R_2)$ and $< Sq^9Sq^3,\omega > = <Sq^7Sq^5,\omega> = <Sq^6Sq^6,\omega > = 1$. Note that by the Steenrod Adem relations $Sq^7Sq^5 = Sq^9Sq^3$ and $Sq^6Sq^6 = Sq^{11}Sq^1 + Sq^{10}Sq^2 + Sq^9Sq^3$. This process allows us to compute Adem relations "backwards." For example, to find out which nonadmissible operations Sq^aSq^b contain Sq^9Sq^3 in their admissible expansion, first expand Q^8Q^2 to get $Q^6Q^4 + Q^5Q^5$ and then replace these summands by Sq^7Sq^5 and Sq^6Sq^6.

2. Let X be an infinite loop space. Since $H^*(X; Z/p)$ is a module over A_p, $H_*(X; Z_p)$ is a module over A_p^{op} with adjoint action given by the formula $<u,\ P_*^a\beta^{\varepsilon}x > = <\beta^{\varepsilon}P^a u,x >$. Nishida derived the following relations for the interaction of A_p^{op} and the geometric Dyer Lashof algebra $R_p(0)$ (see [M4] and [N] for proofs).

$$\begin{cases} P_*^aQ^cx = \Sigma\ (-1)^{a+i} \binom{(c-a)(p-1)}{a-pi} Q^{c-a+i}P_*^ix \\ P_*^a\beta Q^cx = \Sigma\ (-1)^{a+i} \binom{(c-a)(p-1)-1}{a-pi} \beta Q^{c-a+i}P_*^ix \\ \qquad + \Sigma\ (-1)^{a+i} \binom{(c-a)(p-1)}{a-pi-1} Q^{c-a+i}\beta P_*^ix. \end{cases} \tag{2.1}$$

REMARK 2.2. For p=2 if we identify $\beta^{\varepsilon}P^a$ with $Sq^{2a+\varepsilon}$ and, as in section 1, $\beta^{\delta}Q^c$ with $Q^{2c-\delta}$, then the standard Nishida relations follow:

$$Sq_*^aQ^c = \Sigma \binom{c-a}{a-2i} Q^{c-a+i}Sq_*^i .$$

These formulae are not sufficient to make the algebraic Dyer-Lashof algebra into an A_p^{op} module. For example with p=2 and using standard notation $Q^5Q^0 = Q^1Q^4$ but $Sq^2_*(Q^5Q^0) = Q^3Q^0 = Q^1Q^2 \neq 0 = Sq^2_*(Q^1Q^4)$. It is probably possible to extend the Nishida relations so that R_p does become an A_p^{op} module. Indeed there is evidence to believe that this will happen if we make the conventions that $Q^i = 0$ if $i < 0$ and $\binom{-m}{-n} = \binom{m}{n}$ if $m, n \geq 0$. Rather than pursue this avenue, we opt for the following simpler and geometrically more natural approach.

Let $\bar{B}_{k,*}(0)$ be the subcomplex of $\bar{B}_{k,*}(R_p)$ generated by elements $Q^{I_1}\otimes\ldots\otimes Q^{I_k}$ so that $(I_1,\ldots,I_k)$ can be written (J,K,L) with $e(K) < 0$. If we assume that I_j is (Dyer Lashof) admissible for each j, then this condition is equivalent to saying that for some $j, e(I_j) < \sum_{t>j} \deg(Q^{I_t})$. Alternatively $Q^{I_1} \ldots Q^{I_k}$ vanishes on all homology classes purely for excess reasons. In analogy with Miller's definition of UnTor [M5], we define $\mathrm{UnExt}_{R_p}^{*,*}(Z/p, Z/p)$ to be the cohomology of the Z/p dual of $\bar{B}_{*,*}(R_p)/\bar{B}_{*,*}(0)$.

PROPOSITION 2.3. $\mathrm{UnExt}_{R_p}^{k,*}(Z/p, Z/p) \approx \bar{K}^{k,*}(R_p) \approx A_p(L)^{k,*}$ as Z/p modules.

Proof: As noted by Miller in Proposition 3.1.2 [M5], Priddy's proof of our Theorem 1.6 extends to show that $\bar{B}_{*,*}(R_p)/\bar{B}_{*,*}(0)$ is chain equivalent to $\bar{K}_{*,*}(R_p)/\bar{K}_{*,*}(R_p) \cap \bar{B}_{*,*}(0)$. Since (Dyer Lashof) Adem relations involve elements $Q^a Q^b$ with $a > pb$, in particular of positive excess, it is easy to see that elements of $\bar{K}_{k,*}(R_p)$ are sums with at least one summand not in $\bar{B}_{*,*}(0)$. Thus $\bar{K}_{*,*}(R_p)/(\bar{K}_{*,*}(R_p) \cap \bar{B}_{*,*}(0))$ is isomorphic to $\bar{K}_{*,*}(R_p)$.

This proposition says that $A_p(L)$ may be considered to be either the module generated by all sequences $\beta^{\varepsilon_1} P^{a_1} \ldots \beta^{\varepsilon_k} P^{a_k}$ subject to the Adem relations or by the subset of those sequences satisfying $2a_j + \varepsilon_j \geq \sum_{t>j} \deg \beta^{\varepsilon_t} P^{a_t}$ subject to the same relations. Since each module has the set of admissible sequences as basis, this is obvious.

REMARK 2.4. Consider the quotient complex $\bar{B}_{*,*}(R_p)/\bar{B}_{*,*}(m)$ "generated" by $Q^{I_1} \otimes \ldots \otimes Q^{I_k}$ with I_j admissible and with $e(I_j) \geq (\sum_{t>j} \deg Q^{I_t}) + m$ for all j. Let T_*^m be the graded module $T_n^m = Z/p$ if $m = n$ and 0 if $m \neq n$. Then, following Miller, we can define $\mathrm{UnExt}_{R_p}(Z/p, T_*^m)$ to be the cohomology of $(\bar{B}_{*,*}(R_p)/\bar{B}_{*,*}(m))^*$. It can be shown that this cohomology is isomorphic to the submodule of $A_p(L)$ generated by admissible sequences $\beta^{\varepsilon_1} P^{a_1} \ldots \beta^{\varepsilon_k} P^{a_k}$ with $2a_k + \varepsilon_k > m$. This construction is important in studying the Miller spectral sequence [M5], [KL1], [KL2].

The (Steenrod) Adem relations induce a left A_p module structure on $A_p(L)^{k,*}$ as follows. Assume that $\beta_L^{\varepsilon} P_L^a P_L^I \in A_p(L)^{k,*}$ is admissible where $\varepsilon = 0$ or 1 and we distinguish elements of $A_p(L)$ by using the subscript L. Then $\beta(\beta_L^{\varepsilon} P_L^a P_L^I) = (1-\varepsilon)\, \beta_L P_L^a P_L^I$, $P^a P_L^I = 0$ if $a \neq 0$ and $I = \emptyset$, and

$$P^a P_L^b P_L^I = \begin{cases} \sum (-1)^{a+i} \binom{(p-1)(b-i)-1}{a-pi} P_L^{a+b-i} P^i P_L^I & \text{if } a < pb \\ 0 & \text{if } a \geq pb \end{cases} \tag{2.5}$$

$$P^a \beta_L P_L^b P_L^I = \begin{cases} \sum (-1)^{a+i} \binom{(p-1)(b-i)}{a-pi} \beta_L P_L^{a+b-i} P^i P_L^I & \\ + \sum (-1)^{a+i-1} \binom{(p-1)(b-i)-1}{a-pi-1} P^{a+b-i} \beta_L P^i P_L^I & \text{if } a \leq pb \\ 0 & \text{if } a > pb. \end{cases} \tag{2.6}$$

REMARK 2.7. The assumption that $P^b_L P^I_L$ be admissible is necessary. For example, $Sq^5 Sq^2_L = 0$ while $Sq^5(Sq^2_L Sq^5_L) = Sq^5(Sq^6_L Sq^1_L) = Sq^{11}_L Sq^1_L$.

THEOREM 2.8. $\mathrm{UnExt}^{*,*}_{R_p}(Z/p, Z/p) \approx A_p(L)^{*,*}$ as bigraded A_p modules.

Proof: We must show that $\langle P^a P^J_L, \omega\rangle = \langle P^J_L, P^a_* \omega\rangle$ where $P^J_L = \beta^\varepsilon P^b_L\ z$ is admissible in $A_p(L)$ and where $\omega \in \bar{K}_{k,*}(R_p)$. Assume inductively that this equation holds for sequences of length $< k$. Write $\omega = \sum_{\delta,c} \beta^\delta Q^c \otimes \omega'$ for $\omega' = \omega'_{\delta,c} \in \bar{K}_{k-1,*}(R_p)$.

For p=2, this theorem is essentially proven in Section 4 of [M5]. The general proof breaks up into four cases depending on ε and δ.

CASE 1. $\varepsilon = \delta = 0$. Then by (1.13)

$$\begin{aligned}\langle P^a P^b_L\ z, Q^c \otimes \omega' \rangle &= \Sigma\, \gamma_t \langle P^{a+b-t}_L P^t z, Q^c \otimes \omega'\rangle \\ &= 0\end{aligned}$$

and

$$\begin{aligned}\langle P^b_L\ z, P^a_*(Q^c \otimes \omega') \rangle &= \Sigma\, \eta_t \langle P^b_L\ z, Q^{c-a+t} \otimes P^t_*\ \omega' \rangle \\ &= 0\end{aligned}$$

CASE 2. $\varepsilon = 0$ and $\delta = 1$.

We first expand the left side to obtain

$$\begin{aligned}\langle P^b_L z, P^a_*\ \beta Q^c \otimes \omega\rangle &= \Sigma\, (-1)^{a+i} \binom{(c-a)(p-1)-1}{a-pi} \langle P^b_L z, \beta Q^{c-a+i} \otimes P^i_*\ \omega' \rangle \\ &\quad + \Sigma\, \gamma_i \langle P^b_L z, Q^{c-a+i} \times \beta P^i_*\ \omega' \rangle \\ &= \begin{cases} (-1)^{a+i} \binom{(c-a)(p-1)-1}{a-pi} \langle z, P^i_*\ \omega'\rangle & \text{if } b = c-a+i \\ 0 & \text{otherwise.}\end{cases}\end{aligned}$$

In particular, this is 0 if $(c-a)(p-1)-1 < a-pi$, i.e. if $b=c-a+i \le \frac{c}{p}$.

If $a \ge pb$ so that $P^a P^b_L\ z = 0$ by (2.5), then either $a > c$ so that $P^a_*\ \beta Q^c = 0$, or $c \ge a \ge pb$ so that the binomial coefficient above is 0. If $a < pb$, then

$$\begin{aligned}\langle P^a P^b_L\ z, \beta Q^c \otimes \omega'\rangle &= \Sigma\, (-1)^{a+i} \binom{(p-1)(b-i)-1}{a-pi} \langle P^{a+b-i}_L P^i z, \beta Q^c \otimes \omega'\rangle \\ &= \begin{cases} (-1)^{a+i} \binom{(p-1)(b-i)-1}{a-pi} \langle P^i z, \omega' \rangle & \text{if } a+b-i=c \\ 0 & \text{if } a+b-i \ne c.\end{cases}\end{aligned}$$

Thus the two coefficients agree and $\langle P^i z, \omega'\rangle = \langle z, P^i_*\ \omega'\rangle$ by the induction hypothesis.

The cases where $\varepsilon = 1$ are almost identical arguments using (2.6) and will be left to the reader.

BIBLIOGRAPHY

[B1] Bisson, T. Divided sequences and bialgebras of homology operations. Thesis, Duke Univ., 1977.

[B2] Browder, W. Homology operations and infinite loop spaces, Ill. J. Math. 4 (1960), 347-357.

[CLM] Cohen, F. R., Lada, T., and May, J. P. The Homology of Iterated Loop Spaces. Springer Lecture Notes in Math. Vol. 533. 1976.

[DL] Dyer, E. and Lashof, R. K. Homology of iterated loop spaces. Amer. J. Math. 84 (1962), 35-88.

[KL1] Kraines, D. and Lada, T. A counter-example to the transfer conjecture. Springer Lecture Notes in Math. Vol 741 (1979), 588-624.

[KL2] ______________________. Applications of the Miller Spectral Sequence, AMS Contemporary Mathematics Series, to appear.

[KA] Kudo, T. and Araki, S. Topology of H_n-sequences and H-squaring operations. Mem. Fac. Sci. Kyusyu Univ. Ser. A 10 (1956), 85-120.

[M1] MacLane, S. Homology. Springer-Verlag, 1963.

[M2] Madsen, I. On the action of the Dyer Lashof algebra on G. Pac. J. Math 60 (1975), 235-275.

[M3] May, J. P. A general algebraic approach to Steenrod operations. Springer Lecture Notes in Math. Vol. 168, 1970.

[M4] __________ Homology operations on infinite loop spaces. Proc. Symp. Pure Math. Vol. 22. AMS 1971, 171-185.

[M5] Miller, H. A spectral sequence for the homology of an infinite delooping. Pac. J. Math., 79 (1978), 139-155.

[N] Nishida, G. Cohomology operations in iterated loop spaces. Proc. Japan Acad. 44 (1968), 104-109.

[P1] Priddy, S. Koszul resolutions. Trans. AMS. 152 (1970), 39-60.

[P2] __________ Primary cohomology operations for simplicial Lie algebras. Ill. J. Math. 14 (1970), 585-612.

[C] Cohen, R. Odd primary infinite families in stable homotopy. AMS Memoirs No. 242, 1981.

DEPARTMENT OF MATHEMATICS
DUKE UNIVERSITY
DURHAM, NC 27706

DEPARTMENT OF MATHEMATICS
NORTH CAROLINA STATE UNIVERSITY
RALEIGH, NC 27650

Contemporary Mathematics
Volume 19, 1983

SPACELIKE RESOLUTIONS OF SPECTRA

Nicholas J. Kuhn*

§1 INTRODUCTION

The primary purpose of this paper is to establish a "homological algebra" of spectra based on the adjoint functors Σ^∞ and Ω^∞. Here Σ^∞ is the suspension functor from the category of spaces to the category of spectra and Ω^∞ is the functor which assigns to a spectrum E, the 0th space of an Ω-spectrum equivalent to E.

In this setting, free objects are just suspension spectra and projectives are wedge summands of suspension spectra. These are the "spacelike" spectra of the title. In this context, the Kahn-Priddy Theorem [10] is shown to be a statement about the beginning of a minimal projective resolution of $H\mathbb{Z}$, the integral Eilenberg-MacLane spectrum. Our recent work [13], affirmatively answering a conjecture of G. Whitehead, can be interpreted as a construction of just such a minimal projective resolution. Results of Adams [2] and Mitchell and Priddy [19] are also appropriately reinterpreted.

In §2 we make our basic definitions in a manner entirely analogous to standard definitions in the homological theory of R-modules. Continuing our analogy, we state basic consequences such as a comparison theorem for projective resolutions. Some proofs consist of standard categorical arguments and these are generally omitted.

In §3 we show how our theory leads to a quick and natural construction of the Miller spectral sequence and its basic properties. This is the spectral sequence defined by Miller in [18] to relate $H_*(E)$ to $H_*(\Omega^\infty E)$ where E is a (-1)-connected spectrum, and subsequently studied by Kraines and Lada in [11, 12]. (See [16, p.155] for yet another discussion of the existence of this spectral sequence, predating Miller's work.) We also prove the Miller spectral sequence analogue of the "geometric boundary theorem" of [6, 9].

The question of minimality among projective resolutions is discussed in

* This research was partially supported by NSF grant MCS-8201652

§4, and a reasonable criterion is established using the notions of §3. This then leads to an easy proof of a "mod 2^m" Whitehead conjecture: the construction of a minimal projective resolution for $H\mathbb{Z}/2^m$.

Finally we end the paper with some further problems.

§2 A HOMOLOGICAL ALGEBRA OF SPECTRA

We first note the categories of spaces and spectra in which we wish to work. Let C be the category of C.W. complexes with basepoint, with morphisms homotopy classes of maps. Let S be Boardman's category of spectra [5] and let S_c be the full subcategory of (-1)-connected spectra. $[X, Y]$ will denote the set of morphisms from X to Y . Then there is the natural adjunction

$$[\Sigma^\infty X, E]_S = [X, \Omega^\infty E]_C .$$

For the remainder of this paper, spaces are to be taken in C and spectra are defined to be objects in S_c .

DEFINITION 2.1

(1) A map $f:B \to C$ between spectra is onto (written $B \xrightarrow{f} C \to 0$) if $\Omega^\infty f:\Omega^\infty B \to \Omega^\infty C$ is a split epimorphism, i.e. has a right inverse.

(2) A sequence of spectra $A \xrightarrow{f} B \xrightarrow{g} C$ is short exact if it is a cofibration sequence and g is onto. Then $\Omega^\infty B \simeq \Omega^\infty A \times \Omega^\infty C$.

(3) A sequence of spectra

$$\cdots \to X_2 \xrightarrow{d_1} X_1 \xrightarrow{d_0} X_0 \to E_0$$

is exact if it is obtained from a diagram of the form

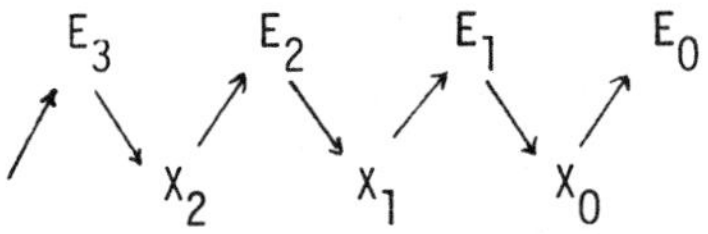

where the sequences $E_{n+1} \to X_n \to E_n$ are short exact.

REMARK 2.2 Suppose $\{F_n\}$ is an increasing filtration of a spectrum E . Let $X_0 = F_0$ and, for $n > 0$, let $X_n = \Sigma^{-n} F_n/F_{n-1}$ and $E_n = \Sigma^{-n} E/F_{n-1}$. One can then form the diagram

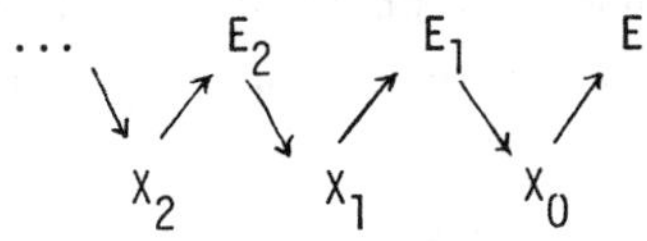

in which the sequences $E_{n+1} \to X_n \to E_n$ are cofibration sequences.

We will refer to such a diagram as a complex (over E). Conversely, via diagram chasing, a complex over E gives rise to a filtration of E.

DEFINITION 2.3

(1) A spectrum E is free if $E = \Sigma^\infty X$ for some space X.

(2) A spectrum P is projective (or spacelike) if every diagram of the form

$$\begin{array}{ccccc} & & P & & \\ & & \downarrow & & \\ B & \longrightarrow & C & \longrightarrow & 0 \end{array} \qquad \text{can be completed to} \qquad \begin{array}{ccccc} & & P & & \\ & \swarrow & \downarrow & & \\ B & \longrightarrow & C & \longrightarrow & 0 \end{array} .$$

(3) A sequence of spectra $\ldots \to P_2 \to P_1 \to P_0 \to E$ is a projective resolution of E if the sequence is exact and each P_i is projective.

Standard arguments now imply the following proposition.

PROPOSITION 2.4 The following statements are equivalent.

(1) P is projective.

(2) P is a wedge summand of $\Sigma^\infty\Omega^\infty P$.

(3) P is a wedge summand of some suspension spectrum.

(4) If $A \to B \to C$ is short exact, so is $0 \to [P,A] \to [P,B] \to [P,C] \to 0$.

COROLLARY 2.5 Suppose that $\ldots \to X_2 \to X_1 \to X_0 \to E$ is exact. Then, for all projective P there is an exact sequence

$$\ldots \to [P, X_2] \to [P, X_1] \to [P, X_0] \to [P, E] \to 0 .$$

In particular, there is an exact sequence of homotopy groups

$$\ldots \to \pi_*(X_2) \to \pi_*(X_1) \to \pi_*(X_0) \to \pi_*(E) \to 0 .$$

EXAMPLE 2.6 Let E be the fiber of the unit $\Sigma^\infty S^0 \xrightarrow{\varepsilon} H\mathbb{Z}$. Note that $\Omega^\infty E = Q_0 S^0$, the component of the identity in $QS^0 = \Omega^\infty\Sigma^\infty S^0$. The Kahn-Priddy Theorem [10] essentially says that, localized at 2, there is a map $f:\Sigma^\infty\mathbb{R}P^\infty \to E$ that is onto. Note that ε is also clearly onto. Thus we have an exact sequence

$$\Sigma^\infty\mathbb{R}P^\infty \to \Sigma^\infty S^0 \to H\mathbb{Z} \to 0 ,$$

the beginning of a projective resolution of $H\mathbb{Z}$.

REMARK 2.7 For any spectrum E, $\Sigma^\infty\Omega^\infty E$ is projective and the evaluation map $\varepsilon:\Sigma^\infty\Omega^\infty E \to E$ is onto, so that projective resolutions always exist. However, it is "small" projective resolutions that are most interesting and much harder to find. The sequence of Example 2.6 is in fact the beginning of a minimal projective resolution of $H\mathbb{Z}$ (see Examples 2.12 and 4.4).

EXAMPLE 2.8 In [2], Adams noted that the Kahn-Priddy Theorem could be strengthened to the following statement: If $\phi:\Sigma^\infty\mathbb{R}P^\infty \to \Sigma^\infty S^0$ is any map inducing an isomorphism on π_1, then, localized at 2, $\Omega^\infty\phi:Q\mathbb{R}P^\infty \to Q_0S^0$ is a split epimorphism and any two such maps differ by a self equivalence

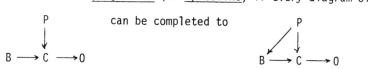

$\alpha:\Sigma^\infty \mathbb{R}P^\infty \to \Sigma^\infty \mathbb{R}P^\infty$.

In our language, Adams' argument is simple and goes as follows. Since $\mathbb{R}P^\infty$ is connected, the composite $\Sigma^\infty \mathbb{R}P^\infty \xrightarrow{\phi} \Sigma^\infty S^0 \xrightarrow{\varepsilon} H\mathbb{Z}$ is zero and thus ϕ lifts to $\phi':\Sigma^\infty \mathbb{R}P^\infty \to E$, where E is the fiber of ε . Recall, as in Example 2.6, that Kahn and Priddy construct a map $f:\Sigma^\infty \mathbb{R}P^\infty \to E$ that is onto. Since $\Sigma^\infty \mathbb{R}P^\infty$ is projective, there is a map $\alpha:\Sigma^\infty \mathbb{R}P^\infty \to \Sigma^\infty \mathbb{R}P^\infty$ making the diagram

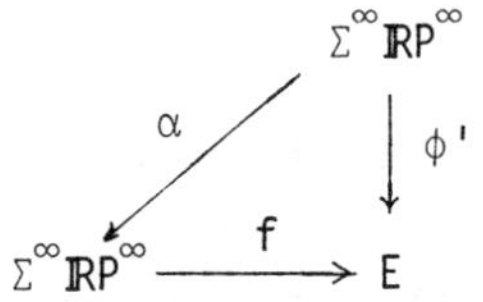

commute.

Adams' argument is completed by noting that α must be an equivalence: α is an isomorphism on π_1 , and any such self map of $\Sigma^\infty \mathbb{R}P^\infty$ must induce the identity in mod 2 homology ([2], lemma 4.5) and thus be an equivalence.

DEFINITION 2.9 Let X_* and Y_* be complexes over spectra A and B , respectively, with boundary maps $d_n:X_{n+1} \to X_n$ and $d'_n:Y_{n+1} \to Y_n$.

(1) A chain map $f_*:X_* \to Y_*$ over $f:A \to B$ is a commutative diagram

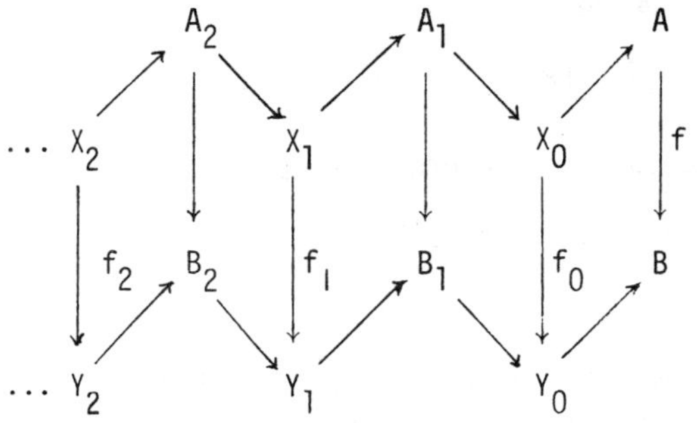

(2) Two chain maps f_* and g_* over $f:A \to B$ are chain homotopic if there are maps $H_n:X_n \to Y_{n+1}$ such that $d'_0 H_0 = f_0 - g_0$ and, for $n \geq 1$, $d'_n H_n - H_{n-1} d_{n-1} = f_n - g_n$.

REMARK 2.10 Under the correspondence of Remark 2.2, a chain map between complexes over A and B is equivalent to a map between the associated filtrations of A and B .

PROPOSITION 2.11 (Comparison Theorem)

If P_* is a projective complex over A and X_* is an exact complex over B then any map $f:A \to B$ lifts to a chain map $f_*:P_* \to X_*$ over f . Any two such liftings are chain homotopic.

EXAMPLE 2.12 Let $F_n = SP^{2^n}(S)$, the 2^n-th symmetric product of the sphere spectrum S . The F_n filter $SP^\infty(S)$ and $SP^\infty(S) \simeq H\mathbb{Z}$ by a theorem of Dold and Thom [7]. Let $L(0) = S$ and, for $n \geq 1$, let $L(n) = \Sigma^{-n} F_n/F_{n-1}$.

Let all spectra and homotopy groups be localized at 2. A conjecture of G. W. Whitehead [17, p.200] is equivalent to the statement that the following sequence is exact:

$$\cdots \to \pi_*(L(2)) \to \pi_*(L(1)) \to \pi_*(L(0)) \to \mathbb{Z} \to 0 .$$

By Corollary 2.5 this would follow from the statement that

$$\cdots \to L(2) \to L(1) \to L(0) \to H\mathbb{Z}$$

is exact. In [14, 19], S. Mitchell and S. Priddy show that $L(n)$ splits off $\Sigma^\infty B(\mathbb{Z}/2)^n$ so that this complex is projective.

The main result of [13] is the construction of an exact resolution of $H\mathbb{Z}$ (which is incidentally also shown to be projective). The two complexes over $H\mathbb{Z}$ can thus be compared and Whitehead's conjecture is then proved by noting that the resulting chain map clearly must be an equivalence. See [13] for details.

Note that $L(1) = \Sigma^\infty \mathbb{R}P^\infty$ [8] so that the sequence of Example 2.6 has been extended to a projective resolution of $H\mathbb{Z}$.

We end this section with two forms of a converse to Corollary 2.5.

PROPOSITION 2.13 Suppose that

$$\cdots \to X_3 \to X_2 \to X_1 \to X_0 \to E$$

is a sequence of spectra having the following exactness property: for all projective spectra P, there is an exact sequence

$$\cdots \to [P, X_2] \to [P, X_1] \to [P, X_0] \to [P, E] \to 0 .$$

If either (1) the sequence is a complex
or (2) the X_i are projective
then the sequence is exact.

PROOF. We sketch the proof. To show that condition (1) implies exactness one lets $P = \Sigma^\infty\Omega^\infty E_n$ where $E_0 = E$ and E_{n+1} is defined to be the fiber of $X_n \to E_n$. Then the exactness property inductively gives liftings

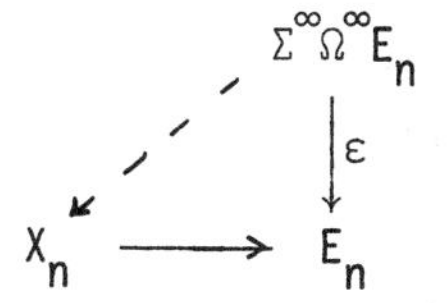

implying that $X_n \to E_n$ is onto.

To show that condition (2) implies exactness one proceeds as above, and also lets $P = X_{n+1}$ so that the exactness property inductively gives liftings

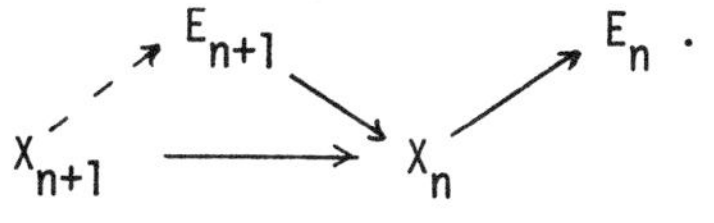

Details are routine and left to the reader.

§3 THE MILLER SPECTRAL SEQUENCE

Let h_* be any homology theory (not necessarily (-1)-connected) and let P_* be any projective resolution of E. Applying h_* to the corresponding filtration of E yields a spectral sequence $\{E^r_{*,*}\}$, called the Miller Spectral Sequence [18].

PROPOSITION 3.1

(1) $E^r_{*,*}$ is well defined, independent of the particular projective resolution, for $r \geq 2$.

(2) The spectral sequence converges to $h_*(E)$ and $E^\infty_{0,*}$ is the image of the homology suspension $\sigma: h_*(\Omega^\infty E) \to h_*(E)$

PROOF. Statement (1) follows in the usual way from the comparison theorem of §2. To prove statement (2), it suffices to prove it for any one particular choice of projective resolution P_*. But if we let P_* be the "bar resolution" of E,

$$\cdots \to \Sigma^\infty QQ\Omega^\infty E \to \Sigma^\infty Q\Omega^\infty E \to \Sigma^\infty\Omega^\infty E \to E,$$

the associated spectral sequence certainly converges to $h_*(E)$ with $E^\infty_{0,*}$ as required (see [16, p.155], for example).

We now show that a short exact sequence of spectra induces a long exact sequence of Miller E^2-terms. We make use of the following mapping cone construction.

Suppose that $A \xrightarrow{f} B \xrightarrow{g} C$ is a fibration sequence of spectra and suppose that there are projective resolutions

$$\cdots \longrightarrow P_2 \xrightarrow{d_1'} P_1 \xrightarrow{d_0'} P_0 \xrightarrow{\varepsilon'} A \quad \text{and}$$

$$\cdots \longrightarrow Q_2 \xrightarrow{d_1''} Q_1 \xrightarrow{d_0''} Q_0 \xrightarrow{\varepsilon''} B.$$

We construct a sequence

$$\cdots \longrightarrow R_2 \xrightarrow{d_1} R_1 \xrightarrow{d_0} R_0 \xrightarrow{\varepsilon} C$$

in the following way. Let $f_*: P_* \to Q_*$ be a lift of f. Let $R_0 = Q_0$ and, for $n \geq 1$, let $R_n = Q_n \vee P_{n-1}$. Let $\varepsilon = g \circ \varepsilon''$, let $d_0 = d_0'' \vee f_0$, and, for $n \geq 1$, let $d_n: Q_{n+1} \vee P_n \to Q_n \vee P_{n-1}$ be $\begin{pmatrix} d_n'' & (-1)^n f_n \\ 0 & d_{n-1} \end{pmatrix}$ in matrix form. It is easy to verify that $d_n \circ d_{n+1} = 0$.

PROPOSITION 3.2 $A \xrightarrow{f} B \xrightarrow{g} C$ is short exact if and only if R_* is a projective resolution of C.

PROOF. Note that we have a commutative diagram

$$\begin{array}{ccccccccc}
\cdots \longrightarrow & Q_3 & \xrightarrow{d''_2} & Q_2 & \xrightarrow{d''_1} & Q_1 & \xrightarrow{d''_0} & Q_0 & \longrightarrow 0 \\
& \downarrow & & \downarrow & & \downarrow & & \| & \\
\cdots \longrightarrow & R_3 & \xrightarrow{d_2} & R_2 & \xrightarrow{d_1} & R_1 & \xrightarrow{d_0} & R_0 & \longrightarrow 0 \\
& \downarrow & & \downarrow & & \downarrow & & \downarrow & \\
\cdots \longrightarrow & P_2 & \xrightarrow{d'_1} & P_1 & \xrightarrow{d'_0} & P_0 & \longrightarrow & 0 &
\end{array}$$

in which all the vertical sequences are split cofibrations and the composite of any two consecutive horizontal maps is 0.

If X is any spectrum, application of the functor $[X,\ ]$ to the above diagram will yield a short exact sequence of chain complexes of abelian groups

$$0 \to Q_*(X) \to R_*(X) \to P_{*-1}(X) \to 0$$

where $Q_n(X) = [X, Q_n]$, $R_n(X) = [X, R_n]$, and $P_n(X) = [X, P_n]$.

This in turn induces a long exact sequence of the corresponding homology groups:

$$\cdots \to H_n(P_*(X); d') \to H_n(Q_*(X); d'') \to H_n(R_*(X); d) \to H_{n-1}(P_*(X); d') \to \cdots$$

By Proposition 2.13, R_* is exact if and only if, for all projective spectra X, $H_n(R_*(X); d) = 0$ for $n > 0$ and $H_0(R_*(X); d) = [X, C]$. Similarly, the fact that P_* and Q_* are exact implies that, for any projective X, $H_n(P_*(X); d') = H_n(Q_*(X); d'') = 0$ if $n > 0$, $H_0(P_*(X); d') = [X, A]$, and $H_0(Q_*(X); d'') = [X, B]$. Then the exact homology sequence implies that $H_n(R_*(X); d) = 0$ for $n > 1$ and that there is an exact sequence:

$$0 \to H_1(R_*(X); d) \to [X, A] \xrightarrow{f_*} [X, B] \to H_0(R_*(X); d) \to 0 .$$

If $A \xrightarrow{f} B \xrightarrow{g} C$ is short exact then f_* will be injective so that $H_1(R_*(X); d) = 0$ and $H_0(R_*(X); d) = [X, C]$ as needed. Conversely, if R_* is exact then, for all projective X, $H_1(R_*(X); d) = 0$ and $H_0(R_*(X); d) = [X, C]$ so that

$$0 \to [X, A] \to [X, B] \to [X, C] \to 0$$

will be exact. It follows then that $A \xrightarrow{f} B \xrightarrow{g} C$ is short exact.

<u>COROLLARY 3.3</u> If $A \xrightarrow{f} B \xrightarrow{g} C$ is short exact and h_* is any homology theory, then there is a long exact sequence of Miller spectral sequence E^2-terms:

$$\cdots \to E^2_{s,t}(A) \xrightarrow{f_*} E^2_{s,t}(B) \xrightarrow{g_*} E^2_{s,t}(C) \xrightarrow{\delta} E^2_{s-1,t}(A) \to \cdots .$$

<u>PROOF</u>. This follows immediately upon applying the functor $h_*(\ )$ to the diagram in the proof of the last proposition. The resulting long exact

sequence of homology groups is, up to sign, the sequence of the corollary.

We now strengthen this result by proving the Miller spectral sequence analogue of the "geometric boundary theorem" for the Adams spectral sequence [6,9]. We need the following lemma.

LEMMA 3.4

Suppose a diagram

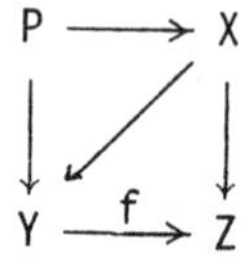

satisfies

(1) P is projective.

(2) f is the fiber of an onto map.

(3) The square commutes.

If the lower triangle commutes, then so does the upper triangle.

PROOF. This is straightforward, noting that conditions (1) and (2) imply that

$$f_*: [P, Y] \to [P, Z] \text{ is monic.}$$

PROPOSITION 3.5 (Geometric Boundary Theorem)

If $A \overset{f}{\to} B \overset{g}{\to} C$ is short exact and h_* is any homology theory, then the connecting homomorphism of Corollary 3.3,

$$\delta: E^2_{s,t}(C) \to E^2_{s-1,t}(A)$$

commutes with differentials in the spectral sequence and converges to the map $\delta: h_*(C) \to h_*(\Sigma A)$ induced by the geometric connecting map $\delta: C \to \Sigma A$.

PROOF. Using the lemma, we can inductively construct maps $t_1: C_1 \to A$ and $t_n: C_n \to A_{n-1}$ making the following diagram of exact sequences commute.

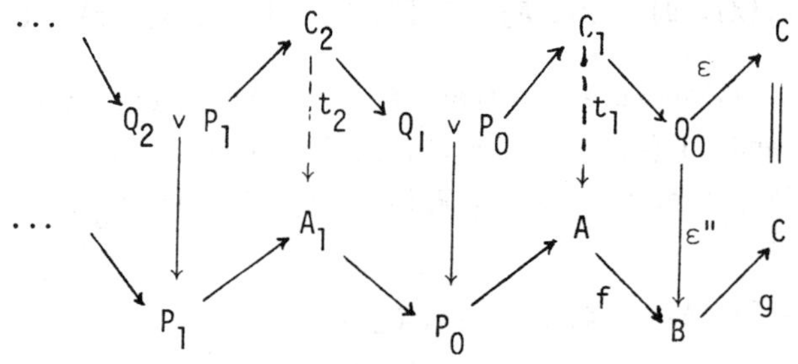

Here, the unlabeled vertical maps are projections and C_n is defined inductively to be the fiber of $Q_n \vee P_{n-1} \to C_{n-1}$, as usual.

We also have a chain map of complexes:

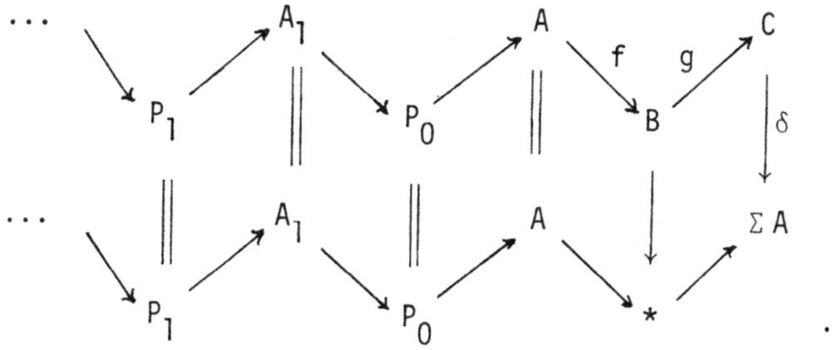

The composition of these two chain maps then induces a map of Miller spectral sequences converging to $\delta_*: h_*(C) \to h_*(\Sigma A)$ such that the induced map on E^2-terms is precisely the map δ of Corollary 3.3.

REMARK 3.6 As a final observation, we note that the Miller spectral sequence is not suspension invariant, i.e. for a homology theory h_* and a spectrum A, the spectral sequence for A and ΣA are in general different. This is because the suspension of an exact resolution need not be exact (and dually, the desuspension of a projective need not be projective). However, the suspension of a projective resolution will at least be projective, and the comparison theorem then yields a map of Miller spectral sequences:

$$\sigma_*: E^r_{s,t}(A) \to E^r_{s,t+1}(\Sigma A) \quad .$$

EXAMPLE 3.7 Letting h_* be ordinary $\mathbb{Z}/2$ homology, Proposition 3.1 implies that $E^\infty_{0,*}(\Sigma^n H\mathbb{Z}/2)$ is isomorphic to the dual of the subspace of the Steenrod algebra with basis $\{Sq^I | I$ admissible and $e(I) \leq n\}$, where $e(I)$ = excess of I. Thus as n increases, more and more of A^* appears in filtration 0. We note that, by calculations of Kraines and Lada [12], $E^2 = E^\infty$ in these cases.

EXAMPLE 3.8 The projective resolution of example 2.12

$$\cdots \to L(2) \to L(1) \to L(0) \to H\mathbb{Z}$$

in fact remains exact after a single suspension (see [13]). Thus for all homology theories, $\sigma_*: E^r_{s,t}(H\mathbb{Z}) \approx E^r_{s,t+1}(\Sigma H\mathbb{Z})$, $r \geq 2$.

§4 MINIMAL RESOLUTIONS

DEFINITION 4.1 A projective resolution P_* of E is minimal if P_* splits off any other resolution of E. That is, if Q_* is any other projective resolution of E there are chain maps $f: P_* \to Q_*$ and $g: Q_* \to P_*$ over $1: E \to E$ such that $g \circ f = 1: P_* \to P_*$.

To establish a workable condition guaranteeing minimality we make the following definition.

DEFINITION 4.2 Let $\cdots \to P_2 \xrightarrow{d_1} P_1 \xrightarrow{d_0} P_0 \to E$ be a projective resolution of E, localized at 2. P_* is H_*-minimal if $d_{n*}: H_*(P_{n+1}; \mathbb{Z}/2) \to H_*(P_n; \mathbb{Z}/2)$ is zero for all n.

REMARK 4.3 In the language of §3, P_* is H_*-minimal if it geometrically realizes the Miller spectral sequence E^2-term for $\mathbb{Z}/2$ homology.

PROPOSITION 4.3 Suppose P_* is H_*-minimal, with each P_n a 2-local spectrum of finite type. Then P_* is minimal.

PROOF If Q_* is any other projective resolution of E, the comparison theorem implies that there are chain maps $f:P_* \to Q_*$, $g:Q_* \to P_*$, and a chain homotopy $g\circ f \sim 1:P_* \to P_*$. H_*-minimality then implies that $(gf)_* = 1: H_*(P_*;\mathbb{Z}/2) \to H_*(P_*;\mathbb{Z}/2)$ so that gf is a homotopy equivalence. Standard arguments then imply that g (or f) can be modified so that $gf = 1$, as needed.

EXAMPLE 4.4 The sequence of Example 2.12,

$$\cdots \quad L(3) \xrightarrow{d_2} L(2) \xrightarrow{d_1} L(1) \xrightarrow{d_0} L(0) \to H\mathbb{Z}$$

is H_*-minimal and thus minimal, localized at 2.

EXAMPLE 4.5 $H\mathbb{Z} \xrightarrow{2^m} H\mathbb{Z} \to H\mathbb{Z}/2^m$ is clearly a short exact sequence. Thus by Proposition 3.2 the mapping cone sequence

$$\cdots \to L(2) \vee L(1) \xrightarrow{D_1} L(1) \vee L(0) \xrightarrow{D_0} L(0) \to H\mathbb{Z}/2^m$$

is exact. Recalling that $D_n = \begin{pmatrix} d_n & (-1)^n 2^m \\ 0 & d_{n-1} \end{pmatrix}$ in matrix form, it is clear that the sequence is H_*-minimal.

EXAMPLE 4.6 Let F'_n be the cofiber of the diagonal map $\Delta: SP^{2^{n-1}}(S) \to SP^{2^n}(S)$. The F'_n filter $H\mathbb{Z}/2$, and, if we let $M(n) = \Sigma^{-n}F'_n/F'_{n-1}$ for $n > 0$ and $M(0) = S$, there are natural maps $\delta_n: M(n+1) \to M(n)$, $\varepsilon: M(0) \to H\mathbb{Z}/2$.

S. Mitchell and S. Priddy show that $M(n)$ is projective. (and then deduce that $M(n) = L(n) \vee L(n-1)$) [14]. As in [13, §6], we can then immediately conclude that the sequence

$$\cdots \to M(3) \xrightarrow{\delta_2} M(2) \xrightarrow{\delta_1} M(1) \xrightarrow{\delta_0} M(0) \xrightarrow{\varepsilon} H\mathbb{Z}/2$$

is equivalent to the sequence of Example 4.5, and is thus exact.

EXAMPLE 4.7 $\Sigma H\mathbb{Z}/2$ has no H_*-minimal resolution because the Miller spectral sequence $E^2_{0,*}$ term cannot be realized as the homology of a spectrum. By Example 3.7, $E^2_{0,*}$ is dual to the quotient of the Steenrod algebra $A/\{Sq^I \mid I \text{ admissible and } e(I) > 1\}$, and the nonexistance of Hopf invariant one elements [1] precludes the existence of a spectrum with this cohomology structure.

EXAMPLE 4.8 (Suggested by J. F. Adams, in correspondance.) Given a projective resolution, one can pass in the usual way to a free resolution. For example, Mitchell and Priddy show that $L(n)$ splits off $\Sigma^\infty B(\mathbb{Z}/2)^n$. Let $R(n)$ be the "n^{th} remainder", so that $L(n) \vee R(n) = \Sigma^\infty B(\mathbb{Z}/2)^n$. Let $f_n\colon B(\mathbb{Z}/2)^{n+1} \vee B(\mathbb{Z}/2)^n \to B(\mathbb{Z}/2)^n \vee B(\mathbb{Z}/2)^{n-1}$ be the stable map defined, using the sequence of Example 4.5, by

$$\Sigma^\infty B(\mathbb{Z}/2)^{n+1} \vee B(\mathbb{Z}/2)^n \approx$$

$$R(n+1) \vee L(n+1) \vee L(n) \vee R(n) \xrightarrow{0 \vee D_n \vee 1} R(n-1) \vee L(n) \vee L(n-1) \vee R(n)$$

$$\approx \Sigma^\infty B(\mathbb{Z}/2)^n \vee B(\mathbb{Z}/2)^{n-1} .$$

Then the sequence of stable maps

$$\ldots \to B(\mathbb{Z}/2)^3 \vee B(\mathbb{Z}/2)^2 \xrightarrow{f_2} B(\mathbb{Z}/2)^2 \vee B(\mathbb{Z}/2) \xrightarrow{f_1} B(\mathbb{Z}/2) \vee S^0 \xrightarrow{f_0} S^0 \longrightarrow H\mathbb{Z}/2^m$$

forms a "small" free resolution of $H\mathbb{Z}/2^m$. There is a similar resolution of $H\mathbb{Z}$.

§5 FURTHER PROBLEMS

PROBLEM 1 Find interesting examples of minimal or near minimal projective resolutions.

For example, let $bu < 0 >$ and $bo < 0 >$ be the usual spectra with 0th-spaces BU and BO respectively. Results of G. B. Segal [20] and J. C. Becker and D. H. Gottlieb [4] imply that there are onto maps $\Sigma^\infty \mathbb{C}P^\infty \to bu < 0 >$ and $\Sigma^\infty BO(2) \to bo < 0 >$.

PROBLEM 2 Extend these to interesting small projective resolutions.

PROBLEM 3 Identify the maps f_n of Example 4.8 in terms of the recent description by J. Adams, J. Gunawardena, and H. Miller of all stable maps between spaces of the form BV , where V is a finite vector space. This might lead to a more conceptual proof of G. W. Whitehead's symmetric products of spheres conjecture.

PROBLEM 4 Can anything substantive be said about the Grothendieck group of projective spectra modulo free spectra?

DEFINITION 5.1 If E is a spectrum, define hom dim E to be s if E has a projective resolution of length s and no shorter projective resolution.

PROBLEM 5 Is hom dim E = 0 or ∞ ?

REMARK 5.2 This would be analogous to algebraic results of T. Y. Lin [15] for π_*^S and Adams and Margolis [3] for the Steenrod algebra. Note that this problem is equivalent to the following. Given $f\colon P \to Q$ with P and Q projective, if $\Omega^\infty f$ is split monic must f be split monic also?

REFERENCES

1. J.F. Adams, On the non-existance of elements of Hopf invariant one, Ann. of Math. 72 (1960), 20-104.

2. ————————, The Kahn-Priddy Theorem, Proc. Cambridge Phil. Soc. 73 (1973), 45-55.

3. J.F.Adams and H.R. Margolis, Modules over the Steenrod algebra, Topology 10 (1971), 271-282.

4. J.C. Becker and D.H. Gottlieb, Characteristic classes and K-theory, Springer Lecture Notes in Math., Vol. 428, 1974, 132-143.

5. J.M. Boardman, Stable homotopy theory, mimiographed notes, University of Warwick, 1966.

6. R. Bruner, Algebraic and Geometric connecting homomorphisms in the Adams spectral sequence, Springer Lecture Notes in Math., Vol. 658, 1978, 131-13 .

7. A. Dold and R. Thom, Quasifaserungen und unendliche symmtriche Produkte, Ann. of Math. 67 (1958), 239-281.

8. I.M. James, E. Thomas, H. Toda, and G.W. Whitehead, On the symmetric square of a sphere, J. of Math. and Mech. 12 (1963), 771-776.

9. D.C. Johnson, H.R. Miller, W.S. Wilson, R.S. Zahler, Boundary homomorphisms in the generalized Adams spectral sequence and the nontriviality of infinitely many γ_t in stable homotopy, Notas de Matematicas y Simposia 1, Soc. Mat. Mex., 1975, 47-63.

10. D.S. Kahn and S.B. Priddy, Applications of the transfer to stable homotopy theory, Bull. Amer. Math. Soc. 76 (1972), 981-987.

11. D. Kraines and T.J. Lada, A Counterexample to the Transfer Conjecture, Springer Lecture Notes in Math., Vol. 741, 1979, 588-624.

12. ————————, The Miller spectral sequence, preprint.

13. N.J. Kuhn, A Kahn-Priddy Sequence and a Conjecture of G.W. Whitehead, Math. Proc. Cambridge Phil. Soc., to appear.

14. N.J. Kuhn, S. Mitchell and S.B. Priddy, The Whitehead conjecture and splitting $B(\mathbb{Z}/2)^k$, Bull. Amer. Math. Soc. 7, no. 1 (1982), 255-258.

15. T.Y. Lin, Homological Algebra of the stable homotopy ring π_* of spheres, Pacific J. Math. 38 (1971), 117-143.

16. J.P. May, The Geometry of Iterated Loop Spaces, Springer Lecture Notes in Math., Vol. 271, 1972.

17. R.J. Milgram, ed. Problems presented to the 1970 AMS Summer Colloquium in Algebraic Topology, in Algebraic Topology, Proc. Symp. Pure Math. XXII, AMS (1971), 187-201.

18. H.R. Miller, A spectral sequence for the homology of an infinite delooping, Pacific J. Math. 79 (1978), 139-155.

REFERENCES (cont'd)

19. S. Mitchell and S.B. Priddy, in preparation.

20. G.B. Segal, The stable homotopy of complex projective spaces, Quart. J. Math. 24 (1973), 1-5.

Department of Mathematics
University of Washington
Seattle, WA 98195

Current address:

Department of Mathematics
Princeton University
Princeton, NJ 08544

Contemporary Mathematics
Volume 19, 1983

EQUIVARIANT BUNDLES WITH ABELIAN STRUCTURAL GROUP

by R. K. Lashof, J. P. May, and G. B. Segal

Let G and A be compact Lie groups and recall that a principal (G,A)-bundle is a principal A-bundle $p\colon D \to X$ such that X and D are G-spaces, p is a G-map, and the actions of G and A on D commute. For a G-space X of the homotopy type of a G-CW complex, define $\mathcal{B}(G,A)(X)$ to be the set of equivalence classes of principal (G,A)-bundles over X. For a space Y of the homotopy type of a CW-complex, define $\mathcal{B}(A)(Y)$ to be the set of equivalence classes of principal A-bundles over Y. Let $X_G = EG \times_G X$, where EG is a contractible and G-free G-CW complex. Define a natural transformation

$$\Phi\colon \mathcal{B}(G,A)(X) \longrightarrow \mathcal{B}(A)(X_G)$$

by sending a (G,A)-bundle $p\colon D \to X$ to the A-bundle $p_G\colon D_G \to X_G$. We shall give an elementary proof of the following result.

<u>Theorem A.</u> If A is Abelian, then Φ is an isomorphism.

A choice of basepoint in EG determines a natural injection $\iota\colon X \to X_G$ and thus a natural transformation

$$\iota^*\colon \mathcal{B}(A)(X_G) \to \mathcal{B}(A)(X).$$

If $\pi\colon EG \times X \to X_G$ is the quotient map and $\varepsilon\colon EG \times X \to X$ is the projection, then $\iota\circ\varepsilon \simeq \pi$ and thus ι^* agrees with the composite

$$\mathcal{B}(A)(X_G) \xrightarrow{\ \pi^*\ } \mathcal{B}(A)(EG \times X) \xrightarrow{\ (\varepsilon^*)^{-1}\ } \mathcal{B}(A)(X).$$

0271-4132/83 \$1.00 + \$.25 per page

The composite $\iota^*\Phi: \mathcal{B}(G,A)(X) \to \mathcal{B}(A)(X)$ coincides with the forgetful transformation Ψ from (G,A)-bundles to A-bundles. Its image consists of those A-bundles over X which admit a structure of (G,A)-bundle, that is, for which the action of G on X lifts appropriately to the total space.

<u>Corollary B.</u> If A is Abelian, then the image of ι^* is the set of A-bundles over X which admit a structure of (G,A)-bundle.

When A is a torus and X is connected and locally finite, this is the main theorem of Hattori and Yoshida [1,1.1].

Since the product and inverse maps of an Abelian group are homomorphisms, they induce natural internal operations which make $\mathcal{B}(G,A)(X)$ and $\mathcal{B}(A)(Y)$ Abelian groups. When $A = S^1$, the product may be viewed as the tensor product of complex line bundles and these are known as Picard groups. Clearly Φ and ι^* are homomorphisms.

<u>Corollary C.</u> If A is Abelian, then the (G,A)-bundle structures (if any) on a given A-bundle over X are in bijective correspondence with the elements of the kernel of ι^*.

Again, when A is a torus and X is connected and locally finite, essentially this enumeration was given by Hattori and Yoshida [1,4.1].

Of course, ι is the inclusion of a fibre in the natural bundle $\gamma: X_G \to BG$ and we can use the Serre spectral sequence of γ to compute ι^*. We assume that X is connected throughout the following discussion.

Since A is isomorphic to the product of a torus T^n and a finite Abelian group F, each (G,A)-bundle decomposes uniquely into the Whitney sum of a (G,T^n)-bundle and a (G,F)-bundle. Thus we can discuss these cases separately.

Of course, we have

$$\mathcal{B}(F)(X) = [X,BF] = H^1(X;F),$$

an F-bundle ξ being given by an F-characteristic class $f_1(\xi)$. An immediate calculation gives that the bottom row is exact in the commutative diagram

$$\begin{array}{ccccccccc}
 & & \mathcal{B}(G,F)(pt) & \xrightarrow{\varepsilon^*} & \mathcal{B}(G,F)(X) & \xrightarrow{\Psi} & \mathcal{B}(F)(X) & & \\
 & & \Phi\downarrow\cong & & \cong\downarrow\Phi & & \cup & & \\
0 & \longrightarrow & H^1(BG;F) & \xrightarrow{\gamma^*} & H^1(X_G;F) & \xrightarrow{\iota^*} & H^1(X;F)^G & \xrightarrow{d_2} & H^2(BG;F).
\end{array}$$

Thus ξ lifts to a (G,F)-bundle if and only if $f_1(\xi)$ is G-invariant and annihilated by d_2, and there is then one lift for each element of $H^1(BG;F)$.

In the torus case, we have

$$\mathcal{B}(T^n)(X) = [X,BT^n] = H^2(X;Z^n),$$

a T^n-bundle ξ being given by a Z^n-characteristic class $c_1(\xi)$. We consider $E_2^{p,q} = H^p(BG;H^q(X;Z^n))$. The corollaries concern

$$\iota^*: H^2(X_G;Z^n) \to E_\infty^{0,2} \subset E_2^{0,2} = H^2(X;Z^n)^G \subset H^2(X;Z^n),$$

where $E_\infty^{0,2} = \mathrm{Ker}(d_3:E_3^{0,2} \to E_3^{3,0}) \subset \mathrm{Ker}(d_2:E_2^{0,2} \to E_2^{2,1})$, and we have the short exact sequence

$$(\alpha) \qquad 0 \longrightarrow E_\infty^{2,0} \longrightarrow \mathrm{Ker}\ \iota^* \longrightarrow E_\infty^{1,1} \longrightarrow 0,$$

where $E_\infty^{2,0} = \mathrm{Coker}(d_2:E_2^{0,1} \to E_2^{2,0})$ and $E_\infty^{1,1} = \mathrm{Ker}(d_2:E_2^{1,1} \to E_2^{3,0})$. We conclude that <u>$\xi$ lifts to a (G,T^n)-bundle if and only if $c_1(\xi)$ is G-invariant and killed by d_2 and d_3, and the exact sequence (α) then determines the number of liftings</u>. For example, if G is simply connected, then BG is 3-connected and every ξ lifts uniquely, this being a result of Stewart [8]. Now assume that $H^1(X;Z) = 0$. Then the bottom row is exact in the commutative diagram

$$\begin{array}{ccccccccc} & & \mathcal{B}(G,T^n)(pt) & \xrightarrow{\varepsilon^*} & \mathcal{B}(G,T^n)(X) & \xrightarrow{\Psi} & \mathcal{B}(T^n)(X) & & \\ & & \Phi\downarrow\cong & & \cong\downarrow\Phi & & \cup & & \\ 0 & \to & H^1(BG;Z^n) & \xrightarrow{\gamma^*} & H^2(X_G;Z^n) & \xrightarrow{\iota^*} & H^2(X;Z^n)^G & \xrightarrow{d_3} & H^3(BG;Z^n). \end{array}$$

Note that $H^3(BG;Z^n) = 0$ if G is Abelian and $H^2(X;Z^n)^G = H^2(X;Z^n)$ if G is connected. Thus every ξ lifts uniquely if G is a torus, this being a result of Su [9]. If $n = 1$, the top row is the exact sequence of Picard groups discussed by Liulevicius [3,Thm 2].

To prove Theorem A, we first interpret Φ on the classifying space level; A need not be Abelian for this part. There is a classifying G-space $B(G,A)$ such that

$$\mathcal{B}(G,A)(X) = [X,B(G,A)]_G,$$

where $[X,X']_G$ denotes the set of homotopy classes of G-maps $X \to X'$. We may take $B(G,A)$ to be a G-CW complex. Of course, we also have

$$\mathcal{B}(A)(Y) = [Y,BA] = [Y,BA]_G,$$

where Y and BA are regarded as G-trivial G-spaces. As a nonequivariant space, $B(G,A)$ is itself a classifying space for A-bundles. Moreover, there is a map $\zeta\colon BA \to B(G,A)$ which takes values in $B(G,A)^G$ (hence may be regarded as a G-map) and is a nonequivariant homotopy equivalence. On the level of represented functors on spaces, ζ corresponds to the transformation which sends an A-bundle to the same A-bundle regarded as a G-trivial (G,A)-bundle. The following diagram of functors clearly commutes.

$$\text{(I)}\qquad \begin{array}{ccc} \mathcal{B}(G,A)(X) & \xrightarrow{\Phi} & \mathcal{B}(A)(X_G) \\ \downarrow\varepsilon^* & & \downarrow\zeta_* \\ \mathcal{B}(G,A)(EG \times X) & \xleftarrow{\pi^*} & \mathcal{B}(G,A)(X_G) \end{array}$$

We also have the obvious commutative diagram

(II)
$$\begin{array}{ccccc}
[EG \times X, BA]_G & \xleftarrow{\pi^*} & [X_G, BA]_G & = & [X_G, BA] \\
\downarrow \zeta_* & & \downarrow \zeta_* & & \\
[EG \times X, B(G,A)]_G & \xleftarrow{\pi^*} & [X_G, B(G,A)]_G & = & [X_G, B(G,A)^G]
\end{array}$$

Here the upper map π^* is clearly a bijection and will be regarded as an identification. We shall see in a moment that the left map ζ_* is also a bijection. This implies that $\pi^*\zeta_*$ is a bijection in both diagrams and that Φ may be regarded as the composite

$$[X, B(G,A)]_G \xrightarrow{\varepsilon^*} [EG \times X, B(G,A)]_G \xrightarrow{\zeta_*^{-1}} [EG \times X, BA]_G.$$

For G-spaces X and X', let $M(X,X')$ denote the function G-space of continuous maps $X \to X'$, with G acting by conjugation. Then ε^* and ζ_* are obtained by application of the functor $[X,?]_G$ to the G-maps

(*) $$B(G,A) = M(pt, B(G,A)) \xrightarrow{\varepsilon^*} M(EG, B(G,A)) \xleftarrow{\zeta_*} M(EG, BA).$$

Recall that a G-map $f\colon D \to E$ is said to be a weak G-equivalence if its fixed point map $f^H\colon D^H \to E^H$ is an ordinary weak equivalence for each closed subgroup H of G. By the G-Whitehead theorem [5,10],

$$f_*\colon [X,D]_G \to [X,E]_G$$

is then a bijection for any G-space X of the homotopy type of a G-CW complex. Since we have restricted ourselves to such X, we may as well regard classifying G-spaces as defined only up to weak G-homotopy type. The point is that such function G-spaces as $M(EG,BA)$ will generally fail to have the homotopy types of G-CW complexes. Our assertion above that ζ_* is a bijection was proven by obstruction theory in [2,1.4], but we give the following simple argument to illustrate the convenience of using function G-spaces in this context.

Lemma 1 ζ_*: M(EG,BA) $\to$ M(EG,B(G,A)) is a weak G-equivalence.

Proof: Via $f \leftrightarrow \sigma$ if $\sigma(x) = (x,f(x))$, $M(EG,B(G,A))^H$ may be identified with the space of sections of the natural fibration

$$EG \times_H B(G,A) \longrightarrow EG/H = BH.$$

Similarly, $M(EG,BA)^H = M(BH,BA)$ is the space of sections of

$$EG \times_H BA = BH \times BA \longrightarrow BH.$$

Since $1 \times \zeta$: $EG \times BA \to EG \times B(G,A)$ is a G-map and a nonequivariant homotopy equivalence between free G-spaces of the homotopy type of G-CW complexes, it is a G-homotopy equivalence by the G-Whitehead theorem. Therefore $1 \times_H \zeta$ is a homotopy equivalence over BH and thus a fibre homotopy equivalence (by a standard elementary argument). The induced homotopy equivalence between the respective spaces of sections coincides with the fixed point map $(\zeta_*)^H$.

Henceforward, we assume that A is Abelian. It is clear from the discussion above that Theorem A is an immediate consequence of the following result, which implies that (*) displays a weak G-equivalence between B(G,A) and M(EG,BA).

Theorem 2. ε^*: B(G,A) $\to$ M(EG,B(G,A)) is a weak G-equivalence when A is Abelian.

For the proof, we note first that ε^* and ζ_* in (*) are Hopf G-maps between Hopf G-spaces. Indeed, our G-spaces have equivariant sums which make them Abelian topological G-groups up to homotopy. This is clear for M(EG,BA), which inherits a structure of Abelian topological G-group from the structure of Abelian topological group on BA. For B(G,A) and M(EG,B(G,A)), it follows from the fact that, up to G-homotopy, B(G,A) is a product-preserving functor of A. The zero of $B(G,A)$ is the image of the point $B(G,\{0\})$. We shall use the following triviality.

Lemma 3. Let Y be a homotopy associative and commutative Hopf space such that $\pi_0 Y$ is a group and let Y_0 be the basepoint component of Y. Then Y is naturally equivalent as a Hopf space to $Y_0 \times \pi_0 Y$.

Proof: Choose a point a in each component Y_a, writing a^{-1} for the chosen point in the inverse component. Define $\alpha: Y \to Y_0 \times \pi_0 Y$ by $\alpha(y) = (a^{-1}\cdot y, Y_a)$ for $y \in Y_a$ and define $\beta: Y_0 \times \pi_0 Y \to Y$ by $\beta(z, Y_a) = a \cdot z$ for $z \in Y_0$. Homotopy associativity ensures that α and β are inverse equivalences; homotopy commutativity ensures that they are Hopf maps.

Thus to prove Theorem 2 it suffices to show that $(\varepsilon^*)^H$ restricts to an equivalence on basepoint components and induces an isomorphism on π_0 for each $H \subset G$.

The basepoint component of $B(G,A)^H$ classifies H-trivial (H,A)-bundles and is thus a copy of BA. Indeed, ζ may be regarded as the inclusion of the basepoint component in $B(G,A)^H$ for any H. In view of Lemma 1 and the obvious commutative diagram

$$\begin{array}{ccc} BA & \xrightarrow{\varepsilon^*} & M(BH,BA) = M(EG,BA)^H \\ \downarrow{\scriptstyle\zeta} & & \downarrow{\scriptstyle(\zeta_*)^H} \\ B(G,A)^H & \xrightarrow{(\varepsilon^*)^H} & M(EG,B(G,A))^H \end{array},$$

$(\varepsilon^*)^H$ will be a weak equivalence on basepoint components provided that ε^* is a weak equivalence from BA to the basepoint component $M_0(BH,BA)$ of $M(BH,BA)$, the basepoint being the trivial map. Now A has the form $F \times T^n$, where F is finite, and we have the commutative diagram

$$\begin{array}{ccc} \pi_q BA & = & H^1(S^q;F) \oplus H^2(S^q;Z^n) \\ \Big\downarrow \varepsilon^* & & \Big\downarrow \varepsilon^* \oplus \varepsilon^* \\ \pi_q M_0(BH,BA) & = & H^1(\Sigma^q BH^+;F) \oplus H^2(\Sigma^q BH^+;Z^n), \end{array}$$

where BH^+ is the union of BH and a disjoint basepoint. If $q = 1$, the first summand ε^* is clearly an isomorphism and the second summands are zero since $H^1(BH;Z) = 0$. If $q = 2$, the first summands are clearly zero and the second summand ε^* is clearly an isomorphism. If $q \geq 3$, all groups are zero.

It remains to consider $(\varepsilon^*)^H$ on π_0. For any G-space X, $\pi_0(X^H) = [G/H,X]_G$. By Lemma 1 and diagrams (I) and (II), $(\varepsilon^*)^H$ will induce an isomorphism on π_0 provided that

$$\Phi\colon \mathcal{B}(G,A)(G/H) \longrightarrow \mathcal{B}(A)(BH)$$

is an isomorphism. We claim that Φ here may be identified with the homomorphism

$$B\colon \mathrm{Hom}(H,A) \longrightarrow [BH,BA]$$

given by the classifying space functor, where $\mathrm{Hom}(H,A)$ denotes the Abelian group of continuous homomorphisms $\rho\colon H \to A$. Indeed, we obtain an isomorphism from $\mathrm{Hom}(H,A)$ to $\mathcal{B}(G,A)(G/H)$ by sending ρ to the natural (G,A)-bundle $\xi_\rho\colon G \times_H A_\rho \to G/H$, where A_ρ denotes A regarded as an H-space via ρ, and Φ carries ξ_ρ to the natural A-bundle $EG \times_H A_\rho \to BH$. It is classical bundle theory that the latter is classified by $B\rho$. Thus the following result completes the proof of Theorem 2.

<u>Proposition 4.</u> $B\colon \mathrm{Hom}(G,A) \to [BG,BA]$ is an isomorphism when A is Abelian.

Proof: If A is finite, elementary calculations show that π_0 and π_1 are isomorphisms in the commutative diagram

$$\begin{array}{ccc} \mathrm{Hom}(G,A) & \xrightarrow{B} & [BG,BA]. \\ \pi_0 \downarrow & & \downarrow \pi_1 \\ \mathrm{Hom}(\pi_0 G, \pi_0 A) & = & \mathrm{Hom}(\pi_1 BG, \pi_1 BA) \end{array}$$

In general, $A = F \times T^n$ where F is finite, hence it suffices to prove the result when A is the circle group S^1. This is very easy if G is finite (by group cohomology [4,IV.5.5]), if G is a torus (by inspection), or if G is connected (by use of a maximal torus). The general case is easily handled by use of the third author's continuous group cohomology theory [7]. For topological G-modules A, there are cohomology groups $H^*(G;A)$ (denoted $R^*\Gamma^G A$ in [7]). We shall only be concerned with trivial G actions. Here $H^*(G;A)$ is the ordinary cohomology $H^*(BG;A)$ if A is discrete [7,3.3], and $H^1(G;A) = \mathrm{Hom}(G,A)$ in general [7,4.3]. If A is contractible, then $H^*(G;A)$ can be calculated by continuous cochains [7,3.1], and it follows from Mostow [6,2.5 and 2.14] that $H^q(G;\mathbf{R}) = 0$ for $q > 0$. Suitable topological short exact sequences in A give rise to long exact sequences of cohomology groups [7,1.3]. In particular, the extension $Z \to \mathbf{R} \to S^1$ gives rise to a connecting isomorphism $\delta: H^1(G;S^1) \to H^2(G;Z)$. A comparison of definitions shows that δ coincides with B.

Bibliography

1. A Hattori and T. Yoshida. Lifting compact group actions in fibre bundles. Japan J. Math. 2 (1976), 13-25.

2. R. K. Lashof. Obstructions to equivariance. Springer Lecture Notes in Mathematics Vol 763, 1979, p. 476-503.

3. A. Liulevicius. Homotopy rigidity of linear actions: characters tell all. Bull. Amer. Math. Soc. 84 (1978), 213-221.

4. S. MacLane. Homology. Springer-Verlag. 1963.

5. T. Matumoto. On G-CW complexes and a theorem of J.H.C. Whitehead. J. Fac. Sci. Tokyo 18 (1971), 363-374.

6. G. D. Mostow. Cohomology of topological groups and solvmanifolds. Annals Math. 73 (1961), 20-48.

7. G. B. Segal. Cohomology of topological groups. Symposia Mathematica, vol IV (INDAM, Rome, 1968/69), p. 377-387.

8. T. E. Stewart. Lifting group actions in fibre bundles. Annals Math. 74 (1961), 192-198.

9. J. C. Su. Transformation groups on cohomology projective spaces. Trans. Amer. Math. Soc. 106 (1963), 305-318.

10. S. Waner. Equivariant homotopy theory and Milnor's theorem. Trans. Amer. Math. Soc. 258 (1980), 351-368.

DEPARTMENT OF MATHEMATICS
UNIVERSITY OF CHICAGO
CHICAGO, IL 60637

and

DEPARTMENT OF MATHEMATICS
UNIVERSITY OF CHICAGO
CHICAGO, IL 60637

and

UNIVERSITY OF OXFORD - MATH INSTITUTE
24-29 ST. GILES
OXFORD
ENGLAND

Contemporary Mathematics
Volume 19, 1983

COHOMOLOGY EXTENSIONS IN CERTAIN 2-STAGE POSTNIKOV SYSTEMS

Wen-Hsiung Lin*

Let $i \geqq 4$. Consider the following 2-stage Postnikov system

$$(1)\qquad \begin{array}{ccc} \Sigma^{-1}L & \xrightarrow{j} & E \\ & & \downarrow p \\ & & K \xrightarrow{f=(Sq^2, Sq^2 Sq^1, Sq^4, \cdots, Sq^{2^i})} L \end{array}$$

where K denotes the Eilenberg-Maclane spectrum $K(\mathbb{Z}_2)$ and $L = \Sigma^2 K \times \Sigma^3 K \times \Sigma^4 K \times \cdots \times \Sigma^{2^i} K$. It is easy to see that $H^*(K)/\mathrm{im} f^*$ is generated by 1, Sq^1, $Sq^{2^{i+1}}$ and $Sq^{2^{i+1}+1}$ for $* \leqq 2^{i+1} + 1$ where $H^*(\ )$ is the mod 2 cohomology functor. This group is embedded in $H^*(E)$ via p^*. It is also easy to see that any minimal set of generators of $H^*(E)$ over the mod 2 Steenrod algebra A for $3 \leqq * \leqq 2^{i+1} - 1$ is one-to-one correspondent with a $\mathbb{Z}_2$-basis of $\mathrm{Ext}_A^{2,*+1}(H^*(S^0 \cup_{2\iota} e^1), \mathbb{Z}_2)$ where S^0 is the sphere spectrum and $S^0 \cup_{2\iota} e^1$ is the mapping cone of $2\iota : S^0 \to S^0$. Let $h_k \in \mathrm{Ext}_A^{1,k}(H^*(S^0), \mathbb{Z}_2)$ be the classes corresponding to the generators $Sq^{2^k} \in A$; their images in $\mathrm{Ext}_A^{1,*}(H^*(S^0 \cup_{2\iota} e^1), \mathbb{Z}_2)$ are also denoted by h_k. Adams ([1]) has shown that $h_i^2 h_1 \neq 0$ in $\mathrm{Ext}_A^{3,2^{i+1}+2}(H^*(S^0 \cup_{2\iota} e^1), \mathbb{Z}_2)$

AMS Subject Classifications: 55S45, 55T15.

* Supported by the National Science Council, Republic of China.

(since $i \geqq 4$). This implies

$$j^*(Sq^2\psi_{i,i} + Sq^{2^i}\psi_{i,1} + \cdots) = 0$$

where $\psi_{m,\ell}$ are the classes in $H^*(E)$ corresponding to the Adams invariants $h_m h_\ell \in \mathrm{Ext}_A^{2,*}(H^*(S^0 \cup_{2\iota} e^1), \mathbb{Z}_2)$ (in dimensions $\leqq 2^{i+1} - 1$). So

$$(2) \qquad Sq^2\psi_{i,i} + Sq^{2^i}\psi_{i,1} + \cdots = \lambda Sq^{2^{i+1}+1}$$

for some coefficient λ which is 0 or 1. In this note we prove

<u>Theorem A</u>. $\lambda = 1$ for all $i \geqq 4$.

This extension has been known to Mark Mahowald for years if the strong Kervaire invariant conjecture is true (see Proposition B). This result gives a strong evidence for the conjecture that $h_i^2 h_1$ survives the Adams spectral sequence for $S^0 \cup_{2\iota} e^1$.

The result <u>A</u> is analogous to the one proved by Adams in [1]. He considered, for each $i \geqq 3$, a 2-stage Postnikov system obtained by replacing f (and hence L accordingly) in (1) by $f' = (Sq^1, Sq^2, \cdots, Sq^{2^i})$ and from this system got an equation similar to (2) which is obtained by replacing Sq^2, $\psi_{i,1}$ and $Sq^{2^{i+1}+1}$ in (2) by Sq^1, $\psi_{i,0}$ and $Sq^{2^{i+1}}$ respectively, and proved $\lambda = 1$ in this equation for all $i \geqq 3$. Adams' result is of course the well known theorem on the Hopf invariant. The consideration of the extensions (2) comes from the Kervaire invariant conjecture for which we refer to J. Jones' talk at this conference or to [4]. In particular we recall that the strong Kervaire invariant conjecture asserts that $h_{i+1} \in \mathrm{Ext}_A^{1,2^{i+1}}(H^*(S^0 \cup_{2\iota} e^1), \mathbb{Z}_2)$ detects homotopy elements in $\pi_{2^{i+1}-1}(S^0 \cup_{2\iota} e^1)$ for each $i \geqq 3$. This conjecture is known to be true for $i = 3$ and for $i = 4$ ([3]).

Proposition B. If $h_{i+1} \in Ext_A^{1,2^{i+1}}(H^*(S^0 \cup_{2\iota} e^1)$ detects elements in $\pi_{2^{i+1}-1}(S^0 \cup_{2\iota} e^1)$, where $i \geq 4$, then the coefficient λ is 1.

Our work is to show $\lambda = 1$ without the assumption on h_{i+1}. But let us prove B first.

Proof: Let $\{h_{i+1}\} : S^{2^{i+1}-1} \to S^0 \cup_{2\iota} e^1$ be a homotopy class detected by $h_{i+1} \in Ext_A^{1,2^{i+1}}(H^*(S^0 \cup_{2\iota} e^1), \mathbb{Z}_2)$. Barratt and Mahowald ([4]) have shown that the composite

$$S^{2^{i+1}-1} \xrightarrow{\{h_{i+1}\}} S^0 \cup_{2\iota} e^1 \longrightarrow S^1 \text{ is a } \theta_i \in \pi_{2^{i+1}-1}(S^1)$$

with $2\theta_i = 0$ where θ_i is detected by h_i^2 in $Ext_A^{2,2^{i+1}+1}(H^*(S^1), \mathbb{Z}_2)$. Let $\bar{\theta}_i$ be the image in $\pi_{2^{i+1}-1}(S^0 \cup_{2\iota} e^1)$ of the desuspension of θ_i under the inclusion $S^0 \to S^0 \cup_{2\iota} e^1$. Then $\eta\bar{\theta}_i = 2\{h_{i+1}\}$; this is due to the fact that $\eta\theta_i$ is contained in the Toda bracket $\langle 2\iota, \theta_i, 2\iota \rangle$ ([5], 3.7). Note that $\eta\bar{\theta}_i$ is detected by $h_i^2 h_1$ in $Ext_A^{3,2^{i+1}+2}(H^*(S^0 \cup_{2\iota} e^1), \mathbb{Z}_2)$.

Consider the mapping cone $X = S^0 \cup_{2\iota} e^1 \cup e^{2^{i+1}}$ of $\{h_{i+1}\}$. The Steenrod operation $Sq^{2^{i+1}}$ is non-zero in $H^*(X)$. From this one verifies that there is only one non-zero class in $Ext_A^{1,2^{i+1}+1}(H^*(X), \mathbb{Z}_2)$ and that this class corresponds to $Sq^{2^{i+1}+1} \in H^*(E)$ in the system (1); we denote it by γ. It is easy to see that the image $\widetilde{h_i^2 h_1}$ of $h_i^2 h_1$ in $Ext_A^{3,2^{i+1}+2}(H^*(X), \mathbb{Z}_2)$ is non-zero and is a permanent cycle in the Adams spectral sequence for X (since $h_i^2 h_1$ is for $S^0 \cup_{2\iota} e^1$). If this class is non-zero in E_∞ then the image in $\pi_{2^{i+1}-1}(X)$ of $\eta\bar{\theta}_i$ projects to it. This is impossible since $\eta\bar{\theta}_i = 2\{h_{i+1}\}$. Thus $\widetilde{h_i^2 h_1}$ is a boundary, and the only possibility is that $d_2(\gamma) = \widetilde{h_i^2 h_1}$ which means exactly that λ in (2) is 1. Q. E. D.

The remainder of this note is devoted to the proof of Theorem A. Mahowald and Tangora ([3]) have shown that $h_5 \in \mathrm{Ext}_A^{1,32}(H^*(S^0 \cup_{2\iota} e^1), \mathbb{Z}_2)$ detects homotopy elements in $\pi_{31}(S^0 \cup_{2\iota} e^1)$. So, by Proposition B, $\underline{A}$ is true for $i = 4$. We may thus assume $i \geqq 5$. In this case $h_i^2 h_1^2 \neq 0$ in $\mathrm{Ext}_A^{4,2^{i+1}+4}(H^*(S^0), \mathbb{Z}_2)$ which can be proved by using the method of Mahowald and Tangora in [2]. Using Lemma 2.6.1 of [1] one easily calculates $\mathrm{Ext}_A^{s,t}((S^0 \cup_{2\iota} e^1), \mathbb{Z}_2)$ for $2^{i+1}-2 \leqq t-s \leqq 2^{i+1}+1$, $1 \leqq s \leqq 4$ to be

(3)

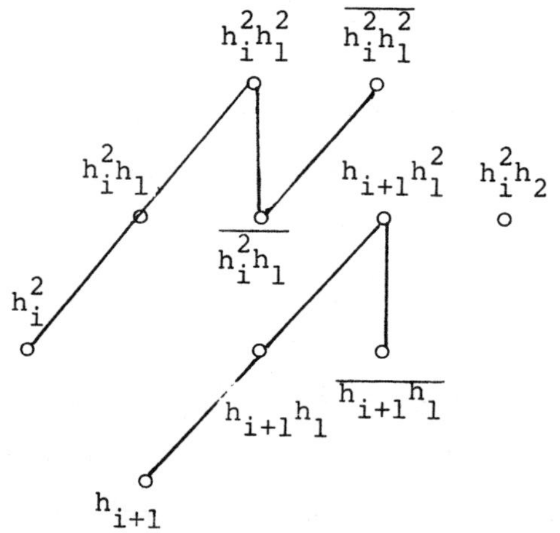

which excludes elements not related to the h_j's (if there are any). Note that $h_0\overline{h_i^2h_1} = h_i^2h_1^2$.

<u>Proposition C</u>. $d_2(\overline{h_{i+1}h_1}) = h_i^2h_1^2$ in the Adams spectral sequence for $S^0 \cup_{2\iota} e^1$.

<u>Proof</u>. Let η be the generator of $\pi_1(S^0) = \mathbb{Z}_2$; it is detected by $h_1 \in \mathrm{Ext}_A^{1,2}(H^*(S^0),\mathbb{Z}_2)$ and has order 2. Let $\tilde{\eta} : S^2 \to S^0 \cup_{2\iota} e^1$ be a coextension of $\Sigma\eta$ and Y the fiber of the nontrivial map $S^0 \cup_{2\iota} e^1 \to K = K(\mathbb{Z}_2)$. There is a unique lifting $\bar{\eta} : S^2 \to Y$ of $\tilde{\eta}$. The induced homomorphism

$$\bar{\eta}_*: \mathrm{Ext}_A^{*,*}(H^*(S^2),\mathbb{Z}_2) \longrightarrow \mathrm{Ext}_A^{*,*}(H^*(Y),\mathbb{Z}_2)$$

maps h_{i+1} to $\overline{h_{i+1}h_1}$ and h_i^2 to $\overline{h_i^2h_1}$. Adams theorem on the Hopf invariant ([1]) implies $d_2(h_{i+1}) = h_i^2h_0$. Since $\bar{\eta}_*$ commutes with d_2 it follows that $d_2(\overline{h_{i+1}h_1}) = h_0\overline{h_i^2h_1} = h_i^2h_1^2$.

Q. E. D.

We will interpret the result of *Proposition C* in the following 3-stage Postnikov system

$$\begin{array}{ccccc}
\Sigma^{-1} L_1 & \xrightarrow{j_2} & E_2 & & \\
 & & \downarrow p_2 & & \\
\Sigma^{-1} L_0 & \xrightarrow{j_1} & E_1 & \xrightarrow{f_1} & L_1 \\
 & & \downarrow p_1 & & \\
 & & K & \xrightarrow{f_0=(Sq^2,Sq^2Sq^1,Sq^4,\cdots Sq^{2^i},Sq^{2^{i+1}})} & L_0
\end{array} \tag{4}$$

which corresponds to a minimal Adams resolution of $S^0 \cup_{2\iota} e^1$ in a range of dimensions and is described as follows.

The set of the classes defining f_0 corresponds to the $\mathbb{Z}_2$- basis $\{h_1,\bar{h}_1,h_2,\cdots,h_i,h_{i+1}\}$ of $\mathrm{Ext}_A^{1,*}(H^*(S^0 \cup_{2\iota} e^1),\mathbb{Z}_2)$ for $* \leqq 2^{i+1}$. Let B_1 be a $\mathbb{Z}_2$-basis of $\mathrm{Ext}_A^{2,*}(H^*(S^0 \cup_{2\iota} e^1),\mathbb{Z}_2)$ for $* \leqq 2^{i+1} + 3$, of which h_i^2, $h_{i+1}h_1$ and $\overline{h_{i+1}h_1}$ in diagram (3) are the last three basis elements. Then B_1 corresponds to a minimal set $\bar{B}_1$ of generators of the A-module $H^*(E_1)$ for $3 \leqq * \leqq 2^{i+1} + 2$. Let $\psi_{i,i}$, $\psi_{i+1,1}$ and $\bar{\psi}_{i+1,1}$ be the classes in $\bar{B}_1$ corresponding to h_i^2 , $h_{i+1}h_1$ and $\overline{h_{i+1}h_1}$ respectively.

Let $\tilde{h}_{i+1}$ be the generator of $H^{2^{i+1}-1}(\Sigma^{2^{i+1}-1}K) = \mathbb{Z}_2$ where $\Sigma^{2^{i+1}-1}K$ is the last factor of $\Sigma^{-1}L_0$. Let $\bar{j}_1 : \Sigma^{2^{i+1}-1}K \to E_1$ be the composite $\Sigma^{2^{i+1}-1}K \to \Sigma^{-1}L_0 \xrightarrow{j_1} E_1$. Then

$$\bar{j}_1^*(\bar{\psi}_{i+1,1}) = Sq^2Sq^1\tilde{h}_{i+1} \tag{5}$$

which is easily verified.

The f_1 in (4) is defined to be the map which kills all the classes in $\bar{B}_1$ except $\bar{\psi}_{i+1,1}$. Then the group $H^*(E_1)/\mathrm{im} f_1^*$, which is embedded in $H^*(E_2)$ via p_2^*, is generated by 1, Sq^1 and $\bar{\psi}_{i+1,1}$ for $* \leq 2^{i+1}+3$. Let B_2 be a $\mathbb{Z}_2$-basis of $\mathrm{Ext}_A^{3,*}(H^*(S^0 \cup_{2\iota} e^1),\mathbb{Z}_2)$ for $* \leq 2^{i+1}+3$ of which $h_i^2h_1$ and $\overline{h_i^2h_1}$ in diagram (3) are the last two basis elements. Then B_2 corresponds to a minimal set $\bar{B}_2$ of generators of the A-modules $H^*(E_2)$ for $3 \leq * \leq 2^{i+1}+1$. Let $\psi_{i,i,1}$ and $\bar{\psi}_{i,i,1}$ be the classes in $\bar{B}_2$ corresponding to $h_i^2h_1$ and $\overline{h_i^2h_1}$ respectively.

Since $h_i^2h_1^2 \neq 0$ in $\mathrm{Ext}_A^{4,2^{i+1}+4}(H^*(S^0 \cup_{2\iota} e^1),\mathbb{Z}_2)$, there is a relation

$$j_2^*(Sq^2\psi_{i,i,1} + Sq^1\bar{\psi}_{i,i,1} + \cdots) = 0$$

which implies that there is an equation

$$Sq^2\psi_{i,i,1} + Sq^1\bar{\psi}_{i,i,1} + \cdots = \lambda' p_2^*(\bar{\psi}_{i+1,1}) \tag{6}$$

in $H^*(E_2)$. The interpretation of the result of *Proposition* C in system (4) is that the coefficient λ' in equation (6) is 1.

To prove *Theorem* A we consider the following Postnikov system

$$
(7)\qquad
\begin{array}{ccccccc}
\Sigma^{-1} L_1' & \xrightarrow{j_2'} & E_2' & & & & \\
 & & \downarrow & & & & \\
\Sigma^{-1} L_0' & \xrightarrow{j_1'} & E_1' & \xrightarrow{f_1'} & L_1' & & \\
 & & \downarrow & & & & \\
 & & K & & \xrightarrow{f_0'=(Sq^2, Sq^2Sq^1, Sq^4, \cdots, Sq^{2^i}, Sq^{2^{i+1}+1})} & & L_0'
\end{array}
$$

which is described as follows.

Let M be the A-module

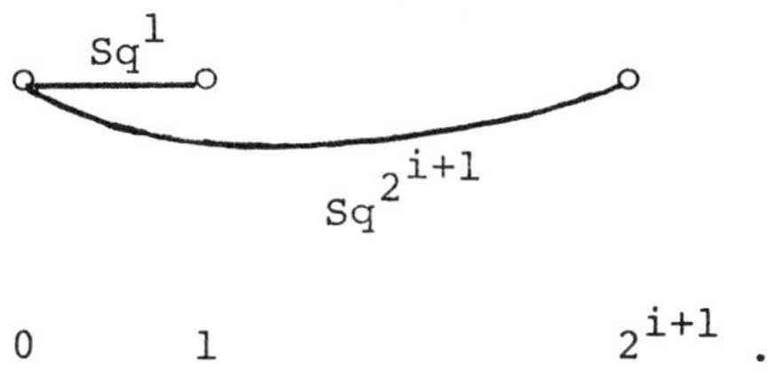

So $M \cong H^*(X)$ where X is as in the proof of *Proposition B*. The set of the classes defining f_0' corresponds to the $\mathbb{Z}_2$-basis $\{h_1, \bar{h}_1, h_2, \cdots, h_i, \gamma\}$ of $\mathrm{Ext}_A^{1,*}(M, \mathbb{Z}_2)$ for $* \leqq 2^{i+1} + 3$ and in this range of dimensions there is a decomposition of A-modules $H^*(E_1') = H^*(S^0 \cup_{2\iota} e^1) \oplus \mathrm{Ker}\ \tau \oplus \Sigma^{2^{i+1}} \mathbb{Z}_2$ where $\tau : H^*(\Sigma^{-1} L_0') \to H^*(K)$ is the transgression and $Sq^{2^{i+1}}$ generates $\Sigma^{2^{i+1}} \mathbb{Z}_2$. That the last factor splits off $H^*(E_1')$ is due to the fact that, in the Adams spectral sequence for $S^0 \cup_{2\iota} e^1$, $d_2(h_{i+1}) = 0$ since $\mathrm{Ext}_A^{3, 2^{i+1}+1}(H^*(S^0 \cup_{2\iota} e^1), \mathbb{Z}_2) = 0$.

It is easy to see that $\mathrm{Ext}_A^{2,*}(M, \mathbb{Z}_2) \cong \mathrm{Ext}_A^{2,*}(H^*(S^0 \cup_{2\iota} e^1), \mathbb{Z}_2)$ for $* \leqq 2^{i+1} + 3$ except $* = 2^{i+1} + 2$ in which case $\mathrm{Ext}^{2, 2^{i+1}+2}(M, \mathbb{Z}_2) = 0$; thus $B_1' = B_1 - \{h_{i+1}h_1\}$ is a $\mathbb{Z}_2$-basis of $\mathrm{Ext}_A^{2,*}(M, \mathbb{Z}_2)$ for $* \leqq 2^{i+1} + 3$, of which h_i^2 and $\overline{h_{i+1}h_1}$ are the last two basic elements. B_1' corresponds to a minimal set $\bar{B}_1'$ of generators of the A-module $(\mathrm{Ker}\ \tau)^*$ for $* \leqq 2^{i+1} + 2$.

Let $\psi'_{i,i}$ and $\bar{\psi}'_{i+1,1}$ be the classes in $\bar{B}'_1 \subset H^*(E'_1)$ corresponding to h_i^2 and $\overline{h_{i+1}h_1}$ respectively.

The f'_1 in (7) is defined to be the map which kills all the classes in $\bar{B}'_1$ except $\bar{\psi}'_{i+1,1}$. Then $H^*(E'_1)/\mathrm{im}(f'_1)^*$ is generated by 1, Sq^1 and $\bar{\psi}'_{i+1,1}$ for $* \leqq 2^{i+1} + 2$. It is easy to see that $\mathrm{Ext}_A^{3,*}(M,\mathbb{Z}_2) \cong \mathrm{Ext}_A^{3,*}(H^*(S^0 \cup_{2\iota} e^1),\mathbb{Z}_2)$ for $* \leqq 2^{i+1} + 3$; so $B'_2 = B_2$ is a $\mathbb{Z}_2$-base for $\mathrm{Ext}_A^{3,*}(M,\mathbb{Z}_2)$ in this range of dimensions. B'_2 corresponds to a minimal set $\bar{B}'_2$ of generators of the A-modules $H^*(E'_2)$ for $3 \leqq * \leqq 2^{i+1} + 1$. Let $\psi'_{i,i,1}$ and $\bar{\psi}'_{i,i,1}$ be the classes in $\bar{B}'_2$ corresponding to $h_i^2h_1$ and $\overline{h_i^2h_1}$ respectively.

Let $\tilde{\gamma}$ be the generator of $H^{2^{i+1}}(\Sigma^{2^{i+1}}K) = \mathbb{Z}_2$ where $\Sigma^{2^{i+1}}K$ is the last factor of $\Sigma^{-1}L'_0$. It is clear that the composite $\Sigma^{2^{i+1}}K \xrightarrow{j} \Sigma^{-1}L'_0 \xrightarrow{j'_1} E'_1 \xrightarrow{f'_1} L'_1$ is trivial; so there is a lifting $j''_1 : \Sigma^{2^{i+1}}K \to E'_2$ of $\bar{j}'_1 = j'_1 \circ j$ as indicated in the following portion of diagram (7).

(8)
$$\begin{array}{ccccc} & & E'_2 & & \\ & \overset{j''_1}{\nearrow} & \downarrow p'_2 & & \\ \Sigma^{2^{i+1}}K & \xrightarrow{\bar{j}'_1} & E'_1 & \xrightarrow{f'_1} & L'_1 \end{array}$$

To prove that the coefficient λ in equation (2) is 1 is equivalent to proving

$$(j''_1)^*(\psi'_{i,i,1}) = \tilde{\gamma}.$$

For this purpose we map diagram (4) to diagram (7) by defining $\phi_0 : K \to K$ to the identity map and $g : L_0 \to L'_0$ to be the map $g = (\mathrm{id},\mathrm{id},\cdots,\mathrm{id},Sq^1)$:

$$L_0 = \Sigma^2 K \times \Sigma^3 K \times \Sigma^4 K \times \cdots \times \Sigma^{2^i} K \times \Sigma^{2^{i+1}} K$$

$$\longrightarrow L_0' = \Sigma^2 K \times \Sigma^3 K \times \Sigma^4 K \times \cdots \times \Sigma^{2^i} K \times \Sigma^{2^{i+1}+1} K.$$

Then $(\Sigma^{-1}g)^*(\tilde{\gamma}) = Sq^1 \tilde{h}_{i+1}$. The maps ϕ_0 and g determine a map from diagram (4) to diagram (7). Let $\phi_i : E_i \to E_i'$, $i = 1,2$, be the induced maps. Then $\phi_1^*(\bar{\psi}'_{i+1,1}) = \bar{\psi}_{i+1,1}$, $\phi_2^*(\psi'_{i,i,1}) = \psi_{i,i,1}$ and $\phi_2^*(\bar{\psi}'_{i,i,1}) = \bar{\psi}_{i,i,1}$.

From the commutative diagram

$$\begin{array}{ccc} \Sigma^{2^{i+1}-1} K & \xrightarrow{\ \bar{j}_1\ } & E_1 \\ \Big\downarrow {\scriptstyle \Sigma^{-1} g} & & \Big\downarrow {\scriptstyle \phi_1} \\ \Sigma^{2^{i+1}} K & \xrightarrow{\ \bar{j}_1'\ } & E_1' \end{array}$$

and equation (5) we get

$$\begin{aligned} (\Sigma^{-1}g)^*[(\bar{j}_1')^*(\bar{\psi}'_{i+1,1})] &= \bar{j}_1^*[\phi_1^*(\bar{\psi}'_{i+1,1})] \\ &= \bar{j}_1^*(\bar{\psi}_{i+1,1}) \\ &= Sq^2 Sq^1 \tilde{h}_{i+1} \\ &= Sq^2 (\Sigma^{-1}g)^*(\tilde{\gamma}) \\ &= (\Sigma^{-1}g)^*(Sq^2 \tilde{\gamma}). \end{aligned}$$

Since $(\Sigma^{-1}g)^* : H^{2^{i+1}+2}(\Sigma^{2^{i+1}} K) \longrightarrow H^{2^{i+1}+2}(\Sigma^{2^{i+1}-1} K)$ is a monomorphism it follows that

(9) $$(\bar{j}_1')^*(\bar{\psi}'_{i+1,1}) = Sq^2 \tilde{\gamma}.$$

We recall from (6) that

(6)' $$Sq^2 \psi_{i,i,1} + Sq^1 \bar{\psi}_{i,i,1} + \cdots = p_2^*(\bar{\psi}_{i+1,1}) \in H^{2^{i+1}+2}(E_2).$$

Here any term after $Sq^1\,\bar{\psi}_{i,i,1}$ on the left side of the equation is of the form $a_\mu\beta_\mu$ with $\beta_\mu \in \bar{B}_2$, $a_\mu \in A$ and $\dim(a_\mu) > 2$. Let $\beta'_\mu \in \bar{B}'_2 \subset H^*(E'_2)$ be the basic element corresponding to β_μ. Applying ϕ_2^* to (6)' and noting that $\phi_2^* : H^{2^{i+1}+2}(E'_2) \cong H^{2^{i+1}+2}(E_2)$ we see

$$Sq^2\psi'_{i,i,1} + Sq^1\bar{\psi}'_{i,i,1} + \sum_\mu a_\mu\beta'_\mu = (p'_2)^*(\bar{\psi}'_{i+1,1}).$$

Applying $(j''_1)^*$ to this equation and noting the commutative diagram (8) we get

$$Sq^2(j''_1)^*(\psi'_{i,i,1}) + Sq^1(j''_1)^*(\bar{\psi}'_{i,i,1}) + \sum_\mu a_\mu(j''_1)^*(\beta'_\mu)$$

$$= (\bar{j}'_1)^*(\bar{\psi}'_{i+1,1}) = Sq^2\,\tilde{\gamma} \quad \text{(by (9))}.$$

It is clear that $Sq^1(j''_1)^*(\bar{\psi}'_{i,i,1}) = 0$. By dimensional reason $(j''_1)^*(\beta'_\mu) = 0$ all μ. Since

$$Sq^2 : H^{2^{i+1}}(\Sigma^{2^{i+1}}K) \longrightarrow H^{2^{i+1}+2}(\Sigma^{2^{i+1}}K)$$

is an isomorphism it follows that

$$(j''_1)^*(\psi'_{i,i,1}) = \tilde{\gamma}.$$

This completes the proof of Theorem A.

References

1. J. F. Adams, On the non-existence of elements of Hopf invariant one, *Ann. Math.* 72(1960), 20-103.

2. M. E. Mahowald and M. C. Tangora, On secondary operations which detect homotopy classes, *Bol. Soc. Math. Mexicana* (2) 12 (1967), 71-75.

3. M. E. Mahowald and M. C. Tangora, Some differentials in the Adams spectral sequence, *Topology* 6(1967), 349-369.

4. M. E. Mahowald, Some remarks on the Kervaire invariant problem from the homotopy point of view, *Proc. Symp. Pure Math. A. M. S.* Vol. 22, (1971) 165-169.

5. H. Toda, Composition methods in homotopy groups of spheres, *Ann. Math. Studies* No. 49, Princeton University Press, Princeton, 1962.

INSTITUTE OF MATHEMATICS
NATIONAL TSING HUA UNIVERSITY
HSINCHU, TAIWAN 300
REPUBLIC OF CHINA

Contemporary Mathematics
Volume 19, 1983

BORSUK-ULAM THEOREMS FOR SPHERICAL SPACE FORMS

Arunas Liulevicius[1]

ABSTRACT. If G is a nontrivial compact Lie group acting freely on S^m and S^n, and $h: S^m \longrightarrow S^n$ is a G-map, then it is shown that m is at most n. The proof is elementary and is based on mod p homology of the classifying space of a cyclic group.

1. STATEMENT OF RESULTS. The classical Borsuk-Ulam theorem [1] asserts that if $f: S^m \longrightarrow S^n$ is an antipodal map (that is, for each x in S^m we have $f(-x) = -f(x)$), then $m \leq n$. Its immediate corollary is that if $h: S^m \longrightarrow R^m$ is a map then there exists an $x_0 \in S^m$ such that $h(-x_0) = h(x_0)$. We generalize these results to topological spherical space forms and give a proof which simplifies the classical one.

THEOREM 1. If G is a nontrivial compact Lie group acting freely on S^m and S^n and $f: S^m \longrightarrow S^n$ is a G-map, then $m \leq n$.

THEOREM 2. If G is a nontrivial compact Lie group acting freely on S^m and freely and orthogonally on S(W), the unit sphere of a representation W of $\dim_R W \leq m$, then given a map $h: S^m \longrightarrow W$ there exists an $x_0 \in S^m$ such that

$$\int_G gh(g^{-1}x_0)\, dg = 0 ,$$

where $\int_G$ denotes the Haar integral on G.

The compact Lie groups G which can act freely on some S^m are now completely known [6]. If G has positive dimension, then G is isomorphic to one of S^1, N, S^3, where N is the normalizer of S^1 in S^3 ([3], p.153). If G is a finite group, then G acts freely on a suitable S^m if and only if for all primes p all subgroups of G of orders 2p and p^2 are cyclic [6]. For free orthogonal actions on spheres necessary and sufficient conditions are given in [9].

1980 Mathematics Subject Classification 55R40, 57S25, 55M20.

[1] Supported by NSF grant MCS 80-02730.

Let p be a prime and let $C = Z/(p)$ be the cyclic group of order p, F_p the field of p elements. Let C act freely on S^m, let $\pi : S^m \longrightarrow S^m/C$ be the quotient map, $c_\pi : S^m/C \longrightarrow BC$ the classifying map [7] of the principal C-bundle π.

THEOREM 3. The map $c_{\pi *} : H_m(S^m/C;F_p) \longrightarrow H_m(BC;F_p)$ is an isomorphism, and both groups are isomorphic to F_p.

The structure of this paper is: in 2. we show how Theorem 1 implies Theorem 2, in 3. we prove Theorem 3, and in 4. we show how Theorem 3 implies Theorem 1.

The simple proofs are a pleasant memento of the Northwestern homotopy theory conference. Special thanks go to H.Miller, L.Taylor, and C.B.Thomas who all contributed key ideas to simplify the argument. The standard proofs of the Borsuk-Ulam theorem (see [5], Theorem 4.3.6 p.152; [4] p.223) rely on the cup product structure of $H^*(RP^m;F_2)$. The approach given here is more elementary in the sense that cohomology is never needed.

2. THEOREM 1 IMPLIES THEOREM 2 . Consider the vector space $Map(S^m,W)$ of all maps $h:S^m \longrightarrow W$, and let G act on it by setting $(g_*h)(x) = g.h(g^{-1}.x)$. The subspace $Map_G(S^m,W)$ of G-maps is precisely the trivial representation in $Map(S^m,W)$, and the norm map $N : Map(S^m,W) \longrightarrow Map_G(S^m,W)$ defined by $Nh = \int_G g_*h\, dg$ is the projection onto the trivial summand. The standard retraction $r: W - \{0\} \longrightarrow S(W)$ given by $r(w) = \|w\|^{-1}.w$ is a G-map since G acts orthogonally on W. Given a map $h:S^m \longrightarrow W$, we claim that its norm Nh has 0 in its image. If not, the composition $r\circ Nh : S^m \longrightarrow S(W)$ would be a G-map of S^m to a sphere of dimension at most m-1 , contradicting Theorem 1.

3. PROOF OF THEOREM 3 . Since S^m is (m-1)-connected, the classifying map $c_\pi : S^m/C \longrightarrow BC^{(m)}$ is a weak homotopy equivalence [7] to the m-skeleton of the classifying space BC. Indeed, the (m+1)-skeleton of BC is weakly homotopy equivalent to the space $(S^m/C)\cup_\pi e^{m+1}$. Now notice that $H_m(S^m/C;F_p) = F_p$, for S^m/C is a connected topological manifold of dimension m (this does the case p=2), and if p is odd, S^m/C is orientable, for in this case m is odd and each element of C preserves the orientation of S^m. We claim: $\pi_* : H_m(S^m;F_p) \longrightarrow H_m(S^m/C;F_p)$ is the zero map - once we show this, Theorem 3 will follow. It is enough to do the case of the standard action of C on S^m (remember the weak

equivalence!). Here S^m/C is a CW complex with one cell in each dimension between 0 and m. The covering map $\pi: S^m \longrightarrow S^m/C$ induces a CW structure on S^m such that [8] for $0 \leq k \leq m$ we have p k-cells of S^m permuted freely by the group $C = Z/(p)$. If we let C_k denote the cellular k-chains of S^m with coefficients F_p, then C_k is isomorphic to the group ring $F_p[C]$, and the chain map induced by $\pi: S^m \longrightarrow S^m/C$ corresponds to the augmentation $\varepsilon: F_p[C] \longrightarrow F_p$. Since the action of C on $H_m(S^m/C;F_p)$ is trivial, a generator of this group is represented by the norm element $N = \sum_{g\in C} g$ in $F_p[C] \approx C_m$. However $\varepsilon N = 0$, so $\pi_*: H_m(S^m;F_p) \longrightarrow H_m(S^m/C;F_p)$ is the zero map, as was to be shown.

4. THEOREM 3 IMPLIES THEOREM 1. Suppose $f: S^m \longrightarrow S^n$ is a G-map, where G is a nontrivial compact Lie group. Let $C \subset G$ be a cyclic group of order p, p a prime. Then of course f is a C-map, so if we let $\pi: S^m \longrightarrow S^m/C$, $\pi': S^n \longrightarrow S^n/C$ be the quotient maps, then f induces a map $\underline{f}: S^m/C \longrightarrow S^n/C$ such that the following diagram of maps commutes up to homotopy:

$$\begin{array}{ccc} S^m/C & \xrightarrow{\ \underline{f}\ } & S^n/C \\ & {\scriptstyle c_\pi}\searrow \quad \swarrow{\scriptstyle c_{\pi'}} & \\ & BC, & \end{array}$$

where c_π and $c_{\pi'}$ are classifying maps for π and π', respectively. We claim: $m \leq n$. If not, $H_m(S^n/C;F_p) = 0$, for $m > n$ and S^n/C is a topological n-manifold. This is ridiculous, since according to Theorem 3 the map $c_{\pi *}: H_m(S^m/C;F_p) \longrightarrow H_m(BC;F_p)$ is an isomorphism onto a group isomorphic to F_p.

BIBLIOGRAPHY

1. K.Borsuk, "Drei Sätze über die n-dimensionale Euklidische Sphäre," Fund. Math. 20 (1933), 177-190.

2. K.Borsuk, "An application of the theorem on antipodes to the measure theory," Bull. Acad. Polon. Sci. Cl III, 1 (1953), 87-90.

3. G.E.Bredon, Introduction to compact transformation groups, Academic Press, New York and London, 1972.

4. M.J.Greenberg and J.R.Harper, Algebraic Topology, A First Course, The Benjamin/Cummings Publishing Company, Reading, 1981.

5. P.J.Hilton and S.Wylie, Homology Theory, Cambridge University Press, Cambridge, 1960.

6. I.Madsen, C.B.Thomas and C.T.C.Wall, "The topological sphe-

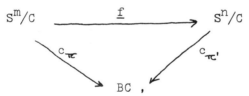

rical space form problem - II. Existence of free actions", Topology 15 (1976), 375-382.

7. N.Steenrod, The Topology of Fibre Bundles, Princeton University Press, Princeton, 1951.

8. J. H.C.Whitehead, "Combinatorial homotopy, I ", Bull. Amer. Math. Soc. 55 (1949), 213-245.

9. J.A.Wolf, Spaces of constant curvature, McGraw-Hill, New York, 1967.

DEPARTMENT OF MATHEMATICS
THE UNIVERSITY OF CHICAGO
CHICAGO, ILLINOIS 60637

Contemporary Mathematics
Volume 19, 1983

ON THE CLASSIFICATION OF G-SPHERES

(An outline)

Ib Madsen and Mel Rothenberg*

§0. Introduction

Let G be a cyclic group of odd order. This outline contains classification results for locally linear G-manifolds in the piecewise linear and topological categories. In one direction, we show that topological conjugate linear representations of groups of odd order are linearly conjugate. In another direction, our results imply a classification of 'fake' PL G-spheres S(V) when V is a topological stable respresentation (that is, when $10 < 2 \dim V^H < \dim V^K$ for $K \subsetneq H \subset G$), generalizing C.T.C. Wall's classification of fake Lens spaces.

Stated in homotopy theoretic terms our key results are stability theorems for the groups $PL_G(V)$ and $Top_G(V)$ of G-homomorphisms of linear representations in the two categories. The results roughly correspond to the stability one would expect from the orthogonal case, but with the change that characters of the same period are considered equal. As a consequence we deduce a (stable) G-transversality result in the Pl category. In the TOP category, G-transversality fails in general (even stably). There are obstructions connnected with K_{-1}-groups, but they are G-cobordism invariants (because $K_{-i}(\mathbb{Z}G) = 0$ for $i > 1$!) and are 2-primary.

*Partially supported by NSF MCS-8002730A03

Generalizing methods used by D. Sullivan in the G-trivial case (cf. [MM]) we show for G cyclic of odd order:

Theorem A. Locally linear PL or TOP G-bundles*) are oriented with respect to $KO_G(-;\mathbb{Z}[\frac{1}{2}])$.

Standard methods (cf. [SC] and §5) then lead to

Theorem B. Topological conjugate representations of groups of odd order are linearly conjugate.

Special cases of Theorem B, namely for p-groups, were shown earlier by D. Sullivan and R. Schultz. They used the G-trivial version of Theorem A and the fact that for p-groups the representation ring injects into its completion. For groups of order 4n with n > 1, Theorem B is false by results due to S. Cappell and J. Shaneson [CS]. Theorem B has been independently proved, using completely different methods by W. C. Hsiang and W. Pardon [HP].

The classification of fake S(V) uses a G-version of the surgery exact sequence [DR]. The PL-case is treated in §3 but we have not included the TOP-case where there are added difficulties, related to the complicated homotopy structure of $TOP_G(V)/PL_G(V)$. In the PL case, normal invariants can be identified with G-homotopy classes $[M/\partial M, F/PL]_G$, and can be calculated from

Theorem C. For cyclic groups of odd order,

$$[M/\partial M, F/PL]_G \otimes \mathbb{Z}[\tfrac{1}{2}] = \sum_{H \subseteq G}^{\oplus} KO_{G/H}(M^H, \partial M^H) \otimes \mathbb{Z}[\tfrac{1}{2}]$$

Here F/PL is the fiber over appropriate G fixed point of natural G map BPL → BF, the classifying spaces of locally linear G fibrations in the

---

*G-bundle means stable G-$\mathbb{R}^N$-bundle, cf. [LR].

respective categories

We shall not give the precise classification result fo 'fake' G-spheres. But we describe a qualitative result which simultaneously generalizes earlier results from [W] and [R]. Let S be a PL G-manifold, G-homotopy equivalent to the unit sphere of a (complex) representation. There are generalizations $\tau(S)$ and $\rho(s)$ of the classical Reidemeister torsion and ℓ-invariant (cf. [R] and §3).

We call the PL G-sphere stable if it satisfies the strong gap condition: For $H_1 \subsetneq H_2 \subset G$ and $S^{H_1} \neq \emptyset$, $S^{H_2} \neq \emptyset$, $\dim S^{H_1} > 2 \dim S^{H_2} \geq 10$.

<u>Theorem D</u>. Stable PL G-spheres are determined up to PL G-homeomorphism by the invariants $\tau(S)$ and $\rho(S)$.

The above results are linked to equivariant transversality in the involved categories, and in geometric terms our basic new result is a stable transversality theorem. In the piecewise linear category transversality holds in the following stable form: Let ξ be a PL G-bundle over Y and $f: M \to \xi^+$ a G-map. It is called <u>stable</u> if for all $H \subset G$ and $x \in M^H$, $y \in Y^H$, $T_xM = V_{x,y} \oplus \xi_y$ where $V_{x,y}$ and ξ_y are stable representations of H, and $\xi_y^H \neq 0$ implies $V_{x,y}^H \neq 0$.

<u>Theorem E</u>. In the locally linear PL-category a stable G-map $f: M \to \xi^+$ is G-homotopic to a map transverse to the zero section.

In the locally linear topological category stable transversality fails in general, but an obstruction theory is developed.

At present we know only very little about these questions when G has even order except for the Cappell-Shaneson results about topological similarity of linear representations. One obvious question would be if there is stable transversality in the PL G-category for all groups G.

At several places below we use not only that G has odd order but also tha it is cyclic. However, this second condition is probably not so serious. In fact, we believe that our results could well be true for odd order groups in general.

Our debt to the work of Dennis Sullivan in the G-trivial case is obvious. We have also been much inspired by the work of D. Anderson and W. C. Hsiang on pseudoisotopies and of course by C. T. C. Wall's work on fake Lens spaces [cf. [AH] and [W]).

In this paper, G will always denote a cyclic group of odd order. The terms G-manifold, G-bundle always mean locally linear G-manifold, G-bundle.

§1. Equivariant bordism and K-theory

We begin with a reformulation of the G-signature theorem (for G cyclic of odd order). Let ξ be an oriented 2ℓ-dimensional G-vector bundle over X. Let $u(\xi)$ be the universal symbol class of the D^+ operator. Given ξ a Riemannian metric, then $u(\xi) \in \tilde{K}_G(\xi^+)$ is the class of

$$0 \longrightarrow \pi^*\Lambda_+(\xi) \longrightarrow \pi^*\Lambda_-(\xi) \longrightarrow 0.$$

Here $\pi:\varepsilon \to X$ and $\Lambda_\pm(\xi)$ = (± 1)-eigenspaces of the operator τ, and $\varphi_v = F_v + F_v^*$ where $F_v(w) = v \wedge w$. ([ASIII, §6]).

Let η be a complex G-vector bundle over X. The total exterior power $\lambda_1(\eta) \in K_G(X)$ is invertible away fromm 2 (because $\lambda_1(\mathbf{C}[G])$ is invertible in $R(G) \otimes \mathbf{Z}[\frac{1}{2}]$).

Definition 1.1. Let ξ be a 2ℓ-dimensional oriented G-vector bundle. Define $\Delta(\xi)$ $u(\xi)/\lambda_1(\varepsilon \otimes \mathbf{C}) \in K_G(\xi^+) \otimes \mathbf{Z}[\frac{1}{2}]$.

The class $\Delta(\xi)$ is a Thom class. It can be considered to lie in $\widetilde{KO}_G^{2\ell}(\xi^+) \times \mathbf{Z}[\frac{1}{2}]$. In fact,

1.2 $$KO_G^{2k}(Y) \otimes \mathbf{Z}[\tfrac{1}{2}] = (1 + (-1)^k\psi^{-1})K_G^0(Y) \otimes \mathbf{Z}[\tfrac{1}{2}],$$

and ψ^{-1} is induced from conjugation. Thus $KO_G^0(pt)$ is the real representation ring and $KO_G^2(pt)$ is the purely imaginary representations (away from 2).

If ξ is the underlying real bundle of a complex bundle η then $\Delta(\xi) = \lambda(\eta)/\lambda_1(\eta)$ where $\lambda(\eta)$ is the usual Thom class. In particular, the Euler class $e(\xi) \in K_G(X)$ associated with $\Delta(\xi)$ is given as $e(\xi) = \prod(1 - L_i)/(1 + L_i)$ when $\xi = \sum L_i$ is the sum of complex line bundles. For a representation W,

1.3 $$e(W) = \prod(1 - \chi_i)/(1 + \chi_i), \quad W = \sum \chi_i.$$

Suppose M is an oriented smooth G-manifold of even dimension 2n. Let $M \subset E$ be G-embedding in a represenation space with normal bundle η. Consider the composition (of degree -2n)

$$\varphi_*:K_G^*(M)_{od} \xrightarrow{\cdot\Delta(\eta)} \tilde{K}_G^*(\eta^+)_{od} \longrightarrow \tilde{K}_G^*(E^+)_{od} \xleftarrow[\cong]{\cdot\Delta(E)} K_G^*(pt)_{od}$$

where $K_G(M)_{od} = K_G(M) \otimes \mathbf{Z}[\frac{1}{2}]$.

We shall use the following variant of the G-signature theorem, easily derived from its usual form as given in [ASI], [ASIII, §6].

<u>Theorem 1.4</u>. For a smooth, oriented G-manifold M without boundary, $\mathrm{Sign}_G(M) = \varphi_*(1)$.

We now turn to equivariant bordism. Let V be a representation of G. Let $BO_k(V)$ denote the Grassman manifold of k-planes in $V \oplus \mathbf{R}^\infty$. It admits an action of G, and the usual bundle $\gamma_k(V)$ over $BO_k(V)$ is a G-vector bundle whose Thom space is denoted $MO_k(V)$. The obvious maps

$$i_k(V,W):W^+\wedge MO_k(V) \to MO_{k+\ell}(V \oplus W), \quad \ell = |W|$$

make $\{MO_k(V)\}$ into a G-spectrum. Let X be a G-space. Define

(1.5
$$\Omega_k^G(X) = \lim_{\to V} [V^+ \wedge S^k, MO_{|V|}(V) \wedge X^+]_G$$
$$\Omega_G^k(X) = \lim_{\to V} [V^+ \wedge X^+, MO_{|V|+k}(V)]_G.$$

For a pointed G-space Y and a representation V we write $\Omega^V Y$ for the pointed maps from V^+ to Y with its usual G-structure, $(g\cdot f)(v) = gf(g^{-1}v)$. Then $MO_k(G) = \lim_{\to} \Omega^V MO_{k+|V|}(V)$ classifies $\Omega_G^k(X)$.

Geometrically, $\Omega_k^G(X,A)$ consists of G-bordism classes of pairs (M,f) where M is an oriented smooth G-manifold and $f:(M,\partial M) \to (X,A)$ is a G-map, cf. [Wa].

The Thom class of 1.1 implies a natural transformation

$$\Delta:\Omega_*^G(X) \to KO_*^G(X)_{od}$$

which for X = pt reduces to the G-signature.

Lemma 1.6. In degrees at least 4, $\Delta:\Omega_*^G \to KO_*^G \otimes \mathbf{Z}[\frac{1}{2}]$ is surjective.

The proof is based upon explicit use of the G-signature theorem for certain G-surfaces, cf. [K]. For G of prime order, 1.6 can be found in [E], and the general case is similar.

Let $BO(G)_{od}$ be the classifying space for $KO_G(-)_{od}$. There is a fibration (of G-spectra)

1.7
$$F \longrightarrow MO_{-4}(G)_{od} \xrightarrow{\Delta} BO(G)_{od}$$

where Δ represents $\Delta(\gamma) \in K_G(MO_{|V|}(V))$. Here is the key lemma for relating equivariant bordism to equivariant K-theory, cf. [CF1] for the G-trivial case.

Lemma 1.8. The map Δ in 1.7 has a section, and

$$\Omega_{*+4}^G(Z)_{od} \cong KO_*^G(Z)_{od} \oplus F_*^G(Z)$$

where $F_*^G(Z)$ is the homology theory defined from the fibre in 1.7.

Proof (outline). Let $\pi: BO(G)_{od} \to BF$ classifying 1.7. Consider the usual stratification $X_1 \subset X_2 \subset \cdots$ of $X = BO(G)_{od}$ with $X_1 = X^G$ and $X_i - X_{i-1} = X_{(H)}$, $H = H_i$ (see e.g. [tD, §8]). Assuming $\pi | X_{i-1} \simeq 0$ the obstructions to deform π over H_i to 0 lies in the Bredon cohomology groups of the $\bar{G}$-pair (X_i^H, X_{i-1}^H) with coefficients in the $\bar{G}$-coefficient system $\pi_*(BF^H)$, $\bar{G} = G/H$. Using the $\bar{G}$-homotopy equivalence $X_i^H/X_{i-1}^H \simeq E\bar{G} \times X_i^H / E\bar{G} \times X_{i-1}^H$, we can calculate the Bredon cohomology groups in question to be

$$\sum_{j=0}^{n} H^j(B\bar{G}) \otimes H^{n-j}(X_i^H, X_{i-1}^H; \pi_n(BF^H)).$$

From 1.7 it follows that $\pi_n(BF^H) = 0$ for n even, so the obstructions to a G-homotopy $\pi | X_i \simeq 0$ lie in $H^{2k+1}(X_i^H, X_{i-1}^H; \pi_n(BF^H))$. This group maps injectively under Δ^*. But the obstructions also lie in $\operatorname{Ker}(\Delta^*)$, hence they must vanish. The rest of 1.8 is obvious.

Theorem 1.9. The homomorphism $\Delta: \Omega_*^G(X) \to KO_*^G(X) \otimes \mathbf{Z}[\frac{1}{2}]$ factors to give an isomorphism $\Omega_*^G(X) \otimes_{\Omega_*^G} (KO_*^G \otimes \mathbf{Z}[\frac{1}{2}]) \xrightarrow{\cong} KO_*^G(X) \otimes \mathbf{Z}[\frac{1}{2}]$.

Proof (outline). The argument is similar to the G-trivial case from [CF1]. First note by equivariant S-duality ([Wi]): to each $\alpha \in \Omega_*^G(X)$ there exists a pair (Y, f) where Y is a Thom space over a Grassmanian, and $f: Y \to X$ is a G-map such that $\alpha \in \operatorname{Image}(\Omega_*^G(Y) \to \Omega_*^G(X))$. Thus it suffices to prove 1.9 for such Y.

Consider the G-Atiyah-Hirzebruch spectral sequence

$$H_*^G(Y; \underline{\underline{\Omega}}_*^G) \Longrightarrow \Omega_*^G(Y)$$

where $\underline{\underline{\Omega}}_*^G$ is the covariant coefficient system with $\underline{\underline{\Omega}}_*^G(G/H) = \Omega_*^H$. Let $Y_1 \subset Y_2 \subset \cdots$ be the stratification (with $Y_i - Y_{i-1} = Y_{(H_i)}$ as in 1.8). Since Y is a Thom space as described it is "even-dimensional" in the following sense:

(i) $H_*(Y_i^{H_i})$ is concentrated in even dimensions

(ii) $Y_{i-1}^{H_i} \to Y_i^{H_i}$ has a (non-equivariant) section.

Using reduction to one-strata spaces (as in the proof of 1.8) one finds $H_{2n+1}(Y_i^H/G) = 0 \quad (H = H_i)$, and the push-out diagram

$$\begin{array}{ccc} C_*^G(Y_{i-1};\ \underline{\underline{\Omega}}_*^G) & \longrightarrow & C_*^G(Y_i;\ \underline{\underline{\Omega}}_*^G) \\ \uparrow & & \uparrow \\ C_*(Y_{i-1}^H/G;\ \Omega_*^H) & \longrightarrow & C_*(Y_i^H/G;\Omega_*^H) \end{array}$$

gives inductively $H_{2n+1}^G(Y_i,\ \underline{\underline{\Omega}}_*^G) = 0$ for all i. Thus the spectral sequence collapses for degree reasons. Moreover, inductive use of the exact sequences

$$H_{2n}(Y_{i-1}^H/G) \otimes \Omega_*^H \to H_{2n}^G(Y_{i-1};\underline{\underline{\Omega}}_*^G) \oplus H_{2n}(Y_i^H/G) \otimes \Omega_*^H \to H_{2n}^G(Y_i;\underline{\underline{\Omega}}_*^G) \to 0$$

shows that $H_*^G(Y;\underline{\underline{\Omega}}_*^G) \otimes_{\Omega_*^G} KO_*^G \cong H_*^G(Y;\underline{\underline{KO}}^G{}_*)$ (away from 2). This completes the outline.

From [B], we recall the universal coefficient theorem for equivariant K-theory:

<u>Theorem 1.10</u>. For a finite G-CW complex, there is an exact sequence

$$0 \to \mathrm{Ext}^1_{K_G^*}(K_*^G(X), K_*^G) \to K_G^*(X) \to \mathrm{Hom}_{K_G^*}(K_*^G(X), K_*^G) \to 0$$

Away from 2, $KO_G^*(X)$ is a factor in $K_G^*(X)$ and 1.10 gives an analogous exact sequence for $KO_G^*(X) \otimes \mathbf{Z}[\frac{1}{2}]$. Note also the isomorphism

$$\mathrm{Ext}^1_{K_G^*}(K_*^G(X),\ K_*^G) \cong \mathrm{Ext}^1_{\mathbf{Z}}(K_G^*(X);\mathbf{Z}).$$

This is a finite group. For a homomorphism $\sigma:\Omega_*^G(X) \to KO_*^G \subset R(G)$ which satisfies

$$\sigma(\alpha \cdot [M]) = \sigma(\alpha)\ \mathrm{sign}_G(M), \quad M \in \Omega_*^G \text{ and } \alpha \in \Omega_*^G(X)$$

we get from 1.9 and 1.10 an element $\sigma \in K_G^*(X)$, well-defined up to finite ambiguity. We shall use this procedure also for infinite G-CW complexes, and must then appeal to an equivariant version of Milnor's $\varprojlim$-sequence, [Mil].

We close with a convenient 'adjoining' result. For $H \subseteq G$, let $H^{\perp} = \{K \subset G \mid K \not\supset H\}$. Let $E(H^{\perp})$ be the terminal object in the category of spaces with isotropy subgroups in $H^{\perp}$, i.e., spaces X such that $X^H = 0$. Explicitly $E(H^{\perp}) = \ast E(G/K)$, $K \in H^{\perp}$. Using equivariant obstruction theory one can easily prove

Lemma 1.11]. Let X be a based G-CW complex and Y a based G/H CW-complex. Then $[X^H, Y]_{G/H} \cong [X, E(H^{\perp}) * Y]_G$.

§2. Equivariant transversality.

We consider the G-transversality question in the locally linear category of PL or TOP G-manifolds. The general theory is the same in the two cases, so to avoid unnecessary repetition we write CAT for either case. In particular, if V is a representation CAT(V) will denote the (simplicial) space of CAT-automorphisms. It has the usual G-action and $CAT_G(V)$ is the fixed set.

Let ξ be a CAT G-bundle over a CAT C-manifold Y, and assume for convenience that Y^H is connected for all H. Let ξ^+ denote the Thom space.

A singular G-manifold $f: M \to \xi^+$ is called transverse to Y if for each $x \in f^{-1}(Y)$ there is a decomposition of G_x-representations $T_xM = T_x^b \oplus T_x^v$, a G_x-embedding $(T_xM, 0) \subset (M, x)$, and a CAT G-bundle map isomorphic on fibers

$$\begin{array}{ccc} T_xM & \xrightarrow{\hat{f}} & \xi \\ \downarrow & & \downarrow \\ T_x^b & \xrightarrow{f|} & Y \end{array}$$

with $\hat{f} = f$ near the zero section. (Here $T_x^b \subset T_xM \subset M$.)

When f is transversal to Y then $f^{-1}(Y) = N$ is a CAT G-manifold, and there exists a CAT G-bundle $\nu(M,N)$ over N together with an embedding $\nu(M,N) \subset M$

(rel N) and a bundle map $\hat{f}:\nu(M,N) \to \xi$ (over f) agreeing with f near the zero section

The existence of sufficiently many transverse maps (under suitable stability conditions) is based on the equivariant submersion theorem, which we now recall.

Let $T = W \oplus V$ be representations. The $CAT(T) \times CAT(W)$ orbit of the linear projection $p:T \to W$ is denoted $Epi(T,W)$. Its G-fixed points are determined (up to homotopy) by

(2.1) $$Epi_G(T,W) \simeq CAT_G(T)/CAT_G(V).$$

Similarly, for CAT G-bundles τ_i over U_i $(i = 1,2)$, $EPI(\tau_1,\tau_2)$ denotes the space of G-bundle maps $\hat{f}:\tau_1 \to \tau_2$, $f:U_1 \to U_2$, with $\hat{f}_u \in Epi((\tau_1)_u(\tau_2)_{f(u)})$. The component over a fixed homotopy class f can be considered as the space of sections in a G-bundle over U_1 with fibres $Epi_{G_u}((\tau_1)_u,(\tau_2)_{f(u)})$.

Suppose U_1,U_2 are CAT G-manifolds. Let $Sub(U_1,U_2)$ denote the space of G-submersions. There is a natural map (the 'differential')

$$d:Sub(U_1,U_2) \to EPI(TU_1,TU_2).$$

In analogy with [LR, §5 Theorem 2] Lashof [L] has proved the following submersion theorem.

<u>Theorem 2.2</u>. For CAT G-manifolds U_1, U_2 with U_1^H nonclosed for each $H \subseteq G$, the map d is a homotopy equivalence.

Given $f:U_1 \to U_2$, it can be deformed into a G-submersion if and only if a certain bundle section can be constructed. The obstructions to such a section lie in the Bredon-cohomology groups:

2.3 $$\sigma_n(f) \in H^n(U_1; \{\pi_{n-1}(Epi_{G_x}(T_xU_1,T_{f(x)}U_2))\})$$

(cf. [Br. §2]). We need stability conditions in order to calculate these groups.

A representation V is called topological stable if

2.4 $$10 < 2 \dim V^H < \dim V^K$$

for each pair of subgroups $H \supsetneq K$ of G with $V^H \neq V^K$. Write $|V^H|$ for the real dimension of V^H. For CAT = PL we have

Theorem 2.5. Suppose $V \subsetneq T$ are topological stable representations of G. If V and T have the same isotropy subgroups then $PL_G(T)/PL_G(V)$ is $(|V^G|+1)$-connected.

The situation in the locally linear topological category is more complicated as $K_{-1}(\mathbf{Z}G)$ enters into the picture, which means there is no assumption in terms of dimension, which assures stable topological transversality obstructions vanish. To express the results conveniently we introduce the homogeneous space

$$R_G(V) = TOP_G(V)/PL_G(V).$$

Homotopy theoretic properties of this quotient were originally considered by D. Anderson and W. C. Hsiang.

At this point we must assume that V is so-called superstable; this is a technical concept defined in 4.7 below. To every representation V there exist sufficiently large representations W so that $V \oplus W$ is superstable, but at present we don't know the exact conditions on V, say in terms of the dimensions of the fixed points sets V^H to guarantee superstability.

Let V be a representation with $|V^G| = k$. for $i \leq k-2$, $\pi_i(R_G(V \oplus \mathbf{R}), R_G(V)) = 0$. In the range $i \leq k+1$ the groups depend only on the lattice of subgroups of G, not on the actual V, provided V is superstable.

For superstable representations we write

$$\mathcal{K}_{-1}(G) = \pi_{k-1}(R_G(V \oplus \mathbf{R}), R_G(V)).$$

If V is semi-free then $\mathcal{K}_{-1}(G) = K_{-1}(\mathbf{Z}G)$. In any case, consider the chain complex

$$2.6 \quad \cdots \to \pi_k(R_G(V \oplus \mathbf{R}^2), R_G(V \oplus \mathbf{R})) \xrightarrow{d_k} \pi_{k-1}(R_G(V \oplus \mathbf{R}), R_G(V)) \xrightarrow{d_{k-1}} \cdots$$

where d_k is the boundary map in the homotopy exact sequence of the involved triple.

The proof of the following is sketched in Section 4.

Theorem 2.7. (i) Let $V \subseteq T$ be super-stable representations satisfying the conditions in 2.5. If $i < |V^G| - 1$ then $\pi_i(R_G(T), R_G(V)) = 0$.

(ii) The homology groups $H^*(\mathcal{K}_{-1}(G), d)$ of the complex in 2.6 are finite 2-groups.

We return to the question of equivariant transversality. First we treat CAT = PL. Let ξ be a PL G-bundle over Y and let $f: M \to \xi^+$ be a singular PL G-manifold in its Thom space.

Call the pair (M,f) topological stable if for each $H \subseteq G$ and $x \in M^H$, $y \in Y^H$, $T_xM \cong V_{x,y} \oplus \xi_y$ where $V_{x,y}$ and ξ_y are topological stable representations such that $\xi_y^H \neq 0$ implies $V_{x,y}^H \neq 0$ for all $H \subseteq G$. There is a similar concept of superstable pairs.

Theorem 2.8. In the locally PL-category, each topological stable singular G-manifold $f: M \to \xi^+$ is G-homotopic to a map transverse to the zero section. the reasonable realtive version is also valid.

This result is an application of 2.2 and 2.5 and can be outlined as follows. Suppose $f^G: M^G \to (\xi^+)^G$ is transverse to Y^G. One attempts to extend transversality to an open tube around $(f^G)^{-1}(Y^G)$. By 2.2 this translates into a bundle section problem. Using 2.5 the obstructions to a section (cf. 2.3) vanish. The process can be continued up the stratification as in the proof of 1.8 and 1.9.

For CAT = TOP there is a global obstruction to transversality due to 2.7(ii). One proceeds as in the proof above but there are obstructions to extending transversality to a tube around $(f^G)^{-1}(Y^G)$. The obstruction is concentrated in the top-dimension, hence is a cobordism invariant, and it lies in $H^*(\mathcal{K}_{-1}(G);d)$ so is 2-primary. Climbing up the stratifiction we meet obstructions in $H^*(\mathcal{K}_{-1}(H),d)$ for all the isotropy subgroups.

Let $H_1 \subset H_2 \subset \cdots$ be an admissible ordering of the orbit types: $H_j \subset H_i$ only if $j < i$. We formulate our results about stable topological transversality in

Theorem 2.9. Let ξ be a TOP G-bundle. Let $f:M \to \xi^+$ be a superstable singular G-manifodl (in TOP) and suppose that each component of M^H is simply connected for all $H \subseteq G$. There exists a hierarchy of obstructions

$$\pitchfork_i(M,f) \in H^*(\mathcal{K}_{-1}(H_i),d).$$

Suppose $\pitchfork_i(M,f) = 0$ for $i < j$. Then $\pitchfork_j(M,f)$ is a G-bordism invariant relative to smaller fixed sets, and is additive with respect to bordism addition. If $\pitchfork_i(M,f) = 0$ for all i then f is G-homotopic to a map which is transverse to the zero section of ξ.

The reasonable relative version is also valid.

§3. Equivariant surgery exact sequence: PL case

Let $\widetilde{PL}_G(V)$ be the Δ-group of G-equivariant block PL-homeomorphisms, cf. [RS]. The next lemma is essentially equivalent to the uniqueness of regular neighbourhood, and is well-known.

Lemma 3.1. (i) The inclusion $PL_G(V) \subset \widetilde{PL}_G(V)$ is $(|V^G|+1)$-connected.

(ii) Let V_G be a complement to V^G in V. There is a homotopy equivalence $\widetilde{PL}_G(S(V_G)) \to \widetilde{PL}_G(V)$ where $S(V_G)$ is the sphere in V_G.

We can refer to [AH] for a proof. Indeed, there are principal convering maps $PL_G(V) \to PL(c(S^{k-1} * L))$ and $\widetilde{PL}_G(V) \to \widetilde{PL}_G(c(S^{k-1} * L))$ with fiber G. Here $k = |V^G|$, $L = S(V_G)/G$ with its natural stratification, c( ) denotes the open cone and PL( ) denotes the stratum preserving PL-homeomorphisms. Now use 11.3 and (the proof of) 8.1 in [AH] together with the easy fact that the space of block pseudo isotopies is contractible: $\widetilde{PL}_G(V) \simeq \widetilde{PL}_G(V \oplus \mathbf{R})$.

Let Y be a based G-space and V any representation of G. Let χ be a complex irreducible representation of G or $\chi = \mathbf{R}$ $(S(\chi) = S^1$ or $S^0)$. Consider the natural map

$$\Sigma_\chi : Map_G(S(V),Y) \to Map_G(S(V \oplus \chi), Y * S(\chi))$$

and the corresponding non-equivariant maps on the fixed sets

$$\Sigma_\chi(H) : Map(S(V^H),Y^H) \to Map(S(V^H \oplus \chi^H),\ Y^H * S(\chi^H))$$

<u>Theorem 3.2.</u> Σ_χ is k-connected if and only if $\Sigma_\chi(H)$ is k-connected for all $H \subseteq G$.

The proof follows inductively from the two fibrations below where $V = V_0 \oplus \psi$, ψ a complex character of G with isotropy subgroup $G_0 \neq G$.

$$Map_G(DV_0 \times S(\psi) rel\ \partial, Y) \to Map_G(SV, Y) \to Map_G(SV_0, Y)$$

$$Map_{G_0}(D(V_0 \oplus \mathbf{R}) rel\ \partial, Y) \to Map_G(DV_0 \times S(\psi) rel\ \partial, Y) \to Map_{G_0}(DV_0 rel\ \partial, Y).$$

The space of G-maps, $Map^G(S(V),S(V))$, is easily related to the space (of Δ-space) $F_G(S(V))$ of G-homotopy equivalences of S(V). Moreover, $F_G(S(V))$ is contained as a deformation retract in the corresponding block version $\widetilde{F}_G(S(V))$. From 3.2 we get

Corollary 3.3. Let χ be a complex (1-dimensional) character or $\chi = \mathbf{R}$. Then

$$\Sigma_{\chi}: F_V(S(V)) \to \tilde{F}_G(S(V \oplus \chi))$$

is (k-2)-connected where $k = \mathrm{Min}\{|V^H| \mid H \subseteq G, \chi^H = \chi\}$.

The Δ-space $\tilde{F}_G(S(V))/\widetilde{PL}_G(S(V))$ classified equivalence classes or $PL_G(S(V))$ block bundles with block homotopy trivializations, cf. [RS, 3.21]. Alternatively, it can be considered as the Δ-space of simple G-manifold structures on S(V). In particular,

3.4 $$\pi_i(\tilde{F}_G(S(V))/PL_G(S(V))) = \mathcal{S}_G(D^i \times S(V), \partial).$$

In the G-trivial case 3.4 can be found in [ABK]. The argument is similar in general, using the equivariant s-cobordism theorem (cf. [R]). The right-hand side in 3.4 consists of simple G-homotopy equivalences $f:(M,\partial) \to (D^i \times S(V), \partial)$ with ∂f a PL-homeomorphism. In a range, $\mathcal{S}_G(D^i \times S(V), \partial)$ can be analysed by equivariant surgery. This requires, on the one hand, transversality (to identify the normal invariants) and on the other hand, stability results for PL G-bundles, hence it depends on 2.5 which is what we attempt to prove. Thus we must proceed by induction and assume 2.5 for smaller order groups. First we list the results.

Let F/PL = F/PL(G) be the G-space which classifies stable equivalence classes of homotopy trivialized G-bundles. It is the fibdre of the mapping $BPL(G) \to BF(G)$, and

$$F/PL \simeq_G \lim_{\substack{\to \\ V}} F(S()V))/PL(S(V))$$

where V runs over all representations of G.

In the setup of equivariant surgery, one starts out with a given set $\pi = \{H_\alpha\}$ of (isotropy) subgroups of G. For example, in calculating the structure set $\mathcal{S}_G(X, \partial X)$ the relevant set of subgroups is the set of isotropy subgroups of points of X. The corresponding version of F/PL to consider is

$$(F/PL)_\pi = \lim_{\substack{\to \\ V\in\pi}} F(S(V))/PL(S(V))$$

where $V \in \pi$ if $S(V_G)$ has isotropy subgroups belonging to π.

Consider PL G-manifolds of the form $X = Y \times D^1$ which satisfy the stability conditions

3.5 $\dim X^H > 5$, $2 \dim X^H < \dim X^K - 1$ for $X^H \neq X^K$ and $K \subset H$

Theorem 3.6 (see [D-R]. In the setting of 3.5 there is an exact sequence of abelian groups

$$\longrightarrow [X \times D^1/\partial(X \times D^1), (F/PL_\pi]_G \xrightarrow{\lambda} \sum_{H\in\pi}{}^{\oplus} L^s_{d(H)+1}(G/H) \longrightarrow$$

$$\xrightarrow{\alpha} \mathcal{J}_G(X, \partial X) \xrightarrow{\eta} [X/\partial X, (F/PL)_\pi]_G \xrightarrow{\lambda} \sum_{H\in\pi}{}^{\oplus} L^s_{d(H)}(G/H)$$

where $d(H) = \dim X^H$ and H runs over all isotropy subgroups.

Using the fact that $X = Y \times D^1$ and pasting the objects along boundary components, the geometrically defined sets are groups.

Our next result identifies the pointed G-homtopy classes $[X/\partial X, (F/PL)_\pi]_G$ away from 2. It generalizes to the equivariant case [S] a well known theorem of Sullivan.

Theorem 3.7. There is an isomorphism

$$[X/\partial X, (F/PL)_\pi]_G \otimes \mathbf{Z}[\tfrac{1}{2}] \cong \sum_{H\in\pi}{}^{\oplus} KO_{G/H}(X^H, \partial X^H) \otimes \mathbf{Z}[\tfrac{1}{2}].$$

In the rest of the section we shall suppress the set π of subgroups of G and write F/PL istead of $(F/PL)_\pi$ etc.

Let M be an even dimensional PL G-manifold. It can be embedded in a representation space: $M \subset U$ with PL normal bundle ξ (cf. [Wa], Proposition 1.9). By Theorem A(PL), ξ is $KO_G(\ )_{od}$-oriented and we can form

$$\varphi_*: KO^*_G(M)_{od} \to KO_G(\xi^+)_{od} \to KO^*_G(U^+)_{od} \xrightarrow{\cong} KO^*_G(pt)_{od} \subset R(G) \otimes \mathbf{Z}[\tfrac{1}{2}].$$

We write $\operatorname{sign}_G(M)$ for the G-signature of M defined from the cup product pairing in $H^*(M,\mathbb{R})$. If M is topologically stable, then we have PL normal G-bundles $\nu(M,M^g)$ for every $g \in G$. They are $KO_G(-) \otimes \mathbb{Z}[1/2]$ oriented by Theorem AZ, and in particular they have Euler classes $e(M,M^g) \in KO_G(M^g)_{od}$. Here is a PL-version of the G-signature theorem.

<u>Theorem 3.8.</u> Let M be a closed PL G-manifold

(i) $\operatorname{sign}_G(M) = \varphi_*(1)$ in $R(G)$

(ii) $e(M,M^g)$ is invertible in $KO(M^g)_{od}$ localized at the prime ideal in $R(G)$ defined by g.

(iii) $\operatorname{sign}_G(M)(g) = \varphi(M^g)_*(e^{-1}(M,M^g))(g)$.

In the rest of this paragraph, W will be a topological stable representation with $W^G = 0$. Suppose i is an integer such that $D^{2i} \times S(W)$ satisfies 3.5. We define a modified ρ-invariant (cf [W], §14)

$$\rho(G): \mathscr{S}_G(D^{21} \times S(W),\partial) \to R(G) \otimes \mathbb{Q}$$

as follows. Let $(\tilde{E},t)$ be a homotopy trivialized block bundle over D^{2i} (rel ∂), representing an element of $\mathscr{S}_G(D^{2i} \times S(W),\partial)$. There exists an integer r and a PL-manifold Y_r with $\partial Y_r = r \cdot (\tilde{E} \cup_\partial D^{2i} \times S(W))$ and such that Y_r and ∂Y_r have the same orbit types. Define

3.9 $$\rho(G)(\tilde{E},t) = \frac{1}{r} \operatorname{sign}_G(Y_r)e(W) \in R(G) \otimes \mathbb{Q}$$

where $e(W)$ is the Euler class from §1.

It appears that $\rho(G)(\tilde{E},t)$ depends on the choice of Y_r, but in fact it does not. Indeed, if Y_r' is a second choice then 3.8(ii) applied to $Y_r \cup_\partial Y_r'$ shows that $\operatorname{sign}_G(Y_r) - \operatorname{sign}_G(Y_r') \in \sum \operatorname{Image}(\operatorname{Ind}_H^G(R(H)))$ where H runs over the subgroups with $H^H \neq 0$; and $e(W) \cdot \operatorname{Ind}_H^G(R(H)) = 0$ by Frobenius reciprocity.

For each H (with $W^H \neq 0$) we get an invariant $\rho(G/H)$, and consider the direct sum

$$\rho_W : \mathcal{S}_G(D^{2i} \times S(W), \partial) \to \sum\nolimits^{\oplus} R(G/H) \otimes \mathbb{Q}$$

<u>Theorem 3.11.</u> Suppose $X = D^{2u} \times S(W)$ satisfies 3.5. Away from the prime 2, the exact sequence in 3.6 can be identified with the sequence

$$r \to \sum\nolimits^{\oplus} KO^{m-1}_{G/H}(S(W^H)) \xrightarrow{\delta^*} \sum\nolimits^{\oplus} KO^{m}_{G/H}(D(W^H), S(W^H)) \to$$
$$\to \sum\nolimits^{\oplus} KO^{m}_{G/H}(DW^H) \to \sum\nolimits^{\oplus} KO^{m}_{G/H}(S(W^H)) \to 0$$

where $m = |W| - 2i$.

<u>Theorem 3.12.</u> Suppose $V \subset T$ are representations of G satisfying the requirements of 2.5. Then $\widetilde{PL}_G(T)/\widetilde{PL}_G(V)$ is $(|V^G|+1$-connected.

We now outline the proofs. First observe that Theorem A(PL), 2.8, 3.6, 3.7 and 3.8 all have "restricted" versions where we assume G does not occur as isotropy subgroup. They will be denoted $A(G^{\perp})$, $2.8(G^{\perp})$, $3.6(G^{\perp})$, $3.7(G^{\perp})$, and $3.8(G^{\perp})$. For example, $3.6(G^{\perp})$ has the same conclusion as 3.6 (namely exactness of the listed sequence) but under the extra assumption that $X^G = \emptyset$. The restricted versions of 2.8, 3.7, 3.8 and Theorem A are similar.

The proofs proceed by induction over the order of the group, and starts with 2.5. Thus we fix the group G and suppose 2.4 is true for groups of smaller order. Here is a chart of total argument

$$\begin{array}{ccccccccc}
A(G^{\perp}) & \overset{(4)}{\Longrightarrow} & 3.8(G^{\perp}) & & & & & & \\
\Uparrow{\scriptstyle (2)} & & & & 3.11 & \overset{(7)}{\Longrightarrow} & 3.12 & \overset{(8)}{\Longrightarrow} & 2.5(\leq |G|) \\
 & & & & & & \Uparrow{\scriptstyle (6)} & & \\
2.5(<|G|) & \overset{(1)}{\Longrightarrow} & 2.8(G^{\perp})) & \overset{(3)}{\Longrightarrow} & 3.6(G^{\perp}) & \overset{(5)}{\Longrightarrow} & 3.7(G^{\perp}) & &
\end{array}$$

Finally, 2.5 ($\leq |G|$) implies the unrestricted versions of the results via a similar chain of arguments, outlined at the end of this section.

(1): If $M^G = \emptyset$ in 2.8 then the argument outlined in §2 starts with making f transverse in a neighbourhood of a minimal stratum M^H (on which G/H acts freely). The obstructions to extend transversality to an open neighbourhood depends on $\pi_i(PL_K(T)/PL_K(V))$ but only for proper subgroups K. They vanish by 2.5 ($< |G|$).

(2): Let ξ be a topological stable PL G-bundle over Y with $Y^G = \emptyset$. Let $f:(M,\partial M) \to (D\xi, S\xi)$ represent an element of $\Omega_*^G(\xi^+)$. There exists a projective space **P** with G-action so that $f \circ \mathrm{proj}: M \times \mathbf{P} \to D\xi$ is topological stable, and hence is G-homotopic to a transversal map f_0. Let $N = f_0^{-1}(Y)$ and define

$$\sigma_G : \tilde{\Omega}_*^G(\xi^+) \to R(G)$$

by $\sigma_G([M,f]) = \mathrm{sign}_G(N)$, with the convention that $\mathrm{sign}_G(N) = 0$ if dim N is odd. Multiplicativity of the signature shows that σ_G factors over $\Omega^G(\xi^+) \otimes_{\Omega_*^G} R(G)$. The results of §1 implies a Thom class,

$$\Delta_G(\xi) \in KO_G^*(\xi^+) \otimes \mathbf{Z}[\tfrac{1}{2}].$$

We have (at least rationally) $\Delta_G(\xi) \oplus \eta) = \Delta(\xi)\Delta(\eta)$.

(3): See the remarks following 3.4.

[4]: The first part of $3.8(G^{\perp})$ follows from the definition above of $\Delta_G(\xi)$. The second part follows as in [AS II, Lemma 27].

(5): Let $\varphi_H: F/PL^H \to F/PL(G/H)$ classify the functor of taking H-fixed points: $(\eta,t) \to (\eta^H, t^H)$. for each $\bar{G} = G/H$ of order less than $|G|$, define

$$\lambda(\bar{G}) : \tilde{\Omega}_*^{\bar{G}}(F/PL(\bar{G})) \to R(\bar{G})$$

in analogy with σ_G of (2): to a singular $\bar{G}$-manifold $f:(M,\partial M) \to (F/PL(\bar{G}),*)$, let (η,t) be the $F/PL(\bar{G})$ bundle over $M \times \mathbf{P}$ represented by $f \circ \mathrm{proj}$, where **P** is a suitable projective $\bar{G}$-space. We make t transverse to the 0-section to obtain $\hat{M} = t^{-1}(M \times P)$ and define

$$\lambda(\overline{G})[M,f] = \operatorname{sign}_G(\hat{M})(- \operatorname{sign}_G(M \times \mathbf{P}).$$

As in (2) we get a $\overline{G}$-homotopy class $\lambda(\overline{G}):F/PL(\overline{G}) \to BO(\overline{G})$. We compose $\lambda(G/H)$ with φ_H and use 1.11 to get a G-homotopy class

$$\varphi(G/H):F/PL \to BO(G/H) * E(H^{\perp}).$$

If $H = 1$ the necessary transversality is only avialable if $M^G = \emptyset$, so instead of $\Omega_*^G(X,A)$ we consider the functor

$$\Omega_*^G[G^{\perp}](X,A) = \Omega_*^G(X \wedge E(G^{\perp})^+, A \wedge E(G^{\perp})^+).$$

Geometrically, this is generated by G-bordism classes with $M^G = \emptyset$ (cf. [CFII], [tD]). Altogether, we have a G-map

$$\varphi:F/PL \quad E(G^{\perp})^+ \to \prod_{H \subseteq G} BO(G/H) * E(H^{\perp}).$$

$(E(1^{\perp}) = \emptyset$, $BO(G) * \emptyset = BO(G))$. We can assume inductively that φ^H is a homotopy equivalence for each proper subgroup H; this follows from the isomorphism in 3.7 and for the group H, cf. the proof of 3.7 below. We then use that

$$[X, Y \wedge E(G^{\perp})]_G \cong [X,Y]_G$$

when $X^G = *$. This completes the proof of $3.7(G^{\perp})$.

(6): There is a commutative diagram

$$\begin{array}{ccccccc}
\cdots \to \sum^{\oplus} L^s_{d(H)+1}(G/H) & \longrightarrow & \mathcal{S}_G(X,\partial X) & \longrightarrow & [X/\partial X, F/PL]_G & \longrightarrow & 0 \\
\downarrow \sigma & & \downarrow \hat{\rho}_W & & \downarrow \varphi_* & & \\
\cdots \to \sum^{\oplus} RO_*(G/H) & \xrightarrow{\sum^+ e(W^H)} & \sum^{\oplus} RO_*(G/G) & \longrightarrow & \sum^{\oplus} KO_{G/H}(X^H, \partial X^H) & \longrightarrow & 0
\end{array}$$

where $X = D^{2i} \times S(W)$ and $RO_*(G/H) = KO^*_{G/H}(pt) \otimes \mathbf{Z}[\frac{1}{2}]$. Here $\hat{\rho}_W$ is the sum of the invariant ρ_W defined in the paragraph following 3.9 twisted by certain normal invariants (on proper subgroups). These are invariants coming from $[X/\partial X, F/PL]_H$, $H \subset G$. Commutativity of the left-hand square follows directly from the definitions; the right-hand square is more delicate. Since σ and φ_* are isomorphisms away from 2, so is $\hat{\rho}_W$.

(7): Let $V = W \oplus \mathbf{R}^k, W^G = 0$ and $T = W \oplus U \oplus \mathbf{R}^\ell$ with $(W \oplus U)^G = 0$. By 3.1 it suffices to examine the inclusion $\widetilde{PL}_G(S(W)) \to \widetilde{PL}_G(S(W \oplus U))$, and using 3.3 it suffices to examine the map

$$\Sigma(U):\widetilde{F}_G(S(W))/PL_G(S(W)) \to \widetilde{F}_G(S(W \oplus U))/PL_G(S(W \oplus U))$$

which clasifies the operation $(\eta,t) \to (\eta \oplus W, t \oplus 1)$. The ρ-invariant of 3.9 is stable:

$$\begin{array}{ccc} \mathcal{S}_G(D^{2i} \times S(W),\partial) & \xrightarrow{\rho(G)} & \\ \downarrow{\scriptstyle \Sigma(U)} & & R(G) \\ \mathcal{S}_G(D^{2i} \times S(W \oplus U),\partial) & \xrightarrow{\rho(G)} & \end{array}$$

is a commutative square. (In the special case of a free representation this is equivalent to [W], Theorem 14.A.1). It follows from 3.11 that $\Sigma(U)$ induces isomorphism on $\pi_i(-) \otimes \mathbf{Z}[1/2]$ in the required range.

We have left to show that $\Sigma(U)$ is also a 2-local isomorphism. This requries a partial calculation of $\mathcal{S}_G(X,\partial X) \otimes \mathbf{Z}_{(2)}$ for $X = D^i \times S(W)$. We outline the calculation when $S(W)$, whence X, has two strata, say X^H and $X - X^H$.

Without special indication in the notation we assume all functors are localized at 2. Let $Y =- X/\partial X$. Since $Y/Y^H \simeq_G EG \times Y/EG \times Y^H$ we have

$$[(Y,Y^H),F/PL]_G \cong [(EG \times_G Y,EG \times_G Y^H),F/PL] \cong [(Y,Y^H),F/PL].$$

A similar argument shows that

$$[Y^H,F/PL^H]_{G/H} \cong [Y^H,F/PL^H] \cong [Y^H,F/PL]_H .$$

Using the exact sequence for the pair (Y,Y^H) it follows that

$$\mathrm{Res}:[Y,F/PL]_G \to [Y,F/PL]_H$$

is a (2-local) isomorphism. On the other hand,

$$[Y,F/PL]_H \cong [S(\mathbf{R}^i + W),F/PL]_H \oplus \pi_i(F/PL^H).$$

In the relevant range $\mathcal{S}_H(S(\mathbf{R}^i \oplus W)) = 0$, so

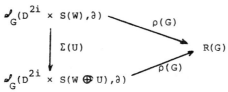

$\lambda:[S(R^i + W),F/PL]_H \to L^s_*(H) \oplus L^s_*(1)$ is an isomorphism. Hence η maps $\mathscr{S}_G(X,\partial X)$ onto $\pi_i(F/PL^H)$, and 2-locally 3.6 reduces to the exact sequence

$$0 \to L^s_*(G)/L^s_*(H) \oplus L^s_*(G/H)/L^s_*(1) \xrightarrow{\alpha} \mathscr{S}_G(X,\partial X) \xrightarrow{\eta} \pi_i(F/PL^H) \to 0.$$

The suspension $\Sigma(U)$ clearly maps $\pi_i(F/PL^H)$ isomorphically. We use the ρ-invariant from 3.9 and [W, Theorem 13.A.4] to see that it also maps the image of α isomorphically.

The general case with more complicated orbit structure is similar. In particular, the image of η only depends on the isotropy subgroups of G in W, not on the actual representation.

(8): This follows from 3.1(i) and 3.12.

We have left to see that 2.5 implies the unrestricted results 2.8, 3.6, 3.7, 3.8 and Theorem A(PL). This follows as above except for 3.7 where extra information is needed. So we end this section with

Proof of 3.7. We now have sufficient transversality to get a G-homotopy class

$$\varphi:F/PL \to \prod BO(G/H) * E(H^\perp),$$

which we may assume induces isomorphisms $\pi_i(\varphi^H)$ for $i > 0$ and $H \in G^\perp$. We have

$$\tilde{F}_G/\widetilde{PL}_G = \lim_{\to} \tilde{F}_G(S(W))/PL_G(S(W))$$

where W runs over representations with $W^G = 0$. There is a fibration,

$$\tilde{F}_G/PL_G \xrightarrow{j} (F/PL)^G \xrightarrow{\lambda(G/G)} F/PL(G/G),$$

cf. (5) above, and $F/PL(G/G) \simeq BO$ (away from 2). Let

$$\rho = \lim \rho_W: \pi_{2i}(\tilde{F}_G/\widetilde{PL}_G) \to \sum{}^{\oplus} R(G/H).$$

There is a commutative diagram

$$\begin{array}{ccc} \pi_{2i}(\widetilde{F}_G/\widetilde{PL}_G) & \longrightarrow & \sum_{H \neq G} {}^{\oplus} R(G/H) \\ \downarrow j_* & & \uparrow \\ \pi_{2i}((F/PL)^G) & \xrightarrow{\varphi_*^G} & \sum_{H \neq G} {}^{\oplus} KO_{G/H}^{-2i}(pt), \end{array}$$

and we can apply 3.11.

§4. Topological stability

In this section we use results from [AH] and [LR] to examine stability questions for the homogeneous spaces $TOP_G(V)/PL_G(V)$. In particular, we outline the proof of Theorem 2.7. When these results are combined with the results from Section 3 one obtains the necessary stability properties for $TOP_G(V)$, used in the obstruction theory for topological transversality contained in Theorem 2.9.

We assume throughout that V is a representation of G with $\dim V^G \geq 5$, but not that V is topologically stable (in the sense of 2.4).

Let $R(V) = TOP(V)/PL(V)$, considered as a G-space with fixed sets

$$R(V)^H = R_H(V) = TOP_H(V)/PL_H(V).$$

The corresponding space of pseudo isotopies can be identified with the fibre in the G-fibration

4.1 $$\mathcal{P}R(V) \to R(V) \to R(V \oplus \mathbf{R}).$$

There are G-homotopy equivalences

$$\mathcal{P}R(V) \simeq R(V,V^G) \times R(V^G),$$

$$R(V) \simeq \mathcal{P}R(V,V^G) \times \mathcal{P}R(V^G),$$

and $\mathcal{P}R(V^G)$ is $(\dim V^G + 1)$-connected.

We shall continuously use the following standard homotopy equivalences, all based upon the Alexander isotopy (cf. [AH] or [LR]):

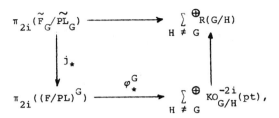

$$\mathcal{P}R_G(V,V^G) \simeq \mathcal{P}R_G(c(S^{r-1} * S(W)),cS^{r-1}) \simeq \mathcal{P}R_G(S^r * S(W),S^r);$$

4.2 $$\Omega^k \mathcal{P}R_G(V,V^G) \simeq \mathcal{P}R_G((W \oplus \mathbf{R}^{r-k},\mathbf{R}^{r-k}) \times (D^k,\partial D^k)), \quad k \leq r;$$

$$\Omega^{r+1}\mathcal{P}R_G(V,V^G) \simeq \mathcal{P}R_G(S(W) \times (D^{r+1}, \partial D^{r+1})) ,$$

where $r = \dim V^G$ and $V = V^G \oplus W$. There are corresponding equivalences when $\mathcal{P}R_G$ is replaced with R_G.

Let M be a (not necessarily compact) PL G-manifold. We write $\mathcal{P}R_G(M) = \mathcal{P}TOP_G(M)/\mathcal{P}PL_G(M)$ for the space of topological G pseudo isotopies modulo the PL ones. This is in agreement with the above notation when M = V,cf. [LR, §8, Proposition 1].

Our next result is a pseudo isotopy version of [LR, §6, Theorem 9] and the equivariant s-cobordism theorem. In the compact case the equivariant s-cobordism theorem can be found in [R]. In the noncompact case we need a relatively straightforward generalization of the proper s-cobordism theorem from [S]. The obstruction group is in the compact case the equivariant Whitehead group $Wh_G(\mathbf{Z})$ from [R]. In the non-compact situations where we shall use the theorem the obstruction group is an equivariant K_{-i}-group, $i \geq 0$. In order to keep a uniform notation we write the obstruction group as $\mathcal{W}_G(\mathbf{Z})$ in all cases.

To ease the formulation we assume that M has trivial tangent bundle, $\tau_M = M \times T$ for some representation T.

The following is based on techniques of [LR] appolied to the problem of classifying equivariant triangulation rather than equivariant smoothings.

<u>Theorem 4.3.</u>. Let M be as above and suppose no component of the M^H has dimension 4. Then there is an exact sequence

$$0 \longrightarrow \pi_0(\mathcal{P}R_G(M,\partial M)) \longrightarrow [(M,\partial M), \mathcal{P}R(T)]_G \longrightarrow \mathcal{W}_G(\mathbf{Z})$$

In analogy with the notation from [AH] we let $\mathcal{P}R_G^+(V,V^G)$ denote the space of pseudo isotopies which are PL outside V^G. The equivalences of 4.2 are valid for $\mathcal{P}R_G^+(\ )$ as well, so we have fibrations (up to homotopy) for $k \leq r$,

4.4 $$\Omega^k\mathcal{P}R_G^+(V,V^G) \longrightarrow \Omega^k\mathcal{P}R_G(V,V^G) \longrightarrow \mathcal{P}R_G(Y_{r,k} \times (D^k,\partial D^k)),$$

with $Y_{r,k} = (W - 0) \times R^{r-k}$, $r = \dim V^G$.

<u>Proposition 4.5</u>. Let V be a representation with $r = \dim V^G \geq 5$. Then $\pi_k(\mathcal{P}R_G(V)) = 0$ for $k < r - 2$.

<u>Proof</u>. Since $R(V^G)$ is $(r+1)$-connected we have $\pi_k(\mathcal{P}R_G(V)) = \pi_k(\mathcal{P}R_G(V,V^G))$. We prove 4.5 by induction on the order of G.

The tangent bundle of $Y_{r,k} \times D^k$ is G-trivial with fibre V, so 4.3 gives

$$\pi_0(\mathcal{P}R_G(Y_{r,k} \times (D^k,\partial D^k))) \subseteq [Y_{r,k} \times (D^k,\partial D^k), \mathcal{P}R(V)]_G.$$

There is a G-homotopy equivalence $Y_{r,k} \simeq S(W)$ and $[S(W) \times (D^k,\partial D^k), \mathcal{P}R(V)]_G$ can be calculated from G-obstruction theory. Indeed, for each proper subgroup H and each component C of $S(W^H)$, the Bredon cohomology groups

(*) $$H^{i-k}_{G/H}(C,C^S;\pi_i(\mathcal{P}R_H(V))) = 0 \quad \text{for } i < r-2,$$

by the inductive assumption; (C^S is the singular set). Hence $\pi_0(\mathcal{P}R_G(Y_{r,k} \times (D^k,\partial D^k))) = 0$ for $k < r-2$. Since $\pi_k(\mathcal{P}R_G^+(V,V^G)) = 0$ by [AH,11.2], 4.4 implies the result.

We next consider the stabilization. First, we stabilize with **R** and consider the maps coming from identification 4.2.

$$i_k : \Omega^k\mathcal{P}R_G(V,V^G) \to \Omega^{k+1}(\mathcal{P}R_G(V \oplus \mathbf{R}, V^G \oplus \mathbf{R}))$$

and the similar maps i_k^+ for $\mathcal{P}R_G^+(V,V^G)$.

Proposition 4.6. For sufficiently large $r = \dim V^G$

$$(i_k)_*:\pi_k(\mathcal{P}R_G(V,V^G)) \to \pi_{k+1}(\mathcal{P}R_G(V \oplus \mathbf{R}, V^G \oplus \mathbf{R}))$$

is an isomorphism for all $k \leq r$.

Proof. Consider the fibration diagram,

$$\begin{array}{ccccc}
\Omega^k \mathcal{P}R_G^+(V,V^G) & \longrightarrow & \Omega^k \mathcal{P}R_G(V,V^G) & \longrightarrow & \mathcal{P}R_G(Y_{r,k} \times (D^k, \partial D^k)) \\
\downarrow{\scriptstyle i_k^+} & & \downarrow{\scriptstyle i_k} & & \downarrow{\scriptstyle \bar{i}_k} \\
\Omega^{k+1} \mathcal{P}R_G^+(V_1,V_1^G) & \longrightarrow & \Omega^{k+1} \mathcal{P}R_G(V_1,V_1^G) & \longrightarrow & \mathcal{P}R_G(Y_{r,k} \times (D^{k+1}, \partial D^{k+1}))
\end{array}$$

where $V_1 = V \oplus \mathbf{R}$. It follows from [AH] that i_k^+ induces isomorphisms on π_0. From 4.3 and the obvious inductive asssumption we have that $\bar{i}_k$ induces isomorphism on π_0.

The fibration in 4.4 is not onto, so the 5-lemma does not apply directly. However, the image of $\pi_k(\mathcal{P}R_G(V,V^G))$ in $\pi_0(\mathcal{P}R_G(Y_{r,k} \times (D^k, \partial D^k))$ must stabilize under iterated use of $(i_k)_*$. Since $(i_k^+)_*$ is an isomorphism the result follows.

At present we do not have an estimate of size of $\dim V^G$ needed in 4.6. We introduce the concept superstability.

Definition 4.7. A representation V is called super-stable if it is topological stable and if

$$(i_k)_*:\pi_k(\mathcal{P}R_H(V,V^H)) \to \pi_{k+1}(\mathcal{P}R_H(V \oplus \mathbf{R}, V^H \oplus \mathbf{R}))$$

is an isomorphism for all $k \leq \dim V^H$.

Suppose χ is a representation with isotropy groups C those of V and $\chi^G = 0$. Geometrically, one defines a suspension mapping

$$\Sigma_\chi : \mathcal{P}R(V) \to \Omega^\chi \mathcal{P}R(V + \chi)$$

which on H-fixed sets reduces to an iteration of the i_k when $\chi|H$ is trivial. (Here $\Omega^\chi(X)$ means maps of $\dot{\chi}$ into X.) Using an argument similar to the one proving 4.6 one gets:

Proposition 4.8. If V is super-stable and $\chi^G = 0$ then Σ_χ induces a map $(\Sigma_\chi)_*$

$$(\Sigma_\chi)_* : \pi_k(\mathcal{P}R_G(V,V^G)) \to \pi_k(\mathcal{P}R_G(V \oplus \chi, V^G))$$

which is an isomorphism for $k \leq \dim V^G$.

We next consider the exact couple of homotopy groups associated to the fibrations

$$\mathcal{P}R_G(\mathbf{R}^r \oplus W) \to R_G(\mathbf{R}^r \oplus W) \to R_G(\mathbf{R}^{r+1} \oplus W)$$

where $W^G = 0$. We truncate at π_{r+1}, and get

4.9
$$\begin{aligned} E^1_{p,q} &= \pi_{p+q-1}(\mathcal{P}R_G(\mathbf{R}^p + W)), \quad q \leq 1; \\ E^1_{p,2} &= \pi_{p+2}(\tilde{R}_G(S(W))); \\ E^1_{p,q} &= 0 \quad \text{for } q > 2. \end{aligned}$$

Here $\tilde{R}_G$ denotes the block version of R_G; it enters into the exacty couple via the exact sequence

$$\pi_{r+1}(\mathcal{P}R_G(\mathbf{R}^r \oplus W)) \to \pi_{r+1}(R_G(\mathbf{R}^r \oplus W)) \to \pi_{r+1}(\tilde{R}_G(S(W))) \to 0,$$

cf. [LR, p.258].

The other terms in the exact couple are given by

4.10
$$\begin{aligned} D^1_{p,q} &= \pi_{p+q}(R_G(\mathbf{R}^{p+1} \oplus W)), \quad q \leq 1; \\ D^1_{p,2} &= \pi_{p+2}(\tilde{R}_G(S(W))); \\ D^1_{p,q} &= 0 \quad \text{for } q > 2. \end{aligned}$$

The associated spectral sequence converges to $\lim_{\overrightarrow{N}} \pi_{p+q}(R_G(\mathbf{R}^N + W))$, and the limit is attained for $p+q < N+2$.

<u>Proposition 4.11.</u> Let χ be as in 4.8. The suspension mapping

$$\Sigma_\chi : \pi_k(\tilde{R}_G(S(W))) \to \pi_k(\tilde{R}_G(S(W + \chi)))$$

is an isomorphism for $k \geq 5$ when W is superstable.

<u>Proof.</u> Elements of $\pi_k(\tilde{R}_G(S(W)))$ are represented by G-homeomorphisms

$$f : D^k \times S(W) \to D^k \times S(W)$$

which are PL on the boundary. Let

$$I_k(W) : [D^k \times S(W)/\partial, \mathcal{P}R(R^{k-1} \oplus W]_G \to [D^k \times S(W)/\partial, R(R^{k-1} \oplus W)]_G$$

be the natural map. Then 4.3 and an easy calculation yields an exact sequence

$$\cdots \longrightarrow \hat{H}^k(\mathbf{Z}/2; Wh_G(\mathbf{Z})) \longrightarrow \pi_k(\tilde{R}(SW)) \longrightarrow \text{Coker } I_k(W) \longrightarrow \hat{H}^{k-1}(\mathbf{Z}/2; Wh_G(\mathbf{Z})) \longrightarrow$$

where $Wh_G(\mathbf{Z})$ denotes the equivariant Whitehead group w.r.t. the isotropy subgroups of S(W).

The algebraic terms $\hat{H}^*(\mathbf{Z}/2; Wh_G(\mathbf{Z}))$ map isomorphically under Σ_χ so it suffices to compare $I_k(W)$ and $I_k(W \oplus \chi)$.

The calculation of $I_k(W)$ is done inductively over the stratification. first suppose SW has only one non-trivial orbit type, and consider the sequence

$$[X/X^H, R]_G \longrightarrow [X, R]_G \longrightarrow [X^H, R]_G \longrightarrow [\Sigma^{-1}(X/X^H), R]_G \longrightarrow \cdots$$

where $X = D^k \times SW/\partial$ and $R = R(\mathbf{R}^{k-1} + W)$. Since X/X^H is free outside the base point and, non-equivariantly $Top/PL = K(\mathbf{Z}/2, 3)$ we have

$$[X/X^H, R]_G = [(X/X^H)/G, K(\mathbf{Z}/2,3)] = [X/X^H \; K(\mathbf{Z}/2,3)] = 0.$$

Similarly, $[\Sigma^{-1}(X/X^H), R]_G = 0$, so that

$$[X, R]_G = [X^H, R]_G = [X^H/G, R_H(\mathbf{R}^{k-1} \oplus W)].$$

It follows from the spectral sequence listed in 4.9 and 4.10 applied to

$R_H(\mathbf{R}^{k-1} \oplus W)$ that

$$\operatorname{coker} I_k(W) = \hat{H}^{k-1}(\mathbf{Z}/2;Wh(\mathbf{Z}H)).$$

There is a similar calculation for coker $I_k(W \oplus \chi)$ and the suspension Σ_χ is an isomorphism.

In the general case, let $X_1 \subset X_2 \subset \cdots \subset X_n$ be a stratification of $X = D^k \times S(W)/\partial'$ such that $X_1 = X^{H_1}$, $X_i - X_{i-1} = X_{(H_i)}$ and such that $H_i \supseteq H_j$ only if $i \leq j$. The corresponding exact couple together with 4.6 and 4.8 shows that $[D^{k-1} \times S[W]/\partial, \; R(\mathbf{R}^{k-2} + W)]_G$ maps isomorphically onto the corresponding group with W replaced by $W \oplus \chi$ when W is superstable. Hence the cokernel of

$$J_k(W):[D^k \times S(W)/\partial, \mathcal{P}R(\mathbf{R}^{k-1} \oplus W)]_G \to [D^{k-1} \times S(W)/\partial, \mathcal{P}R(\mathbf{R}^{k-2} \oplus W)]_G$$

suspends isomorphically. Finally, coker $I_k(W)$ surjects onto coker $J_k(W)$ and its kernel can inductively be expressed in terms of $\pi_*(\tilde{R}_K(SW))$ for proper subgroups $K \subset H$. Hence this kernel also suspends isomorphically.

Proof of Theorem 2.7. The first part follows directly from 4.5 and the spectral sequence defined in 4.9. Indeed, suspension with χ induces isomorphism on E^1 by 4.8 and 4.11, where $T = V \oplus \chi$. We have left to prove 2.7(ii).

By 4.6 it follows that

$$\mathcal{K}_{-1}(G) = \pi_{r-1}(R_G(V \oplus \mathbf{R}), R_G(V)) = \pi_{r-2}(\mathcal{P}R_G(V))$$

depends only on V through its isotropy subgroups.

Consider the map

$$j_r: \mathcal{P}R_G(cS(W) \times \mathbf{R}^2 \times D^{r-1}, \bar{\partial}) \to \mathcal{P}R_G(cS(W) \times \mathbf{R}^2 \times D^{r-2}, \bar{\partial})$$

which maps the pseudo isotopy (rel $\bar{\partial}$)

$$f: I \times cS(W) \times \mathbf{R}^2 \times D^{r-1} \to I \times cS(W) \times \mathbf{R}^2 \times D^{r-1}$$

into $f|1 \times cS(W) \times \mathbf{R}^2 \times D^{r-1}$, viewed as a pseudo isotopy by readjusting D^{r-1} to $I \times D^{r-2}$. Here we have written $\bar{\partial}$ to denote

$$\bar{\partial} = cS(W) \times \mathbf{R}^2 \times \partial D^{r-1} \cup * \times \mathbf{R}^2 \times D^{r-1}, \quad * = \text{basepoint of } cS(W)$$

One can chase through the homotopy equivalences in 4.2 to see that $\pi_0(j_r)$ is equal to the 'differential'

$$d_r : \pi_r(R_G(V \oplus \mathbf{R}^2), R_G(V \oplus \mathbf{R})) \to \pi_{r-1}(R_G(V \oplus \mathbf{R}), R_G(V))$$

in the sequence 2.6.

Similarly, the groups

$$\pi_{r-2}(\mathcal{P} R_G^+(V)) = \pi_{r-1}(R_G^+(V \oplus \mathbf{R}), R_G^+(V))$$

depend only on the isotropy subgroups of V, and they are denoted $\mathcal{K}_{-1}^+(G)$. There are maps j_r^+ similar to the maps j_r above and we have a natural homomorphism

$$\alpha : \mathcal{K}_{-1}^+(G) \to \mathcal{K}_{-1}(G)$$

induced by the inclusion of the fibre in 4.4.

We use 4.3 and G-obstruction theory as in the proof of 4.5 to see that α is surjective. It commutes with $\pi_0(j_r^+)$, $\pi_0(j_r)$ and, more importantly, it commutes with the usual (geometrically defined) involutions on $\mathcal{K}_{-1}^+(g)$, $\mathcal{K}_1(G)$.

Finally,

$$\pi_0(j_r^+)(a) = a - (-1)^r \bar{a}$$

where the bar indicates the involution on $\mathcal{K}_{-1}^+(G)$, cf. [AH, §12]. It follows that

$$\pi_0(j_r)(a) = a - (-1)^r \bar{a},$$

and in conclusion, the homology of the complex 2.6 can be identified with the Tate cohomology groups $\hat{H}^*(\mathbf{Z}/2; \mathcal{K}_{-1}(G))$.

Our use of 2.7 to the question of obstructions to stable topological transversality is via the following

<u>Corollary 4.12</u>. Let $V = \mathbf{R}^r \oplus W$ with $W^G = 0$, and suppose $T \supseteq V \oplus \mathbf{R}^2$ is a representation with the same isotropy groups as V, and that all representations

are superstable. Then $\pi_i(TOP_G(T)/TOP_G(V)) = 0$ for $i < r-1$, and

$$\text{image}(\pi_{r-1}(TOP_G(T)/Top_G(V)) \to \pi_{r-1}(TOP_G(T)/TOP_G(V \oplus \mathbf{R})))$$

is a finite 2-group, isomorphic to $\hat{H}^r(\mathbf{Z}/2; \mathcal{K}_{-1}(G))$.

Proof. It follows from section 3 that in the considered range

$$\pi_*(TOP_G(T), TOP_G(V)) \cong \pi_*(R_G(T), R_G(V)),$$

and 4.12 follows from 2.7.

§5. The proofs of Theorems A and B

In section 3 we gave the proof of Theorem A in the PL category. In the TOP category the proof is similar but is now basaed on 2.9 instead of 2.8.

Since $KO_G^*(-)$ satisfies the suspension axiom with respect to an arbitrary representation it suffices to prove A for stable bundles ξ. Let $f: M \to \xi^+$ be a (superstable) singular G-manifold (see Theorem 2.3 and the preceding paragraph). By 2.90 there exists a k such that $2^k \cdot (M,f)$ is cobordant to a map transverse to he zero section. Thus the procedure used in §3 also defines

$$\sigma_G : \tilde{\Omega}^G(\xi^+) \otimes \mathbf{Z}[\tfrac{1}{2}] \to KO_*^G \otimes \mathbf{Z}[\tfrac{1}{2}]$$

and we get a Thom class

$$\Delta(\xi) \in KO_G^*(\xi^+) \otimes Z[\tfrac{1}{2}].$$

This proves Theorem A.

If ξ is a G-vector bundle then the above class coincides with the class from 1.1. By construction they correspond to the same homomorphisms from $KO_*^G(\xi^+)$ to $R(G) \otimes \mathbf{Z}[1/2]$. Thus they are rationally the same. But $KO_G^*(MO_{k+|V|}(V))$ is torsion free so the classes must be equal.

For each integer k which is prime to the order of G, and each TOP G-bundle ξ over X we define

5.1 $$\rho^k(\xi) = \frac{1}{k^n}\Psi^k(\Delta(\xi))/\Delta(\xi) \in KO_G^0(X) \otimes \mathbb{Q}$$

where ψ^k is the Adams operation and 2n is the (fibre) degree of ξ. Rationally, $\Delta(\xi + \eta) = \Delta(\xi)\Delta(\eta)$ and hence ρ^k is also exponential.

Let $KTop_G(X)$ be the Grothendieck group of TOP G-bundles over X. Then we get an exponential homomorphism

$$\rho^k : KTop_G(X) \to KO_G(X) \otimes \mathbb{Q}.$$

Let $J: KO_G(X) \to KTop_G(X)$ be the natural map and consider a (complex) G-line bundle L over X. Then

5.2 $$\rho^k \circ J(L) = \frac{1}{k}\frac{1 - L^k}{1 - L}\left(\frac{1 + L^k}{1 + L}\right)^{-1}$$

(cf. the paragraph preceding 1.3).

<u>Lemma 5.3.</u> For X = pt, the homomorphism

$$(\rho^k \circ J): RO(G) \to \prod_{(k,|G|)=1} RO(G) \otimes \mathbb{Q}$$

is injective.

This follows from 5.2 and Franz' independence lemma, see e.g. [Sc, p.265].

Now Theorem B follows: If W_1 and W_2 are representations of G which are topologically conjugate then $J(W_1) = J(W_2)$ and 5.3 applies.

REFERENCES

[AH] D. Anderson, W. C. Hsiang, Extending combinatorial piecewise linear structures on stratified spaces II, (preprint).

[ABK] P. L. Antonelli, D. Burghelea, P. J. Kahn, The concordance-homotopy groups of geometric automorphism groups, LNM vol. 215, Springer-Verlag.

[ASI,III] M. F. Atiyah, I. M. Singer, The index of elliptic operators I; III, Annals of Math. Vo. 87, no.3 (431-484), (531-546).

[ASII] M. F. Atiyah, G. B. Segal, The index of elliptic operators II, Annals of Math. vol. 87, nr. 3(485-530).

[B] M. Bökstedt, Universal coefficient theorems for equivariant K and KO-theory, Aarhus University (preprint) 1981.

[Br] G. Bredon, Equivariant cohomology theories, LNM vol. 34 (1967), Springer-Verlag.

[CS] S. Cappell, J. Shaneson, Non-linear similarity, Annals of Math. vol. 113(2) (1981), 351-357.

[dR] G. de Rham, Torsion et Type Simple d'Homotopie, LNH vol. 48 (1967), Springer-Verlag.

[DF1] P. Conner, E. E. Floyd, The relation of cobordism to K-theories, LNM vol. 28 (1966), Springer-Verlag.

[CF2] P. Conner, E. E. Floyd, Maps of odd period, Annals of Math. vol. 84 (1966), 132-156.

[tD] T. tom Dieck, Transformation groups and representation theory, LNM vol. 766 (1979), Springer-Verlag.

[E] J. Ewing, The image of the Atiyah-Bott map, Mat. Z. vol. 165 (1979), 53-71.

[K] C. Kosniowski, Generators of the $\mathbb{Z}/p$ bordism ring, Math. Z. 149 (1976), 121-130.

[LR] R. Lashof, M. Rothenberg, G-smoothing theory, Proc. Symp. Pure Mth. vol. 32 (1978), 211-266, AMS, Providence.

[Mil] J. Milnor, On axiomatic homology theory, Pacific J. Math., 12 (1962), 337-341.

[MM] I. Madsen, R. J. Milgram, The classifying spaces for surgery and cobordism of manifolds, Annals of Math. Studies,nr. 92(1979) Princeton University Press.

[R] M. Rothenberg, Torsion invariants and finite transformation groups, Proc. Symp. Pure Math. vol. 32 (1978), 267-312, AMS, Providence.

[RS] C. Rourke, B. Sandeson, Δ-sets II: Block bundles and block fibrations, Quart. J. Math. Oxford (2) 22 (1971) 465-85.

[Sc] R. Schultz, On the topological classification of linear representations, Topology 16 (1977), 263-270.

[W] C. T. C. Wall, Surgery on compact manifolds, Academic Press, London (1970).

[Wa] A. G. Wasserman, Equivariant differential topology, Topology 8 (1969), 127-150.

[Wi] K. Wirthmüller, Equivariant S-duality, Arch. Math. 27(1975), 427-431.

[S] L. C. Siebenmann, Infinite simple homotopy types, Indag. Math. 43 (1979), 502-513.

[DR] K. H. Dovermann, M. Rothenberg, An equivariant surgery sequence and equivariant diffeomorphsim and homeomorphism classification, preprint.

[HP] W. C. Hsiang, W. Pardon, When are topologically equivalent orthogonal representations linearly equivalent? preprint.

[L] R. Lashof, A topological G submersion theorem, to appear.

[S] D. Sullivan, Thesis, Princeton University (1965).

AARHUS UNIVERSITY
NY MUNKEGADE
(8000) AARHUS (C)
DENMARK

and

DEPARTMENT OF MATHEMATICS
UNIVERSITY OF CHICAGO
CHICAGO, IL 60637

Contemporary Mathematics
Volume 19, 1983

Some Root Invariants in $\mathrm{Ext}_A^{*,*}(\mathbf{Z}/2,\mathbf{Z}/2)$.

Mark Mahowald and Paul Shick

It is becoming increasingly clear that the "root invariant", first defined in [M1], is a very important tool for studying v_i-periodicity in stable homotopy theory. A more algebraic definition is given in [DM] and the geometric root invariant defined below appears as the "Mahowald invariant" in [R]. The following table lists the root invariants of elements in the cohomology of the mod 2 Steenrod algebra, A, in the range $t - s \leq 16$. The calculations were carried out using George Whitehead's extensive tables for the mod 2 Lambda algebra for RP^∞ (unpublished).

The root invariant is defined using W.-H. Lin's theorem, [L1], which states that $\lim\limits_{\leftarrow k} \pi_*^S(RP_k^\infty) \simeq \pi_*^S(S^{-1})$. Further, if we denote by P the direct limit of the cohomology groups $H^*(RP_k^\infty;\mathbf{Z}/2)$, then $\mathrm{Ext}_A(P,\mathbf{Z}/2) \simeq \mathrm{Ext}_A(\Sigma^{-1}\mathbf{Z}/2,\mathbf{Z}/2)$. Here, P is isomorphic to the ring of Laurent series, $\mathbf{Z}/2[x,x^{-1}]$, where $|x| = 1$. The Steenrod algebra action is given by $Sq^i x^j = \binom{j}{i} x^{i+j}$.

Let P_m denote the A-module $H^*(RP_m^\infty;\mathbf{Z}/2)$, where m is any integer. We have a map of A-modules $j_m: P_m \to \Sigma^m\mathbf{Z}/2$, induced from the map generating $\pi_m(RP_m^\infty)$. There is also a map $k_m: P_m \to P$, given by the system of maps $RP_{m-k}^\infty \to RP_m^\infty$ which collapse the bottom k cells of RP_{m-k}^∞. With these conventions, we define the root invariant as follows: for $\underline{a} \in \mathrm{Ext}_A^{s,t}(\mathbf{Z}/2,\mathbf{Z}/2)$, we may regard $\underline{a}$ as living in $\mathrm{Ext}_A^{s,t-1}(P,\mathbf{Z}/2)$. There will exist a maximal integer N such that $k_N^*(\underline{a}) \neq 0$ in $\mathrm{Ext}_A^{s,t-1}(P_N,\mathbf{Z}/2)$. We then define the root invariant of $\underline{a}$, $R(\underline{a})$, to be the coset given by $R(\underline{a}) = \{y \in \mathrm{Ext}_A^{s,t-1}(\Sigma^N\mathbf{Z}/2,\mathbf{Z}/2): j_N^*(y) = k_N^*(\underline{a})\}$. Note that N will always be negative, for $s > 0$, by the proof of the

0271-4132/83 $1.00 + $.25 per page

algebraic Kahn-Priddy theorem [L2], so that R will preserve the s-filtration and raise the (t-s)-filtration of an element. The diagram that one should have in mind is:

$$\begin{array}{ccc}
\mathrm{Ext}_A^{s,t}(\mathbf{Z}/2,\mathbf{Z}/2) & \xrightarrow{\ \cong\ } & \mathrm{Ext}_A^{s,t-1}(P,\mathbf{Z}/2) \\
\big\downarrow R & & \big\downarrow k_N^* \quad \text{largest } N \text{ s.t. } k_N^*(\underline{a}) \neq 0 \\
\mathrm{Ext}_A^{s,t-N-1}(\mathbf{Z}/2,\mathbf{Z}/2) & & \\
\big\downarrow \cong & & \\
\mathrm{Ext}_A^{s,t-1}(\Sigma^N \mathbf{Z}/2,\mathbf{Z}/2) & \xrightarrow[\text{same } N]{\ j_N^*\ } & \mathrm{Ext}_A^{s,t-1}(P_N,\mathbf{Z}/2)
\end{array}$$

For $\alpha \in \pi_*^S(S^0)$, we define the geometric root invariant of α, $R_G(\alpha)$, in a similar manner. We may regard α as living in $\pi_{j-1}^S(RP_{-2}^{\infty}L)$, where L is sufficiently large. Then $R_G(\alpha)$ is given by:

$$\begin{array}{ccc}
\pi_j^S(S^0) & \xrightarrow[\text{this range}]{\text{iso in}} & \pi_{j-1}^S(RP_{-2}^{\infty}L) \\
\big\downarrow R & & \big\downarrow k_N^* \quad \text{largest } N \text{ s.t. } k_N^*(\underline{a}) \neq 0 \\
\pi_{j-N-1}^S(S^0) & & \\
\big\downarrow \cong & & \\
\pi_{j-1}^S(S^N) & \xrightarrow[\text{same } N]{\ j_N^*\ } & \pi_{j-1}^S(RP_N^{\infty})
\end{array}$$

It is easy to check that if $\underline{a} \in \mathrm{Ext}_A^{s,t}(\mathbf{Z}/2,\mathbf{Z}/2)$ is an Adams spectral sequence representative for $\alpha \in \pi_{t-s}^S(S^0)$, then if $R(\underline{a})$ represents a homotopy class, that class will be $R_G(\alpha)$.

The table is to be read as follows: "$a \xrightarrow[N]{} b$" means that the root invariant of a is b, and that $k_N^*(a) \neq 0$, but $k_{N+1}^*(a) = 0$. If the element a is a representative for $\alpha \in \pi_*^S(S^0)$, then $R(a)$ is a representative for $R_G(\alpha)$, unless noted with an asterisk. The root invariant for an element a will appear at the usual (t-s,s) coordinates for a. Finally, one should note that $R(h_0^k)$ is equal to the element of least (t-s)-filtration on the line $s=k$, with $t-s > 0$. These represent the Adams-Barratt elements given in [A] or [B].

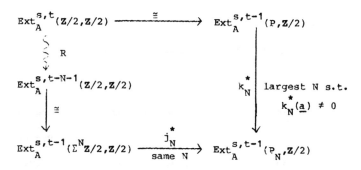

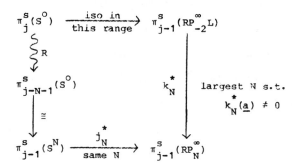

R(a) for $a \in \mathrm{Ext}_A^{**}(\mathbf{Z}/2,\mathbf{Z}/2)$

13	$h_0^{13} \xrightarrow[-26]{} P^3h_1$								
12	$h_0^{12} \xrightarrow[-24]{} h_0^5 i$								
11	$h_0^{11} \xrightarrow[-20]{} P^2h_1^3$								
10	$h_0^{10} \xrightarrow[-19]{} P^2h_1^2$								
9	$h_0^{9} \xrightarrow[-18]{} P^2h_1$								
8	$h_0^{8} \xrightarrow[-16]{} h_0^7h_4$								
7	$h_0^{7} \xrightarrow[-12]{} P^1h_1^3$								
6	$h_0^{6} \xrightarrow[-11]{} P^1h_1^2$								
5	$h_0^{5} \xrightarrow[-10]{} P^1h_1$								
4	$h_0^{4} \xrightarrow[-8]{} h_0^3h_3$							$h_0^3h_3 \xrightarrow[-12]{} h_1^3h_4$	
3	$h_0^{3} \xrightarrow[-4]{} h_1^3$			$h_1^3 \xrightarrow[-7]{} h_2^3$				$h_0^2h_3 \xrightarrow[-11]{} h_1^2h_4$	$c_0 \xrightarrow[-12]{} c_1$
2	$h_0^{2} \xrightarrow[-3]{} h_1^2$		$h_1^2 \xrightarrow[-5]{} h_2^2$	$h_0h_2 \xrightarrow[-6]{} h_1h_3$			$h_2^2 \xrightarrow[-9]{} h_3^3$	$h_0h_3 \xrightarrow[-10]{} h_1h_4$	$h_1h_3 \xrightarrow[-11]{} h_2h_4$
1	$h_0 \xrightarrow[-2]{} h_1$	$h_1 \xrightarrow[-3]{} h_2$		$h_2 \xrightarrow[-5]{} h_3$				$h_3 \xrightarrow[-9]{} h_4^*$	
s=0	$1 \xrightarrow[-1]{} 1$								
t-s =	0	1	2	3	4	5	6	7	8

R(a) for $a \in \mathrm{Ext}_A^{**}(\mathbf{Z}/2,\mathbf{Z}/2)$, continued

s	9	10	11	12	13	14	15	16
8							$h_0^7h_4 \xrightarrow[-28]{} P^1h_1^3h_5$	
7			$P^1h_1^3 \xrightarrow[-23]{} h_1q$				$h_0^6h_4 \xrightarrow[-27]{} P^1h_1^2h_5$	$P^1c_0 \xrightarrow[-24]{} c_1g$
6		$P^1h_1^2 \xrightarrow[-21]{} r^*$	$P^1h_0h_2 \xrightarrow[-22]{} q$			$h_0^2d_0 \xrightarrow[-25]{} y$	$h_0^5h_4 \xrightarrow[-26]{} P^1h_1h_5$	$h_1^2d \xrightarrow[-23]{} h_2^2d_1$
5	$P^1h_1 \xrightarrow[-15]{} h_2g$		$P^1h_2 \xrightarrow[-20]{} h^3h_4^{2*}$			$h_0d_0 \xrightarrow[-20]{} h_1d_1$	$h_0^4h_4 \xrightarrow[-24]{} h_0^3h_3h_5$ $h_1d_0 \xrightarrow[-21]{} h_2d_1$	
4	$h_1c_0 \xrightarrow[-14]{} h_2c_1$					$d_0 \xrightarrow[-19]{} d_1$	$h_0^3h_4 \xrightarrow[-20]{} h_1^3h_5$	
3	$h_2^3 \xrightarrow[-13]{} h_3^3$					$h_0h_3^2 \xrightarrow[-18]{} h_1h_4^2$	$h_0^2h_4 \xrightarrow[-19]{} h_1^2h_5$	
2						$h_3^2 \xrightarrow[-17]{} h_4^2$	$h_0h_4 \xrightarrow[-18]{} h_1h_5$	$h_1h_4 \xrightarrow[-19]{} h_2h_5$
1							$h_4 \xrightarrow[-17]{} h_5$	
s=0								
t−s =	9	10	11	12	13	14	15	16

REFERENCES

[A] J.F. Adams, "On the groups J(X). IV", Topology 5 (1966) 21-71.

[B] M.G. Barratt, "Homotopy operations and homotopy groups", (mimeographed notes), Seattle conference, 1963.

[DM] D. Davis and M. Mahowald, "Ext over the subalgebra A_2 of the Steenrod algebra for stunted projective spaces", to appear in Proc. London, Ontario Conf., Springer Lect. Notes.

[L1] W.-H. Lin, "On conjectures of Mahowald, Segal and Sullivan", Math. Proc. Camb. Phil. Soc. 87 (1980) 449-458.

[L2] ———, "Algebraic Kahn-Priddy Theorem", Pac. J. Math 96 (1981) 435-455.

[M] Mark Mahowald, "The metastable homotopy of S^n", Memoirs of the A.M.S. 72, 1967.

[R] D. Ravenel, "The Segal conjecture for cyclic groups and its consequences", to appeaar in Am. J. Math.

DEPARTMENT OF MATHEMATICS
NORTHWESTERN UNIVERSITY
EVANSTON, IL 60201

and

DEPARTMENT OF MATHEMATICS
NORTHWESTERN UNIVERSITY
EVANSTON, IL 60201

Contemporary Mathematics
Volume 19, 1983

On the Segal Conjecture for Periodic Groups

Haynes Miller and Clarence Wilkerson

Using a hyperext calculation and D. C. Ravenel's "modified Adams spectral sequence" it is shown that the Segal conjecture for $D2^n$ implies the Segal conjecture for $Q2^{n+1}$; in particular, it holds for Q8. Also, a homotopy theoretic argument is provided for Ravenel's theorem that the Segal conjecture for $\mathbf{Z}/p^n$ implies the Segal conjecture for $\mathbf{Z}/p^{n+1}$.

In a talk at the Northwestern University Homotopy Theory Conference in March, 1982, Gunnar Carlsson announced a complete affirmative resolution of the Burnside ring conjecture of G. B. Segal. This represented a tremendous advance over previous knowledge, as the reader will see by glancing at J. F. Adams' report [3] at the Adem conference the preceding August. We refer the reader to that report, or to Adams' earlier report [2], for an account of the conjecture itself. The first author had been analyzing the proof by D. C. Ravenel [13] of the implication, "Segal conjecture for $\mathbf{Z}/p^n \Longrightarrow$ Segal conjecture for $\mathbf{Z}/p^{n+1}$," in hopes of constructing an analogous proof for more general group extensions. The starting point for an induction was to have been the validity of the conjecture for elementary Abelian groups, due to Adams, Gunawardena, and Miller [4], as the starting point for Ravenel's induction was its validity for $\mathbf{Z}/p$, due to W. H. Lin [11], [12], and J. H. C. Gunawardena [9]. (Carlsson's proof in fact uses [4] to ground an induction.) In particular, he noticed that Ravenel actually provided the techniques for two distinct proofs of his result, though, at least in the first version of [13] (available in September, 1981), the two were mixed together. One proof relied on a modification of the Adams spectral sequence, designed to make the algebraic result underlying the theorems of Lin and Gunawardena applicable in the presence of higher torsion. The other was entirely homotopy-theoretic, depending instead on a modification of a certain "Atiyah-Hirzebruch-Serre" spectral sequence. Naturally, the latter appeared to the first author more

---

Both authors were Alfred P. Sloan Fellows, and were supported in part by the NSF.

promising as a pattern for generalization. However, during a visit in November, 1981, the second author convinced him that the modified Adams spectral sequence approach was worth a try, at least for groups over which one has homological control - such as the quaternion groups. Joint work soon led to

Theorem A. The Segal conjecture is valid for the quaternion group $Q_{2^{n+1}}$ provided it is valid for the dihedral group D_{2^n}.

Since the Segal conjecture was known for $D_4 = \mathbf{Z}/2 \times \mathbf{Z}/2$ (a proof was in fact announced by the first author at the Winter AMS meeting in January, 1981), this constituted a proof of it for Q_8, the first non Abelian p-group for which it had been checked. Of course, this result in itself is now of mainly historical interest, and probably only to us, at that. However, we feel that the techniques used merit an audience.

Recall that if DX denotes the Spanier-Whitehead dual of a spectrum X - i.e., the spectrum representing the contravariant functor $W \longmapsto \pi^0(X \wedge W)$ - then the Segal conjecture describes a map to DBG, for any finite group G, which should be a homotopy equivalence. If G is a p-group (which suffices, by independent work of May and McClure, of Laitinen, and of Segal) then DBG is p-adically complete. Our technique yields a general theorem concerning the p-adic completion of DBG for compact Lie groups G which admit a representation which is free away from 0. The only nondiscrete examples of such groups are $SO(2) = S^1$, $Spin(3) = S^3$, and the normalizer N_2 of the maximal torus in S^3. If G is discrete, it must have periodic cohomology. We restrict attention to p-groups; so [8] the only examples are the cyclic p-groups and quaternion 2-groups. Each of these groups has a unique normal subgroup C of order p. Write $\bar{G}$ for the quotient G/C. In the statement of the following theorem, α denotes the adjoint representation of G on its Lie algebra. Generally, we shall denote by B^λ the Thom space of the vector bundle formed over the base B of a principal G-bundle by mixing with a representation λ of G.

Theorem B. Assume G is a periodic p-group, S^1, S^3, or N_2. There is a canonical map

$$DB\bar{G} \vee BG^{\alpha} \longrightarrow DBG$$

which is an equivalence after p-adic completion.

In Section 1 of this paper we recall Ravenel's method of proof and his "modified Adams spectral sequence," and reduce Theorem B to an algebraic calculation which we carry out in Section 2. In particular, we prove the following Theorem in which A is the Steenrod algebra.

Theorem C. Make $\mathbf{F}_2[z^{\pm 1}]$, $|z| = 4$, into an A-algebra by declaring $Sq\, z = z + z^2$. Make $\mathbf{F}_2[w_2, w_3]$, $|w_i| = i$, into an A-algebra by declaring $Sq\, w_2 = w_2 + w_3$, $Sq\, w_3 = w_3$, and let $\mathbf{F}_2[w_2, w_3]_n$ denote the sub A-module of elements of degree n in the variables. Then there is a canonical isomorphism

$$\mathrm{Ext}_A^{s,t}(\mathbf{F}_2, \mathbf{F}_2[z^{\pm 1}]) \cong \bigoplus_{n \geq 0} \mathrm{Ext}^{s-n,t-n}(\mathbf{F}_2, \mathbf{F}_2[w_2, w_3]_n) \quad .$$

The paper ends with an account (due to the first author) of a proof of the inductive step for cyclic groups which does not use Ext.

We wish to thank Doug Ravenel for letting us see an early draft of [13], and the first author thanks Northwestern University for its support and hospitality during the 1981-82 academic year.

Section 1.

We shall set up some machinery valid for any compact Lie group G with a real representation V such that G acts freely off 0. Such groups, and, in fact, such representations, have of course been completely classified [15]. Since G is compact, we are free to give V an equivariant inner product. Let S be the unit sphere and $B_V = G \backslash S$ the orbit space. We begin with a lemma; the notation is as in the introduction.

Lemma 1.1. (i) There is a cofibration sequence

$$B_V^{-V} \longrightarrow BG^{-V} \longrightarrow BG^0$$

in which the first map is induced from the classifying map of $S \longrightarrow B_V$ and the second by including a complement of V (over a finite skeleton of BG) into a trivial bundle.

(ii) ΣB_V^{-V} is Spanier-Whitehead 0-dual to B_V^{α}.

Proof. (i) Ravenel has observed [13] that for any CW complex X and vector bundles α and β over X there is a cofibration sequence

$$S(\alpha)^{\beta} \longrightarrow B(\alpha)^{\beta} \longrightarrow X^{\alpha\oplus\beta}$$

where $S(\alpha)$ and $B(\alpha)$ are the sphere- and disk-bundles of α. Since

$B(\alpha) \cong X$, this may be rewritten, up to homotopy, as

$$(1.2) \qquad S(\alpha)^{\beta} \longrightarrow X^{\beta} \longrightarrow X^{\alpha \oplus \beta}$$

where the second map is induced from the inclusion of 0 into α. By restricting to finite subcomplexes and considering complementary bundles, we obtain the same statement for β virtual. Now take $X = BG$, $\alpha = V$, $\beta = -V$, and notice that $S(\alpha) \cong B$.

(ii) Recall (e.g. from [6]) that if G acts freely on a smooth manifold S with orbit manifold B, then $\tau(B) \oplus \alpha \cong G \backslash \tau(S)$. In our situation, $\tau(S) \oplus \underline{1} \cong S \times V$ as G-vector bundles. To see this, identify $\tau(S)_s$ with $\{v \in V : v \perp s\}$; then $g \in G$ acts by $g_*(s, v) = (gs, gv)$. The isomorphism is then $(s, v, t) \longmapsto (s, v + ts)$. Dividing by G we find

$$G \backslash (S \times V) \cong G \backslash \tau(S) \oplus \underline{1} \cong \tau(B) \oplus \alpha \oplus \underline{1} \ .$$

The result is thus Atiyah duality [5] in this instance. □

Let G and $\overline{G}$ be as in Theorem B. Part (i) of Lemma 1.1 gives the top cofibration sequence in the following important diagram, in which X^0 denotes the Thom space of the zero bundle over X - i.e., X with a disjoint base point adjoined.

$$(1.3) \qquad \begin{array}{ccccc} BG^{-V} & \longrightarrow & BG^{0} & \longrightarrow & \Sigma B_V^{-V} \\ & & \big\downarrow & & \\ & & B\overline{G}^{0} & & \end{array}$$

Our main technical result is

<u>Theorem</u> 1.4. The composite $BG^{-V} \longrightarrow B\overline{G}^{0}$ induces an isomorphism in $\hat{\pi}^q$ for $q > 1 - \dim V$, where $\hat{\pi}^*$ denotes p-adic cohomotopy.

We pause to explain about p-adic completion. Let $\tilde{S}^0$ be the fiber of $S^0 \longrightarrow S^0[\frac{1}{p}]$; it is a Moore spectrum with $H_{-1}(\tilde{S}^0) = \mathbf{Z}[\frac{1}{p}]/\mathbf{Z}$. Following A. K. Bousfield [7], we define the p-adic completion of a spectrum X as the spectrum $\hat{X}$ representing the contravariant functor $W \longmapsto [\tilde{S}^0 \wedge W, X]$. It comes equipped with a canonical map $X \longrightarrow \hat{X}$; and it clearly satisfies

$$D(\tilde{S}^0 \wedge X) \cong (DX)\hat{}$$

where D is Spanier-Whitehead duality as in the introduction. Define

$\hat{\pi}^q(X) = [X, \hat{S}^q]$; then

$$\hat{\pi}^{-q}(X) = \pi_q((DX)\hat{\ }) \quad . \tag{1.5}$$

Now D carries cofibration sequences to cofibration sequences [1], so by Lemma 1.1(ii) we obtain from (1.3) the diagram

$$\begin{array}{ccccc} DBG^{-V} & \longleftarrow & DBG^0 & \longleftarrow & B_V^{\alpha} \\ & & \uparrow & & \\ & & DB\bar{G}^0 & & \end{array}$$

We may now p-adically complete and take a direct limit over increasing V, using the maps induced by the inclusions $V \longrightarrow V \oplus W$. We get:

$$\begin{array}{ccccc} \varinjlim_V (DBG^{-V})\hat{\ } & \longleftarrow & (DBG^0)\hat{\ } & \longleftarrow & (BG^{\alpha})\hat{\ } \\ & \nwarrow & \uparrow & & \\ & & (DB\bar{G}^0)\hat{\ } & & \end{array}$$

Theorem 1.4 implies that the diagonal map is a homotopy equivalence, using (1.5) and the Whitehead theorem. Consequently the cofibration splits, and we have obtained Theorem B.

We remark that Segal's conjecture specifies a map $BG^{\alpha} \longrightarrow DBG^0$, constructed from the transfer, which should enter into a description of the homotopy-type of DBG^0. The appendix to [13] shows that the present map agrees with Segal's; so Theorem A follows.

The proof of Theorem 1.4 is achieved by exhibiting spectral sequences converging to the two completed cohomotopy groups involved, and checking that the map at E_2 is an isomorphism in a suitable range of dimensions. Following Ravenel, we use a modification of the Adams spectral sequence, constructed as follows. Assume given a diagram of cofibration sequences

$$\begin{array}{ccccc} X = X_0 & \longrightarrow & X_1 & \longrightarrow & X_2 \cdots \\ \nwarrow & & \nwarrow & & \\ E_0^0X & & E_1^0X & & \end{array} \tag{1.6}$$

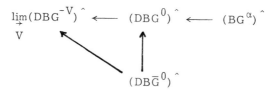

The fiber terms are named E_s^0X because we choose to think of $X \longrightarrow X_s$ as the quotient of X by the sth stage in an increasing filtration, and of E_s^0X as the associated quotient. Pick an Adams resolution [1]

$$\begin{array}{ccccc} \hat{S}^0 = Y^0 & \longleftarrow & Y^1 & \longleftarrow & Y^2 \ \ldots \\ \downarrow & & \downarrow & & \\ K^0 & & K^1 & & \end{array}$$

for the p-adic sphere. By means of mapping telescopes (as in [13]) we may take $F(X_m, Y^n)$ to be a subspectrum of $F(X, \hat{S}^0)$. Filter $F(X, \hat{S}^0)$ by

$$F_s = \bigcup_{m+n\leq s} F(X_m, Y^n) \quad .$$

To describe the E_2-term of the associated homotopy spectral sequence, let $\bar{H}^*(E_\bullet^0X)$ denote the cochain complex of A-modules obtained from (1.6) using the boundary maps. Then [13]

$$E_2 = \mathbf{Ext}_A(\mathbf{F}_2, \bar{H}^*(E_\bullet^0X)) \quad .$$

Here $\mathbf{Ext}_A(\mathbf{F}_2, C^\bullet)$ denotes the hyperext module associated to a cochain complex $C^\bullet$ over A; it is computed by means of a projective resolution $P_\bullet$ of $\mathbf{F}_2$:

(1.7) $$\mathbf{Ext}_A(\mathbf{F}_2, C^\bullet) \cong H(\text{tot } \mathrm{Hom}_A(P_\bullet, C^\bullet)) \quad .$$

To prove Theorem 1.4 we construct filtrations of BG^{-V} and of $B\bar{G}^0$, compatible with the map in question, and verify:

(1.8) $$\mathrm{Ext}_A(\mathbf{F}_p, \bar{H}^*(BG^{-V})) \xrightarrow{\cong} \mathbf{Ext}_A(\mathbf{F}_p, \bar{H}^*(E_\bullet^0BG^{-V}))$$

(1.9) $$\mathrm{Ext}_A(\mathbf{F}_p, e^{-1}\bar{H}^*(BG^0)) \xrightarrow{\cong} \mathbf{Ext}_A(\mathbf{F}_p, \bar{H}^*(E_\bullet^0B\bar{G}^0))$$

where e is the mod p reduction of the period of G. Notice that since the mod p Euler class of V is a nonzero scalar multiple of some power of e, we have

(1.10) $$\bar{H}^q(BG^{-V}) \xrightarrow{\cong} e^{-1}\bar{H}^q(BG^0) \quad \text{iso for } q \geq -\dim V \quad .$$

Using (1.8) and (1.9) it follows that the map $BG^{-V} \longrightarrow B\bar{G}^0$ at $E_2^{s,t}$ is iso for $s-t > -\dim V$. According to [13], the two spectral sequences are convergent in the sense that such an E_2-iso induces an isomorphism in $\hat{\pi}^q$ for $q > 1 - \dim V$; and Theorem 1.4 follows.

Section 2.

In this section we construct filtrations satisfying (1.8) and (1.9). We actually do this explicitly only for $p = 2$ and $G = Q_2n+1$. The modifications required for the other cases may safely be left to the reader, and of course the cases $G = \mathbf{Z}/p^n$ and $G = S^1$ were done by Ravenel [13], though from a slightly different perspective. We shall begin with a proof of Theorem C of the introduction.

If $C^\bullet$ is a cochain complex over A, there are two spectral sequences useful in studying $\mathbf{Ext}_A(\mathbf{F}_p, C^\bullet)$:

$$'E_2 = \mathrm{Ext}_A(\mathbf{F}_p, H(C^\bullet)) \Longrightarrow \mathbf{Ext}_A(\mathbf{F}_p, C^\bullet)$$

$$''E_1 = \mathrm{Ext}_A(\mathbf{F}_p, \natural C^\bullet) \Longrightarrow \mathbf{Ext}_A(\mathbf{F}_p, C^\bullet) \quad .$$

Here $\natural C^\bullet$ denotes $C^\bullet$ without its differential. These are obtained from the usual filtrations of the double complex defining $\mathbf{Ext}_A(\mathbf{F}_p, C^\bullet)$, and result in the following:

Lemma 2.1. Let $f : C^\bullet \longrightarrow D^\bullet$ be a map of bounded below cochain complexes over A. Then $\mathbf{Ext}_A(\mathbf{F}_p, f)$ is an isomorphism if either $H(f)$ or $\mathrm{Ext}_A(\mathbf{F}_p, \natural f)$ is an isomorphism. □

Consider the Laurent series algebra $\mathbf{F}_2[z^{\pm 1}]$ on a generator z of dimension 4, with its natural A-algebra structure: $Sq\, z = z + z^2$. We begin by replacing $\mathbf{F}_2[z^{\pm 1}]$ by a homologically equivalent cochain complex. Consider the bigraded polynomial algebra $\mathbf{F}_2[w_2, w_3]$, $|w_i| = (1, i-1)$, as a cochain complex over A by declaring $d = 0$ and $Sq\, w_2 = w_2 + w_3$, $Sq\, w_3 = w_3$. Let $|x| = 1$ and consider $\mathbf{F}_2[x]$ with $Sq\, x = x + x^2$, and its quotient Hopf algebra over A, $\mathbf{F}_2[x]/x^4$. The map $\tau : \mathbf{F}_2[x]/x^4 \longrightarrow \mathbf{F}_2[w_2, w_3]$ sending $1 \longmapsto 0$, $x \longmapsto w_2$, $x^2 \longmapsto w_3$, $x^3 \longmapsto 0$, is a twisting morphism [10] which is A-linear and has acyclic total complex. For any $\mathbf{F}_2[x]/x^4$-comodule K over A, τ induces an A-linear differential in $\mathbf{F}_2[w_2, w_3] \otimes K$, indicated by decorating the tensor symbol with a superscript τ. An easy calculation shows that $z \longmapsto x^4$ induces an A-linear homology isomorphism

$$\mathbf{F}_2[z] \longrightarrow \mathbf{F}_2[w_2, w_3] \otimes^\tau \mathbf{F}_2[x]$$

where $\mathbf{F}_2[x]$ is an $\mathbf{F}_2[x]/x^4$-comodule via the quotient map. Now x^4 acts on $\mathbf{F}_2[x]$ by comodule maps, so we get a structure of $\mathbf{F}_2[x]/x^4$-comodule over A on $\mathbf{F}_2[x^{\pm 1}]$ and on the $\mathbf{F}_2[x]$-submodule $x^{-4k}\mathbf{F}_2[x]$ generated by x^{-4k}, for any k. Thus we obtain A-linear homology isomorphisms

$$\mathbf{F}_2[z^{\pm 1}] \longrightarrow \mathbf{F}_2[w_2, w_3] \otimes^{\tau} \mathbf{F}_2[x^{\pm 1}] \tag{2.2}$$

$$z^{-k}\mathbf{F}_2[z] \longrightarrow \mathbf{F}_2[w_2, w_3] \otimes^{\tau} x^{-4k}\mathbf{F}_2[x] \quad . \tag{2.3}$$

On the other hand, Lin's theorem [11] asserts that $\mathbf{F}_2 \longrightarrow \mathbf{F}_2[x^{\pm 1}]$ is an isomorphism in $\mathrm{Ext}_A(\mathbf{F}_2, -)$. Since each degree of $\mathbf{F}_2[w_2, w_3]$ is a finite A-module, it follows that

$$\mathbf{F}_2[w_2, w_3] \longrightarrow \Sigma\mathbf{F}_2[w_2, w_3] \otimes^{\tau} \mathbf{F}_2[x^{\pm 1}] \tag{2.4}$$

is an isomorphism in $\mathrm{Ext}_A(\mathbf{F}_2, -)$. Combining (2.2) and (2.4) with Lemma 2.1 yields Theorem C. Notice also that we may tensor everything with a finite A-module without altering the proof. □

Given a coalgebra C, there is a universal twisting morphism $\theta : \Omega^{\bullet}C \longrightarrow C$ from the cobar construction [10]. If C is a coalgebra over A then, using the diagonal tensor product to make $\Omega^{\bullet}C$ an algebra over A, θ becomes A-linear. The acyclicity of the total complex of τ thus implies that the natural map

$$\Omega^{\bullet}(\mathbf{F}_2[x]/x^4) \otimes^{\theta} M \longrightarrow \mathbf{F}_2[w_2, w_3] \otimes^{\tau} M \tag{2.5}$$

is an A-linear homology isomorphism, for any $\mathbf{F}_2[x]/x^4$-comodule M over A.

Now all the above algebra is closely modelled on geometric constructions. Thus, if G is a compact Lie group, BG comes equipped with a canonical filtration, due to Milnor. By [14],

$$H^*(E^0_{\bullet}BG) \cong \Omega^{\bullet}(H^*G) \quad ,$$

using the Pontrjagin coalgebra structure of H^*G, as cochain complexes over A. Moreover, given a G-space X, the pullback of the Milnor filtration filters $EG \times_G X$, and

$$H^*(E^0_{\bullet}(EG \times_G X)) \cong \Omega^{\bullet}(H^*G) \otimes^{\theta} H^*X \quad . \tag{2.6}$$

Suppose we have a pullback diagram of compact Lie groups and homomorphisms

(2.7)
$$\begin{array}{ccc} H & \longrightarrow & G \\ \downarrow & & \downarrow \\ \bar{H} & \longrightarrow & \bar{G} \end{array}$$

in which the vertical maps are epic and the horizontal maps are monic. Let

$$N = \ker(H \longrightarrow \bar{H}) = \ker(G \longrightarrow \bar{G})$$
$$M = G/H = \bar{G}/\bar{H} \quad .$$

M is a smooth $\bar{G}$-manifold via left translation. On the other hand, recall that in general, the fiber of a fibration $E \longrightarrow B\bar{G}$ is canonically homotopy-equivalent to a principal $\bar{G}$-bundle over E, namely, the pullback over E of $E\bar{G} \longrightarrow B\bar{G}$. Thus we may take BN to be a $\bar{G}$-space, by virtue of the fibration sequence $BN \longrightarrow BG \longrightarrow B\bar{G}$.

Lemma 2.8. There are canonical homotopy equivalences

$$BH \cong E\bar{G} \times_{\bar{G}} (M \times BN)$$
$$B\bar{H} \cong E\bar{G} \times_{\bar{G}} M \quad ,$$

under which the map $BH \longrightarrow B\bar{H}$ is induced by $pr_1 : M \times BN \longrightarrow M$.

Proof. If we arrange that $BG \longrightarrow B\bar{G}$ and $B\bar{H} \longrightarrow B\bar{G}$ are fibrations, then BH may be obtained as a pullback in

$$\begin{array}{ccc} BH & \longrightarrow & BG \\ \downarrow & & \downarrow \\ B\bar{H} & \longrightarrow & B\bar{G} \end{array}$$

Thus in the three-dimensional diagram

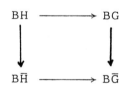

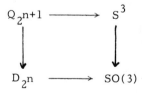

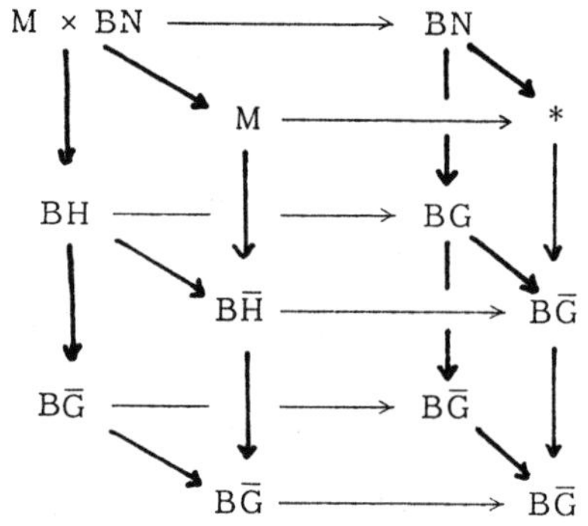

each horizontal square is a pullback and the right-most three columns are fibration sequences. It follows that the left-hand column is a fibration sequence too, and this gives the result. □

Using this lemma we obtain filtrations of BH and $B\bar{H}$ from the Milnor filtration of $B\bar{G}$, and the map $BH \longrightarrow B\bar{H}$ is compatible with them. Moreover, if V is a representation of H, then the Thom spectrum BH^{-V} is filtered by the Thom spectra of the pullback of $-V$ to the filtration degrees of BH. Of course the "transfer" map $BH^{-V} \longrightarrow BH^0$ respects filtrations. We obtain:

(2.9) $$\bar{H}^*(E^0_\cdot BH^{-V}) \cong \Omega^\cdot(H^*G) \otimes^\theta (H^*M \otimes \bar{H}^*BN^{-V})$$

(2.10) $$\bar{H}^*(E^0_\cdot B\bar{H}^0) \cong \Omega^\cdot(H^*G) \otimes^\theta H^*BN \ .$$

We now specialize by taking for (2.7) the diagram

$$\begin{array}{ccc} Q_2n+1 & \longrightarrow & S^3 \\ \downarrow & & \downarrow \\ D_2n & \longrightarrow & SO(3) \end{array}$$

Thus $N = \mathbf{Z}/2$, and M is a certain 3-manifold. Take for V a 4k-dimensional free representation of Q_2n+1; then we claim that (1.8) and (1.9) hold for the filtrations analyzed above in (2.9) and (2.10). We require only the following Lemma whose proof is left to the reader.

Lemma 2.11. (i) The coaction of $H^*SO(3)$ on H^*M is trivial.

(ii) $H^*(BQ_2n+1) \cong \mathbf{F}_2[z] \otimes H^*M$ with the diagonal A-action; $|z| = 4$.

Assembling (2.11), (2.3), (2.5), and (2.9),

$$\begin{aligned}\bar{H}^{*}BQ_{2n+1}^{-V} &\cong z^{-k}\mathbf{F}_2[z] \otimes H^{*}M \\ &\longrightarrow \mathbf{F}_2[w_2, w_3] \otimes^{\tau} x^{-4k}\mathbf{F}_2[x] \otimes H^{*}M \\ &\longleftarrow \Omega^{\bullet}(H^{*}SO(3)) \otimes^{\theta} \bar{H}^{*}BN^{-V} \otimes H^{*}M \\ &\cong \bar{H}^{*}(E_{\bullet}^{0}BQ_{2n+1}^{-V})\end{aligned}$$

where both arrows are homology-isomorphisms and hence, by Lemma 2.1, $\mathbf{Ext}_A(\mathbf{F}_2, -)$-isomorphisms. This gives (1.8). Equation (1.9) is immediate from Theorem C tensored with $H^{*}M$. The compatibility of everything is immediate from the close parallel between the algebra and the geometry.

Section 3.

In this section we give a proof of Theorem 1.4 (and hence of Theorem B) in case $G = \mathbf{Z}/p^n$ or $G = S^1$ which, while using ideas from [13], is free of Ext calculations. To begin with, we outline a faulty proof; then we indicate how to fix it up. For notational convenience, take p to be 2.

Let G be any compact Lie group and C a normal subgroup of order 2, with quotient group $\bar{G}$. Let V be a representation of G which restricts to $n = \dim V$ times the sign representation of C. Consider the composite

$$BG^{-V} \longrightarrow BG^{0} \longrightarrow B\bar{G}^{0} .$$

Pseudoproposition 3.1. This composite induces an isomorphism in $\hat{\pi}^q$ for $q > 1-n$.

Pseudoproof. If ξ is a (virtual) vector bundle over the total space E of a fibration $p : E \longrightarrow B$, and h^{*} is any cohomology theory, then we have a relative Atiyah-Hirzebruch-Serre spectral sequence

$$H^{*}(B;\, h^{*}(F^{\xi})) \Longrightarrow h^{*}(E^{\xi})$$

where F is the fiber of p. Apply this with $p : BG \longrightarrow B\bar{G}$ and $\xi = -V$, so that F^{ξ} is the "stunted real projective space" P_{-n}, and with $h^{*} = \hat{\pi}^{*}$. There is a map of spectral sequences

$$\begin{array}{ccc} H^*(B\bar{G};\ \hat{\pi}^*) & \Longrightarrow & \hat{\pi}^*(B\bar{G}^0) \\ \big\downarrow & & \big\downarrow \\ H^*(BG;\ \hat{\pi}^* P_{-n}) & \Longrightarrow & \hat{\pi}^*(BG^{-V}) \end{array}$$

Lin's theorem asserts that $\hat{\pi}^q \longrightarrow \hat{\pi}^q P_{-n}$ is iso for $q > 1-n$, so the map is iso at E_2 in a range, and hence in the abutment. □

The first author was very happy to have found such a short proof, till the next morning, when Mark Mahowald pointed out that the result contradicted the validity of the Segal conjecture for $\mathbf{Z}/2 \times \mathbf{Z}/2$. Some comfort was gained from the later realization that he was not the first to fall into this trap, which was dubbed the "canonical error."

The trouble, of course, lies in the fact that Lin's theorem gives us an isomorphism above a horizontal line, while what is needed to conclude anything about the abutment is an isomorphism above a line of slope -1. In order to explain how to achieve this, in case G is cyclic or $G = S^1$, we first express the "Serre filtration," inducing the spectral sequence of the pseudo-proof, in terms of Thom spectra. For clarity, we deal only with $G = S^1$; the case of a cyclic p-group follows similar lines.

Let $X = \mathbf{C}P^\infty$, and consider the map $p : X \longrightarrow X$ induced by squaring in S^1. Let λ be the canonical one-dimensional complex representation of S^1. The Serre filtration of $X^{-n\lambda}$, viewed as a succession of quotients, is equivalent to the system of cofibration sequences

$$\begin{array}{ccccccc} X^{-n\lambda} & \longrightarrow & X^{\lambda^2 - n\lambda} & \longrightarrow & X^{2\lambda^2 - n\lambda} & \longrightarrow & \cdots \\ \nwarrow & & \nwarrow & & \nwarrow & & \\ P_{-2n} & & \Sigma^2 P_{-2n} & & \Sigma^4 P_{-2n} & & \end{array}$$

The horizontal maps are induced from the inclusion $0 \hookrightarrow \lambda^2$, and the identification of the cofibers results from (1.2) together with the observation that $S(\lambda^2) = \mathbb{R}P^\infty$.

We propose to alter this system so as to obtain stunted projective spaces with bottom cell in lower and lower dimension. To achieve this, we employ a device which Ravenel used (needlessly, as shown above), in his $\mathbf{Ext}$ proof. Let λ be any complex line bundle over an arbitrary CW complex X. The map $S(\lambda) \longrightarrow S(\lambda^2)$ by $z \longmapsto z^2$ induces a map $\psi : X^\lambda \longrightarrow X^{\lambda^2}$ of Thom spaces which is trivial in mod 2 cohomology (since it has degree 2 on the Thom class). We may also twist with a virtual vector bundle α to obtain a map $\psi : X^{\alpha+\lambda} \longrightarrow X^{\alpha+\lambda^2}$ of Thom spectra.

Apply this with X and λ as above, and $\alpha = -(k+i+1)\lambda + i\lambda^2$, to obtain a map

$$X^{-(k+i)\lambda+i\lambda^2} \longrightarrow X^{-(k+i+1)\lambda+(i+1)\lambda^2} .$$

It is not hard to check that the cofiber of this map is $\Sigma^{2i-1}P_{-2(k+i)+1}$, and that we receive a commutative diagram

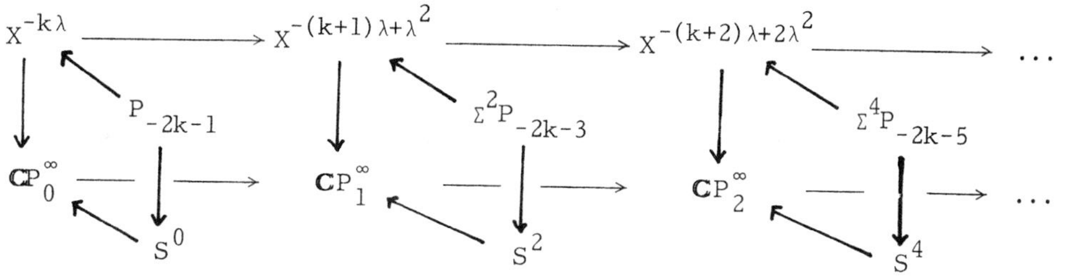

in which the vertical fiber maps are exactly those guaranteed by Lin's theorem to induce isomorphisms in $\hat{\pi}^q$ for $q > -2k$.

What remains is to prove convergence of each spectral sequence: i.e., that $\varprojlim \hat{\pi}^* = \varprojlim^1 \hat{\pi}^* = 0$ in both cases.

The $\varprojlim^1 \hat{\pi}^*$ terms vanish in both cases since these are inverse systems of compact Abelian groups. Since each horizontal map in the top "exact couple" is trivial in mod 2 cohomology, any string in $\varprojlim \hat{\pi}^* X^{-(k+i)\lambda+i\lambda^2}$ consists entirely of elements of infinite Adams filtration. But the Adams spectral sequences converge [11], [4], so this $\varprojlim$ is zero. For the bottom "exact couple," notice that $\varprojlim \hat{\pi}^*(\mathbf{C}P_i^\infty)$ is contained in the group of phantom completed cohomotopy classes in $\mathbf{C}P^\infty$, i.e., $\varprojlim^1 \hat{\pi}^*(\mathbf{C}P^{i-1})$, and this group is zero, again by compactness.

References

1. J. F. Adams, Stable Homotopy and Generalized Homology, Univ. of Chicago Press, Chicago, 1974.

2. ———, Graeme Segal's Burnside ring conjecture, Bull. (N.S.) Amer. Math. Soc. 6 (1982), 201-210.

3. ———, Graeme Segal's Burnside ring conjecture, Symposium on Algebraic Topology in Honor of José Adem, Contemporary Mathematics 12, Amer. Math. Soc., 1982, 9-18.

4. ———, J.H.C. Gunawardena, and H. R. Miller, in preparation.

5. M. F. Atiyah, Thom complexes, Proc. Lon. Math. Soc. 11 (1961), 291-310.

6. J. C. Becker and R. E. Schultz, Equivariant function spaces and stable homotopy theory, Com. Math. Helv. 49 (1974), 1-34.

7. A. K. Bousfield, The localization of spectra with respect to homology, Topology 18 (1979), 257-281.

8. H. Cartan and S. Eilenberg, Homological Algebra, Princeton Univ. Press, Princeton, 1956.

9. J. H. C. Gunawardena, Segal's Burnside Ring Conjecture for Cyclic Groups of Odd Prime Order, J. T. Knight Prize Essay, Cambridge, 1980.

10. D. Husemoller, J. C. Moore, and J. Stasheff, Differential homological algebra and homogeneous spaces, J. Pure and Appl. Alg. 5 (1974), 113-185.

11. W. H. Lin, On conjectures of Mahowald, Segal, and Sullivan, Math. Proc. Camb. Phil. Soc. 87 (1980), 449-458.

12. ———, D. M. Davis, M. E. Mahowald, and J. F. Adams, Calculation of Lin's Ext groups, Math. Proc. Camb. Phil. Soc. 87 (1980), 459-469.

13. D. C. Ravenel, The Segal conjecture for cyclic groups and its consequences, Am. J. Math., to appear.

14. M. Rothenberg and N. E. Steenrod, The cohomology of classifying spaces of H-space, Bull. Amer. Math. Soc. 71 (1965), 872-875.

15. J. A. Wolf, Spaces of Constant Curvature, McGraw Hill, 1967.

INSTITUTE FOR ADVANCED STUDY
DEPARTMENT OF MATHEMATICS
PRINCETON UNIVERSITY
PRINCETON, N.J. 08544

DEPARTMENT OF MATHEMATICS
WAYNE STATE UNIVERSITY
DETROIT, MI 48202

Contemporary Mathematics
Volume 19, 1983

POWER SERIES METHODS IN UNORIENTED COBORDISM

Stephen A. Mitchell

ABSTRACT. Formal group laws are used to give simple and conceptual proofs of classical theorems of Thom, Milnor, and Landweber on the unoriented cobordism ring and its relation to the complex cobordism ring. Elementary power series methods are used to obtain various formulae of Brown, Davis and Peterson on the coaction in H_*MO and the canonical antiautomorphism of the dual Steenrod algebra.

By a famous theorem of Quillen (see [1] and [9]) the complex cobordism ring MU_* is naturally isomorphic to the Lazard ring L - that is, the universal ring for commutative one-dimensional formal groups. Quillen's work also applied to unoriented cobordism; he showed that MO_* is the universal ring for "formal $\mathbb{Z}/2$-modules" (see Section One below). In this note we show first of all how formal groups can be used to give very simple and conceptual proofs of the following theorems:

THEOREM A (Thom [10]). The unoriented cobordism ring MO_* is a polynomial algebra $\mathbb{Z}/2\,[x_i\colon\ i \neq 2^k-1]$ on generators x_i of dimension i .

Let $r\colon\ MU_* \to MO_*$ denote the forgetful map.

THEOREM B (Milnor [7]). The image of r is the subring of squares in MO_* .

THEOREM C (Landweber [5]). The kernel of r is the ideal C generated by weakly complex manifolds with a free conjugation (a conjugation is an involution reversing the complex structure).

The proof of Theorem A (given in Section 2) combines formal groups with the standard method of first identifying H_*MO as a certain extended A-comodule $A \otimes L_2$. The isomorphism $H_*MO \cong A \otimes L_2$ is viewed here as the "universal example" of an isomorphism between a certain pair of functors

1980 Mathematics Subject Classification 55N22

on the category of commutative rings. Theorems B and C are proved simultaneously in Section three, using only the formal groups; in particular no use is made of characteristic numbers.

Formal groups aside, it has long been known that the coaction maps of the dual Steenrod algebra A and the A-comodule H_*MO are given by simply composing certain universal power series. This means that various classical methods of calculus (some dating back to the 17th century) can be applied. In Section four we use this approach to give complete and extremely elementary proofs of the results of Brown, Davis and Peterson ([2], [3]) on the coaction in H_*MO and the canonical anti-automorphism in A .

Many of the proofs given here have probably been discovered before. Be that as it may, it is the author's opinion that these methods deserve to be more widely appreciated. We have included a summary of the relevant properties of formal $\mathbb{Z}/p$-modules; for general information about formal groups the reader should consult [1] or Appendix 2 of a forthcoming book by Doug Ravenel.

1. FORMAL $\mathbb{Z}/p$-MODULES.

Let R denote a commutative ring with identity. We will call a formal group F over R "p-simple" if the p-fold formal sum $[p]_F(x)$ is zero (i.e., F is a formal $\mathbb{Z}/p$-module). Let G_a denote the additive formal group.

(1.1) THEOREM (Lazard) F is p-simple if and only if Char R = p and F is (strictly) isomorphic to G_a .

In other words, if $[p]_F(x) = 0$ then F admits a logarithm $\log_F(x)$. Of course this logarithm is not unique, since we may compose with any automorphism of G_a ; note that when Char R = p , $\mathrm{End}_R\, G_a = \{f(x)\colon\ f(x) = \sum_{i=0}^{\infty} a_i x^{p^i}\}$. Nevertheless, a canonical log is defined by the following proposition ([9])

(1.2) PROPOSITION A p-simple formal group F admits a unique log of the form $\log_F^c(x) = \sum_{i=0}^{\infty} a_i x^{i+1}$ with $a_0 = 1$ and $a_{p^i-1} = 0$ for $i > 0$.

PROOF: Given $f(x) = \sum_{i=0}^{\infty} a_i x^{i+1}$, define $\overline{f}(x) = \sum_{i=0}^{\infty} a_{p^i-1} x^{p^i}$.

Then if $\phi \in \mathrm{Aut}_R G_a$ we have $\overline{\phi f} = \phi \overline{f}$. Hence, taking f to be any log of F and ϕ to be the composition inverse of $\overline{f}$, the desired canonical log is defined by $\log_F^c(x) = \phi f$. A similar argument shows $\log_F^c(x)$ is unique.

Let F_p denote the functor (Rings) $\to$ (Sets) given by $F_p(R) =$ {p-simple formal groups over R}. Then F_p is corepresented by a certain universal ring L_p.

(1.3) <u>COROLLARY</u> $L_p = \mathbb{Z}/p\,[y_n: \ n \geq 1 \ , \ n \neq p^i-1]$, where the y_n are the coefficients of the universal canonical log.

(1.4) <u>REMARKS</u> (a) Let L denote the usual Lazard ring; L corepresents the functor {formal groups over R}. For any positive integer d, let $I_d \subseteq L$ denote the ideal generated by the coefficients of the universal d-sequence. Then obviously $L_p = L/I_p$.

(b) The inverse of a log is called an exponential; in particular the canonical exponential $\exp_F^c$ is the inverse of $\log_F^c$.

(c) Another construction of the canonical log is the following: Any formal group F over a $\mathbb{Z}_{(p)}$-algebra is canonically isomorphic to a p-typical formal group $F^{(p)}$. If F is p-simple, then $F^{(p)}$ is both p-simple and p-typical, and so must equal the additive group G_a (a p-typical formal group is uniquely determined by its p-sequence). Hence we obtain an isomorphism $F \to G_a$, which is easily seen to coincide with $\log_F^c$.

(d) If $f(x) = \sum_{i=0}^{\infty} a_{p^i-1}x^{p^i}$, it is easy to show that the coefficients of its composition inverse are $c_{p^i-1} = s_{p^i-1}(0,\ldots,0, a_{p-1}, 0,\ldots,0, a_{p^i-1})$, where s_{p^i-1} is the Newton polynomial. Alternatively, if $\frac{x}{f(x)} = \sum d_i x^i$, one has $c_{p^i-1} = -d_{p^i-1}$ (cf. 4.4).

2. <u>THE STRUCTURE OF MO_*</u>. Let $A = \mathbb{Z}/2\,[\xi_1, \xi_2, \ldots]$ be the dual Steenrod algebra. Recall that the coproduct $A \xrightarrow{\eta} A \otimes A$ can be written $\eta(\xi(x)) = (1 \otimes \xi)((\xi \otimes 1)(x))$ where $\xi(x)$ is the formal power series $\sum_{i=0}^{\infty} \xi_i x^{2^i}$. Similarly, the coaction Ψ: $H_*MO \to A \otimes H_*MO$ can be written $\Psi(b(x)) = (1 \otimes b)((\xi \otimes 1)(x))$ (or simply $\Psi b = b(\xi)$), where $b(x) = \sum_{i=0}^{\infty} b_i x^{i+1}$ and the b_i are the usual generators of H_*MO.

Functors $\mathcal{A}$ and $\mathcal{E}$ are defined on our category of rings by $\mathcal{A}(R) = \mathrm{Aut}_R G_a$ and $\mathcal{E}(R)$ = set of exponentials $\exp_F$ of 2-simple formal groups over R . Moreover it is clear that $\mathcal{A}$ is corepresented by A and $\mathcal{E}$ is corepresented by H_*MO . The group structure on $\mathcal{A}(R)$ (composition of power series) yields a Hopf algebra structure on A which is clearly the usual one. This implies, for example, that the conjugate $\zeta(x)$ of $\xi(x)$ is just the composition inverse of $\xi(x)$. Similarly, the right action of $\mathcal{A}(R)$ on $\mathcal{E}(R)$ defined by $f \cdot \exp_F = \exp_F(f)$ yields a left coaction $H_*MO \to A \otimes H_*MO$, which again coincides with the usual one.

We may now prove:

(2.1) PROPOSITION There is an isomorphism of A-comodule algebras

$H_*MO \xrightarrow{\cong} A \otimes L_2$ (here $A \otimes L_2$ is the extended A-comodule on L_2) .

PROOF: $\mathcal{A}(R)$ acts freely on $\mathcal{E}(R)$, with orbit space $\mathcal{F}(R)$ = {2-simple formal groups over R} . The projection $\mathcal{E}(R) \to \mathcal{F}(R)$ has a section defined by the canonical exponential; this induces in the usual way an $\mathcal{A}$-equivariant isomorphism of functors $\mathcal{F} \times \mathcal{A} \xrightarrow{\cong} \mathcal{E}$. The resulting map $H_*MO \to A \otimes L_2$ of universal rings is then an isomorphism of comodule algebras.

The usual argument then shows that the Hurewicz map $\pi_*MO \to H_*MO$ is an isomorphism onto L_2 , proving Theorem A (and providing a canonical set of generators).

(2.2) REMARKS (a) The formal group over MO_* obtained from RP^∞ is 2-simple, so there is an induced ring homomorophism $L_2 \to MO_*$. The argument of ([1], 6.5) shows this is an isomorphism inverse to the one described above. (b) At odd primes $\mathcal{A}$ is corepresented by $S = \mathbb{Z}/p\,[\xi_1, \xi_2, \ldots]$. We obtain as in (2.1) an isomorphism of S-comodule algebras $H_*(MU; \mathbb{Z}/p) \cong S \otimes L_p$. (c) One advantage of the proof of (2.1) given here is that it easily yields many explicit formulae. For example, one obtains a canonical splitting s: $A \to H_*MO$ of the Thom homomorphism $H_*MO \to A$, such that $s(\zeta_i) = m_{2^i-1}$ ($m(x) = \sum_0^\infty m_i x^{i+1}$ is the composition inverse of $b(x)$). To see this, consider the morphism of functors $\mathcal{E} \to \mathcal{A}$ given by $\exp_F \longrightarrow \log_F^c(\exp_F)$; this is one factor of the isomorphism $\mathcal{E} \to \mathcal{F} \times \mathcal{A}$ given above. The resulting map s of universal rings maps $\zeta(x)$ to $\overline{m}(x)$, by the proof of (1.2). S. Bullett has shown [4] that this splitting coincides with the one obtained geometrically from "braid bordism". (See also (4.6) below.)

3. THE RELATION BETWEEN MU_* AND MO_* . In this section we make use of the theorem of Quillen (cf. [1]) asserting that the natural map $L \to MU_*$ is an isomorphism. We also recall:

(3.1) PROPOSITION Let J_d denote the ideal in MU_* generated by the Fermat hypersurfaces $X^n(d) = \{y \in CP^{n+1} : \sum y_i^d = 0\}$. Then $J_d = I_d$.

PROOF: $X^n(d)$ has normal bundle λ^d in CP^{n+1} ; hence $[d]_{MU}(z) \cap [CP^{n+1}] = [X^n(d)] \in MU_* CP^{n+1}$, by the definition of cap product. Hence if

$$CP(z) = \sum_{i=0}^{\infty} [CP^i] z^i \quad \text{and} \quad X(z) = \sum_{i=0}^{\infty} [X^i] z^i \text{ , we have } ([d]_{MU}(z))(CP(z)) = X(z) \text{ .}$$

Hence $J_d \subseteq I_d$. But $CP(z)$ is invertible, so also $I_d \subseteq J_d$.

Now suppose E is a cohomology theory with a real orientation class $x \in E^2 CP^{\infty}$. Then E_* receives three formal groups: F_R (obtained from x) , F_C (obtained from z) and F_R^σ (induced from F_R by the Frobenius or squaring map $\sigma: E_* \to E_*$). Let $j: RP^{\infty} \to CP^{\infty}$ be the nontrivial map. Since j is a map of H-spaces we have

(3.2) PROPOSITION If $j^* z = x^2$, then $F_C = F_R^\sigma$.

Now MO, with its usual orientations, satisfies the hypothesis of (3.2). Hence there is a commutative diagram

(3.3)

$$
\begin{array}{ccc}
MU_* & \xrightarrow{\quad r \quad} & MO_* \\
& {\scriptstyle f} \searrow \quad \nearrow {\scriptstyle \sigma} & \\
& MO_* &
\end{array}
$$

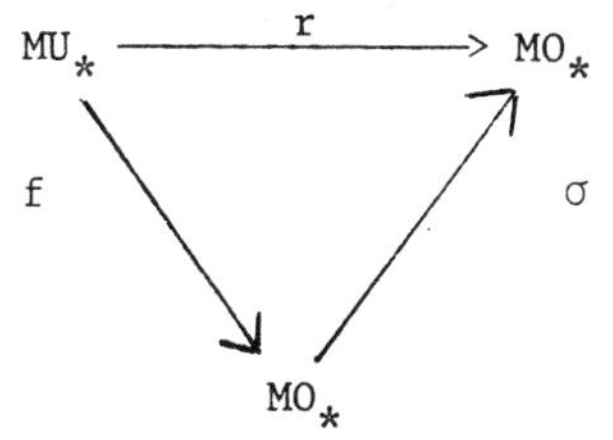

where f is the map inducing F_R . Thus f is the canonical surjection $L \to L_2$. In particular Im r = Im σ , which proves Theorem B. Since σ is injective, Ker r = Ker f = I_2 = J_2 (using 3.1). But J_2 is generated by the Fermat quadrics, which admit the free conjugation defined by $[y_0, \ldots, y_{n+1}] \longrightarrow [\bar{y}_0, \ldots, \bar{y}_{n+1}]$. Hence Ker r $\subseteq$ C . Since obviously C $\subseteq$ Ker r (any manifold M with free involution T bounds the evident one-disc bundle over M/T), the proof of Theorem C is complete.

4. POWER SERIES METHODS FOR H_*MO . Let $f(x) = \sum_{i\geq N} a_i x^i$ be a Laurent series over a commutative ring R . Then we can define $d\log f = \frac{f'}{f}\,dx$. The residue of f , denoted $\oint f(x)dx$, is the coefficient a_{-1} . The following was used in ([1], 7.5):

(4.1) SUBSTITUTION LEMMA Suppose $g(y) = \sum_{i=0}^{\infty} c_i y^{i+1}$ with $c_o = 1$. Then $\oint f(x)dx = \oint f(g(y))g'(y)dy$.

For any $f(x)$, we let $\hat{f}(x) = f(x)/x$. The coproduct or diagonal on $H_*BO \cong H_*MO$ can then be written $\Delta\hat{b} = (\hat{b}\otimes 1)(1\otimes\hat{b})$. Now let

$d\log b(x) = \sum_{i\geq 0} p_i x^{i-1} dx$ (note $p_i = b_1^i \bmod(b_2, b_3, \ldots)$) .

(4.2) PROPOSITION The p_i are nonzero coalgebra primitives.

PROOF: $\Delta(d\log\hat{b}) = d\log(\Delta\hat{b}) = d\log(\hat{b}\otimes 1) + d\log(1\otimes\hat{b})$.

Theorem 1.1 of [3] can now be restated as

(4.3) PROPOSITION $\Psi(\frac{d}{dx}\log b(x)) = (\frac{d}{dx}\log b)(\xi(x))$. (Composition is to be understood here.)

PROOF: $\Psi(\frac{d}{dx}\log b) = \frac{d}{dx}\log\Psi(b) = b'(\xi)\cdot\xi'/b(\xi) = \frac{d}{dx}\log b)(\xi)$, since $\xi' = 1$.

(4.4) PROPOSITION ([3], Theorem 1.2) $(\hat{\zeta})_n^k = (\hat{\xi})_n^{-(n+k+1)}$ (The subscript refers to the coefficient of x^n .)

PROOF: $(\hat{\xi})_n^{-(n+k+1)} = \oint \frac{1}{x^{n+1}\hat{\xi}^{n+k+1}}\,dx = \oint \frac{x^k}{\xi^{n+k+1}}\,dx$. Now let $x = \zeta(y)$, so that dx = dy . We obtain $\oint \frac{\zeta(y)^k}{y^{n+k+1}}\,dy = (\hat{\zeta})_n^k$.

Theorem 1.4 of [3] (see also [2], Theorem 2.7) follows immediately from:

(4.5) PROPOSITION (a) $m_{2^i-1} = p_{2^i-1}$ (b) Recall $\overline{m}(x) = \sum_0^{\infty} m_{2^i-1} x^{2^i}$. Then $\Psi\overline{m} = \zeta(\overline{m})$.

PROOF: (a) $p_n = \oint \frac{d\log b}{x^n}$. Now let $x = m(y)$, $dx = m'(y)dy$.

We obtain $\oint \frac{b'(m(y))m'(y)}{(m(y))^n y} dy = \oint \frac{dy}{(m(y))^n y} = (\hat{m}(y))^{-n}_n$. Now let $n = 2^i-1$.

Then $\hat{m}(y)^{-n} = \hat{m}(y)/(\hat{m}(y))^{2^i} = \hat{m}(y) \bmod y^{2^i}$. Hence $p_{2^i-1} = m_{2^i-1}$.

(b) $\Psi b = b(\xi)$, so $\Psi m = \zeta(m)$. Hence $\Psi\overline{m} = \overline{\zeta(m)} = \zeta(\overline{m})$.

REMARK Of course (b) also follows from the fact that $s(\zeta) = \overline{m}$, where s is the canonical splitting of the Thom homomorphism discussed in Section 1. Moreover we can now easily show that this splitting (obtained from formal groups) agrees with the splitting obtained geometrically from Mahowalds Ω^2S^3 construction [6]: Recall there is a map $\Omega^2S^3 \xrightarrow{\gamma} BO$ such that the composite ϕ: $T(\gamma) \xrightarrow{\gamma'} MO \xrightarrow{\alpha} K\mathbb{Z}/2$ is an equivalence, where $T(\gamma)$ is the Thom spectrum over Ω^2S^3 obtained from γ , γ' is the Thomification of γ , and α is the Thom class.

(4.6) PROPOSITION $s = \gamma'_*\phi_*^{-1}$.

PROOF: $H_*T(\gamma) = H_*\Omega^2S^3$ is a polynomial algebra on certain generators x_{2^i-1} , $i \geq 1$, such that $\gamma'_*x_{2^i-1} = p_{2^i-1}$ ([8]) . But $p_{2^i-1} = m_{2^i-1}$, and it follows that $\phi_*(x_{2^i-1}) = \zeta_i$. Hence $\gamma'_*\phi_*^{-1}(\zeta_i) = m_{2^i-1} = s(\zeta_i)$.

Our last example is another simple residue calculation, and is left to the reader.

(4.7) PROPOSITION $m_{2^i-2} = b_{2^i-2}$.

All the results of this section (except 4.6) have analogues for $H_*(MU; \mathbb{Z}/p)$. Details are left to the reader.

BIBLIOGRAPHY

1. J. F. Adams, Stable Homotopy and Generalized Homology, University of Chicago Press, Chicago, 1974.

2. E. H. Brown and F. P. Peterson, H^*MO as an algebra over the Steenrod algebra, Notas de Mat. y Simposia, Vol. 1, 11-21, (Soc. Mat. Mexico, 1975).

3. E. H. Brown, D. M. Davis and F. P. Peterson, The homology of BO and some results about the Steenrod algebra, Math. Proc. Camb. Phil. Soc. 81 (1977), 393-398.

4. S. R. Bullett, Permutations and braids in cobordism theory, Proc. London Math. Soc. 38 (1979), 517-531.

5. P. S. Landweber, Fixed point free conjugations on complex manifolds, Annals of Math. 86 (1967).

6. M. Mahowald, A new infinite family in ${}_2\pi_*^S$, Topology 16 (1977) 249-256.

7. J. Milnor, On the Stiefel-Whitney numbers of complex manifolds and of spin manifolds, Topology 3 (1965), 223-230.

8. S. B. Priddy, $K\mathbb{Z}/2$ as a Thom spectrum, Proc. Amer. Math. Soc. 70 (1978) 207-208.

9. D. Quillen, Elementary proofs of some results of cobordism theory using Steenrod operations, Advances in Math. 7 (1971), 29-56.

10. R. Thom, Quelques propriétés globales des variétés differentiables, Comm. Math. Helv. 28 (1954), 17-86.

DEPARTMENT OF MATHEMATICS
MASSACHUSETTS INSTITUTE OF TECHNOLOGY
CAMBRIDGE, MASSACHUSETTS 02139

Contemporary Mathematics
Volume **19**, 1983

ROTHENBERG SEQUENCES AND THE ALGEBRAIC S^1-BUNDLE TRANSFER

Hans J. Munkholm[1]

ABSTRACT. It is shown that the algebraic S^1-bundle transfer maps on Whitehead groups and Wall groups defined by the author and Erik K. Pedersen in [1] and [2] commute with the maps in the exact sequences of Rothenberg, [4].

0. INTRODUCTION. In [1] and [2] the author and Erik K. Pedersen gave algebraic descriptions of the homomorphisms

$$\varphi^{\#} : K_1(\mathbb{Z}\pi_1(B)) \to K_1(\mathbb{Z}\pi_1(E)), \text{ respectively}$$

$$\varphi^{!} : L^{\varepsilon}_{\ell}(\mathbb{Z}\pi_1(B)) \to L^{\varepsilon}_{\ell+1}(\mathbb{Z}\pi_1(E)), \quad (\varepsilon = h \text{ or } s)$$

defined by pull back in a given S^1-bundle $S^1 \to E \to B$. The algebraically defined maps (which we call "algebraic S^1-bundle transfers") are defined in greater generality (i.e. for rings which are not integral group rings). It is the purpose of this note to prove that the maps $\varphi^{\#}$ and $\varphi^{!}$ fit into the Rothenberg exact sequence in the way expected from geometry. For the case of integral group rings $\mathbb{Z}\pi_1(E)$ and $\mathbb{Z}\pi_1(B)$ this follows from the geometric description of the maps involved. However, in direct computations concerning these maps one cannot stay with integral group maps. Thus an algebraic proof of the commutativity of the diagrams in question is well motivated.

Our basic set up and the results obtained are given in §1. In §2 we recall the various definitions (in matrix terms). Finally §§3-5 contain the proofs.

Ideally the present note should have been §2E of [2]. However, when we wrote that paper we could not quite prove what we wanted.

1980 Mathematics Subject Classification. 57R67, 57Q10, 18F25.

[1]Supported by the Danish Natural Science Research Counsil.

1. STATEMENT OF RESULTS. Let R be an associative ring with unit equipped with the following data

- an anti involution $r \to r^*$,
- an automorphism $r \to r^t$,
- an invertible element t.

Assume that the data satisfy the following identities (for all $r \in R$)

$$(1.1) \qquad \begin{aligned} &r^{*t} = r^{t*},\ (t-1)r = r^t(t-1),\ trt^{-1} = r^{t^2} (=(r^t)^t) \\ &t^* = t^{-1},\ t^t = t. \end{aligned}$$

The ideal $(t-1)R$ is then 2-sided. Let $\varphi : R \to \overline{R} = R/(t-1)R$ be the projection. The quotient ring inherits an automorphism $\overline{r} \to \overline{r}^t$ with $\varphi(r)^{\overline{t}} = \varphi(r^t)$ and an anti involution $\overline{*}$ with $\varphi(r)^{\overline{*}} = \varphi(r^*)$. To make notation fit with the motivating example below we write $*$ for the composite anti involution $\overline{*}\overline{t}$.

Note that then $\varphi(r)^* = \varphi(r^{t*})$.

MOTIVATING EXAMPLE. Let $0 \to \mathbb{Z}/n\mathbb{Z} \to \pi \to \rho \to 1$ be a group extension with the action of ρ on $\mathbb{Z}/n\mathbb{Z}$ given by a map $\omega : \rho \to \{\pm 1\}$. Let t be a generator for $\mathbb{Z}/n\mathbb{Z}$ and let $w_\pi : \pi \to \{\pm 1\}$, $w_\rho : \rho \to \{\pm 1\}$ be homomorphisms having $w_\pi(g) = w_\rho(\varphi(g))\,\omega(\varphi(g))$, $g \in \pi$. For any commutative ring A one may take $R = A\pi$ with its usual anti involution ($g^* = w_\pi(g)g^{-1}$, $g \in \pi$) and with

$$g^t = \begin{cases} -gt^{-1} & , \omega(\varphi(g)) = -1 \\ g & , \omega(\varphi(g)) = 1 \end{cases}$$

Then $(A\pi, *, t)$ fits into the above pattern and $\varphi : A\pi = R \to \overline{R} = A\rho$ becomes the ring homomorphism induced by $\pi \to \rho$. Also the anti involution on $A\rho$ becomes the usual one $(\overline{g} \to w_\rho(\overline{g})\overline{g}^{-1})$.

In the above situation there is defined, see [1], an <u>algebraic S^1-transfer map</u> $\varphi^\# : K_1(\overline{R}) \to K_1(R)$ (the definition will be recalled below). It induces a map $\widetilde{\varphi}^\# : \widetilde{K}_1(\overline{R}) = K_1(\overline{R})/\{\pm 1\} \to \widetilde{K}_1(R)$. For any subgroup $\overline{X}$ of $\widetilde{K}_1(\overline{R})$ we let X be the subgroup of $\widetilde{K}_1(R)$ generated by $\widetilde{\varphi}^\#(\overline{X})$ and (the images of) t^i, $i \in \mathbb{Z}$.

Now let $\overline{X} \subseteq \overline{Y}$ be $*$-invariant subgroups of $\widetilde{K}_1(\overline{R})$. It is our purpose to show the following

PROPOSITION A. $X \subseteq Y$ *are* $*$*-invariant subgroups of* $K_1(R)$, *and the map* $\widetilde{\varphi}^{\#} : \overline{Y}/\overline{X} \to Y/X$ *induced by* $\widetilde{\varphi}^{\#}$ *anti commutes with* $*$ *so that it induces a map of Tate cohomology*

$$\hat{\varphi}^{\#} : \hat{H}^{\ell}(\mathbb{Z}/2\mathbb{Z};\overline{Y}/\overline{X}) \to \hat{H}^{\ell+1}(\mathbb{Z}/2\mathbb{Z};Y/X)$$

PROPOSITION B. The formulas for the algebraic S^1*-transfer map* $\varphi^! : L^s_{\ell}(\overline{R}) \to L^s_{\ell+1}(R)$ *define a map* $\varphi^! : L^{\overline{X}}_{\ell}(\overline{R}) \to L^X_{\ell+1}(R)$.

PROPOSITION C. The following diagram, the rows of which are Rothenberg type sequences, commutes

$$\begin{array}{ccccccccc}
\cdots \to & L^X_{2k+2}(R) & \xrightarrow{a} & L^Y_{2k+2}(R) & \xrightarrow{b} & \hat{H}^{2k+2}(\mathbb{Z}/2\mathbb{Z};Y/X) & \xrightarrow{c} & L^X_{2k+1}(R) & \xrightarrow{d} \\
 & \uparrow\varphi^! & (1) & \uparrow\varphi^! & (2) & \uparrow\hat{\varphi}^{\#} & (3) & \uparrow\varphi^! & (4) \\
\cdots \to & L^{\overline{X}}_{2k+1}(\overline{R}) & \xrightarrow[d]{} & L^{\overline{Y}}_{2k+1}(\overline{R}) & \xrightarrow[e]{} & \hat{H}^{2k+1}(\mathbb{Z}/2\mathbb{Z};\overline{Y}/\overline{X}) & \xrightarrow[f]{} & L^{\overline{X}}_{2k}(\overline{R}) & \xrightarrow[a]{}
\end{array}$$

$$\begin{array}{ccccccc}
\to & L^Y_{2k+1}(R) & \xrightarrow{e} & \hat{H}^{2k+1}(\mathbb{Z}/2\mathbb{Z};Y/X) & \xrightarrow{f} & L^X_{2k}(R) & \to \cdots \\
 & \uparrow\varphi^! & (5) & \uparrow\hat{\varphi}^{\#} & (6) & \uparrow\varphi^! & \\
\xrightarrow[a]{} & L^{\overline{Y}}_{2k}(\overline{R}) & \xrightarrow[b]{} & \hat{H}^{2k}(\mathbb{Z}/2\mathbb{Z};\overline{Y}/\overline{X}) & \xrightarrow[c]{} & L^{\overline{X}}_{2k-1}(\overline{R}) & \to \cdots
\end{array}$$

2. DEFINING FORMULAS. In this section we present the matrix formulas for the maps $\varphi^{\#}: K_1(\overline{R}) \to K_1(R)$ and $\varphi^!: L^{\overline{X}}_{\ell}(\overline{R}) \to L^X_{\ell+1}(R)$

We extend $t,*$ to the matrix ring $M_n(R)$ by $(a_{ij})^t = (a^t_{ij})$ and $(a_{ij})^* = a_{ji}{}^*)$. Then the identities (1.1) still hold when $r \in M_n(R)$. We write $\varphi : M_n(R) \to M_n(\overline{R})$ for the map of matrix rings induced by $\varphi : R \to \overline{R}$.

We shall have occasion to use the inverse of t both as element and as automorphism. For brevity we let $t^{-1} = s$. We then have the following identities (for any $r \in M_n(R)$)

(2.1) $r^{*s} = r^{s*}$, $(s-1)r = r^s(s-1)$, $srs^{-1} = r^{s^2} (=(r^s)^s)$

$s^* = t\ ,\ s^s = s$

(2.2) $\varphi(r)^* = \varphi(r^s{}^*)$

(2.3) $s^t = s\ ,\ t^s = t\ ,\ st = ts$ (in two senses)

Also we shall write 1 for the identity matrix the size of which must often be inferred from the context. Similarly if $r \in R$ then r can also denote r times the identity matrix.

Recall that $K_1(R) = G\ell(R)/[G\ell(R), G\ell(R)] = G\ell(R)_{ab}$ where $G\ell(R) = \bigcup_n G\ell_n(R)$. Write $[A] \in K_1(R)$ for the element represented by $A \in G\ell_n(R) \subseteq G\ell(R)$.

DEFINITION 2.4. For any $\bar{A} \in G\ell_r(\bar{R})$ *there exist matrices* $A,B,C \in M_r(R)$ *such that* $BA = 1 + (1-s)C$. *For any such choice one has*

$$\varphi^{\#}([\bar{A}]) = \begin{bmatrix} A & s-1 \\ -C & B^t \end{bmatrix} \in K_1(R)$$

REMARK. This formula differs slightly from the one given in [1]. The difference is due to the fact that there we used left modules, here we use right modules.

DEFINITION 2.5. Let $\bar{\alpha} \in M_r(\bar{R})$ *represent an element* $[\bar{\alpha}] \in L_{2k}^{\bar{X}}(\bar{R})$ *and let* $\alpha \in M_r(R)$ *have* $\varphi(\alpha) = \bar{\alpha}$. *Then there is a unitary* $2r \times 2r$ *matrix over* R *of the form*

$$\begin{pmatrix} ? & \eta \\ ? & \beta \end{pmatrix} = \begin{pmatrix} ? & (-1)^k(s-1) \\ ? & \alpha^t + (-1)^k\alpha^* s \end{pmatrix}$$

(the form of the $r \times r$ *matrices called* ? *is immaterial) and*

$$\varphi^!([\bar{\alpha}]) = \begin{bmatrix} ? & \eta \\ ? & \beta \end{bmatrix} \in L_{2k+1}^X(R)$$

REMARK. In section 4 we recall how the Wall groups are defined in matrix terms.

DEFINITION 2.6. Let $\bar{A} \in M_{2r}(\bar{R})$ *be a unitary matrix representing an element* $[\bar{A}] \in L_{2k+1}^{\bar{X}}(\bar{R})$, *and let* $A \in M_{2r}(R)$ *have* $\varphi(A) = \bar{A}$. *Then there exist matrices* Θ *and* Z *over* R *such that*

(2.7) $A^{s^*}JA = s\Theta - (-1)^k\,\Theta^{s^*} - (1-s)Z + J$

For any such choice the matrix

$$W = \begin{pmatrix} J + (-1)^k J^* s & A \\ 0 & Z \end{pmatrix}$$

has $W + (-1)^{k+1}W^*$ *nonsingular and*

$$\varphi^!([A]) = [W] \in L^X_{2k+2}(R) .$$

§3. PROOF OF PROPOSITION A.

Since $\varphi^\#([-1]) = \begin{bmatrix} -1 & s-1 \\ 0 & -1 \end{bmatrix} = 0$ we do get $\widetilde{\varphi}^\#\colon \widetilde{K}_1(\overline{R}) \to \widetilde{K}_1(R)$.

If $\overline{A} \in G\ell_r(\overline{R})$ and $A,B,C \in M_r(R)$ have $BA = 1 + (1-s)C$ as in definition 2.4 then $\varphi(B) = \overline{A}^{-1}$ so there also exists $D \in M_r(R)$ with $AB = 1 + (1-s)D$, and hence $B^{t^*}A^{t^*} = 1 - (1-s)tD^*$. Since $\varphi(A^{t^*}) = \overline{A}^*$ it follows that

$$\varphi^\#([\overline{A}]^*)^* = \begin{bmatrix} A^{t^*} & s-1 \\ tD^* & B^{t^2 *} \end{bmatrix}^* = \begin{bmatrix} A^t & Ds \\ t-1 & B^{t^2} \end{bmatrix}$$

$$= \left[\begin{pmatrix} 0 & s \\ 1 & 0 \end{pmatrix} \begin{pmatrix} A^t & Ds \\ t-1 & B^{t^2} \end{pmatrix} \begin{pmatrix} 0 & 1 \\ t & 0 \end{pmatrix}\right]$$

$$= \begin{bmatrix} B & 1-s \\ D & A^t \end{bmatrix}$$

Since

$$\begin{pmatrix} B & 1-s \\ D & A^t \end{pmatrix} \begin{pmatrix} A & s-1 \\ -C & B^t \end{pmatrix} = \begin{pmatrix} 1 & 0 \\ DA-A^tC & 1 \end{pmatrix}$$

is simple we see that

$$\varphi^\#([\overline{A}]^*)^* = -\varphi^\#([\overline{A}]),$$

i.e. $\varphi^\#$ anticommutes with $*$ as claimed. The rest of proposition A is now very easy.

§4. PROOF OF PROPOSITION B.

First we recall the definition of $L^X_\ell(R)$ (for any ring R with an anti involution $*$, and for any $*$-invariant subgroup $X \subseteq \tilde{K}_1(R)$).

Elements $[A] \in L^X_{2k}(R)$ are represented by matrices $A \in M_r(R)$, some $r \geq 1$, having the following properties

(4.1) $A + (-1)^k A^* \in G\ell_r(R)$ and $[A + (-1)^k A^*] \in X \subseteq \tilde{K}_1(R)$.

The addition is given by

(4.2) $[A_1] + [A_2] = [A_1 \oplus A_2]$

And the relations are generated by the following identities

(4.3) $[A] = [A + B - (-1)^k B^*] = [\Sigma^* A \Sigma]$

for any $B \in M_r(R)$ and any $\Sigma \in G\ell_r(R)$ with $[\Sigma] \in X \subseteq \tilde{K}_1(R)$.

(4.4) $\left[\begin{pmatrix} 0 & 1 \\ 0 & 0 \end{pmatrix}\right] = 0$

To describe $L^X_{2k+1}(R)$ let $J_r = \begin{pmatrix} 0 & I_r \\ (-1)^k I_r & 0 \end{pmatrix}$

and call a matrix $A \in M_{2r}(R)$ <u>unitary</u> if there exists $\Theta \in M_{2r}(R)$ with

(4.5) $A^* J_r A = J_r + \Theta - (-1)^k \Theta^*$.

Now elements $[A] \in L^X_{2k+1}(R)$ are represented by matrices $A \in M_{2r}(R)$, some $r \geq 1$, having the following properties

(4.6) A is unitary and $[A] \in X \subseteq \tilde{K}_1(R)$.

We write A in the form $\begin{pmatrix} \alpha_{11} & \alpha_{12} \\ \alpha_{21} & \alpha_{22} \end{pmatrix}$ with $\alpha_{ij} \in M_r(R)$.

Then the addition is given by

(4.7) $\begin{bmatrix} \alpha_{11} & \alpha_{12} \\ \alpha_{21} & \alpha_{22} \end{bmatrix} + \begin{bmatrix} \beta_{11} & \beta_{12} \\ \beta_{21} & \beta_{22} \end{bmatrix} = \begin{bmatrix} \alpha_{11} \oplus \beta_{11} & \alpha_{12} \oplus \beta_{12} \\ \alpha_{21} \oplus \beta_{21} & \alpha_{22} \oplus \beta_{22} \end{bmatrix}$

Finally the relations are generated by the following identities

(4.8) $\begin{bmatrix} 0 & 1 \\ (-1)^k & 0 \end{bmatrix} = 0$

(4.9) $[A] = \left[\begin{pmatrix} (\sigma^*)^{-1} & 0 \\ 0 & \sigma \end{pmatrix} A\right] = \left[\begin{pmatrix} I_r & 0 \\ \nu - (-1)^k \nu^* & I_r \end{pmatrix} A\right]$

$$= \left[A\begin{pmatrix}(\sigma^*)^{-1} & 0\\ 0 & \sigma\end{pmatrix}\right] = \left[A\begin{pmatrix}I_r & 0\\ \nu-(-1)^k\nu^* & I_r\end{pmatrix}\right]$$

for any $\sigma \in G\ell_r(R)$ with $[\sigma] \in X \subseteq \tilde{K}_1(R)$ and for any $\dot{\nu} \in M_r(R)$.

$$(4.10)\qquad \begin{bmatrix}\alpha_{11} & \alpha_{12}\\ \alpha_{11} & \alpha_{22}\end{bmatrix} = \begin{bmatrix}\alpha_{21} & \alpha_{22}\\ (-1)^k\alpha_{11} & (-1)^k\alpha_{12}\end{bmatrix}$$

We start with the formula from definition 2.5. In the proof of theorem 2C.1 of [2] it is shown that there is indeed a unitary matrix of the form $\begin{pmatrix}\sigma & \eta\\ \rho & \beta\end{pmatrix}$ with η,β given in the definition quoted. The unitaricity implies that $\sigma^*\beta + (-1)^k\rho^*\eta = 1$, i.e. $\sigma^{*s}\beta^s = 1 + (1-s)\rho^*$. Now $\varphi(\beta^s) = \bar{\alpha} + (-1)^k\bar{\alpha}^*$ so $[\varphi(\beta^s)] \in \bar{X}$ by assumption. Consequently X contains the element

$$\varphi^{\#}\left(\left[\varphi(\beta^s)\right]\right)^* = \begin{bmatrix}\beta^s & s-1\\ -\rho^* & \sigma^*\end{bmatrix}^* = \begin{bmatrix}\beta^{s*} & -\rho\\ t-1 & \sigma\end{bmatrix}$$

$$= \begin{bmatrix}(-1)^k\beta t & -\rho\\ -(-1)^k\eta t & \sigma\end{bmatrix}$$

But $\begin{pmatrix}(-1)^k\beta t & -\rho\\ -(-1)^k\eta t & \sigma\end{pmatrix} = \begin{pmatrix}0 & 1\\ -1 & 0\end{pmatrix}\begin{pmatrix}\sigma & \eta\\ \rho & \beta\end{pmatrix}\begin{pmatrix}0 & -1\\ (-1)^k t & 0\end{pmatrix}$, so $\begin{bmatrix}\sigma & \eta\\ \rho & \beta\end{bmatrix} \in X$ as desired. Note that here we needed to include the element of $\tilde{K}_1(R)$ represented by t in X.

We leave it to the reader to check that $\begin{bmatrix}? & \eta\\ ? & \beta\end{bmatrix} \in L^X_{2k+1}(R)$ depends only on $[\bar{\alpha}] \in L^{\bar{X}}_{2k}(\bar{R})$. Essentially the proof is identical to the one for the case $\bar{X} = \tilde{K}_1(\bar{R})$ given in section 2C of [2].

Now to the formula given in definition 2.6. We start by showing that $W + (-1)^{k+1}W^*$ <u>is</u> an isomorphism with torsion $[W + (-1)^{k+1}W^*] \in X$. In fact if one applies $(-1)^k(\cdot)^{s^*}$ to (2.7), adds the result to (2.7) and multiplies on the right by K^* $\left(K = J + (-1)^kJ^*s = \begin{pmatrix}0 & 1\\ (-1)^k s & 0\end{pmatrix}\right)$ one gets

$$(4.11)\qquad A^{s^*}KAK^* = 1 + (1-s)(Z + (-1)^{k+1}Z^*)K^*$$

Now $[\bar{A}\,\varphi(K)] = [\bar{A}] \in \bar{X}$ so X contains the element

$$\varphi^{\#}\left(\left[\bar{A}\varphi(K)\right]\right) = \begin{bmatrix}AK^* & s-1\\ -(Z+(-1)^{k+1}Z^*)K^* & A^*K\end{bmatrix}$$

Since $K^*K = KK^* = 1$ and $[K] \in X$ we can multiply on the right by $\begin{pmatrix} 0 & K \\ (-1)^k K^* & 0 \end{pmatrix}$ and on the left by $\begin{pmatrix} 1 & 0 \\ 0 & -1 \end{pmatrix}$ to conclude that $\left[W + (-1)^{k+1} W^*\right] \in X$ as desired.

Once again we leave it to the reader to check that $[W] \in L^X_{2k+2}(R)$ depends only on $[A] \in L^{\overline{X}}_{2k+1}(\overline{R})$. This can be done by adapting the arguments which are given for the case of $\overline{X} = \widetilde{K}_1(\overline{R})$ in section 2D of [2].

§5. PROOF OF PROPOSITION C. We recall from Shaneson [4], see also Ranicki [3], how the maps a,b,c,d,e,f are defined:

(5.1) $a[\alpha] = [\alpha] \in L^Y_{2k+2}(R)$, $[\alpha] \in L^X_{2k+2}(R)$

(5.2) $b([\alpha]) = [\alpha + (-1)^k \alpha^*] \in \hat{H}^{2k+2}(\mathbb{Z}/2\mathbb{Z}, Y/X)$, $[\alpha] \in L^Y_{2k+2}(R)$

(5.3) $c([\varphi]) = \begin{bmatrix} \varphi & 0 \\ 0 & (\varphi^*)^{-1} \end{bmatrix} \in L^X_{2k+1}(R)$, $[\varphi] \in \hat{H}^{2k+2}(\mathbb{Z}/2\mathbb{Z}, Y/X)$

(5.4) $d([A]) = [A] \in L^Y_{2k+1}(R)$, $[A] \in L^X_{2k+1}(R)$

(5.5) $e([A]) = [A] \in \hat{H}^{2k+1}(\mathbb{Z}/2\mathbb{Z}; Y/X)$, $[A] \in L^Y_{2k+1}(R)$

(5.6) $f([\varphi]) = \begin{bmatrix} 0 & \varphi \\ 0 & 0 \end{bmatrix} \in L^X_{2k}(R)$, $[\varphi] \in \hat{H}^{2k+1}(\mathbb{Z}/2\mathbb{Z}, Y/X)$

The squares labelled (1) and (4) commute for trivial reasons.

<u>Square 2</u>: If $[\overline{A}] \in L^{\overline{Y}}_{2k+1}(\overline{R})$ then (since $K^{-1} = K^*$) $e([\overline{A}]) = [\overline{A}] = \left[\varphi(K)\overline{A}\,\varphi(K^*)\right]$ so from (4.11) we see that

$$\hat{\varphi}^{\#} e\left([\overline{A}]\right) = \begin{bmatrix} KAK^* & s-1 \\ -(Z+(-1)^{k+1}Z^*)K^* & A^* \end{bmatrix}$$

Now multiply from the left and the right by $\begin{pmatrix} K^* & 0 \\ 0 & -1 \end{pmatrix}$ and $\begin{pmatrix} 0 & K \\ (-1)^k & 0 \end{pmatrix}$ (note that these matrices represent 0 in $\widetilde{K}_1(R)$) to see that

$$\hat{\varphi}^{\#} e\left([\overline{A}]\right) = \left[\begin{pmatrix} (-1)^k K^*(s-1) & A \\ (-1)^{k+1} A^* & Z + (-1)^{k+1} Z^* \end{pmatrix}\right]$$

Since

$$(-1)^k K^*(s-1) = (J + (-1)^k J^*s) + (-1)^{k+1}(J + (-1)^k J^*s)^*$$

the above matrix also represents $b\varphi^!([\overline{A}])$.

<u>Square 5</u>: If $[\bar{\alpha}] \in L_{2k}^{\bar{Y}}(\bar{R})$ then for suitable matrices ρ, σ over R we have

$$e\varphi^!([\bar{\alpha}]) = \begin{bmatrix} \rho & (-1)^k (s-1) \\ \sigma & \alpha^t + (-1)^k \alpha^* s \end{bmatrix}$$

where $\varphi(\alpha) = \bar{\alpha}$. The unitaricity of the matrix here implies that

$$\rho^*(\alpha^t + (-1)^k \alpha^* s) + \sigma^*(s-1) = 1$$

Applying $(\cdot)^s$ to this equation and noticing that $\varphi(\alpha + (-1)^k \alpha^{*s} s) = \bar{\alpha} + (-1)^k \bar{\alpha}^*$ which represents $b([\bar{\alpha}])$ we see that

$$\hat{\varphi}^\# b[\alpha]) = \begin{bmatrix} \alpha + (-1)^k \alpha^{*s} s & s-1 \\ -\sigma^* & \rho^* \end{bmatrix}$$

Since $*$ acts trivially on $\hat{H}^{2k+1}(\mathbb{Z}/2\mathbb{Z}, Y/X)$ it easily follows that $\hat{\varphi}^\# b = e\varphi^!$

<u>Square 3</u>: Let $[\bar{A}] \in \hat{H}^{2k+1}(\mathbb{Z}/2\mathbb{Z}; \bar{Y}/\bar{X})$, so that $[\bar{A}] = -[\bar{A}]^*$ in $\bar{Y}/\bar{X}$. Let $BA = 1 + (1-s)C$ and $AB = 1 + (1-s)W$. Then

$$\hat{\varphi}^\#([\bar{A}]) = \begin{bmatrix} A & s-1 \\ -C & B^t \end{bmatrix}$$

and

$$\left(\begin{pmatrix} A & s-1 \\ -C & B^t \end{pmatrix}^*\right)^{-1} = \begin{pmatrix} B^* & W^*-B^*U \\ 1-t & A^{t*} \end{pmatrix}$$

where $U = A^*W^* - C^*A^{t*}$. Hence

$$c\hat{\varphi}^\#([\bar{A}]) = \begin{bmatrix} A & s-1 & 0 & 0 \\ -C & B^t & 0 & 0 \\ 0 & 0 & B^* & W^*-B^*U \\ 0 & 0 & 1-t & A^{t*} \end{bmatrix}$$

Now note that

$$\Sigma_1 = \begin{pmatrix} 1 & 0 & 0 & 0 \\ 0 & 0 & 0 & (-1)^k \\ 0 & 0 & 1 & 0 \\ 0 & 1 & 0 & 0 \end{pmatrix} \qquad \Sigma_2 = \begin{pmatrix} 0 & 1 & 0 & 0 \\ 0 & 0 & (-1)^k & 0 \\ 0 & 0 & 0 & 1 \\ 1 & 0 & 0 & 0 \end{pmatrix}$$

represent 0 in $L_{2k+1}^{\{0\}}(\mathbb{Z})$; Hence they also represent 0 in $L_{2k+1}^X(R)$. When we multiply the above representative of $c\hat{\varphi}^\#([\bar{A}])$ by Σ_1 on the left and Σ_2 on the right and ignore the two first columns we get

$$c\hat{\varphi}^{\#}([\bar{A}]) = \begin{bmatrix} (-1)^k(s-1) & 0 \\ 0 & (-1)^k(1-t) \\ 0 & B^* \\ (-1)^k B^t & 0 \end{bmatrix} = \begin{bmatrix} (-1)^k(s-1) & 0 \\ 0 & (-1)^k(s-1) \\ 0 & B^*s \\ (-1)^k B^t & 0 \end{bmatrix}$$

(the last equality is obtained by right multiplication by $\begin{pmatrix} T & 0 \\ 0 & (T^*)^{-1} \end{pmatrix}$ where $T = \begin{pmatrix} 1 & 0 \\ 0 & s \end{pmatrix}$)

On the other hand $[\bar{A}] = -[\bar{A}]^* = [\varphi(B)^*]$ so $f([\bar{A}]) = \begin{bmatrix} 0 & \varphi(B)^* \\ 0 & 0 \end{bmatrix}$ which lifts to $\begin{bmatrix} 0 & B^{s^*}s \\ 0 & 0 \end{bmatrix}$ over R so (definition 2.5)

$$\varphi^! f([\bar{A}]) = \begin{bmatrix} (-1)^k(s-1) & 0 \\ 0 & (-1)^k(s-1) \\ 0 & B^*s \\ (-1)^k B^t & 0 \end{bmatrix}$$

coincides with $c\hat{\varphi}^{\#}([\bar{A}])$.

<u>Square 6</u>: Let $[\bar{A}] \in \bar{Y} \subseteq \tilde{K}_1(\bar{R})$ represent an element of $\hat{H}^{2k}(\mathbb{Z}/2\mathbb{Z}, \bar{Y}/\bar{X})$. Thus $[\bar{A}] = [\bar{A}^*]$ in $\bar{Y}/\bar{X}$. If $BA = 1 + (1-s)C$ as in definition 2.4 then

$$\hat{\varphi}^{\#}([\bar{A}]) = \begin{bmatrix} A & s-1 \\ -C & B^t \end{bmatrix}$$

and

$$f\hat{\varphi}^{\#}([\bar{A}]) = \begin{bmatrix} 0 & 0 & A & s-1 \\ 0 & 0 & -C & B^t \\ 0 & 0 & 0 & 0 \\ 0 & 0 & 0 & 0 \end{bmatrix}$$

Also $c([\bar{A}]) = \begin{bmatrix} \bar{A} & 0 \\ 0 & \bar{B}^* \end{bmatrix}$. This matrix lifts to $\begin{pmatrix} A & 0 \\ 0 & B^* \end{pmatrix}$ which has

$$\begin{pmatrix} A & 0 \\ 0 & B^* \end{pmatrix}^{s^*} \begin{pmatrix} 0 & 1 \\ 0 & 0 \end{pmatrix} \begin{pmatrix} A & 0 \\ 0 & B^* \end{pmatrix} = \begin{pmatrix} 0 & 1 + (1-s)(-C^*t) \\ 0 & 0 \end{pmatrix}$$

Hence, by definition 2.6

$$\varphi^! c([\bar{A}]) = \begin{bmatrix} 0 & 1 & A & 0 \\ (-1)^k s & 0 & 0 & B^* \\ 0 & 0 & 0 & -C^*t \\ 0 & 0 & 0 & 0 \end{bmatrix}$$

$$= \begin{bmatrix} 0 & 1-t & A & 0 \\ 0 & 0 & 0 & 0 \\ 0 & 0 & 0 & 0 \\ 0 & (-1)^k B & (-1)^{k+1} sC & 0 \end{bmatrix}$$

(the last equality by a straightforward application of (4.3)).
Finally change basis by means of

$$\Sigma = \begin{pmatrix} 1 & 0 & 0 & 0 \\ 0 & 0 & 0 & s \\ 0 & 0 & 1 & 0 \\ 0 & (-1)^k s & 0 & 0 \end{pmatrix}$$

to see that $\varphi^! c([\bar{A}]) = f\hat{\varphi}^{\#}([\bar{A}])$ as desired.

§ 6 REFERENCES.

1. H.J. Munkholm and E.K. Pedersen, Whitehead transfers for S^1-bundles. An algebraic description, Comm. Math. Helv. 56 (1981), 404-430.

2. H.J. Munkholm and E.K. Pedersen, The S^1-transfer in Surgery theory, to appear Trans. Amer. Math. Soc.

3. A.A. Ranicki, Algebraic L-theory, III. Twisted Laurent Extensions. Algebraic K-theory III. Batelli Inst. Conf. 1972, Lecture Notes in Math., vol. 343, Springer Verlag, Berlin 1973.

4. J.L. Shaneson, Wall's surgery obstruction groups for $G \times \mathbb{Z}$, Ann. of Math. 90 (1969), 296-334.

ODENSE UNIVERSITY
DEPARTMENT OF MATHEMATICS
ODENSE (DK5230)
DENMARK

Contemporary Mathematics
Volume 19, 1983

SMALL RING SPECTRA AND p-RANK OF THE STABLE HOMOTOPY OF SPHERES

Shichiro Oka

INTRODUCTION AND SUMMARY OF RESULTS

In recent years, infinite families of elements of the stable homotopy groups π_*^S of spheres have been discovered aside from the image of the J-homomorphism. L. Smith [25] proved that for every prime $p \geqq 5$ the elements β_t, $t \geqq 1$, in the p-primary component of π_*^S, which are constructed from a stable self-map of Adams-Toda's 4-cell complex V(1) [2], [30], cf.[32], are all non-trivial, generalising the ealier elements β_t, $t < p$, of Toda [27]. He investigated the image of the MU-theory Hurewicz homomorphism for a skeleton of V(1) to detect the β-family, the first example of infinite family in Coker J at $p \geqq 5$. Following his method, he [26] and the author [17], [19] discovered infinite families closely related to the β-family, extending the elements ϵ_j of P. May [11] and Toda [28], cf.[15]. R. Zahler [33] obtained those families by showing that they are detected in the second line $\mathrm{Ext}^2_{BP_*BP}(BP_*, BP_*)$ of the E_2 term of the Adams-Novikov spectral sequence based on the Brown-Peterson theory.

The BP-method seemed to be powerful, and in fact it is, as the splendid work of Miller, Ravenel and Wilson [12] indicates. They gave a complete structure of Ext^2 including description of generators. Moreover their description of generators suggests that if we once construct an appropriate 4-cell complex K with $BP_*(K) = BP_*/(p^i, v_1^s)$ and a stable self-map of K realising multiplication by v_2^t, for some i, s, t, where v_i is the generator of the coefficient ring BP_*, then the construction of an infinite family and its detection are immediate. In fact, the n-th element of the infinite family constructed is detected by the element in Ext^2 called $\beta_{tn/s,i}$ by them [12]. In this way, one can easily confirm the detection of the aforementioned β-family and its modification, and the author [21] gave more infinite families including the family of elements of order p^2.

1980 Mathematics Subject Classification. 55Q45, 55T15.

Now the structure of Ext^1 is very simple. Novikov [14] proved that if p is odd prime Ext^1 is cyclic or zero and it is isomorphic to the p-primary component of the image of J in the canonical way in the spectral sequence. According to [12], however, Ext^2 consists of quite lot of elements of various orders ; $\mathrm{Ext}^{2,t}$ contains a direct sum of arbitrarily many $\mathbb{Z}/p$ and an element of arbitrarily large order for sufficiently large t.

In this article, I construct much larger family of elements detected in Ext^2 so that it still holds in π_t^S that, for $p \geq 5$, the p-rank becomes arbitrarily large for sufficiently large t.

THEOREM I. *Let* $p \geq 5$. *The elements* $\beta_{tp^n/s}$ *in* $\mathrm{Ext}^2_{BP_*BP}(BP_*, BP_*)$ *of* [12] *survive non-trivially to* π_*^S, *if*

$$t \geq 2,\ n \geq 3,\ 1 \leq s \leq 2^{n-2}p \quad \text{or} \quad t = 1,\ n \geq 3,\ 1 \leq s \leq 2^{n-3}p.$$

The corresponding homotopy elements are of order p *and linearly independent. They generate direct summands in* π_*^S *if* $s \not\equiv 0 \mod p$, *and they are divisible by* p *if* $s \equiv 0 \mod p$.

Here, regarding "linearly independent", we exclude the obvious relation : $\beta_{tp^n/s} = \beta_{t'p^{n'}/s}$ if $tp^n = t'p^{n'}$. The corresponding results are already known for $n = 0$ [25], $n = 1$ [17],[19] and $n = 2$ [21] with better (best for $n = 0, 1$) range of s, t, as mentioned above (see also Remark after Proof of Theorem I in section 7). The dimension of $\beta_{tp^n/s}$ in π_*^S is $2tp^n(p^2-1) - 2s(p-1) - 2$, and hence, Theorem I allows arbitrarily many β's in the same dimension.

THEOREM II. *If* $p \geq 5$, *the p-rank of* π_t^S, *i.e., the number of independent generators of the p-component of* π_t^S, *becomes arbitrarily large for sufficiently large* t.

Theorem I produces little more : in Theorem II, generators may be replaced by either those which generate direct summand $\mathbb{Z}/p$ or those of order $\geq p^2$.

The growth of the p-rank is very slow. The known computations of π_*^S [18], [13], which are, in case $p = 5$, up to dimension 561, indicate that the p-rank is 0 or 1 in most dimensions and there is no example of p-rank ≥ 4. The first dimension of p-rank 3 is 551 at $p = 5$, which is more than a half of the range of recent computation of Ravenel. The range of such computations might be too low to imagine Theorem II.

It would be reasonable to conjecture Theorem II at smaller p, at least at $p = 3$. A part of my method still works for $p = 3$. However, the non-associativity of the mod 3 Moore space [29] causes much difficulity (see Remark after Theorem 7.1).

Theorem I is proved by the induction given in Theorem 7.1. Each stage of the induction constructs a 4-cell complex and a self-map of the complex, and it increases the indices t, s of $\beta_{t/s,i}$, fixing i. There would be a similar induction raising t and i, hence the order of the homotopy elements. However there is a gap in selecting the complex constructed. We merely have the first step of the second induction, which produces elements of order p^2. Completing the second induction suggests much stronger version of Theorem II.

The mod p^n Moore space M (p odd prime) admits a decomposition of $M \wedge M$, which defines a derivation on the graded ring $[M, M]_*$ of stable self-maps of M. The effectiveness of the derivation in studying $[M, M]_*$ is shown by P. Hoffman [8]. Extending his method leads to an induction which constructs a self-map of M inducing the multiplication by $v_1^{sp^{n-1}}$ at the n-th stage. This gives an alternative proof of the result of Novikov on Ext^1 mentioned above, which is our motivation of the induction giving Theorem I.

The alternative proof includes a construction of a complex K with $BP_*(K) = BP_*/(p^n, v_1^{sp^{n-1}})$. In contrast with $BP_*/(p^n)$, $BP_*(K)$ does not determine the homotopy type of K. In any case, K is a 4-cell complex. There is a choice of K which is a ring spectrum [10], [23]. As against M, this is not enough to decompose $K \wedge K$ into 4 copies of K. We shall call K admitting such a decomposition of $K \wedge K$ a <u>split ring spectrum</u> (see Definition 2.1). K is a split ring spectrum over $\mathbb{Z}/t$ if and only if K is a ring spectrum and the attaching map of the third cell to the second is divisible by t in π_*^S (Proposition 2.2). The graded ring $[K, K]_*$ is then possessed of two "derivations" d, d' with the properties:

$$d(fg) = (-1)^m d(f)g + fd(g),$$

$$d'(fg) = (-1)^m d'(f)g + fd'(g) + d(f)(\phi' \wedge 1_K)d(g)$$

for $f, g \in [K, K]_*$, $m = \deg g$,

$$d^2 = 0, \quad dd' + d'd = 0; \text{ and } \quad (d')^2 = 0 \quad \text{when} \quad t \not\equiv 0 \mod 3$$

(Proposition 1.1, Corollary 5.2, Theorem 3.6, Proposition 5.1). Here ϕ' is an element in π_*^S such that $t\phi'$ is the attaching map in K.

There are elements δ, $\delta' \in [K, K]_*$ which satisfy

$$d(\delta) = -1, \quad d'(\delta) = 0, \quad d(\delta') = 0, \quad d'(\delta') = -1,$$

$$(\delta')^2 = 0, \quad \delta\delta' + \delta'\delta = 0, \quad \delta^2 = \delta'(\phi' \wedge 1)$$

(see (2.3),(2.4),(2.5),(2.6),(2.6)',(4.7),(4.9)). Let $\mu_i : K \wedge K \longrightarrow \Sigma^n K$ be the i-th projection (n is the dimension of the i-th cell); μ_1 is the multiplication as ring spectrum. With suitable choice of μ_i, we will conclude, in section 4, the commutativity of μ_i which describe $\mu_i T$ in terms of μ_j, δ, δ' (T is the twisting map), and the associativity describing $\mu_i(1 \wedge \mu_j)$ in terms of $\mu_k(\mu_m \wedge 1)$, $\phi' \wedge 1$, and in case $t \equiv 0 \mod 9$ one more element which is the obstruction to the associativity (Theorems 4.2, 4.12). From these results we will conclude the structure theorems of $[K, K]_*$, Theorems 5.5, 5.6. An essential part of them, which leads to the induction proving Theorem I, is stated as follows:

THEOREM III. Let K be a split ring spectrum over $\mathbb{Z}/p^i$ with p prime ≥ 5. Then $\mathscr{C}_* = \operatorname{Ker} d \cap \operatorname{Ker} d'$ is a commutative subring, and

$$[K, K]_* = \mathscr{C}_* \oplus \mathscr{C}_*^1 \quad \text{(direct sum as additive group)},$$

where the complementary summand $\mathscr{C}_*^1$ has the property that every element induces the trivial homomorphism of BP-homology.

A modification in case $p = 3$, $i \geq 2$ will be also obtained as an analogue of modified Hoffman decomposition for non-associative Moore space discussed in [20].

This paper is divided into seven sections. In section 1, we quote several facts on module spectra over the Moore spectrum, from [20],[31]. They work basically everywhere in this paper. In sections 2-5, we discuss the structure of a split ring spectrum. In section 6, we link the split ring spectrum with homotopy elements detected in Ext^1. The proof of Theorems I and II is given in section 7.

Ending this introduction, I would like to thank Professor I. M. James and the staff of the Mathematical Institute, Oxford for providing a congenial atmosphere and pleasant surroundings during my stay in Oxford.

§1. MODULE SPECTRA OVER THE MOORE SPECTRUM

We shall recall the structure of homotopy group of stable maps between module spectra over the Moore spectrum discussed in [20], [31].

Let t be an odd integer and M_t (or briefly M) the Moore spectrum $S^0 \cup_t e^1$ of the cyclic group $\mathbf{Z}/t$. There is a cofibration

$$S^0 \xrightarrow{t} S^0 \xrightarrow{i} M \xrightarrow{j} S^1,$$

and M is a commutative ring spectrum having unique multiplication

$$\mu_M : M \wedge M \longrightarrow M,$$

which is associative if and only if $t \not\equiv \pm 3 \mod 9$ [4],[20], [29].

Let X be a finite CW spectrum such that $t \cdot 1_X = 0$ in $[X, X]$, where 1_X is the stable class of the identity map of X. This condition is equivalent to the existence of an element

$$m_X : M \wedge X \longrightarrow X \quad \text{such that} \quad m_X(i \wedge 1_X) = 1_X,$$

and we call such a spectrum X or precisely a pair (X, m_X) an <u>M-module spectrum</u>. Here we do not assume associativity of the module multiplication m_X. There is then an element

$$\bar{m}_X : \Sigma X \longrightarrow M \wedge X,$$

uniquely associated with m_X, such that

$$(j \wedge 1_X)\bar{m}_X = 1_X, \quad m_X\bar{m}_X = 0, \quad (i \wedge 1_X)m_X + \bar{m}_X(j \wedge 1_X) = 1_{M \wedge X},$$

which give a decomposition

$$M \wedge X = X \vee \Sigma X.$$

We put $[X, Y]_k = [\Sigma^k X, Y]$ and $[X, Y]_* = \underset{k}{\oplus}[X, Y]_k$ (direct sum), and grade it by defining $\deg f = k$ for $f \in [X, Y]_k$. The composition of maps makes $[X, X]_*$ a graded ring and $[X, Y]_*$ a right $[X, X]_*$-left $[Y, Y]_*$ bimodule. If X and Y are M-module spectra, $[X, Y]_*$ has a derivation

$$d = d_{m_X, m_Y} : [X, Y]_k \longrightarrow [X, Y]_{k+1}$$

defined by $d(f) = m_Y(1_M \wedge f)\bar{m}_X$. The decompositions of $M \wedge X$ and $M \wedge Y$ give an embedding of $[M \wedge X, M \wedge Y]_*$ into 2x2 matrices over $[X, Y]_*$, preserving product, with diagonal entries preserving grade, (1,2) entries raising it by one and (2,1) entries lowering it by one. In particular, $1_M \wedge f$ corresponds with the matrix

$$\begin{pmatrix} f & d(f) \\ 0 & (-1)^k f \end{pmatrix}, \quad k = \deg f.$$

PROPOSITION 1.1. ([20], [31]) <u>For</u> $f \in [Y, Z]_k$, $g \in [X, Y]_*$,

$$d(gf) = (-1)^k d(g)f + gd(f).$$

<u>If</u> m_X <u>and</u> m_Y <u>are associative, then</u> $d^2 = 0$ <u>in</u> $[X, Y]_*$.

Here by associativity of m_X we mean the usual one

$$m_X(1_M \wedge m_X) = m_X(\mu_M \wedge 1_X).$$

DEFINITION 1.2. ${}_t\mathscr{M}[X, Y]_k$ = kernel of $d : [X, Y]_k \longrightarrow [X, Y]_{k+1}$.

We shall omit the subscript t in ${}_t\mathscr{M}$ unless it leads to confusion. Notice that the different notations are used for d and $\mathscr{M}$ in [20] and [31].

PROPOSITION 1.3. <u>Let</u> X <u>and</u> Y <u>be associative M-module spectra. If</u> X <u>admits an element</u> $\delta_X \in [X, X]_{-1}$ <u>such that</u> $d(\delta_X) = -1_X$, <u>then the homologies of</u> $[X, Y]_*$ <u>and of</u> $[Y, X]_*$ <u>with respect to</u> d <u>are trivial, and</u>

$$[X, Y]_k = \mathscr{M}[X, Y]_k \oplus (\mathscr{M}[X, Y]_{k+1})\delta_X,$$

$$[Y, X]_k = \mathscr{M}[Y, X]_k \oplus \delta_X(\mathscr{M}[Y, X]_{k+1}).$$

This is proved in [20], Theorem 7.5, under an additional condition on δ_X, but we can remove it without alteration in the proof. The Moore spectrum M is itself an M-module spectrum with $m_M = \mu_M$. Moreover it is possessed of the element δ_M as in Proposition 1.3. In fact, put

$$\delta_M = ij \in [M, M]_{-1},$$

then the commutativity of μ_M (and of $\bar{\mu}_M$, i.e., $T\bar{\mu}_M = -\bar{\mu}_M$) implies

$$d(\delta_M) = -1_M.$$

Therefore, if $t \not\equiv \pm 3 \mod 9$,

(1.1)([8]) $\quad [M, M]_* = E(\delta_M) \otimes \mathscr{M}[M, M]_* = \mathscr{M}[M, M]_* \otimes E(\delta_M)$

as additive group, where E denotes an exterior algebra. In non-associative case, i.e., $t \equiv \pm 3 \bmod 9$, this still holds if we replace $\mathscr{M}$ by a larger subgroup ([20], §8)

$$\bar{\mathscr{M}}[M, M]_* = \{f \mid d(f)\delta_M = 0\} = \{f \mid d(f) = \pm(\alpha_1 \wedge 1_M)\delta_M f \delta_M\}.$$

The following result is obtained by P. Hoffman [8] together with (1.1).

PROPOSITION 1.4. $\mathscr{M}[M, M]_*$ <u>is a commutative subring of</u> $[M, M]_*$:

$$gf = (-1)^{km} fg \qquad \text{for} \quad f \in \mathscr{M}[M, M]_k, \quad g \in \mathscr{M}[M, M]_m,$$

<u>and any element</u> f <u>in</u> $\mathscr{M}[M, M]_*$ <u>with</u> $\deg f$ <u>even is self-dual in the sense of Spanier and Whitehead.</u>

Let r be a divisor of t. The natural homomorphisms $\mathbb{Z}/r \longrightarrow \mathbb{Z}/t$ and $\mathbb{Z}/t \longrightarrow \mathbb{Z}/r$ are realised by the unique maps

$$\lambda = \lambda_{r,t} : M_r \longrightarrow M_t \quad \text{and} \quad \rho = \rho_{t,r} : M_t \longrightarrow M_r,$$

respectively, such that $\lambda i_r = (t/r)i_t$, $j_t\lambda = j_r$, $\rho i_t = i_r$, $j_r\rho = (t/r)j_t$, $\lambda_{s,t} = \lambda_{s,r}\lambda_{r,t}$, $\rho_{t,s} = \rho_{t,r}\rho_{r,s}$, where i, j for M_t are denoted by i_t, j_t. Moreover there is a cofibration

$$(1.2) \qquad \Sigma^{-1}M_s \xrightarrow{i_r j_s} M_r \xrightarrow{\lambda_{r,rs}} M_{rs} \xrightarrow{\rho_{rs,s}} M_s.$$

If (X, m_X) is an M_t-module spectrum, it is an M_r-module spectrum with multiplication $m_X(\rho \wedge 1_X)$, for any divisor r of t. If Y is also an M_t-module spectrum which is considered as an M_r-module spectrum in this way, then ([20])

$$(1.3) \qquad {}_r\mathscr{M}[X, Y]_k \subset {}_t\mathscr{M}[X, Y]_k.$$

The M_t-module multiplication on M_r is unique and given in this way.

THEOREM 1.5. <u>Let</u> t <u>be odd. For any</u> $f \in {}_t\mathscr{M}[M_t, M_t]_{2k}$, <u>there exist elements</u>

$$f_n \in {}_{t^n}\mathscr{M}[M_{t^n}, M_{t^n}]_{2kt^{n-1}}, \quad n = 1, 2, \ldots,$$

<u>such that</u>

$$f_1 = f, \ \lambda(f_n)^t = f_{n+1}\lambda, \quad (f_n)^t\rho = \rho f_{n+1},$$

<u>where</u> $\lambda = \lambda_{t^n, t^{n+1}}$, $\rho = \rho_{t^{n+1}, t^n}$.

PROOF. We construct f_n by the induction on n. Assume that f_m, $m \leq n$, exist. Since $d(f_n\delta_M - \delta_M f_n) = -f_n + f_n = 0$ by Proposition 1.1, $f_n\delta_M - \delta_M f_n$ commutes with f_n by Proposition 1.4, $M = M_{t^n}$. Therefore,

$$(f_n)^2\delta_M - 2f_n\delta_M f_n + \delta_M(f_n)^2 = 0, \text{ and}$$

$$(f_n)^s\delta_M - \delta_M(f_n)^s = s((f_n)^s\delta_M - (f_n)^{s-1}\delta_M f_n), \quad s \geq 2,$$

by induction on s. We write λ', δ' for λ_{t,t^n}, $i_{t^n}j_t$. Then $(f_n)^t\lambda' = \lambda' f^{t^n}$, $\delta_M\lambda' = \delta'$ and $t\lambda' = 0$. Hence $(f_n)^t\delta' = \delta' f^{t^n}$. By (1.2), there exists $g \in [M', M']_{2kt^n}$ with $g\lambda = \lambda(f_n)^t$, $M' = M_{t^{n+1}}$. By (1.1), we may put $g = g' + \delta_{M'}g''$, $d'(g') = d'(g'') = 0$, where d' is the derivation of M'-module spectrum. Since $[M, M']_1 = 0$, $d'(\lambda) = 0$. By (1.3), $g''\lambda = d'(g'\lambda + \delta_{M'}g''\lambda) = d'(\lambda(f_n)^t) = 0$. Then $\lambda(f_n)^t = g'\lambda$ and we can put $f_{n+1} = g'$. The commutativity with ρ follows from the duality in Proposition 1.4. □

If (X, m_X) and (Y, m_Y) are M-module spectra, both m_X and m_Y define multiplications $m^{(1)}$ and $m^{(2)}$ on the smash product $X \wedge Y$:

$$m^{(1)} = m_X \wedge 1_Y : M \wedge X \wedge Y \longrightarrow X \wedge Y, \quad \text{with } \bar{m}^{(1)} = \bar{m}_X \wedge 1_Y,$$

$$m^{(2)} = (1_X \wedge m_Y)(T \wedge 1_Y) : M \wedge X \wedge Y \longrightarrow X \wedge M \wedge Y \longrightarrow X \wedge Y$$

$$\text{with } \bar{m}^{(2)} = (T^{-1} \wedge 1_Y)(1_X \wedge \bar{m}_Y),$$

where T is the switching map. We sometimes denote $(X \wedge Y, m^{(1)})$, $(X \wedge Y, m^{(2)})$ by $X^* \wedge Y$, $X \wedge Y^*$, when multiplications on X and Y are understood without mention. Similar notations are used for smash products of more factors.

LEMMA 1.6. *Let X_i, Y_i be M-module spectra, $X = X_1 \wedge \cdots \wedge X_n$, $Y = Y_1 \wedge \cdots \wedge Y_n$, and Z be an associative M-module spectrum.*

(i) *Let $d^{(i)}$ be the derivation on $[X, Y]_*$ with respect to $X_1 \wedge \cdots \wedge X_i^* \wedge \cdots \wedge X_n$, $Y_1 \wedge \cdots \wedge Y_i^* \wedge \cdots \wedge Y_n$. Then*

$$d^{(i)}(f_1 \wedge \cdots \wedge f_n) = f_1 \wedge \cdots \wedge f_{i-1} \wedge d(f_i) \wedge f_{i+1} \wedge \cdots \wedge f_n,$$

for $f_j \in [X_j, Y_j]_*$, and $d^{(i)}d^{(j)} = -d^{(j)}d^{(i)}$, $i \neq j$.

(ii) *Let d_i be the derivation on $[X, Z]_*$ or on $[Z, X]_*$ with respect to the i-th factor X_i of X. Then $d_i d_j = -d_j d_i$ for $i \neq j$. In case $X_1 = M$, $X_2 = Z$, $n = 2$,*

$$d_1(i \wedge 1_Z) = \bar{m}_Z, \quad d_1(j \wedge 1_Z) = -m_Z,$$

$$d_2(i \wedge 1_Z) = 0, \quad d_2(j \wedge 1_Z) = 0,$$

$$d_s(m_Z) = 0, \quad d_s(\bar{m}_Z) = 0, \quad s = 1, 2.$$

PROOF. Since $\mu_M(1_M \wedge i) = 1_M$, $(1_M \wedge j)\bar{\mu}_M = -1_M$, the calculation for $d_s(i \wedge 1_Z)$, $d_s(j \wedge 1_Z)$ is easy. Without assuming the associativity of m_Z, there is an element $a(m_Z) \in [Z, Z]_2$ uniquely associated to m_Z such that

$$m_Z(1_M \wedge m_Z) = m_Z(\mu_M \wedge 1_Z) - a(m_Z)(j \wedge j \wedge 1_Z),$$

$$(1_M \wedge \bar{m}_Z)\bar{m}_Z = -(\bar{\mu}_M \wedge 1_Z)\bar{m}_Z + (i \wedge i \wedge 1_Z)a(m_Z),$$

by [20], Theorem 5.2 (We notice that there is an error in the sign of the first formula in [20],(5.1)). Let $T : M \wedge M \longrightarrow M \wedge M$ be the switching map. Then $\mu_M T = \mu_M$, $T\bar{\mu}_M = -\bar{\mu}_M$ [4] and $T(i \wedge i) = i \wedge i = (i \wedge 1_M)i = (1_M \wedge i)i$, $(j \wedge j)T = -(j \wedge j) = j(j \wedge 1_M) = -j(1_M \wedge j)$ by the commutativity of smash products of stable maps [3], cf. [31], Theorem 1.1. Then the calculation using these relations leads to

$$d_1(m_Z) = a(m_Z)(j \wedge 1_Z), \quad d_1(\bar{m}_Z) = (i \wedge 1_Z)a(m_Z),$$

$$d_2(m_Z) = 2a(m_Z)(j \wedge 1_Z), \quad d_2(\bar{m}_Z) = 2(i \wedge 1_Z)a(m_Z).$$

If m_Z is associative, $a(m_Z) = 0$ [20] and $d_s(m_Z) = 0$, $d_s(\bar{m}_Z) = 0$.

The remaining part is proved by an easy diagram chasing. □

REMARK. If $t \equiv \pm 3 \mod 9$, the element $a(m_Z)$ is always non-trivial (unless Z is mod 3 contractible), that is, no (finite) M_t-module spectrum is associative [20], Theorem 6.3. Therefore $d_s(m_Z) \not\equiv 0$, $d_s(\bar{m}_Z) \not\equiv 0$.

§2. 4-CELL RING SPECTRA WHOSE SQUARES SPLIT

Let K be a finite CW spectrum which has the homology groups:

$$H_0(K) = H_{k+1}(K) = \mathbb{Z}/t \quad (k>0), \quad H_i(K) = 0 \quad (i \neq 0, k+1),$$

where t is odd. K is then a cofibre of a map $\phi : \Sigma^k M \longrightarrow M$, $M = M_t$, so that there is a cofibration

$$(2.1) \qquad \Sigma^k M \xrightarrow{\phi} M \xrightarrow{i'} K \xrightarrow{j'} \Sigma^{k+1} M.$$

The attaching map ϕ is unique for K up to constant relatively prime to t, and K has a cell decomposition

$$K = S^0 \cup e^1 \cup e^{k+1} \cup e^{k+2}.$$

To have a decomposition $K \wedge K = \vee \Sigma^i K$, where i runs over all dimensions of cells in K, we introduce the following

DEFINITION 2.1. K is said to be a <u>split ring spectrum</u> (<u>over</u> $\mathbb{Z}/t$), if $t \cdot 1_K = 0$ in $[K, K]_0$ and $\phi \wedge 1_K = 0$ in $[M \wedge K, M \wedge K]_k$.

The first condition means that K is an M-module spectrum, and hence, it provides us with the decomposition $M \wedge K = K \vee \Sigma K$. Similarly the second condition provides $K \wedge K = (M \wedge K) \vee \Sigma^{k+1}(M \wedge K)$ so that a split ring spectrum K admits the decomposition of $K \wedge K$ as above.

PROPOSITION 2.2. <u>If</u> k <u>is odd, then</u> K <u>is a split ring spectrum only if</u> $\phi = 0$. <u>Assume</u> $\phi \neq 0$. <u>Then,</u> K <u>is a split ring spectrum if, and only if,</u>

(A) $d(\phi) = 0$,

(B) k <u>is even,</u>

(C) $o(\phi) \equiv 0 \mod J_*(\phi)$, <u>where</u> $o(\phi)$ <u>is the coset in the group</u> $[M, M]_{2k+1}$ <u>defined in</u> [23], <u>Definition</u> 4.1, <u>and</u> $J_*(\phi)$ <u>is the two sided ideal of</u> $[M, M]_*$ <u>generated by</u> ϕ; <u>and</u>

(D) <u>the element</u> $j\phi i \in \pi^S_{k-1}$ <u>is divisible by</u> t.

LEMMA 2.3. <u>If</u> k <u>is odd, then the condition</u>

(E) <u>there is a map</u> $\mu_K : K \wedge K \longrightarrow K$ <u>such that</u> $\mu_K(i'i \wedge 1_K) = 1_K$ <u>is satisfied only if</u> $\phi = 0$. <u>If</u> k <u>is even,</u> (E) <u>is equivalent to</u> (A) and (C).

PROOF. Immediate from [23], Theorem 4.4. □

LEMMA 2.4. (A) *is equivalent to*

(A_1) *There is an* M-*module multiplication* m_K, *which is associative in case* $t \not\equiv \pm 3 \mod 9$, *such that the derivation given by* m_K *satisfies* $d(i') = 0$ *and* $d(j') = 0$.

PROOF. Immediate from [20], Theorem 4.3 and Lemma 4.7. The associativity follows from [22], Corollary on p.219. □

LEMMA 2.5. *Assume* (A) *and* (B). *Then* (D) *is equivalent to one of the following relations in* $[M, M]_*$:

(D_1) $\phi\delta_M = \delta_M\phi$, *where* $\delta_M = ij$.

(D_2) $\delta_M\phi\,\delta_M = 0$.

(D_3) $(j\phi i)\wedge 1_M = 0$.

PROOF. We have

$$(\phi_1 \wedge 1_M)\delta_M = i\phi_1 j = \delta_M\phi\,\delta_M, \quad \phi_1 = j\phi i.$$

Calculating d of this implies, by the assumption,

$$\phi_1 \wedge 1_M = -d(\delta_M\phi\,\delta_M) = \delta_M\phi - \phi\delta_M.$$

Therefore, (D_1), (D_2) and (D_3) are equivalent to each other. (D) is equivalent to $i\phi_1 = 0$, and to $\delta_M\phi\,\delta_M = i\phi_1 j = 0$, since $j^* : \pi_{k-1}(M) \longrightarrow [M, M]_{k-2}$ is monic. □

PROOF OF PROPOSITION 2.2. As before, $t\cdot 1_K = 0$ is equivalent to the existence of M-module multiplication, which is equivalent to (A_1) [20], and to (A) by Lemma 2.4. By Lemma 2.3, it is enough to show that (A) implies the equivalence:

$$\phi\wedge 1_K = 0 \Longleftrightarrow (D) \text{ and } (E).$$

If $\phi\wedge 1_K = 0$, the cofibration

$$M\wedge K \xrightarrow{i'\wedge 1} K\wedge K \xrightarrow{j'\wedge 1} \Sigma^{k+1}M\wedge K,$$

which is induced by $\phi\wedge 1$, is trivial. Let $r : K\wedge K \longrightarrow M\wedge K$ be a retraction (trivialisation). Then (E) is satisfied with $\mu_K = m_K r$. Since $\phi_1\wedge 1_K = (j\wedge 1_K)(\phi\wedge 1_K)(i\wedge 1_K) = 0$, we have $i'(\phi_1\wedge 1_M) = (\phi_1\wedge 1_K)i' = 0$. (2.1) induces the exact sequence

$$[M, M]_{-1} \xrightarrow{\phi_*} [M, M]_{k-1} \xrightarrow{(i')_*} [M, K]_{k-1},$$

and $[M, M]_{-1}$ is generated by $ij = \delta_M$. Then $(\phi_1 \wedge 1_M)\delta_M = 0$ by $\delta_M\delta_M = 0$, proving (D) by Lemma 2.5.

Conversely, assume (D) and (E) hold. By [23], §4, (E) implies $m_K(\phi \wedge 1_K) = 0$ for some m_K (which may not be associative). Then

$$
\begin{aligned}
(j \wedge 1_K)(\phi \wedge 1_K) &= m_K(\delta_M \phi \wedge 1_K) && (\text{by } m_K(i \wedge 1_K) = 1_K)\\
&= m_K(\phi\delta_M \wedge 1_K) && (\text{by } (D_1))\\
&= 0.
\end{aligned}
$$

Therefore $\phi \wedge 1_K = (i \wedge 1_K)m_K(\phi \wedge 1_K) + \bar{m}_K(j \wedge 1_K)(\phi \wedge 1_K) = 0.$ □

REMARK. If one replaces (D) by $j\phi i = 0$, $\phi = \phi_2 \wedge 1_M$ for some $\phi_2 \in \pi_k^S$ by [16], §3, and $K = M \wedge L$, where L is a cofibre of ϕ_2. The obstruction for L to have a decomposition $L \wedge L = L \vee \Sigma^{k+1}L$, which is a coset in π_{2k+1}^S with indeterminacy $\pi_{k+1}^S\phi_2$, vanishes modulo $\pi_{k+1}^S\phi_2 + t\pi_{2k+1}^S$. The additional term may be negligible, if the composite $(t\pi_k^S) \otimes \pi_{k+1}^S \longrightarrow t\pi_{2k+1}^S$ is epic, for instance. The first case when $j\phi i \neq 0$ happens in dimension $k = (2p-2)p$ with $t = p$ prime, which is, if $p > 3$, the case discussed in [21].

Hereafter the M-module multiplication on a split ring spectrum will be fixed in such a way that it satisfies (A_1) and k will be assumed to be even. We have easily

(2.2) $$[M, M]_j = \begin{cases} \mathbb{Z}/t, \text{ generated by } \delta_M, & \text{for } j = -1,\\ \mathbb{Z}/t, \text{ generated by } 1_M, & \text{for } j = 0,\\ 0 \quad \text{for } j < -1 & \text{and for } j = 1. \end{cases}$$

We put

(2.3) $$\delta_K' = i'j' \in [K, K]_{-k-1}.$$

It clearly satisfies

(2.4) $$\delta_K' i' = 0, \quad j'\delta_K' = 0, \quad (\delta_K')^2 = 0, \quad d(\delta_K') = 0.$$

Exact sequences induced from (2.1) enable us to compute $[K, K]_*$, and we have

LEMMA 2.6. <u>The homomorphism</u>

$$(i')_*(j')^* : [M, M]_{j+k+1} \longrightarrow [K, K]_j$$

<u>is monic if</u> $j \leq -3$ <u>or</u> $j = 0$, <u>and is epic if</u> $j \leq -2$. <u>There is an element</u>

(2.5) $$\delta_K \in [K, K]_{-1}$$

<u>such that</u>

(2.6) $\delta_K i' = i'\delta_M$, $j'\delta_K = -\delta_M j'$, $\delta'_K\delta_K = -\delta_K\delta'_K$,

and in case $t \not\equiv \pm 3 \mod 9$, $d(\delta_K) = -1_K$. *Moreover,*

$[K, K]_{-k-2} = \mathbb{Z}/t$, *generated by* $\delta_K\delta'_K$;

$[K, K]_{-k-1} = \mathbb{Z}/t$, *generated by* δ'_K ;

$[K, K]_j = \mathrm{Im}(i')_*(j')^*$ *for* $-k-1 < j < -1$;

$[K, K]_{-1} = \mathbb{Z}/t \oplus \mathrm{Im}(i')_*(j')^*$,

where the first summand is generated by δ_K ;

$[K, K]_0 = \mathbb{Z}/t \oplus \mathrm{Im}(i')_*(j')^*$,

where the first summand is generated by 1_K.

PROOF. Everything is immediate from (2.2),except for (2.6). By (D_1), there is an element $\delta_0 \in [K, K]_{-1}$ with $\delta_0 i' = i'\delta_M$ and $j'\delta_0 = \pm\delta_M j'$. Then $d(\delta_0)i' = d(\delta_0 i') = -i'$, and hence, $d(\delta_0) = -1_K + i'gj'$ for some $g \in [M, M]_{k+1}$. Then the sign in $j'\delta_0$ must be minus, and we may put $\delta_K = \delta_0$ in case $t \equiv \pm 3 \mod 9$. If $t \not\equiv \pm 3 \mod 9$, $i'd(g)j' = -d(i'gj') = d^2(\delta_0) = 0$ and $d(g) = 0$. By Proposition 1.3, $g = d(h)$ for some h, and we may put $\delta_K = \delta_0 + i'hj'$. □

COROLLARY 2.7. *Let* K *be a split ring spectrum over* $\mathbb{Z}/t$, *where* $t \not\equiv \pm 3 \mod 9$. *Then*

$$[K, K]_* = \mathscr{M}[K, K]_* \oplus (\delta_K)_*(\mathscr{M}[K, K]_*)$$
$$= \mathscr{M}[K, K]_* \oplus (\delta_K)^*(\mathscr{M}[K, K]_*),$$

and similar decompositions hold for $[M, K]_*$ *and* $[K, M]_*$.

PROOF. Immediate from Proposition 1.3 together with (2.6). □

As in (2.4), the square of δ'_K is trivial. However the square of δ_K may not be trivial as the following lemma shows.

LEMMA 2.8. *There is an element* $\phi' \in \pi^S_{k-1}$ *such that* $t\phi' = j\phi i$ *and*

(2.6)' $(\delta_K)^2 = \delta'_K(\phi' \wedge 1_K) = -(\phi' \wedge 1_K)\delta'_K$.

PROOF. Let $M^n = S^n \cup_f e^{n+1}$, where f is a map of degree t, and $i^n : S^n \longrightarrow M^n$ the inclusion, $j^n : M^n \longrightarrow S^{n+1}$ the map collapsing S^n to the base point. Then δ_M is represented by $d^n = i^{n+1}j^n$. Assume n is sufficiently large so that there is a representative $g : M^{n+k} \longrightarrow M^n$ of ϕ such that $d^n g$ is homotopic to $(\Sigma g)d^{n+k}$ with homotopy A_s, $s \in I = [0, 1]$. Then A_s defines a map $h : C(g) \longrightarrow C(\Sigma g)$ as usual by

$$h(m) = d^n(m)$$

$$h(s\wedge m') = \begin{cases} A_{2s}(m') & \text{for } 0 \leq s \leq \frac{1}{2} \\ (2s-1)\wedge d^{n+k}(m') & \text{for } \frac{1}{2} \leq s \leq 1, \end{cases}$$

where $C(g) = M^n \cup_g CM^{n+k}$, $m \in M^n$ and an element in the cone CM^{n+k} is denoted by $s\wedge m'$, $s \in I$, $m' \in M^{n+k}$ with identification $1\wedge m'$ = base point, $0\wedge m' = g(m') \in M^n$ in $C(g)$; similarly for $C(\Sigma g)$. Since $d^{n+1}A_s$ and $(\Sigma A_{1-s})d^{n+k}$ are both homotopies from 0 (the constant map) to $d^{n+1}(\Sigma g)d^{n+k}$, they define a map $A : \Sigma M^{n+k} = M^{n+k+1} \longrightarrow M^{n+2}$, which satisfies $i'Aj' \sim (\Sigma h)h$, $i' : M^{n+2} \longrightarrow C(\Sigma^2 g)$ the inclusion, $j' : C(g) \longrightarrow \Sigma M^{n+k}$ the map collapsing the bottom Moore space to the base point. The homotopy $j^{n+1}A_s i^{n+k}$ from 0 to 0 defines a map $B : \Sigma S^{n+k} \longrightarrow S^{n+2}$, and clearly $Ai^{n+k+1} = i^{n+2}B$ (as a map). Let $p : I \longrightarrow S^1$ be the usual projection, and $-B$ be a map defined by $(-B)(p(u)\wedge x) = B(p(1-u)\wedge x)$, $u \in I$, $x \in S^{n+k}$, as usual. Then we have

$$(-B)(\Sigma^{k+1}f)(p(u)\wedge x) = j^{n+1}A_{1-u}i^{n+k}(\Sigma^k f)(x) = j^{n+1}A_{1-u}(0\wedge x),$$

$$j^{n+1}(\Sigma g)i^{n+k+1}(p(u)\wedge x) = j^{n+1}(\Sigma g)d^{n+k}(u\wedge x) = j^{n+1}A_1(u\wedge x).$$

Let $C : \Sigma S^{n+k} \longrightarrow S^{n+2}$ be defined by $p(u)\wedge x \longmapsto j^{n+1}A_{1-u}(u\wedge x)$. Then $(-B)(\Sigma^{k+1}f) \sim C \sim j^{n+1}(\Sigma g)i^{n+k+1}$ by easy homotopies.

We turn to the stable homotopy class. Let α, β be the stable classes of A, B. The element δ_K is represented by an h with a suitable homotopy A_s. Then we have

$$(\delta_K)^2 = i'\alpha j', \quad \alpha i = i\beta \quad \text{and} \quad -t\beta = j\phi i.$$

Assume first that $t \not\equiv \pm 3$ mod 9. Then $d((\delta_K)^2) = -d(\delta_K)\delta_K + \delta_K d(\delta_K) = 0$ by (2.6), and hence, $d(\alpha) = 0$ by Lemma 2.6 with replacing α by $\alpha + x\phi\delta_M$, if necessary. The sequence

$$\pi^S_{k-1} \xrightarrow{\ t\ } \pi^S_{k-1} \xrightarrow{\ \wedge 1\ } \mathcal{M}[M, M]_{k-1} \xrightarrow{\ j_* i^*\ } \pi^S_{k-2}$$

is exact by [16], §3, so that $\alpha = \alpha' \wedge 1_M$ for some $\alpha' \in \pi^S_{k-1}$. Then $i\alpha' = \alpha i = i\beta$ and $\beta\wedge 1_M = \alpha'\wedge 1_M = \alpha$. Therefore $(\delta_K)^2 = i'(\beta\wedge 1_M)j' = (\beta\wedge 1_K)\partial_K'$. Define $\phi' = -\beta$.

Next consider the case $t \equiv \pm 3$ mod 9. The modified decomposition of $[M, M]_*$ mentioned after (1.1) allows a choice of δ_K satisfying $d(\delta_K) = -1_K + i'g\delta_M j'$, $g \in \mathcal{Z}[M, M]_*$, by a similar replacement to get δ_K from δ_0 in the proof of Lemma 2.6. Then $d(\alpha) = -\delta_M g\delta_M$ and

$\alpha \in \mathcal{X}$. The above exact sequence still holds, if one replaces $\mathcal{M}$ by $\mathcal{X}$, and the above proof does work.([20], §8). □

We shall give sufficient conditions for K to be a split ring spectrum.

PROPOSITION 2.9. Let f be an element in $[M, M]_*$ with $d(f) = 0$, deg f even. Let K be a cofibre of f^n, $n \geq 2$. Then K is a ring spectrum. If moreover $n \equiv 0 \mod t$, it is a split ring spectrum.

PROOF. The first half is immediate from [23], Theorem 5.6. As is seen in the proof of Theorem 1.5, f satisfies $f^t \delta_M = \delta_M f^t$, which implies the condition (D). □

PROPOSITION 2.10. Let K, K_1 be split ring spectra with attaching maps ϕ, ϕ_1. Then the cofibre of the composite $\phi\phi_1$ is again a split ring spectrum.

PROOF. Clearly, $\phi\phi_1$ satisfies (A), (B) and (D_1) in Proposition 2.2 and Lemma 2.5. The condition (C) for it follows from [23], Corollary 5.4, (ii). □

§3. MORE ON SPLITTING

Henceforward the odd integer t will be assumed to satisfy $t \not\equiv \pm 3 \bmod 9$ to avoid the non-associativity of $M = M_t$ and of M-module spectra (see Remark after Lemma 1.6). In this section, we study the decomposition into four copies in more general situation, as Toda [31] generalised the decomposition of $M \wedge M$ [8] to his "$\mathbb{Z}/p$-space".

DEFINITION 3.1. Let ϕ be an element in $[M, M]_k$ with $d(\phi) = 0$, k even, and K be a cofibre of ϕ. A finite CW spectrum X is called a <u>split</u> K-(<u>module</u>) <u>spectrum</u>, if X satisfies the following conditions:

(a) $X = (X, m_X)$ is an associative M-module spectrum;

(b) there is an element $\delta_X \in [X, X]_{-1}$ such that $d(\delta_X) = -1_X$, $d = d_{m_X, m_X}$;

and

(c) $\phi \wedge 1_X = 0$ in $[M \wedge X, M \wedge X]_k$.

As is seen in the proof of Proposition 2.2, the condition (c) is equivalent to the triviality of the cofibration

$$M \wedge X \xrightarrow{i' \wedge 1} K \wedge X \xrightarrow{j' \wedge 1} \Sigma^{k+1} M \wedge X.$$

Regarding the trivialisation, we have the following

LEMMA 3.2. <u>Let</u> X <u>be a split</u> K-<u>module spectrum</u>. <u>Let</u> $d^{(1)}, d^{(2)}$ <u>be the derivations as in Lemma</u> 1.6, (i). <u>Then there is an element</u>

$$m'_X \in [K \wedge X, M \wedge X]_0$$

<u>such that</u>

$$m'_X(i' \wedge 1_X) = 1_{M\wedge X}, \quad d^{(1)}(m'_X) = 0, \quad d^{(2)}(m'_X) = 0,$$

<u>and there is</u>

$$\bar{m}'_X \in [M \wedge X, K \wedge X]_{k+1} \quad \underline{\text{with}} \quad d^{(1)}(\bar{m}'_X) = 0, d^{(2)}(\bar{m}'_X) = 0,$$

<u>uniquely associated to</u> m'_X <u>in such a way that</u>

$$(j' \wedge 1_X)\bar{m}'_X = 1_{M\wedge X}, \quad m'_X \bar{m}'_X = 0,$$

$$(i' \wedge 1_X)m'_X + \bar{m}'_X(j' \wedge 1_X) = 1_{K\wedge X}.$$

PROOF. There are elements $r : K\wedge X \longrightarrow M\wedge X$ and $\bar{r} : \Sigma^{k+1}M\wedge X \longrightarrow K\wedge X$ such that $r(i'\wedge 1_X) = 1_{M\wedge X} = (j'\wedge 1_X)\bar{r}$, $r\bar{r} = 0$, $(i'\wedge 1_X)r + \bar{r}(j'\wedge 1_X) = 1_{K\wedge X}$. The element $\bar{r}$ is uniquely determined for given r with $r(i'\wedge 1_X) = 1_{M\wedge X}$. By Lemma 1.6, (i),

$$d^{(1)}d^{(2)}(\delta_M\wedge\delta_X) = 1_{M\wedge X} = -d^{(2)}d^{(1)}(\delta_M\wedge\delta_X).$$

Also we have $d^{(s)}(i'\wedge 1_X) = 0$, $s = 1, 2$. Then the element

$$m'_X = d^{(1)}d^{(2)}((\delta_M\wedge\delta_X)r)$$

satisfies $d^{(s)}(m'_X) = 0$ (by $(d^{(s)})^2 = 0$) and

$$m'_X(i'\wedge 1_X) = d^{(1)}d^{(2)}((\delta_M\wedge\delta_X)r(i'\wedge 1_X)) = 1_{M\wedge X},$$

as required. The associated element $\bar{m}'_X$ satisfies $m'_X d^{(s)}(\bar{m}'_X) = 0$, $(j'\wedge 1_X)d^{(s)}(\bar{m}'_X) = d^{(s)}(1_{M\wedge X}) = 0$, which imply $d^{(s)}(\bar{m}'_X) = 0$. □

DEFINITION 3.3. Let X be a split K-spectrum. We define

$$\mu_{1,X} = m_X m'_X : K\wedge X \longrightarrow X, \qquad \nu_{1,X} = i'i\wedge 1_X : X \longrightarrow K\wedge X,$$
$$\mu_{2,X} = (j\wedge 1_X)m'_X : K\wedge X \longrightarrow \Sigma X, \qquad \nu_{2,X} = (i'\wedge 1_X)\bar{m}_X : \Sigma X \longrightarrow K\wedge X,$$
$$\mu_{3,X} = m_X(j'\wedge 1_X) : K\wedge X \longrightarrow \Sigma^{k+1}X, \qquad \nu_{3,X} = \bar{m}'_X(i\wedge 1_X) : \Sigma^{k+1}X \longrightarrow K\wedge X,$$
$$\mu_{4,X} = jj'\wedge 1_X : K\wedge X \longrightarrow \Sigma^{k+2}X, \qquad \nu_{4,X} = \bar{m}'_X\bar{m}_X : \Sigma^{k+2}X \longrightarrow K\wedge X.$$

These maps satisfy $\mu_{i,X}\nu_{j,X} = 0$, $\mu_{i,X}\nu_{i,X} = 1_X$, $\sum\nu_{i,X}\mu_{i,X} = 1_{K\wedge X}$ and hence give rise to a decomposition

$$K\wedge X = X\vee\Sigma X\vee\Sigma^{k+1}X\vee\Sigma^{k+2}X.$$

LEMMA 3.4.([21], Lemma 2.2) *As in Lemma 1.6, (ii), let d_1 (resp. d_2) be the derivation in $[K^*\wedge X, X]_*$ or in $[X, K^*\wedge X]_*$ (resp. in $[K\wedge X^*, X]_*$ or in $[X, K\wedge X^*]_*$). Then*

$$d_1(\mu_{1,X}) = 0, \quad d_1(\mu_{2,X}) = -\mu_{1,X}, \quad d_1(\mu_{3,X}) = 0, \quad d_1(\mu_{4,X}) = \mu_{3,X},$$
$$d_1(\nu_{1,X}) = \nu_{2,X}, \quad d_1(\nu_{2,X}) = 0, \quad d_1(\nu_{3,X}) = \nu_{4,X}, \quad d_1(\nu_{4,X}) = 0,$$
$$d_2(\mu_{i,X}) = 0, \quad d_2(\nu_{i,X}) = 0, \quad i = 1, 2, 3, 4.$$

PROOF. Apply Lemma 3.2 and Lemma 1.6,(ii), for example,

$$d_1(\mu_{2,X}) = d_1(j\wedge 1_X)m'_X + (j\wedge 1_X)d^{(1)}(m'_X) = -m_X m'_X = -\mu_{1,X}. \quad \square$$

DEFINITION 3.5. Let X and Y be split K-spectra. For any element $f\in[X, Y]_m$, we define

$$d'(f) = \mu_{1,Y}(1_K \wedge f)\nu_{3,X} \in [X, Y]_{m+k+1}$$

$$d''(f) = \mu_{2,Y}(1_K \wedge f)\nu_{3,X} \in [X, Y]_{m+k}.$$

For $F \in [K \wedge X, K \wedge Y]_*$, $\tau(F)$ denotes a 4x4 matrix over $[X, Y]_*$ with (i, j) entry $\mu_{i,Y}F\nu_{j,X}$.

Clearly d' and d'' are additive homomorphisms, and τ is monic, additive and preserves product. In aid of convenience, we sometimes divide $\tau(F)$ into 2x2 blocks $\tau_i(F)$ as follows :

$$\tau(F) = \begin{pmatrix} \tau_1(F) & \tau_2(F) \\ \tau_3(F) & \tau_4(F) \end{pmatrix},$$

where each block corresponds with $m_Y^! F(i' \wedge 1_X)$, $m_Y^! F \bar{m}_X^!$, $(j' \wedge 1_Y)F(i' \wedge 1_X)$, $(j' \wedge 1_Y)F\bar{m}_X^!$ in order of their indices.

THEOREM 3.6. *Let* X *and* Y *be split* K-*module spectra, and* $f \in [X, Y]_m$. *Then* $\tau(1_K \wedge f)$ *is the following upper triangular matrix* :

$$\begin{pmatrix} f & d(f) & d'(f) & dd'(f) \\ & (-1)^m f & d''(f) & dd''(f) - (-1)^m d'(f) \\ & & (-1)^m f & (-1)^m d(f) \\ & & & f \end{pmatrix},$$

and there hold

$$d'd(f) = -dd'(f), \quad d''d(f) = -dd''(f).$$

PROOF. We have

$$(j' \wedge 1_Y)(1_K \wedge f)(i' \wedge 1_X) = (j' \wedge 1_Y)(i' \wedge 1_Y)(1_M \wedge f) = 0,$$

$$m_Y^!(1_K \wedge f)(i' \wedge 1_X) = m_Y^!(i' \wedge 1_Y)(1_M \wedge f) = 1_M \wedge f,$$

$$(j' \wedge 1_Y)(1_K \wedge f)\bar{m}_X^! = (-1)^m(1_M \wedge f)(j' \wedge 1_X)\bar{m}_X^! = (-1)^m 1_M \wedge f,$$

which determine the blocks τ_3, τ_1, τ_4 of $1_K \wedge f$, respectively. To determine τ_2, it is enough to do the (1,4) and (2,4) entries f_1, f_2 of $\tau(1_K \wedge f)$. We notice the following fact :

(3.1) if A_i, $0 \leqq i \leqq n$, are M-module spectra and $a_i : A_{i-1} \longrightarrow A_i$, the element $d(a_n \cdots a_1)$ is independent of M-module multiplications on the intermediate spectra A_i $(1 \leqq i \leqq n-1)$, and hence, it may be computed in different ways given by varying intermediate multiplications.

In this principle, we shall compute $d(g)$ for $g = d'(f)$, $d''(f)$ in the two ways given by the intermediate multiplications $K^* \wedge X$, $K^* \wedge Y$ and by $K \wedge X^*$, $K \wedge Y^*$. Then, $dd'(f) = \mu_{1,Y}(1 \wedge f)\nu_{4,X} = f_1$ and $dd''(f) = (-1)^m \mu_{1,Y}(1 \wedge f)\nu_{3,X} + \mu_{2,Y}(1 \wedge f)\nu_{4,X} = f_2 + (-1)^m d'(f)$ in the first way, by Lemma 1.6,(i) and Lemma 3.4 ; and $dd'(f) = -\mu_{1,Y}(1 \wedge d(f))\nu_{3,X} = -d'd(f)$ and $dd''(f) = -d''d(f)$ in the second way. □

PROPOSITION 3.7. <u>Let</u> X, Y, Z <u>be split</u> K-<u>module spectra and</u> $f \in [X, Y]_m$, $g \in [Y, Z]_n$. <u>Then</u>

$$d'(gf) = (-1)^m d'(g)f + d(g)d''(f) + gd'(f),$$

$$d''(gf) = (-1)^m d''(g)f + (-1)^n gd''(f).$$

PROOF. Compute (1,3) and (2,3) entries of the product $\tau(1 \wedge g)\tau(1 \wedge f)$. □

We shall see later that the operation d'' can be written in terms of d and the element ϕ' in (2.6)' in case $X = Y = K$, after having associativity of a split ring spectrum K in the next section.

§4. COMMUTATIVITY AND ASSOCIATIVITY

Henceforward let K be a split ring spectrum over $\mathbb{Z}/t$ with $t \not\equiv \pm 3 \mod 9$. The M-module multiplication m_K is the one satisfying (A_1) in Lemma 2.4, which is simply denoted by m. Similarly we abbreviate $\bar{m}_K$, m'_K, $\bar{m}'_K$, $\mu_{i,K}$, $\nu_{i,K}$, δ_K, δ'_K to $\bar{m}$, m', $\bar{m}'$, μ_i, ν_i, δ, δ', respectively.

The element m' in Lemma 3.2 with $X = K$ may not be unique. In a similar way as in [20], Theorem 1.3, we see that if n' is other choice of m', then there are unique elements $\alpha_1 \in \mathscr{M}[K, K]_k$, $\alpha_2 \in \mathscr{M}[K, K]_{k+1}$ such that

$$(4.1) \qquad n' = m' + \alpha(j' \wedge 1_K), \quad \bar{n}' = \bar{m}' - (i' \wedge 1_K)\alpha,$$

where

$$(4.2) \qquad \alpha = \bar{m}\alpha_1 m + (i \wedge 1_K)\alpha_2 m - \bar{m}\alpha_2(j \wedge 1_K) \ ;$$

for, the property $d^{(s)}(m') = 0$ restricts α to satisfy $d^{(s)}(\alpha) = 0$, which is equivalent to (4.2) by Lemma 1.6, (ii).

LEMMA 4.1. <u>Let</u> $T : K \wedge K \longrightarrow K \wedge K$ <u>be the switching map. Then each block of the matrix</u> $\tau(T)$ <u>has the following form</u> :

$$\tau_i(T) = \begin{pmatrix} (-1)^{\epsilon} d(\sigma_i) & 0 \\ \sigma_i & d(\sigma_i) \end{pmatrix}, \quad \epsilon = \begin{cases} 0 & \underline{\text{for}} \ i = 2,3, \\ 1 & \underline{\text{for}} \ i = 1,4, \end{cases}$$

<u>for some</u> $\sigma_i \in [K, K]_*$.

PROOF. Let d_1, d_2 be the derivations as in Lemma 3.4, and $d^{(1,2)}$, $d^{(2,1)}$ be those in $[K^* \wedge K, K \wedge K^*]_*$, $[K \wedge K^*, K^* \wedge K]_*$. Then

$$(4.3) \qquad d^{(1,2)}(T) = 0, \quad d^{(2,1)}(T) = 0.$$

We may calculate $d(\mu_i T \nu_j)$ in the principle (3.1) as follows :

$$\begin{aligned} d(\mu_i T \nu_j) &= \mu_i T d_1(\nu_j) \pm \mu_i d^{(1,2)}(T)\nu_j \pm d_2(\mu_i) T \nu_j \\ &= \mu_i T d_2(\nu_j) \pm \mu_i d^{(2,1)}(T)\nu_j \pm d_1(\mu_i) T \nu_j. \end{aligned}$$

By Lemma 3.4 and (4.3),

$$(4.4)\qquad \begin{aligned} d(\mu_i T\nu_j) &= \mu_i T\nu_{j+1} && \text{for } j = 1,3, \text{ all } i,\\ &= (-1)^{\epsilon'}\mu_{i-1}T\nu_j && \text{for } i = 2,4, \text{ all } j,\\ &= 0 && \text{for } j = 2,4 \text{ and for } i = 1,3,\end{aligned}$$

where $\epsilon' = \deg \mu_i + \deg \nu_j$. This gives $\tau_i(T)$ as required. □

THEOREM 4.2. For a suitable choice of μ_i, ν_i and δ, the matrix $\tau(T)$ of the switching map T is the following lower triangular matrix :

$$\tau(T) = \begin{pmatrix} 1_K & & & \\ \delta & -1_K & & \\ \delta' & 0 & -1_K & \\ \delta\delta' & \delta' & -\delta & 1_K \end{pmatrix}.$$

In other words,

$$(4.5)\qquad \begin{aligned} \mu_1 T &= \mu_1, & T\nu_1 &= \nu_1 + \nu_2\delta + \nu_3\delta' + \nu_4\delta\delta',\\ \mu_2 T &= -\mu_2 + \delta\mu_1, & T\nu_2 &= -\nu_2 + \nu_4\delta',\\ \mu_3 T &= -\mu_3 + \delta'\mu_1, & T\nu_3 &= -\nu_3 + \nu_4\delta,\\ \mu_4 T &= \mu_4 - \delta\mu_3 + \delta'\mu_2 + \delta\delta'\mu_1, & T\nu_4 &= \nu_4.\end{aligned}$$

PROOF. By Lemma 4.1, it is enough to show $\sigma_1 = \delta$, $\sigma_2 = 0$, $\sigma_3 = \delta\delta'$, $\sigma_4 = -\delta$. Since $(jj'\wedge 1_K)(1_K\wedge i'i) = i'ijj' = \delta\delta'$, we have $\sigma_3 = \delta\delta'$.

Let D be the Spanier-Whitehead duality. By Proposition 1.4, K is self-dual. By [8], $D(\delta_M) = -\delta_M$. By Lemma 2.6, the number of possible δ is odd, and hence, there is δ which is self-dual (up to sign). The correspondence $m' \mapsto n'$ with $\bar{n}' = D(m')$ is a bijection of the set of all possible m' with $m'(i'\wedge 1_K) = 1_{M\wedge K}$, and the set consists of odd elements. The construction of m' in Lemma 3.2 (with $X = K$) indicates that there is a choice of m' which satisfies

$$(4.6)\qquad D(m') = \bar{m}'.$$

Also there is an m with $D(m) = \bar{m}$. For these m and m', $D(\mu_i) = \nu_{5-i}$, $D(\nu_i) = \mu_{5-i}$. In (4.1), n' also satisfies (4.6) if and only if α is self-dual, that is, $D(\alpha_1) = -\alpha_1$, $D(\alpha_2) = \alpha_2$ in (4.2). By Lemma 2.6 and (4.4), we may put

$$\mu_2 T\nu_1 = \delta + i'fj', \quad \mu_4 T\nu_3 = -\delta + i'gj', \qquad f, g \in [M, M]_*.$$

By duality, $g = -f$. For the element α in (4.2),

$$(j\wedge 1)\alpha(j'\wedge 1)(1\wedge i'i) = (\alpha_1 m - \alpha_2(j\wedge 1))(j'\wedge 1)(1\wedge i'i)$$
$$= \alpha_1\delta' - \alpha_2\delta\delta'.$$

If we put $f = f_1 + f_2\delta_M$, $f_i \in \mathscr{M}[M, M]_*$, the element σ_1 for n' is calculated as follows :

$$(j\wedge 1)n'T\nu_1 = (j\wedge 1)m'(1\wedge i'i) + (j\wedge 1)\alpha(j'\wedge 1)(1\wedge i'i)$$
$$= \delta + i'f_1j' + i'f_2\delta_M j' + \alpha_1\delta' - \alpha_2\delta\delta'.$$

The element $h_i = m(f_i\wedge 1)(i\wedge 1)$, $i = 1,2$, satisfies $d(h_i) = 0$, $D(h_i) = (-1)^{\epsilon}(j\wedge 1)(f_i\wedge 1)\bar{m}$, $\epsilon = \deg f_i$, and $h_i i' = i'f_i$. The last relation implies

$$(j\wedge 1)n'T\nu_1 = \delta + (h_1+\alpha_1)\delta' - (h_2+\alpha_2)\delta\delta'.$$

Calculating d of $(j\wedge 1)(f_i\wedge 1)(i\wedge 1)$ in the two ways with respect to the coordinates of $M\wedge K$ leads to

$$(j\wedge 1)(f_i\wedge 1)\bar{m} + m(f_i\wedge 1)(i\wedge 1) = 0,$$

that is, h_i is self-dual. We can therefore put $\alpha_1 = -h_1$, $\alpha_2 = -h_2$, for which n' satisfies $(j\wedge 1)n'T\nu_1 = \delta$. Thus, there is a choice of m' for which $\sigma_1 = \delta$, $\sigma_4 = -\delta$.

To make σ_2 trivial, we must make one more replacement. In (4.1), (4.2), if m' and n' both make σ_1, σ_4 as required, then $\alpha_i\delta' = 0$, $\delta'\alpha_i = 0$ (by duality), $d(\alpha_i)$, $D(\alpha_i) = \pm\alpha_i$, and vice versa. Then

(a) $\quad n'T\bar{n}' - m'T\bar{m}' = \bar{m}(\alpha_2\delta - \delta\alpha_2)m - 2(i\wedge 1)\alpha_2 m - 2\bar{m}\alpha_2(j\wedge 1).$

Let σ_2 be given by m'. Calculating the matrix of $T^2 = 1$ gives the relations

$$\sigma_2\delta' = 0, \quad \delta'\sigma_2 = 0, \quad \delta d(\sigma_2) - d(\sigma_2)\delta = 2\sigma_2.$$

By Lemma 4.1, $m'T\bar{m}' = \bar{m}\sigma_2 m + (i\wedge 1)d(\sigma_2)m + \bar{m}d(\sigma_2)(j\wedge 1)$. If we put $\alpha_2 = \frac{1}{2}d(\sigma_2)$, then $n'T\bar{n}' = 0$. Replace m' by n'. □

COROLLARY 4.3. *A split ring spectrum over $\mathbb{Z}/t$ is a commutative ring spectrum, if $t \not\equiv \pm 3 \mod 9$.*

PROOF. The first projection μ_1 behaves as a multiplication as ring spectrum. Its commutativity is the first relation in (4.5). □

LEMMA 4.4. *For given m and δ, there is a unique choice of m' which makes T as given in Theorem 4.2 and defines d' satisfying*

(4.7) $\quad d'(\delta) = 0.$

PROOF. The right hand side of (a) in the proof of Theorem 4.2 is trivial, if and only if, $\alpha_2 = 0$. Therefore, the choice of m' making T as required is given by $\alpha_1\delta' = \delta'\alpha_1 = 0$, $d(\alpha_1) = 0$, $D(\alpha_1) = -\alpha_1$ and $\alpha_2 = 0$ in (4.2). We put

$$\alpha' = d'(\delta),$$

where d' is given by an m' giving Theorem 4.2. If we replace m' by n', the element α' is replaced by $\alpha' + \alpha_1$. Therefore it is enough to show α' satisfies all the properties of α_1 mentioned above. The property $d(\alpha') = 0$ follows from $d'd = -dd'$ in Theorem 3.6, and the duality follows from Theorem 3.6 too. Since $\delta' \wedge 1 = (i' \wedge 1)(j' \wedge 1) = \nu_1\mu_3 + \nu_2\mu_4$ and $\delta\delta' \wedge 1 = (i'i \wedge 1)(jj' \wedge 1) = \nu_1\mu_4$, we have

$$d'(\delta') = \mu_1 T(\nu_1\mu_3 + \nu_2\mu_4)T\nu_3 = -1,$$

$$d''(\delta') = \mu_2 T(\nu_1\mu_3 + \nu_2\mu_4)T\nu_3 = 0,$$

$$d'(\delta\delta') = \mu_1 T\nu_1\mu_4 T\nu_3 = -\delta,$$

by Theorem 4.2. Therefore $\alpha'\delta' = d'(\delta)\delta' = \delta d'(\delta') - d'(\delta\delta') = 0$. By (2.6), $d'(\delta'\delta) = -d'(\delta\delta') = \delta$, which implies $\delta'\alpha' = 0$. □

Henceforward we fix m' (and hence, $\bar{m}'$, μ_i, ν_i) as in Lemma 4.4. From the above discussion together with (2.6)', we obtain immediately the following

LEMMA 4.5. *The element* $\beta = d''(\delta)$ *satisfies*

(4.8) $$d(\beta) = 0,$$

and the matrices for $\delta \wedge 1$ *and* $\delta' \wedge 1$ *are given as follows* :

$$\tau_i(\delta \wedge 1) = \begin{pmatrix} 0 & (-1)^{i-1}1 \\ 0 & 0 \end{pmatrix}, \quad \tau_2(\delta \wedge 1) = \begin{pmatrix} 0 & 0 \\ \beta & 0 \end{pmatrix},$$

$$\tau_3(\delta \wedge 1) = 0, \quad \tau_2(\delta' \wedge 1) = \begin{pmatrix} 1 & 0 \\ 0 & 1 \end{pmatrix}, \quad \tau_j(\delta' \wedge 1) = 0,$$

for $i = 1, 4$, $j = 1, 3, 4$. *Moreover*

(4.9) $$d'(\delta') = -1, \quad d''(\delta') = 0,$$

(4.10) $$\beta\delta' = \delta^2 = -(\phi' \wedge 1)\delta' = \delta'(\phi' \wedge 1) = -\delta'\beta.$$

We shall next study the associativity of a split ring spectrum. The associativity here is the description of $\mu_i(1 \wedge \mu_j)$ in terms of $\mu_k(\mu_m \wedge 1)$'s. We shall abbreviate $\mu_i(\mu_j \wedge 1)$ to $\mu_{i,j}$ and $\mu_i(1 \wedge \mu_j)$ to $\bar{\mu}_{i,j}$.

LEMMA 4.6.

$$\bar{\mu}_{2,2} = -\mu_{2,2} + \beta\mu_{1,3} + \sum_{i,j\in\{3,4\}} \gamma_{i,j}\mu_{i,j}$$

for some elements $\gamma_{i,j}$ *which satisfy*

(4.11) $\gamma_{i,j}\delta' = 0,$

(4.12) $d(\gamma_{i,j}) = 0.$

PROOF. Let $d_k(\mu_{i,j})$, $d_k(\bar{\mu}_{i,j})$ be the derivations of $\mu_{i,j}$, $\bar{\mu}_{i,j}$ with respect to $K^*\wedge K\wedge K$ for $k = 1$, $K\wedge K^*\wedge K$ for $k = 2$, $K\wedge K\wedge K^*$ for $k = 3$. Also let $d_k(\mu_i)$ be the derivation of μ_i with respect to the k-th factor of $K\wedge K$, $k = 1, 2$.

We have $\bar{\mu}_{2,2}(i'\wedge 1\wedge 1) = \mu_2(i'\wedge 1)(1_M\wedge\mu_2) = (j\wedge 1)(1_M\wedge\mu_2) = -\mu_2(j\wedge 1\wedge 1) = -\mu_{2,2}(i'\wedge 1\wedge 1)$, and hence,

$$\bar{\mu}_{2,2} + \mu_{2,2} = \sum_{i=1}^{4}\gamma_{i,3}\mu_{i,3} + \sum_{i=1}^{4}\gamma_{i,4}\mu_{i,4} \quad \text{for some } \gamma_{i,j}.$$

Since $d_3(\mu_{i,j}) = 0$, $d_3(\bar{\mu}_{i,j}) = 0$ by Lemma 3.4, $\sum \pm d(\gamma_{i,3})\mu_{i,3} + \sum \pm d(\gamma_{i,4})\mu_{i,4} = 0$, which implies (4.12).

Let $T' : M\wedge K \longrightarrow K\wedge M$, $T'' : K\wedge M \longrightarrow M\wedge K$ be the switching maps, $T'T'' = 1$, $T''T' = 1$. Then $\mu_2(1\wedge i') = \mu_2 T(i'\wedge 1)T'' = -1\wedge j + \delta m T''$ by (4.5). Therefore $(\bar{\mu}_{2,2} + \mu_{2,2})(1\wedge i'\wedge 1) = \mu_2(\delta\wedge 1)(m\wedge 1)(T''\wedge 1)$ and

$$\begin{aligned}(\textstyle\sum\gamma_{i,3}\mu_{i,3} + \sum\gamma_{i,4}\mu_{i,4})(T\nu_1\wedge 1) &= (\bar{\mu}_{2,2} + \mu_{2,2})(1\wedge i'\wedge 1)(T'\wedge 1)(i\wedge 1\wedge 1)\\ &= \mu_2(\delta\wedge 1),\end{aligned}$$

$$\begin{aligned}(\textstyle\sum\gamma_{i,3}\mu_{i,3} + \sum\gamma_{i,4}\mu_{i,4})(T\nu_2\wedge 1) &= (\bar{\mu}_{2,2} + \mu_{2,2})(1\wedge i'\wedge 1)(T'\wedge 1)(\bar{m}\wedge 1)\\ &= 0.\end{aligned}$$

By (4.5) and Lemma 4.5,

$$\gamma_{1,3}\mu_3 + (\gamma_{2,3} + \gamma_{1,4})\mu_4 = \beta\mu_3,$$
$$\gamma_{1,4}\mu_3 + \gamma_{2,4}\mu_4 = 0,$$

and hence $\gamma_{1,3} = \beta$, $\gamma_{2,3} = \gamma_{1,4} = \gamma_{2,4} = 0$.

A similar calculation for $(\bar{\mu}_{2,2} + \mu_{2,2})(1\wedge 1\wedge i')$ leads (4.11). □

LEMMA 4.7. $\bar{\mu}_{2,4} = \mu_{2,4} - \beta\mu_{3,3} + \mu_{4,2}$, $\bar{\mu}_{4,2} = \mu_{2,4}$, $\bar{\mu}_{4,4} = \mu_{4,4}$.

PROOF. Since $1\wedge\mu_4 = (\mu_4\wedge 1)(T\wedge 1)$, we have

$$\bar{\mu}_{2,4} = \sum_i \mu_2((\mu_4 T\nu_i)\wedge 1)(\mu_i \wedge 1)$$
$$= \mu_2(\delta'\wedge 1)(\mu_2\wedge 1) - \mu_2(\delta\wedge 1)(\mu_3\wedge 1) + \mu_2(\mu_4\wedge 1)$$

by (4.5). Then the first associativity follows from Lemma 4.5. The others are easy (cf. [21], p.435). □

CORRECTIONS TO [21]. Lemma 1.7 in [21] is not correct, regarding the element δ^2. The correct formula was given in Lemma 2.8. Also Lemma 2.4 in [21] should be written in terms of β. Because of these errors, Theorem 2.1 in [21] is not correct in several points. We are giving correct formulas. The discussion in [21] except for Lemma 2.4 does still work, which we will do here. The errors, however, do not affect the main results, Theorems 4.3, 4.4, 5.3, 5.4, in [21].

LEMMA 4.8.

$$\bar{\mu}_{1,1} = \mu_{1,1} + \gamma_{4,4}\mu_{3,3},$$
$$\bar{\mu}_{1,2} = \mu_{2,1} - \mu_{1,2} + (\gamma_{3,4} - \gamma_{4,3})\mu_{3,3} + \gamma_{4,4}\mu_{4,3} + \gamma_{4,4}\mu_{3,4},$$
$$\bar{\mu}_{1,3} = \mu_{3,1} - \mu_{1,3},$$
$$\bar{\mu}_{1,4} = \mu_{4,1} + \mu_{3,2} - \mu_{2,3} + \mu_{1,4},$$
$$\bar{\mu}_{2,1} = \mu_{1,2} + \gamma_{4,3}\mu_{3,3} - \gamma_{4,4}\mu_{3,4},$$
$$\bar{\mu}_{2,3} = -\mu_{3,2} - \mu_{1,4}, \qquad \bar{\mu}_{3,1} = \mu_{1,3},$$
$$\bar{\mu}_{3,2} = -\mu_{2,3} + \mu_{1,4}, \qquad \bar{\mu}_{3,3} = -\mu_{3,3},$$
$$\bar{\mu}_{3,4} = \mu_{4,3} + \mu_{3,4}, \qquad \bar{\mu}_{4,1} = \mu_{1,4}, \qquad \bar{\mu}_{4,3} = \mu_{3,4}.$$

PROOF. The formulas of $\bar{\mu}_{i,j}$ for $(i,j) = (2,2), (2,4), (4,2), (4,4)$ are already given in Lemmas 4.6, 4.7. For such (i,j), by using Lemma 2.3 in [21], $d_1(\bar{\mu}_{i,j})$, $d_2(\bar{\mu}_{i,j})$, $d_1d_2(\bar{\mu}_{i,j})$ give the rest as stated above, where d_1 and d_2 are the same as in the proof of Lemma 4.6. □

Now the smash product $K\wedge K$ is a split K-module spectrum. As mentioned just prior to Lemma 1.6, it is possessed of the M-module multiplication given by each factor K. Similarly we may consider the following two structures of split K-module spectrum :

(i) $m_{K\wedge K} = m\wedge 1$, $\bar{m}_{K\wedge K} = \bar{m}\wedge 1$, $\delta_{K\wedge K} = \delta\wedge 1$,
$m'_{K\wedge K} = m'\wedge 1$, $\bar{m}'_{K\wedge K} = \bar{m}'\wedge 1$;

(ii) $m_{K\wedge K} = (1\wedge m)T'$, $\bar{m}_{K\wedge K} = T''(1\wedge\bar{m})$, $\delta_{K\wedge K} = 1\wedge\delta$,
$m'_{K\wedge K} = (T''\wedge 1)(1\wedge m')(T\wedge 1)$, $\bar{m}'_{K\wedge K} = (T\wedge 1)(1\wedge\bar{m}')(T'\wedge 1)$,

where $1 = 1_K$, and T, T', T'' are the switching maps as in the proof of Lemma 4.6. As before, we call $K \wedge K$ with structures (i) and (ii), respectively, $K^*\wedge K$ and $K \wedge K^*$. The elements in Definition 3.3 are then given by

$$\mu_{i,K\wedge K} = \mu_i \wedge 1, \quad \nu_{i,K\wedge K} = \nu_i \wedge 1 \qquad \text{for } K^*\wedge K,$$
$$\mu_{i,K\wedge K} = (1 \wedge \mu_i)(T \wedge 1), \quad \nu_{i,K\wedge K} = (T \wedge 1)(1 \wedge \nu_i) \quad \text{for } K \wedge K^*.$$

LEMMA 4.9. *Let* $d' : [K \wedge K^*, K]_* \longrightarrow [K \wedge K^*, K]_*$ *be given by the second structure* (ii). *Then*

$$d'(\mu_1) = 2\gamma_{4,4}\mu_3, \qquad d'(\mu_2) = \gamma_{3,4}\mu_3 + 2\gamma_{4,4}\mu_4,$$
$$d'(\mu_3) = 0, \qquad d'(\mu_4) = 0.$$

PROOF. By definition, $d'(\mu_i) = \mu_1(1 \wedge \mu_i)(T \wedge 1)(1 \wedge \nu_3)$. We notice that Lemmas 4.6, 4.7 and 4.8 involve the unique description of $\mu_{i,j}$ in terms of $\bar{\mu}_{k,m}$'s. We call it the reverse formula. We have

$$\begin{aligned}
\mu_{1,1}(T \wedge 1) &= \mu_1(\mu_1 T \wedge 1) + \gamma_{4,4}\mu_3(\mu_3 T \wedge 1) && \text{(by the reverse formula)}\\
&= \bar{\mu}_{1,1} + \gamma_{4,4}\mu_3(\delta' \wedge 1)(\mu_1 \wedge 1) - \gamma_{4,4}\bar{\mu}_{3,3} && \text{(by (4.5))}\\
&= \bar{\mu}_{1,1} - \gamma_{4,4}\bar{\mu}_{3,3} && \text{(by Lemma 4.5)}\\
&= \mu_{1,1} + 2\gamma_{4,4}\mu_{3,3} && \text{(by Lemma 4.8).}
\end{aligned}$$

In a similar way,

$$\mu_{1,2}(T \wedge 1) = \mu_{2,1} + \gamma_{3,4}\mu_{3,3} + 2\gamma_{4,4}\mu_{4,3},$$
$$\mu_{1,3}(T \wedge 1) = \mu_{3,1},$$
$$\mu_{1,4}(T \wedge 1) = \mu_{4,1}.$$

Then the lemma follows at once. □

LEMMA 4.10. *Let* $f \in [K, K]_*$. *Then* $d(f) = 0$ *and* $f\delta' = 0$ *imply* $f = -d'(f)\delta'$. *In particular,* $f = 0$ *if* $d(f) = 0$, $f\delta' = 0$, $d'(f)\delta' = 0$.

PROOF. Immediate from Proposition 3.7 and (4.9). □

We are in a position to determine the "coefficients" in the associativity formulas except for $\gamma_{4,4}$ in case $t \equiv \pm 3 \mod 9$.

LEMMA 4.11.

(4.13) $\gamma_{i,j} = 0$ *for* $(i,j) = (3,3), (3,4), (4,3)$.

(4.14) $3\gamma_{4,4} = 0$, *in particular,* $\gamma_{4,4} = 0$ *if* $t \not\equiv 0 \mod 3$.

(4.15) $d''(\delta) = \beta = -\phi' \wedge 1$.

PROOF. We consider $K \wedge K \wedge K$ to be a split K-module spectrum with structure given by the last factor K, and we compute d' of the equality in Lemma 4.6.

Every term of the equality factors through $K \wedge K$, which we regard as $K \wedge K^*$. In this case $d(\mu_i) = 0$ by Lemma 3.4 and $d'(\mu_i)$ is given in Lemma 4.9. Then

$$d'(\bar{\mu}_{2,2}) = -d'(\mu_2)(1 \wedge \mu_2) + \mu_2(1 \wedge d'(\mu_2))$$

by Lemma 1.6,(i) with d replaced by d' and by Proposition 3.7. By Lemma 4.9, Theorem 3.6 and (4.12),

$$d'(\mu_2)(1 \wedge \mu_2) = \gamma_{3,4}\bar{\mu}_{3,2} + 2\gamma_{4,4}\bar{\mu}_{4,2},$$
$$\mu_2(1 \wedge d'(\mu_2)) = -\gamma_{3,4}\bar{\mu}_{2,3} + d''(\gamma_{3,4})\bar{\mu}_{3,3} + d'(\gamma_{3,4})\bar{\mu}_{4,3} + 2\gamma_{4,4}\bar{\mu}_{2,4} + 2d''(\gamma_{4,4})\bar{\mu}_{3,4} - 2d'(\gamma_{4,4})\bar{\mu}_{4,4}.$$

By Lemmas 4.6 and 4.7,

(a) $$d'(\bar{\mu}_{2,2}) = \gamma_{3,4}\mu_{3,2} + 2\gamma_{4,4}\mu_{4,2} + \gamma_{3,4}\mu_{2,3} - (d''(\gamma_{3,4})+2\gamma_{4,4}\beta)\mu_{3,3} + 2d''(\gamma_{4,4})\mu_{4,3} +(d'(\gamma_{3,4})+2d''(\gamma_{4,4}))\mu_{3,4}-2d'(\gamma_{4,4})\mu_{4,4}.$$

Next we compute the right hand side. We have

$$d'(\mu_{i,j}) = (-1)^{\epsilon} d'(\mu_i)(\mu_j \wedge 1) + \mu_i(\mu_j \wedge d'(1)) = (-1)^{\epsilon} d'(\mu_i)(\mu_j \wedge 1),$$

where $\epsilon = \deg \mu_j$, and hence,

$$d'(\mu_{1,j}) = (-1)^{\epsilon} 2\gamma_{4,4}\mu_{3,j},$$
$$d'(\mu_{2,j}) = (-1)^{\epsilon}\gamma_{3,4}\mu_{3,j} + (-1)^{\epsilon} 2\gamma_{4,4}\mu_{4,j},$$
$$d'(\mu_{i,j}) = 0 \quad \text{for } i = 3,\ 4.$$

Therefore,

(b) $$d'(-\mu_{2,2} + \beta\mu_{1,3} + \sum \gamma_{i,j}\mu_{i,j})$$
$$= \gamma_{3,4}\mu_{3,2} + 2\gamma_{4,4}\mu_{4,2} - d'(\beta)\mu_{1,3} - (2\beta\gamma_{4,4} + d'(\gamma_{3,3}))\mu_{3,3} - d'(\gamma_{4,3})\mu_{4,3} + d'(\gamma_{3,4})\mu_{3,4} + d'(\gamma_{4,4})\mu_{4,4}.$$

Comparing (a) with (b) gives the relations

$$d'(\gamma_{3,3}) = 2(\gamma_{4,4}\beta - \beta\gamma_{4,4}),\quad \gamma_{3,4} = 0,\quad d'(\gamma_{4,3}) = 0,$$
$$d''(\gamma_{4,4}) = 0,\quad 3d'(\gamma_{4,4}) = 0,\quad d'(\beta) = 0.$$

By (4.10) and (4.11), $d'(\gamma_{3,3})\delta' = 0$ and $(\beta + \phi' \wedge 1)\delta' = 0$. Apply Lemma 4.10. □

Summarising Lemmas 4.6, 4.7, 4.8 and 4.11 and Corollary 4.3, we have concluded the following two theorems. We shall abbreviate $\gamma_{4,4}$ to γ.

THEOREM 4.12.

$$\mu_1(1\wedge\mu_1) = \mu_1(\mu_1\wedge 1) + \gamma\mu_3(\mu_3\wedge 1),$$
$$\mu_1(1\wedge\mu_2) = \mu_2(\mu_1\wedge 1) - \mu_1(\mu_2\wedge 1) + \gamma\mu_4(\mu_3\wedge 1) + \gamma\mu_3(\mu_4\wedge 1),$$
$$\mu_1(1\wedge\mu_3) = \mu_3(\mu_1\wedge 1) - \mu_1(\mu_3\wedge 1),$$
$$\mu_1(1\wedge\mu_4) = \mu_4(\mu_1\wedge 1) + \mu_3(\mu_2\wedge 1) - \mu_2(\mu_3\wedge 1) + \mu_1(\mu_4\wedge 1),$$
$$\mu_2(1\wedge\mu_1) = \mu_1(\mu_2\wedge 1) - \gamma\mu_3(\mu_4\wedge 1),$$
$$\mu_2(1\wedge\mu_2) = -\mu_2(\mu_2\wedge 1) - (\phi'\wedge 1)\mu_1(\mu_3\wedge 1) + \gamma\mu_4(\mu_4\wedge 1),$$
$$\mu_2(1\wedge\mu_3) = -\mu_3(\mu_2\wedge 1) - \mu_1(\mu_4\wedge 1),$$
$$\mu_2(1\wedge\mu_4) = \mu_4(\mu_2\wedge 1) + (\phi'\wedge 1)\mu_3(\mu_3\wedge 1) + \mu_2(\mu_4\wedge 1),$$
$$\mu_3(1\wedge\mu_1) = \mu_1(\mu_3\wedge 1),$$
$$\mu_3(1\wedge\mu_2) = -\mu_2(\mu_3\wedge 1) + \mu_1(\mu_4\wedge 1),$$
$$\mu_3(1\wedge\mu_3) = -\mu_3(\mu_3\wedge 1),$$
$$\mu_3(1\wedge\mu_4) = \mu_4(\mu_3\wedge 1) + \mu_3(\mu_4\wedge 1),$$
$$\mu_4(1\wedge\mu_j) = \mu_j(\mu_4\wedge 1) \quad \underline{\text{for}}\ j = 1, 2, 3, 4.$$

Here the element $\gamma\in[K, K]_{2k+2}$ *is trivial if* $t \not\equiv 0 \bmod 3$, *and it satisfies* $3\gamma = 0$, $d(\gamma) = 0$, $\gamma\delta' = 0$ *if* $t \equiv 0 \bmod 9$.

THEOREM 4.13. *A split ring spectrum over* $\mathbb{Z}/t$ *is a commutative, associative ring spectrum, if* $t \not\equiv 0 \bmod 3$.

§5. THE RING OF STABLE SELF-MAPS OF A SPLIT RING SPECTRUM

As we have seen in the previuos sections, the graded ring $[K, K]_*$ of stable self-maps of a split ring spectrum K is an algebra over $\mathbb{Z}/t$ and is equipped with derivations d, d', d''. However the last one d'' can be written in terms of d and the element ϕ' in (2.6)'. Also d' is a differential in associative case, as d is.

PROPOSITION 5.1. *For* $f \in [K, K]_m$,

(5.1) $d''(f) = (\phi' \wedge 1)d(f) = (-1)^{m+1}d(f)(\phi' \wedge 1)$,

(5.2) $d'd'(f) = f\gamma - \gamma f$, *in particular* $(d')^2 = 0$ *if* $t \not\equiv 0 \mod 3$.

PROOF. The formulas in Theorem 4.12 are equivalent to their reverse formulas giving $\mu_i(\mu_j \wedge 1)$ in terms of $\mu_k(1 \wedge \mu_m)$'s. They also induce the formulas giving $(\nu_i \wedge 1)\nu_j$ in terms of $(1 \wedge \nu_k)\nu_m$'s, by picking up all $\mu_m(1 \wedge \mu_k)$'s containing $\mu_j(\mu_i \wedge 1)$ in the right hand side of the formula for $\mu_m(1 \wedge \mu_k)$ in Theorem 4.12. There is one more set of 16 equalities which is equivalent to Theorem 4.12. We omit to make a list of them, because of too many (16x3 = 48) equalities. Here we mention the following two formulas for our necessity:

$$(1 \wedge \nu_2)\nu_3 = -(\nu_3 \wedge 1)\nu_2 - (\nu_1 \wedge 1)\nu_4,$$
$$(1 \wedge \nu_3)\nu_3 = (\nu_1 \wedge 1)\nu_1\gamma + (\nu_2 \wedge 1)\nu_4(\phi' \wedge 1) - (\nu_3 \wedge 1)\nu_3.$$

Then we have

$$\begin{aligned} d'd'(f) &= \mu_1(1 \wedge \mu_1)(1 \wedge 1 \wedge f)(1 \wedge \nu_3)\nu_3 \\ &= \mu_1(1 \wedge \mu_1)(\nu_1 \wedge 1)(1 \wedge f)\nu_1\gamma \\ &\quad +(-1)^m\mu_1(1 \wedge \mu_1)(\nu_2 \wedge 1)(1 \wedge f)\nu_4(\phi' \wedge 1) \\ &\quad - (-1)^m\mu_1(1 \wedge \mu_1)(\nu_3 \wedge 1)(1 \wedge f)\nu_3 \\ &= \mu_1(1 \wedge f)\nu_1\gamma - (-1)^m\gamma\mu_3(1 \wedge f)\nu_3 && \text{(by Theorem 4.12)} \\ &= f\gamma - \gamma f && \text{(by Theorem 3.6)} \end{aligned}$$

as required. Similarly $d''d(f) = \mu_2(1 \wedge \mu_1)(1 \wedge 1 \wedge f)(1 \wedge \nu_2)\nu_3 = 0$. Since $f = -d(f\delta) - d(f)\delta$, we have

$$d''(f) = -d''(d(f)\delta) = (-1)^m d(f)d''(\delta) = (-1)^{m+1} d(f)(\phi'\wedge 1)$$

by Proposition 3.7 and (4.15). □

COROLLARY 5.2. *Let* $f\in[K, K]_m$, $g\in[K, K]_n$. *Then*

(5.3) $$d'(gf) = (-1)^m d'(g)f + gd'(f) + D(g, f),$$ *where*

$$D(g, f) = d(g)(\phi'\wedge 1)d(f) = (-1)^{n+1}(\phi'\wedge 1)d(g)d(f) = (-1)^{m+1}d(g)d(f)(\phi'\wedge 1),$$

and the right upper block of the matrix $\tau(1\wedge f)$ *is given as follows* :

$$\tau_2(1\wedge f) = \begin{pmatrix} d'(f) & dd'(f) \\ (\phi'\wedge 1)d(f) & -(-1)^m d'(f) \end{pmatrix}.$$

PROOF. Immediate from Theorem 3.6 and Proposition 3.7. □

We put

(5.4) ($t \equiv 0 \mod 9$) $$\bar{\gamma} = d'(\gamma)\in[K, K]_{3k+3}.$$

By (5.2), $d'(f\gamma - \gamma f) = d'd'd'(f) = d'(f)\gamma - \gamma d'(f)$, and $d'(f) = 0$ implies $f\gamma = \gamma f$. Therefore

LEMMA 5.3. ($t \equiv 0 \mod 9$) γ *commutes with any element in* $\operatorname{Ker} d'$, *in particular*

$$\gamma\delta = \delta\gamma, \quad \gamma\delta' = \delta'\gamma = 0.$$

The element $\bar{\gamma}$ *commutes with any element in* $[K, K]_*$, *and it satisfies*

$$d(\bar{\gamma}) = 0, \quad d'(\bar{\gamma}) = 0, \quad \bar{\gamma}^2 = 0, \quad \gamma = \delta'\bar{\gamma} = -\bar{\gamma}\delta'.$$

The multiplication μ_1 *is associative if, and only if, one of the elements* γ, $\bar{\gamma}$, $\gamma\delta$, $\delta\gamma$, $\bar{\gamma}\delta$, $\delta\bar{\gamma}$ *vanishes*.

In case $t \not\equiv 0 \mod 3$, $([K, K]_*, d')$ is a differential algebra which is acyclic, because the composition with δ' gives a chain homotopy, by (4.9) and (5.3). In case $t \equiv 0 \mod 9$, we have to modify d' to obtain a differential. So we define

(5.5) $$\bar{d}'(f) = d'(f) + \bar{\gamma}\delta' f\delta' = d'(f) - \gamma f\delta'.$$

Then $\bar{d}'\bar{d}' = 0$, $\bar{d}'(f\delta') = d'(f\delta')$, $\bar{d}'(f)\delta' = d'(f)\delta'$ and $f = -\bar{d}'(f\delta') - \bar{d}'(f)\delta'$, by the relation $(\delta')^2 = 0$ and (5.3), Lemmas 5.2, 4.10. Therefore

$$\operatorname{Ker} \bar{d}' = \operatorname{Im} \bar{d}', \quad [K, K]_* = \operatorname{Ker} \bar{d}' \oplus (\operatorname{Ker} \bar{d}')\delta'.$$

Also $[K, K]_* = \operatorname{Ker} \bar{d}' \oplus \delta'(\operatorname{Ker} \bar{d}')$ by calculating $\bar{d}'(\delta' f)$.

DEFINITION 5.4. We define the following subgroups of $[K, K]_*$:

$$\mathscr{C}_* = \operatorname{Ker} d \cap \operatorname{Ker} d', \quad \bar{\mathscr{C}}_* = \operatorname{Ker} d \cap \operatorname{Ker} \bar{d}'.$$

Clearly $\mathscr{C}_*$ is a subring, $\bar{\mathscr{C}}_*$ contains $\mathscr{C}_*$, and $\mathscr{C}_* = \bar{\mathscr{C}}_*$ if $t \not\equiv 0 \bmod 3$. Together with Corollary 2.7, we then obtain the following decomposition theorem.

THEOREM 5.5. *If* $t \not\equiv 0 \bmod 3$, *then* $\mathscr{C}_*$ *is a direct summand of* $[K, K]_*$, *which decomposes into the direct sums in the following 16 ways*

$$[K, K]_* = \mathscr{C}_* \oplus \mathscr{D}_* \oplus \mathscr{D}'_* \oplus \mathscr{D}''_*$$

for $\mathscr{D}_* = \mathscr{C}_*\delta, \ \delta\mathscr{C}_*$

$\mathscr{D}'_* = \mathscr{C}_*\delta', \ \delta'\mathscr{C}_*$

$\mathscr{D}''_* = \mathscr{C}_*\delta\delta', \ \delta\mathscr{C}_*\delta', \ \delta'\mathscr{C}_*\delta, \ \delta\delta'\mathscr{C}_*.$

If $t \equiv 0 \bmod 9$, $\bar{\mathscr{C}}_*$ *is a direct summand and the same decompositions hold with* $\bar{\mathscr{C}}_*$ *in place of* $\mathscr{C}_*$.

PROOF. As we have seen, $\bar{d}'$ is a differential. Unfortunately it is no longer a derivation. It satisfies (5.3) if either of f and g is δ or δ'. But, that will do, for our purpose. Also it satisfies $d\bar{d}' = -\bar{d}'d$, $\bar{d}'(\delta) = 0$, $\bar{d}'(\delta') = -1$. Then it is easy to show

$$\operatorname{Ker} d = \operatorname{Im} d = \bar{\mathscr{C}}_* \oplus \bar{\mathscr{C}}_*\delta' = \bar{\mathscr{C}}_* \oplus \delta'\bar{\mathscr{C}}_*,$$

$$\operatorname{Ker} \bar{d}' = \operatorname{Im} \bar{d}' = \bar{\mathscr{C}}_* \oplus \bar{\mathscr{C}}_*\delta = \bar{\mathscr{C}}_* \oplus \delta\bar{\mathscr{C}}_*.$$

Combining these with the decompositions of $[K, K]_*$ given in Corollary 2.7 and similar ones in terms of δ' and $\bar{d}'$ produces the 16 decompositions. If $t \not\equiv 0 \bmod 3$, $\bar{d}' = d'$ and $\bar{\mathscr{C}}_* = \mathscr{C}_*$. □

THEOREM 5.6. $\mathscr{C}_*$ *is a commutative subring. For* $f, g \in \bar{\mathscr{C}}_*$,

$$fg - (-1)^{mn}gf = -(-1)^{n}\bar{\gamma}\,\delta' f \delta' g \delta',$$

where $m = \deg f$, $n = \deg g$.

PROOF. If $f \in \mathscr{C}_*$, then the matrix $\tau(1 \wedge f)$ is diagonal with (1,1) entry f, by Theorem 3.6 and Corollary 5.2. Then, $\tau(f \wedge 1) = \tau(T)\tau(1 \wedge f)\tau(T)$ is a lower triangular matrix with (1,1) entry f, by Theorem 4.2. Then the commutativity $fg = (-1)^{mn}gf$, $f, g \in \mathscr{C}_*$, is given by the (1,1) entries of $(1 \wedge f)(g \wedge 1) = (-1)^{mn}(g \wedge 1)(1 \wedge f)$. The relation in $\bar{\mathscr{C}}_*$ follows from a similar calculation. □

COROLLARY 5.7.

$$[[\delta', f], f] = 0 \quad \underline{\text{for}} \quad f \in \mathscr{C}_*,$$
$$[\delta', f^t] = 0 \quad \underline{\text{for}} \quad f \in \mathscr{C}_* \quad \underline{\text{with}} \quad \deg f \quad \underline{\text{even}},$$
$$[[\delta, f], f] = 0 \quad \underline{\text{for}} \quad f \in \mathscr{C}_*,$$
$$[\delta, f^t] = 0 \quad \underline{\text{for}} \quad f \in \mathscr{C}_*, \quad \underline{\text{with}} \quad \deg f \quad \underline{\text{even}}.$$

Here [,] is a (graded) commutator.

PROOF. The triviality of the triple commutators follows immediately from Theorem 5.6. As is seen in the proof of Theorem 1.5, $[\delta', f^s]$ ($f \in \mathscr{C}_*$, deg f even) and $[\delta, f^s]$ ($f \in \mathscr{C}_*$, deg f even) are divisible by s. Since $t \cdot 1_K = 0$, they are trivial when $s = t$.

REMARK. Toda [31] defined another operation

$$\lambda_X : [M, M]_k \longrightarrow [X, X]_{k+1}$$

for X an M-module spectrum, and in case $X = V(1)$ he obtained formulas concerning $\lambda_{V(1)}$ in his Theorem 4.2, Lemma 4.2' and Theorem 4.4. It is not hard to obtain such formulas for a split ring spectrum K in more or general situation. Following him, put $f_{(n)} = j'f^n i'$, $f_n = jf_{(n)}i$, $f' = f_1 \wedge 1$ for $f \in [K, K]_*$. His elements δ_1, δ_0 in $[V(1), V(1)]_*$ correspond with our δ', $\delta\delta'$. Notice that $V(1)$ is not a split ring spectrum and no element in $[V(1), V(1)]_*$ corresponds with $\delta = \delta_K$. Then, for $f \in [K, K]_*$,

$$\lambda_K(f_{(1)}\delta_M) = \mu_3(f \wedge 1)\nu_1, \quad \lambda_K(\delta_M f_{(1)}\delta_M) = \mu_4(f \wedge 1)\nu_1,$$

and hence, the elements corresponding with ϵ, ϵ' in his Theorem 4.2 are written in terms of d, d'. In particular,

$$\lambda_K(f_{(1)}\delta_M) = [f, \delta'], \quad \lambda_K(\delta_M f_{(1)}\delta_M) = [[f, \delta], \delta']$$

if $f \in \mathscr{C}_*$. He also obtained various relations concerning the elements β, $\beta_{(n)}$, β_n in his Lemma 4.2', Theorem 4.4, (4.8)(i)-(ii), (4.8)(v)-(vi), Proposition 4.7 (i), (iii), Theorem 5.1 (iv)-(v), (5.4), Theorem 5.3 (i)-(ii). Following his line, we can easily check them for arbitrary $f \in \mathscr{C}_{2*}$ in place of the specific element β. Therefore we can obtain relations of the homotopy elements in Theorem I similar to Toda's.

§6. REFORMULATION OF ELEMENTS CORRESPONDING WITH Ext^1

We shall begin with the reformulation due to Miller, Ravenel and Wilson [12] of the result of Novikov [14] on Ext^1. Let BP be the Brown-Peterson spectrum at a prime p [5], where p will be assumed to be odd in this paper. The coefficient ring $BP_* = \pi_*(BP)$ is a polynomial ring over the integers localised at p [3]:

$$BP_* = \mathbb{Z}_{(p)}[v_1, v_2, \cdots], \quad \deg v_i = 2(p^i-1),$$

and a description of generators v_i is given by Hazewinkel [7]. For a connective CW spectrum X, $BP_*(X)$ is a BP_*BP-comodule and there is a spectral sequence, due to Adams [1], [3] and Novikov [14], with $E_2^{**} = \mathrm{Ext}^{**}(BP_*, BP_*(X))$, which converges to the stable homotopy $\pi_*(X)_{(p)}$ localised at p, where Ext denotes the BP_*BP-comodule extension Ext_{BP_*BP}.

Let $\delta : \mathrm{Ext}^s(BP_*, BP_*/(p^n)) \longrightarrow \mathrm{Ext}^{s+1}(BP_*, BP_*)$ be the coboundary associated to the short exact sequence of BP_*BP-comodules :

$$(6.1) \qquad 0 \longrightarrow BP_* \xrightarrow{p^n} BP_* \longrightarrow BP_*/(p^n) \longrightarrow 0.$$

The multiplication by $v_1^{sp^{n-1}}$ $(s \geq 1)$:

$$(6.2) \qquad v_1^{sp^{n-1}} : BP_* \longrightarrow BP_*/(p^n)$$

is a BP_*BP-comodule homomorphism and it is an element in $\mathrm{Ext}^0(BP_*, BP_*/(p^n))$. Define

$$(6.3) \qquad \alpha_{sp^{n-1}/n} = \delta(v_1^{sp^{n-1}}) \in \mathrm{Ext}^1(BP_*, BP_*).$$

The internal dimension of this element is $2sp^{n-1}(p-1)$, and

$$(6.4) \qquad p^r\alpha_{sp^{n-1}/n} = \alpha_{sp^{n-1}/n-r} \quad (1 \leq r < n), \quad p^n\alpha_{sp^{n-1}/n} = 0.$$

THEOREM 6.1. (Novikov [14], Miller, Ravenel-Wilson [12]) *In dimension $2sp^{n-1}(p-1)$, $n \geq 1$, $s \geq 1$, $s \not\equiv 0 \bmod p$, $\mathrm{Ext}^1(BP_*, BP_*)$ is a cyclic group of order p^n with generator $\alpha_{sp^{n-1}/n}$; otherwise $\mathrm{Ext}^1 = 0$.*

Let $M(n)$ be the mod p^n Moore spectrum. The exact sequence (6.1) is realised by the BP-homology of the cofibration :

$$(6.5) \qquad S^0 \xrightarrow{p^n} S^0 \xrightarrow{i} M(n).$$

The next map $j : M(n) \longrightarrow S^1$ of (6.5) induces the trivial homomorphism of BP-homology, and composing j corresponds with the coboundary δ in the spectral sequence [9] cf. [12], Lemma 2.10. Therefore if we construct an element in $\pi_*(M(n))$ which realises (6.2), the survival of the element (6.3) is immediate. Because of the multiplicativity of $M(n)$, the existence of such an element in $\pi_*(M(n))$ is equivalent to the existence of a self-map of $M(n)$ which is an extension of the original element in $\pi_*(M(n))$, and such a self-map realises a BP_*BP-homomorphism :

$$(6.6) \qquad v_1^{sp^{n-1}} : BP_*/(p^n) \longrightarrow BP_*/(p^n).$$

The construction of a self-map realising (6.6) here, that is, which does not use the J-homomorphism, was suggested by P. Hoffman[8](see also [6]).

THEOREM 6.2. <u>There exist self-maps</u>

$$A_n : \Sigma^{2p^{n-1}(p-1)}M(n) \longrightarrow M(n), \qquad n \geq 1,$$

<u>which induce</u> (6.6) <u>with</u> $s = 1$ <u>and satisfies</u> $d(A_n) = 0$.

PROOF. A_1 is known as the Adams-Toda map [2]. Let $\alpha_1 \in \pi^S_{2p-3}$ be the first element of order p, and $\bar{\alpha}_1 \in [M(1), S^0]_{2p-3}$ its extension. Then $A_1 = -d(i\bar{\alpha}_1)$ satisfies $jA_1 i = \alpha_1$, $d(A_1) = 0$ (in case $p = 3$, the latter relation follows from $(\alpha_1)^2 = 0$ [20], §8). The Adams invariant of α_1 is $1/p \in \mathbb{Q}/\mathbb{Z}$ and hence A_1 induces isomorphism of the complex K-theory [2]. This means $(A_1)_* = v_1$ [25].

Theorem 1.5 with $t = p$ gives rise to elements

$$f_n : \Sigma^{2p^{n-1}(p-1)}M(n) \longrightarrow M(n), \qquad f_1 = A_1,$$

which induce $(f_n)_* \equiv v_1^{p^{n-1}}$ mod p^{n-1} and satisfies $d(f_n) = 0$ by the commutativity of f_n with λ, ρ.

Put $(f_n)_* = v_1^{p^{n-1}} + p^{n-1}x$. Since both $(f_n)_*$ and $v_1^{p^{n-1}}$ are BP_*BP-comodule homomorphisms, x must be a multiple of $v_1^{p^{n-1}}$ by Theorem 6.1. Thus $(f_n)_* = (1 + ap^{n-1})v_1^{p^{n-1}}$, $a \in \mathbb{Z}/p$. Then we may put $A_n = f_n - a\lambda' A_1 \rho'$ or $A_n = (1 - ap^{n-1})f_n$, where $\lambda' : M(1) \longrightarrow M(n)$ and $\rho' : M(n) \longrightarrow M(1)$ are the canonical maps. □

We notice that A_n in the first definition satisfies

(6.7) $$A_n\lambda = \lambda(A_{n-1})^p, \quad \rho A_n = (A_{n-1})^p\rho.$$

For the latter definition, this also holds up to constant $\not\equiv$ mod p.

THEOREM 6.3 (Novikov). *Every element in* Ext^1 *is a permanent cycle. More precisely, the element* (6.3) *is a permanent cycle and corresponds in* E_∞ *with* $j(A_n)^s i \in \pi_*^S$, *which generates a direct summand* $\mathbb{Z}/p^n$ *if* $s \not\equiv 0$ mod p, *and keeps the relation* (6.4):

$$p^r j(A_n)^s i = j(A_{n-r})^{sp^r} i.$$

PROOF. The last relation follows from (6.7). Apparently $(A_n)^s i$ realises (6.2). Hence the rest is clear from Theorem 6.1 and [9]. □

We define

(6.8) $$M(n, sp^{n-1}) = \text{cofibre of } (A_n)^s : \Sigma^{2sp^{n-1}(p-1)}M(n) \longrightarrow M(n).$$

Notice that A_n is not unique and the definition depends on the choice of A_n. In any case,

(6.9) $$BP_*(M(n,m)) = BP_*/(p^n, v_1^m), \quad n, m \geq 1, \quad m \equiv 0 \mod p^{n-1},$$

and the cofibration

(6.10) $$\Sigma^{2m(p-1)}M(n) \longrightarrow M(n) \xrightarrow{i'} M(n,m)$$

realises the exact sequence of comodules :

(6.11) $$0 \longrightarrow BP_*/(p^n) \xrightarrow{v_1^m} BP_*/(p^n) \longrightarrow BP_*/(p^n, v_1^m) \longrightarrow 0.$$

THEOREM 6.4. $M(n, sp^{n-1})$ *is a split ring spectrum if, and only if,* $s \equiv 0$ mod p^n. $BP_*/(p^n, v_1^{sp^{n-1}})$ *is realised by a split ring spectrum if, and only if,* $s \equiv 0$ mod p^n.

PROOF. The "if" part of both statements is immediate from Prop. 2.9 and (6.9). If $BP_*/(p^n, v_1^m) = BP_*(K)$, $m = sp^{n-1}$, for a split ring spectrum K, the attaching map ϕ in K realises the map (6.6). Therefore $j\phi i$ corresponds with the element $\alpha_{m/n}$. Since $j\phi i$ is divisible by p^n, so is $\alpha_{m/n}$. Hence s must be divisible by p^n. □

§7. AN INDUCTION ; PROOF OF THE MAIN THEOREMS

THEOREM 7.1. *Let* p *be an odd prime and put* $v = \deg v_2 = 2(p^2-1)$.

(a) *Assume, for some* $i, s, t \geq 1$, *where* $i \geq 2$ *if* $p = 3$,

(i) *there is a split ring spectrum* K *over* $\mathbb{Z}/p^i$ *such that*

$$BP_*(K) = BP_*/(p^i, v_1^{sp^{2i-1}}),$$

and

(ii) *there is a self-map* $f : \Sigma^{tv}K \longrightarrow K$ *such that*

$$f_* = v_2^t + v_1^{s'p^{2i-1}}x \quad \bmod (p^i, v_1^{sp^{2i-1}})$$

for some $x \in BP_*$, *where* $s' = s/2$ *if* s *is even,* $s' = (s+1)/2$ *if* s *is odd.*

Then there is a self-map

$$g : \Sigma^{tp^i v}L \longrightarrow L$$

of a split ring spectrum L *over* $\mathbb{Z}/p^i$ *such that*

$$BP_*(L) = BP_*/(p^i, v_1^{2sp^{2i-1}}),$$

$$g_* = v_2^{tp^i} + v_1^{sp^{2i-1}}y \quad \bmod (p^i, v_1^{2sp^{2i-1}})$$

for some $y \in BP_*$.

(b) *Assume moreover*

(iii) *the attaching map* ϕ *in* K *is a power* ψ^u *of some element* $\psi \in \mathcal{M}[M, M]_*$, *where* u *is a divisor of* sp^i, $u \geq 2$, *and* M *is the* mod p^i *Moore spectrum.*

Then there exist M-*module spectra* L_j *and self-maps*

$$g_j : \Sigma^{tp^i v}L_j \longrightarrow L_j, \quad u < j \leq 2u,$$

such that

$$BP_*(L_j) = BP_*/(p^i, v_1^{ju'p^{i-1}}), \quad u' = sp^i/u,$$

$$(g_j)_* = v_2^{tp^i} + v_1^{sp^{2i-1}}y \quad \bmod (p^i, v_1^{ju'p^{i-1}}).$$

PROOF. We shall use the notation $C(h)$ for a cofibre of a map h. Let $K = C(\phi)$. Then $\phi_* = v_1^k$ (up to constant $\not\equiv 0 \bmod p$), where $k = sp^{2i-1}$. We put $L = C(\delta')$ for the δ' of K. Since δ' is a composite $i'j'$ of the maps in the cofibration (2.1), $L = C(\phi^2)$ and δ' induces a cofibration:

$$(7.1) \qquad \Sigma^{2k(p-1)}K \longrightarrow L \longrightarrow K,$$

by a standard method in homotopy (cf. [17], Lemma 1.5). By Prop. 2.10, L is a split ring spectrum and $BP_*(L) = BP_*/(p^i, v_1^{2k})$ because of $(\phi^2)_* = v_1^{2k}$. The cofibration (7.1) induces a short exact sequence

$$0 \longrightarrow BP_*/(p^i, v_1^k) \xrightarrow{v_1^k} BP_*/(p^i, v_1^{2k}) \longrightarrow BP_*/(p^i, v_1^k) \longrightarrow 0.$$

By Theorem 5.5, we may assume $f \in \mathscr{C}_*$ without loss of generality, because the elements δ, δ' induce the trivial homomorphism of the BP-homology. By Corollary 5.7, $f^{p^i}\delta' = \delta' f^{p^i}$, which gives a self-map g of L with commutative diagram :

$$\begin{array}{ccccc}
\Sigma^{tp^iv+2k(p-1)}K & \longrightarrow & \Sigma^{tp^iv}L & \longrightarrow & \Sigma^{tp^iv}K \\
\downarrow f^{p^i} & & \downarrow g & & \downarrow f^{p^i} \\
\Sigma^{2k(p-1)}K & \longrightarrow & L & \longrightarrow & K.
\end{array}$$

By the assumption (ii), $(f^{p^i})_* = v_2^{tp^i}$. Therefore $g_* - v_2^{tp^i}$ is annihilated by v_1^k, completing the proof of (a).

The map ψ induces $v_1^{u'p^{i-1}}$ (up to constant relatively prime to p). We put $L_j = C(\psi^j)$ so that $L_u = K$, $L_{2u} = L$. Then L_j is a ring spectrum by Lemma 2.9, and $BP_*(L_j) = BP_*/(p^i, v_1^{ju'p^{i-1}})$. There is a map $r_j : L \longrightarrow L_j$ which is an extension of the identity of M (this is the same one as the second map in (7.1)). Then the composite $h_j = r_jgi'i : S^{tp^iv} \longrightarrow L_j$ induces $(h_j)_* = g_* = v_2^{tp^i} + v_1^k y$. Extend h_j to $\Sigma^{tp^iv}L_j$ by using the ring structure of L_j. □

The theorem allows us to make an induction with first step giving g from f, and we have got an input f in case $p \geq 5$, $i = s = 1, t = t'p$ with $t' \geq 2$ in [19] and done the first step in [21]. Completing the induction with this input leads to all the statements in Theorem I except for the p-divisibility.

REMARK. The integer t in the theorem may be divisible by a large power of p to have the map f in the assumption (ii). In fact, if f exists, t must be divisible by p^{3i-2+k} [12], [24], where k is the minimal integer such that $s \leq p^k$. The above example of the input is the only one we have known. If $i = 2$, the first possible input happens when $t = p^4$ at least, which gives the beta in dimension $2p^4(p^2-1) - 2p^3(p-1)$. This is too large to get an input by computing the stable homotopy of spheres. Theorem 7.1 would not be effective in practice when $p = 3$.

To show the p-divisibility in Theorem I, we prepare the following

THEOREM 7.2. *Let* $p \geq 5$. *Assume that all the assumptions in Theorem 7.1 (a) hold. Then there are an* M'-*module spectrum* L' *and a self-map*

$$h : \Sigma^{tp^i v} L' \longrightarrow L'$$

such that

$$BP_*(L') = BP_*/(p^{2i}, v_1^{sp^{2i-1}})$$

$$h_* = v_2^{tp^i} + p^i z \qquad \text{mod} \quad (p^{2i}, v_1^{sp^{2i-1}})$$

for some $z \in BP_*$, *where* M' *is the* mod p^{2i} *Moore spectrum.*

PROOF. Let M be the mod p^i Moore spectrum. We put $L' = C(\delta)$ for the δ of K. δ can be obtained from a homotopy giving the relation $\delta_M \phi = \phi \delta_M$. This homotopy also defines a self-map ϕ' of $C(\delta_M)$ and $L' = C(\phi')$ so that there is a cofibration

$$(7.2) \qquad K \longrightarrow L' \longrightarrow K.$$

Since $\delta_M = ij$, $M' = C(\delta_M)$ and the inclusion $M \longrightarrow C(\delta_M)$ is equivalent to the canonical map $\lambda : M \longrightarrow M'$. Therefore $(\phi')_* = v_1^{sp^{2i-1}}$ up to non-zero scalar and $BP_*(L')$ has the required form. By [20], Corollary 3.6, L' is an M'-module spectrum. Moreover (7.2) induces the short exact sequence:

$$0 \longrightarrow BP_*/(p^i, v_1^k) \xrightarrow{p^i} BP_*/(p^{2i}, v_1^k) \longrightarrow BP_*/(p^i, v_1^k) \longrightarrow 0,$$

where $k = sp^{2i-1}$. By Corollary 5.7, $f^{p^i}\delta = \delta f^{p^i}$, giving h. □

PROBLEM. Is there a choice of L' which is a split ring spectrum over Z/p^{2i} when $s \equiv 0 \mod p^{2i}$?

An affirmative answer enables us to make one more induction which construct arbitrarily large order betas.

PROOF OF THEOREM I. Our input of the induction is the elements

$$B_t : \Sigma^{tpv}M(1, p) \longrightarrow M(1, p), \quad t = 2, 3,$$

such that $(B_t)_* = v_2^{tp}$ [19], where $v = 2(p^2-1)$. We mention that, in Theorem 7.1, if $K = M(i, sp^{2i-1})$ and $u = sp^i$, then $L = M(i, 2sp^{2i-1})$ $L_j = M(i, jp^{i-1})$, $sp^i \leq j \leq 2sp^i$. Then, by Theorem 7.1 (a), there are self-maps

$$B_{t,n} : \Sigma^{tp^n v}M(1, 2^{n-1}p) \longrightarrow M(1, 2^{n-1}p), \quad B_{t,1} = B_t,$$

such that

(7.3) $$(B_{t,n})_* = v_2^{tp^n} + v_1^{2^{n-2}p}x, \quad x \in BP_*.$$

We extend the definition of $B_{t,n}$ to include any $t \geq 2$:

$$B_{t,n} = \begin{cases} (B_{2,n})^{t'} & \text{for } t = 2t' \\ (B_{2,n})^{t'-1}B_{3,n} & \text{for } t = 2t'+1. \end{cases}$$

By Theorem 7.1 (b), each stage of the induction produces

$$B_{t,n,s} : \Sigma^{tp^n v}M(1, s) \longrightarrow M(1, s), \quad 2^{n-2}p < s \leq 2^{n-1}p,$$

with the same property as (7.3). For $s \leq 2^{n-2}p$, we may put $B_{1,n,s} = B_{p,n-1,s}$. Now the element $(B_{t,n,s})^p$ induces precisely a power of v_2 so that the homotopy element

(7.4) $$jj'(B_{t,n,s})^p i'i$$

corresponds with $\beta_{tp^{n+1}/s}$ [12], [9], [21].

If $s \equiv 0 \mod p$, $M(1, s)$ is a split ring spectrum by Theorem 6.4. By Theorem 7.2, there is a self-map $B'_{pt,n,s}$ of $M(2, s)$ compatible with $(B_{t,n,s})^p$ through the canonical maps $M(1, s) \longrightarrow M(2, s)$, $M(2, s) \longrightarrow M(1, s)$. Therfore

$$p \cdot jj'B'_{pt,n,s}i'i = jj'(B_{t,n,s})^p i'i. \qquad \square$$

REMARK. If $n, t \geq 2$, $2^{n-2}p < s \leq 2^{n-1}p$, the element $jj'B_{t,n,s}i'i$ is non-zero and of order p. It corresponds with $\beta_{tp^n/s} + \beta'$, where β' is a linear combination of $\beta_{t'/s'}$ with $s' \leq 2^{n-2}p$. By dimensional reason, $p^2 | t'$ and $2p | s'$. Therefore no $\beta_{t'/s'}$ can happen when $n \leq 5$. If $n = 6$, there is only one possible $\beta_{t'/s'}$ at $p = 5$, which is, however, already a permanent cycle by earier stage of the induction. The above proof, therefore, indicates the survival of more betas, for example, $\beta_{tp^n/s}$ with $2 \leq n \leq 6$, $t \geq 2$, $2^{n-2}p < s \leq 2^{n-1}p$, which recover

Theorem 5.4 in [21].

PROOF OF THEOREM II. We define

$$S(n) = \{ m \geqq 3 \mid p^m(p+1) + 2^{m-2}p \leqq 2^{n-2}p \}.$$

If $S(n)$ is non-void, define $a(n) = \max S(n)$. Put

$$b(n, 0) = n, \quad b(n, 1) = a(n).$$

If $S(a(n))$ is non-void, define $b(n, 2) = \max S(a(n)) = a(b(n, 1))$. Similarly we define

$$b(n, k) = \max S(b(n, k-1)) = a(b(n, k-1)) \quad \text{and}$$

$$c(n) = \max \{ k \mid S(b(n, k)) \text{ is non-void} \},$$

that is, $b(n, k)$ $(\geqq 3)$ is defined only if $k \leqq c(n) + 1$. Then $a(n)$, $b(n, k)$, $c(n)$ tend to infinity with n. We put

$$d(n, j) = \begin{cases} \sum_{k=j+1}^{c(n)+1} p^{b(n, k)} & \text{for } 0 \leqq j \leqq c(n), \\ 0 & \text{for } j = c(n) + 1. \end{cases}$$

Let t and s satisfy $t \geqq 2$, $1 \leqq s \leqq 2^{b(n, c(n)+1)-2}p$. Such an s exists for a large n. Since $3 \leqq b(n, k) < b(n, k-1)$,

$$tp^n - (d(n, 0) - d(n, j)) = t_{j,n}p^{b(n, j)}$$

for some $t_{j,n} \geqq 2$. Put

$$s_{j,n} = d(n, j)(p+1) + s, \quad k(t,n,s) = (tp^n - d(n, 0))(p+1) - s.$$

Then

(a) $$t_{j,n}p^{b(n, j)}(p+1) - s_{j,n} = k(t,n,s) \quad \text{for all } j,$$

and by the downward induction on j,

(b) $$1 \leqq s_{j,n} \leqq 2^{b(n,j)-2}p.$$

The inequality (b) indicates that the elements $\beta_{t_{j,n}p^{b(n,j)}/s_{j,n}}$, $0 \leqq j \leqq c(n)+1$, are all permanent cycles, and they are in the same dimension $2k(p-1)$ by (a), $k = k(t,n,s)$. Therefore $\pi^S_{2k(p-1)-2}$ contains a subgroup $G = (\mathbb{Z}/p)^{c(n)+2}$ (direct sum of $c(n)+2$ copies of $\mathbb{Z}/p$). If $s \not\equiv 0 \mod p$, then $s_{j,n} \equiv s \not\equiv 0 \mod p$, and hence, G is a direct summand. If $s \equiv 0 \mod p$, every element in G can be divisible by p and $(\mathbb{Z}/p^2)^{c(n)+2} \subset \pi^S_{2k(p-1)-2}$.

We mention that these facts hold for infinitely many $k = k(t,n,s)$ since $t \geqq 2$ is arbitrary.

□

BIBLIOGRAPHY

1. J.F. Adams, "On the structure and applications of the Steenrod algebra", Comm. Math. Helv. 32 (1958), 180-214.

2. J.F. Adams, "On the groups J(X),IV", Topology, 5 (1966), 21-71.

3. J.F. Adams, Stable Homotopy and Generalised Homology, Univ. Chicago Press, Chicago, 1974.

4. S. Araki and H. Toda, "Multiplicative structures in mod q cohomology theories I, II", Osaka J. Math. 2(1965), 71-115 ; 3 (1966), 81-120.

5. E.H. Brown and F.P. Peterson, "A spectrum whose Z_p-cohomology is the algebra of reduced p^{th} powers", Topology, 5 (1966), 149-154.

6. B. Gray, "On the sphere of origin of infinite families in the homotopy groups of spheres", Topology, 8 (1969), 219-232.

7. M. Hazewinkel, "Constructing formal groups I. Over $Z_{(p)}$ algebras", Report of the Econometric Institute 7119, Netherlands School of Economics, 1971.

8. P. Hoffman, "Relations in the stable homotopy rings of Moore spaces", Proc. London Math. Soc. (3) 18 (1968), 621-634.

9. D.C. Johnson, H.R. Miller, W.S. Wilson and R.S. Zahler, "Boundary homomorphisms in the generalized Adams spectral sequence and the non-triviality of infinitely many γ_t in stable homotopy", in Notas de Matematicas y Simposia, Nr. 1 : Reunion Sobre Teoria de Homotopia, pp. 29-46, Soc, Mat. Mexicana, Mexico City, 1975.

10. M.E. Mahowald, "The construction of small ring spectra", in Lecture Notes in Math. 658 : Geometric Applications of Homotopy Theory II, pp. 234-239, Springer-Verlag, Berlin Heidelberg New York, 1978.

11. J.P. May, "The cohomology of restricted Lie algebras and of Hopf algebras: applications to the Steenrod algebra", Ph.D Thesis, Princeton Univ. Princeton, 1964 ; Bull. Amer. Math. Soc. 71 (1965), 130-142.

12. H.R. Miller, D.C. Ravenel and W.S. Wilson, "Periodic phenomena in the Adams-Novikov spectral sequence", Ann. of Math. 106 (1977), 469-516.

13. O. Nakamura and S. Oka, "Some differentials in the mod p Adams spectral sequence ($p \geq 5$)", Hiroshima Math. J. 6 (1976), 305-330 ; "Corrections", ibid. 7 (1977), 655-656.

14. S.P. Novikov, "The methods of algebraic topology from the viewpoint of cobordism theories", Math. U.S.S.R.-Izvestiia 1 (1967), 827-913.

15. S. Oka, "The stable homotopy groups of spheres II", Hiroshima Math. J. 2 (1972), 99-161.

16. S. Oka, "On the stable homotopy ring of Moore spaces", Hiroshima Math. J. 4 (1974), 629-678.

17. S. Oka, "A new family in the stable homotopy groups of spheres", Hiroshima Math. J. 5 (1975), 87-114.

18. S. Oka, "The stable homotopy groups of spheres III", Hiroshima Math. J. 5 (1975), 407-438.

19. S. Oka, "A new family in the stable homotopy groups of spheres II", Hiroshima Math. J. 6 (1976), 331-342.

20. S. Oka, "Module spectra over the Moore spectrum", Hiroshima Math. J. 7 (1977), 93-118.

21. S. Oka, "Realizing some cyclic BP_*-modules and applications to stable homotopy of spheres", Hiroshima Math. J. 7 (1977), 427-447.

22. S. Oka, "Corrections to "Module spectra over the Moore spectrum" ", Hiroshima Math. J. 8 (1978), 217-221.

23. S. Oka, "Ring spectra with few cells", Japan. J. Math. 5 (1979), 81-100.

24. D.C. Ravenel, "A novice's guide to the Adams-Novikov spectral sequence", in Lecture Notes in Math. 658 : Geometric Applications of Homotopy Theory II, pp.404-475, Springer-Verlag, Berlin Heidelberg New York, 1978.

25. L. Smith, "On realizing complex bordism modules. Applications to the stable homotopy of spheres", Amer. J. Math. 92 (1970), 793-856.

26. L. Smith, "On realizing complex bordism modules IV", Amer. J. Math. 99 (1977), 418-436.

27. H. Toda, "p-Primary components of homotopy groups IV" Mem. Coll. Sci. Univ. Kyoto, Ser.A, 32 (1959), 297-332.

28. H. Toda, "An important relation in homotopy groups of spheres", Proc. Japan Acad. 43 (1967), 839-842.

29. H. Toda, "Extended p-th powers of complexes and applications to homotopy theory", Proc. Japan Acad. 44 (1968), 198-203.

30. H. Toda, "On spectra realizing exterior parts of the Steenrod algebra", Topology, 10 (1971), 53-65.

31. H. Toda, "Algebra of stable homotopy of Z_p-spaces and applications", J. Math. Kyoto Univ. 11 (1971), 197-251.

32. N. Yamamoto, "An application of functional higher operations", Osaka J. Math. 2 (1965), 37-62.

33. R.S. Zahler, "Fringe families in stable homotopy", Trans. Amer. Math. Soc. 224 (1976), 243-254.

DEPARTMENT OF MATHEMATICS
KYUSHU UNIVERSITY 33
FUKUOKA 812 / JAPAN

Current Address:
Mathematical Institute
University of Oxford
Oxford OX1 3LB / England
(until August 1982)
Sonderforschungsbereich
"Theoretische Mathematik"
University of Bonn
5300 Bonn 1 / West Germany
(September 1982——August 1983)

Contemporary Mathematics
Volume 19, 1983

EQUIVARIANT CELL ATTACHMENT AND EQUIVARIANT VECTOR BUNDLES

Ted Petrie

0. INTRODUCTION AND MOTIVATION

One of the basic constructions in homotopy theory is the following: Let $F:W \to Z$ be a map and $\iota':S^k \to W$ be a map such that $F\circ\iota'$ is null homotopic. Use ι' to attach a cell D^{k+1} giving the space $O = W \bigcup_{\iota'} D^{k+1}$. Then there is a map $F':O \to Z$ extending F. This construction is particularly central to the theory of surgery on manifolds. In fact surgery can be viewed as an elaboration of this extension process appropriate to the smooth category. However in surgery theory one also encounters a more subtle problem. The initial data includes bundles over W and Z and a "bundle isomorphism" of the appropriate type over W. One must then extend the "bundle isomorphism" over W to a "bundle isomorphism" over O. Although this point is frequently confusing at first glance, the problem can be handled by standard techniques (see section 1 of Wall's book on surgery). This paper treats the equivariant analog of this extension problem for "bundle isomorphism" of a type appropriate to equivariant surgery. Although one might guess that such questions are word for word generalizations of their non equivariant counterparts, the appropriate concepts of "bundle isomorphism" are considerably more complicated in the presence of a group action and to a large extent the formulation depends on the problem being treated. Nevertheless the final answer to the "bundle isomorphism" extension question is still sufficient for the purposes of equivariant surgery in crucial cases. The purpose of this paper is to establish a result of this type.

We first motivate the main results with a discussion of applications. One of the old problems in transformation groups emphasized by P.A. Smith in [S] is the following: A) Suppose a finite group G acts smoothly on W with fixed set W^G consisting of isolated points. Describe the isotropy representations $\{T_xW \mid x \in W^G\}$ of G when W is a disk or sphere. (Schultz has considered a related question in [Sc].)

0271-4132/83 $1.00 + $.25 per page

The case of a disk is easier and we concentrate on that. In that case the problem has an equivariant homotopy analog: B) Let W be a contractible G c.w. complex with W^G consisting of isolated points and let η be a G vector bundle over W. Describe the G representations $\{\eta_x | x \in W^G\}$. There are two major results which apply: one of Smith about the homology groups of W^H when H is a p subgroup of G for some prime p the other of Atiyah about the equivariant K theory of $E \times W$ where E is an acyclic free G space. Smith's result asserts $\tilde{H}_*(W^H, Z_p) = 0$ if H is a subgroup of p power order. Then W^H is connected and this implies

0.1 η_x and η_y are isomorphic as representations of P for each Sylow subgroup P of G and each $x, y \in W^G$.

Atiyah's results in [A_2] deal with the relation between the complex representation ring $R(G)$ of G and $K_G^*(E)$. In particular there is an obvious homomorphism $\alpha_G : R(G) \to K_G^*(E)$ which sends a representation R to the G vector bundle $E \times R$ over E. Then [A_2] implies

0.2 $\mathrm{Ker}(\alpha_G) = \mathrm{Ker}(R(G) \xrightarrow{\text{res}} \hat{R}(P))$

where the product is over all Sylow P groups and res denotes restriction. It is easy to see that $K_G(E \times W)$ and $K_G(E)$ are naturally isomorphic. Together with 0.1 this implies that if R is any of the representations η_x, for $x \in W^G$,

0.3 $E \times \eta$ and $E \times (W \times R)$ are stably isomorphic G vector bundles over $E \times W$.

It is evident then that 0.1 and 0.3 must appear in the solution of the posed problems. Tentatively let's incorporate 0.1 and 0.3 in this definition: Let η and η' be two G vector bundles over W. A "bundle isomorphism" $\underline{\underline{\beta}} : \eta \to \eta'$ consists of a collection $(B_\infty, B_p | p \in A)$ where A is the set of primes dividing $|G|$ the order of G. Here $B_\infty : E \times \eta \to E \times \eta'$ is a stable G vector bundle isomorphism and $B_p : \eta \to \eta'$ is a P vector bundle isomorphism where P is the p Sylow subgroup. Let now W be some finite set and $\{\eta_x | x \in W\}$ a set of representations of G which satisfy 0.1 for x and $y \in W$. Then $\eta = \{\eta_x | x \in W\}$ is a G vector bundle over W with trivial G action. Let R be η_x for some x. By the results mentioned above there is a bundle isomorphism $\underline{\underline{\beta}} : \eta \to W \times R$. Now the problem can be formulated in equivariant homotopy theoretic terms: C) Does there exist a contractible G c.w. complex W' containing W as fixed set, a G vector bundle Γ over W' extending η and a bundle isomorphism $\underline{\underline{\beta}}' : \Gamma \to W' \times R$ extending $\underline{\underline{\beta}}$? The obvious procedure for treating this is to attach equivariant cells to W to kill the homotopy of W and to extend η and $\underline{\underline{\beta}}$ with each cell attachment.

There are two aspects to this problem: One is to realize W as the fixed set of an action on a disk. This was solved by Oliver [O] and does not concern us. The second is to extend η and $\underline{\underline{\beta}}$ and that is what concerns us here.

With this motivation we can generalize C) to a setting which applies to other manifolds as well as disks and which easily translates into the smooth category. Let X, Y, W and Z be G spaces with $X \subset W$ and $Y \subset Z$. Let $F:W \to Z$ be a G map where $f = F_{|X}:X \to Y$. Let H be a subgroup of G. An element μ of $\pi_{k+1}(f^H)$ is represented by a diagram

$$\begin{array}{ccc} S^k & \xrightarrow{\iota'} & X^H \\ \downarrow & & \downarrow \\ D^{k+1} & \xrightarrow{\kappa'} & Y^H \end{array} \qquad \begin{array}{l} \partial\mu \in \pi_k(X^H) \\ \text{is represented by } \iota'. \end{array}$$

See section 2 for elaboration. Then ι' induces a G map of $G/H \times S^k$ to X which is used to attach a cell $G/H \times D^{k+1}$ to W giving $O = W \cup G/H \times D^{k+1}$. Moreover there is a G map $F':O \to Z$ extending F. Now suppose η is a G vector bundle over W and $\underline{\underline{\beta}}:\eta \to W \times R$ is a "bundle isomorphism" for some representation R of G.

> D) Does there exist a G vector bundle Γ over O and a bundle isomorphism $\underline{\underline{\beta}}':\Gamma \to O \times R$ extending $\underline{\underline{\beta}}$?

By virtue of Smith's Theorem mentioned above, D) must be treated for each subgroup G of prime power order.

Question D) is the homotopy analog of the main geometric step in equivariant surgery. There the spaces W and Z are smooth G manifolds with boundary X resp. Y, η is TW and O is replaced by a smooth G manifold W' having O as a G deformation retract. The data $\mathcal{W} = (W,F,\underline{\underline{\beta}})$ is called a normal map and the process of constructing $\mathcal{W}' = (W',F',\underline{\underline{\beta}}')$ from $\mathcal{W}$ is called surgery on μ. To use these constructions one needs the (compare [D-P_1], [D-P_2], [P_1], [P_2])

<u>SURGERY LEMMA 0.4</u>: Surgery on $(\mathcal{W},\mu)$ $\mu \in \pi_{k+1}(f^H)$ $\mathcal{W} = (W,F,\underline{\underline{\beta}})$ is possible if $k < \frac{1}{2} \dim X^H$. This means $\mathcal{W}' = (W',F',\underline{\underline{\beta}}')$ exists extending $\mathcal{W}$.

The validity of the Surgery Lemma depends fundamentally on the definition of the bundle isomorphism $\underline{\underline{\beta}}$. In the case G acts freely, there is no ambiguity. In fact bundle questions are minor and 0.4 is valid and is one of the fundamental geometric steps in Wall's surgery theory. In the case G does not act freely bundle considerations play a more prominent role and the definition of $\underline{\underline{\beta}}$ depends on the geometric problem to which surgery is applied. In [DP_1] and [DP_2] one definition of $\underline{\underline{\beta}}$ ($\underline{\underline{\beta}} = (b,c)$ as in [DP_2, 2.2]) is given which is relevant to constructing smooth actions of G on spheres with one fixed point $[P_1]$ $[P_2]$. The Surgery Lemma 0.4 is established in $[DP_1, §4]$ in that case. It is one step in the Surgery Induction Theorem of $[DP_2, 2.6]$ which is crucial to $[P_1]$ and $[P_2]$. In this paper we give a definition of $\underline{\underline{\beta}}$ relevant to A). See 3.9. The main results of this paper are 3.12 and 4.12. The Surgery Lemma 0.4 for this definition of $\underline{\underline{\beta}}$ is a formal consequence of 3.12 $[P_3]$. The Induction Theorem of $[DP_2]$ then carries over immediately to this definition of $\underline{\underline{\beta}}$. This Induction Theorem is the main tool in treating A) and gives the first examples of actions of G on spheres and disks with distinct isotropy representations. See $[P_3]$ $[P_4]$. These considerations are the main motivation of 3.12 and 4.12.

In analogy with the smooth case we call $\mathcal{W} = (W, F, \underline{\underline{\beta}})$ $\underline{\underline{\beta}}: \eta \to W\times R$ a normal map. If $\mu \in \pi_{k+1}(f^H)$, let $O = W \cup G/H\times D^{k+1}$ with attaching map determined by $\partial\mu$. If Γ is a G vector bundle over O and if $\underline{\underline{\beta}}':\Gamma \to O\times R$ extends $\underline{\underline{\beta}}$, we say surgery on μ is possible. The analog of 0.4 in this setting is called the Homotopy Surgery Lemma.

One aim of this paper is to give a definition of a "bundle isomorphism" $\underline{\underline{\beta}}: \eta \to W\times R$ which incorporates 0.1 and 0.3 such that D) is always valid and which is compatible with the geometric needs of smooth G surgery (this is 3.9). The results of this paper require the following assumption which is to hold throughout text.

0.5 G is an odd order abelian group.

The main result 3.12 (Homotopy Surgery Lemma) answers D) affirmatively for every subgroup H of prime power order. Then 4.12 establishes the existence of non-trivial bundle isomorphisms which are used in $[P_1]$ and $[P_2]$ to treat A) with the aid of the Induction Theorem mentioned above.

The paper is organized as follows: In section 1 we deal with the relations between representations and topology which are relevant. One new and essential idea is to measure the deviation $\theta(B)$ between B and gBg^{-1} where $B:\eta \to W\times R$ is a P vector bundle isomorphism and $g \in G$. Here η is a G vector bundle and R is a representation of G. This deviation turns out to give a G map $\theta(B):W \to \Delta_p(R)$ (§1). This uses the fact G is abelian. The essential topological property of $\Delta_p(R)$ (1.7) is that it is a $K(\pi,1)$ where π is free abelian. This implies that under rather general circumstances, $\theta(B)$ is G homotopic to a constant map. The homotopy provides the added structure needed to extend $\underline{\underline{\beta}}$ to $\beta':\Gamma \to O\times R$ in D). We remark that this would be rather simple if B were a G vector bundle isomorphism as in $[DP_1, 4.22]$. Section 2 treats D). The main result is 2.19. Section 3 treats D) as H and P are varied. The main result is 3.12 which requires the full data of $\underline{\underline{\beta}}$ while 2.19 does not. Section 4 relates the material to the work of Atiyah, makes use of the functoriality of $\underline{\underline{\beta}}$ and produces non-trivial $\underline{\underline{\beta}}$

relevant to A). In 4.14 we summarized the data involved in the definition of $\underline{\underline{\beta}}$.

1. REPRESENTATIONS AND TOPOLOGY

Here is a rough statement of the content of this section: Let η and η' be G vector bundles over a G space S and let $B:\eta \to \eta'$ be a P vector bundle isomorphism for some subgroup P of G. Since G is abelian B and gBg^{-1} for $g \in G$ are two P vector bundle isomorphisms. Comparing them for all $g \in G$, gives rise to a G map $\theta(B):S\to \Delta_p$ where Δ_p is a G space to be described. If $B':\eta \to \eta'$ is a G vector bundle isomorphism, then B' and B together give a G map $\theta(B',B):S \to U$ where U is a space to be described. In this section we discuss the equivariant homotopy properties of $\theta(B)$ and $\theta(B',B)$.

A G space considered in this paper is to be of the G homotopy type of G c.w. complex of finite type. Complex G vector bundles are defined in [A_1]. From that the definition of a real G vector bundle is immediate. We shall have to treat G vector bundles with infinite dimensional fibers in a mild situation: the base space is $E\times S$ where E an acyclic complex on which G acts freely and S is a compact G space. A G vector bundle ξ over $E\times S$ is a fiber bundle over $E\times S$ with group U and fiber $\mathbb{F}^{\infty}$ together with an action of G on ξ which restricts to an $\mathbb{F}$ linear map on fibers and gives ξ/G the structure of an ordinary $(U,\mathbb{F}^{\infty})$ fiber bundle over $(E\times S)/G$ in the sense of Steenrod. See e.g. [H]. Here $(U,\mathbb{F})$ is $(U_{\infty},\mathbb{C})$ or $(O_{\infty},\mathbb{R})$ accordingly, as ξ is complex or real; $\mathbb{F}^{\infty}$ is the countable infinite direct sum of $\mathbb{F}$ with U acting as a group of $\mathbb{F}$ linear automorphisms and U_{∞} resp. O_{∞} is the infinite unitary resp. orthogonal group. If ξ is a finite dimensional G vector bundle, then $\xi\oplus(E\times S\times\mathbb{F}^{\infty})$ is a G vector bundle in the sense just discussed. Here and throughout the paper $\mathbb{F}^{\infty}$ has trivial G action. If M is a representation of G, then $\underline{M} = S\times M$ is a G vector bundle over S. As the base space is suppressed from notation, S must be understood. Sometimes for clarity we write $\underline{M}$ over S to indicate this.

If ξ is a G vector bundle and M is a representation of G, $s_M(\xi) = \xi\oplus\underline{M}$. If $B:\xi \to \xi'$ is a G vector bundle isomorphism, then $s_M(B):s_M(\xi) \to s_M(\xi')$ is the G vector bundle isomorphism $B\oplus 1_M$. The symbol s will be used to denote s_M for some M. By a G vector bundle isomorphism $B:s(\xi) \to s(\xi')$, we mean B is a G vector bundle isomorphism from $s_M(\xi)$ to $s_M(\xi')$ for some M. When $M = M'\oplus\mathbb{F}^{\infty}$ where M' is a finite dimensional representation and $\mathbb{F}^{\infty}$ has trivial action ($\mathbb{F} = \mathbb{R}$ or $\mathbb{C}$ corresponding to real or complex vector bundles), $s_M(\xi)$ is denoted by $s_{\infty}(\xi)$.

If ξ is a G vector bundle over S with G invariant inner product, then for a subgroup H of G written $H \subset G$

1.1 $\xi_{|S^H} = \xi^H \oplus \xi_H$

where ξ^H is the H fixed set and ξ_H is its orthogonal complement. If $B:\xi \to \xi'$ is an H vector bundle isomorphism, then Schur's lemma implies a splitting

1.2 $B_{|S^H} = B^H \oplus B_H$

where $B_H:\xi_H \to \xi'_H$ is an H vector bundle isomorphism.

We shall now treat several functors of representations of G. Such a functor $\mathfrak{F}$ will associate a G space $\mathfrak{F}(R)$ to a representation R of G and a suspension map $s_M:\mathfrak{F}(R) \to \mathfrak{F}(R\oplus M)$ associated to $R \longmapsto R\oplus M$. Generally the definition of s_M is obvious and is omitted from discussion. This notation holds throughout:

1.3 A denotes the set of primes dividing $|G|$. For $p \in A$, $P \subset G$ is the p Sylow subgroup.

Let R be a representation of G with a G invariant inner product. The group $U(R)$ of isometries of R is a G space via $(gB) = gBg^{-1}$ for $B \in U(R)$ $g \in G$. When R is complex $U(R) = U_n$ $n = \dim_C R$, while in the real case $U(R) = O_n$ $n = \dim_{\mathbb{R}} C$. Let $\Omega(R) \subset U(R)$ be a fixed maximal torus which in the complex case is taken to be the diagonal matrices in U_n. We suppose the matrices representing R lie in $\Omega(R)$. For G abelian this is always possible when R is a complex representation or $|G|$ is odd. There are homomorphisms $s_M:U(R) \to U(R\oplus M)$ and $s_M:\Omega(R) \to \Omega(R\oplus M)$ defined by $s_M(B) = B\oplus 1_M$.

By Schur's Lemma

$$U(R)^H = U(R^H)\times U(R_H)$$

for any subgroup H of G. Since G is abelian and the matrices of the representation R lie in $\Omega(R)$, $\Omega(R) \subset U(R)^H$. In fact since G is abelian of odd order, R_H has the structure of a complex representation of G whether or not R is a complex representation. This means

1.4 $\Omega(R) = \Omega(R^H)\times\Omega(R_H)$

for any subgroup H of G.

The space of maps from S to S' is denoted by (S,S'). If these are G spaces, G acts on this set via $(gf)(x) = gf(g^{-1}x)$ for $f \in (S,S')$. The fixed set $(S,S')^G$ is the space of G maps from S to S' while $[S,S']^G$ is the set of G homotopy classes of G maps.

Let $p \in A$ (1.3) and let $\Delta_p = \Delta_p(R)$ be the space of maps of G/P to $\Omega(R)$ which carry the identity coset to 1 in $\Omega(R)$. Then Δ_p is a G space via

$$(gf)(h) = f(hg)f(g)^{-1}$$

for $g \in G$ and $f \in \Delta_p$. Here as elsewhere we write $f(g)$ for the value at the coset gP. Let $1 \in \Delta_p$ be the map which carries every coset to $1 \in \Omega(R)$ and for any G space S, let

$$* \in (S,\Delta_p)^G \quad \text{resp.} \quad (S,U(R))^G$$

denote the map which carries S to $1 \in \Delta_p$ resp. $1 \in U(R)$. Note that P acts trivially on Δ_p; so

$$(S,\Delta_p)^G = (S/P,\ \Delta_p)^{G/P}.$$

The splitting 1.4 gives a corresponding splitting of Δ_p as a G space

1.5 $\quad \Delta_p = \Delta_p(R^H)\times\Delta_p(R_H)$.

In particular if $\theta \in (S,\Delta_p)^G$, then

1.6 $\quad \theta = \theta^H \oplus \theta_H$ where $\theta^H: S \to \Delta_p(R^H)$ and $\theta_H: S \to \Delta_p(R_H)$ are the components of θ in each factor of Δ_p.

Caution: θ^H is the component of θ in $(S,\Delta_p(R^H))^G$ not the fixed point map from S^H to Δ_p^H; similarly for θ_H.

The most important topological property of the space $\Delta_p(R)$ is

1.7 $\quad \Delta_p(R)$ is a $K(\pi,1)$ where $\pi = \pi_1(\Delta_p(R))$ is free abelian.

If $\gamma \in (S,\Delta_p(R))^G$, then $(S,U(R))$ becomes a G space $(S,U(R))_\gamma$ via

$$(gf)(x) = (gf(g^{-1}x))(\gamma(g^{-1}x)(g))^{-1}$$

for $g \in G$, $x \in S$, $f \in (S,U(R))$. The fixed set $(S,U(R))_\gamma^G$ is called the space of G-γ maps from S to $U(R)$. There is a related G action on $S\times U(R)$ defined by

$$g(x,u) = (gx,(gu)\gamma(x)(g)^{-1}).$$

The resulting space is denoted by $(S\times U(R))_\gamma$. When G acts freely on S, there is a fibration

$$\tilde{\gamma}: U(R) \to (S\times U(R))_\gamma/G \to S/G.$$

LEMMA 1.8. The map ϕ from $(S,U(R))_\gamma^G$ to the space of sections of $\tilde{\gamma}$ defined by $\phi(f)[x] = [x,f(x)]$ for $x \in S$ is a homeomorphism. (Here $[x] \in S/G$ denotes the class defined by $x \in S$ while $[x,f(x)]$ is the class defined by $(x,f(x))$ in $(S\times U(R))_\gamma$.)

Proof: This is left to the reader. See [B, II 2.6] for the case $\gamma = *$; so

$(S,U(R))^G_\gamma = (S,U(R))^G$.

In words 1.8 says that G-γ maps from S to $U(R)$ correspond to sections of $\tilde{\gamma}$.

Let η be a G vector bundle over S and let $B:\eta \to \underline{R}$ be a P vector bundle isomorphism. Then B and gBg^{-1} are two P vector bundle isomorphisms; so for $x \in S$, $B(gx)$ and $gB(x)g^{-1}$ differ by a matrix $\theta(B)(x)(g)$ in $U(R)$

1.9 <u>We suppose $\theta(B)(x)(g)$ lies in $\Omega(R)$ for all $g \in G$ and $x \in S$.</u> Then $\theta(B) \in (S,\Delta_p(R))^G$ is defined by

$$B(gx) = \theta(B)(x)(g)gB(x)g^{-1}.$$

Here $B(x):\eta_x \to \underline{R}_x$ is the map on the fiber over $x \in S$. The assertion in the definition is that $\theta(B)$ defined this way is a G map! The verification is straightforward. We note that $\theta(B)$ splits as

1.10 $\theta(B) = \theta(B)^H \oplus \theta(B)_H$ for $H \subset G$ and for $H \subset P$, $\theta(B^H) = \theta(B)^H|_{S^H}$.
See 1.6.

We should emphasize that the assumption that $\theta(B)$ is defined in $(S,\Delta_p(R))^G$ has important topological implications which derive from 1.7 and 1.14. Its viability rests on the assumption that G is abelian; so the matrices defining R lie in $\Omega(R)$.

Now suppose in addition to the P vector bundle isomorphism B, we have also a G vector bundle isomorphism $B':\eta \to \underline{R}$. Define

1.11 $\theta(B',B) \in (S,U(R))^G_{\theta(B)}$ by $B' = \theta(B',B)\circ B$.

What is asserted by this definition is that if $\theta(B',B)$ is defined this way, then it defines a G-$\theta(B)$ map from S to $U(R)$. A useful converse is true:

<u>LEMMA 1.12</u>. If $\theta' \in (S,U(R))^G_{\theta(B)}$ and $B' = \theta'\circ B$, then $B':\eta \to \underline{R}$ is a G vector bundle isomorphism with $\theta(B',B) = \theta'$.

Proof: This is immediate from the definitions.

1.13 <u>DEFINITION</u>: i) A <u>trivialization of</u> B is a homotopy $\theta_t(B) \in (S,\Delta_p(R))^G$ with $\theta_0(B) = *$ and $\theta_1(B) = \theta(B)$.

ii) A <u>trivialization of</u> B' <u>relative to a trivialization $\theta_t(B)$ of B</u> is a homotopy $\theta_t(B',B) \in (S,U(R))^G_{\theta_t(B)}$ with $\theta_0(B',B) = *$ and $\theta_1(B',B) = \theta(B',B)$.

If B is in fact a G vector bundle isomorphism, then $\theta(B) = *$ and $\theta_t(B) = *$ is a trivialization. A P vector bundle isomorphism with a trivialization is a useful compromise between the weaker notion of a P vector bundle isomorphism and the stronger notion of a G vector bundle isomorphism.

Here is a lemma which tells when trivializations exist.

LEMMA 1.14. i) If G/P acts freely on S/P and $H^1(S/P,\mathbb{Z}) = 0$, then any P vector bundle isomorphism $B:\eta \to \underline{R}$ over S has a trivialization. ii) Suppose G acts freely on S and $[S,U(R)]^G$ has one element - the class of $*$. If $B:\eta \to \underline{R}$ is a P vector bundle with trivialization $\theta_t(B)$ and $B':\eta \to \underline{R}$ is a G vector bundle isomorphism, there is a trivialization of B' relative to $\theta_t(B)$.

Proof: i) As G/P acts freely on S/P, G/P equivariant maps of S/P to Δ_p correspond to sections of a fiber bundle over S/P with fiber Δ_p. See [B]. Homotopy classes of these sections are obstructed by $H^1(S/P, \pi_1(\Delta_p))$ because Δ_p is a $K(\pi,1)$ by 1.7. Since π is free abelian and $H^1(S/P,Z) = 0$, this obstruction group vanishes. This means that any two G maps from S to Δ_p are G homotopic. This guarantees a trivialization of $\theta(B)$. ii) We apply 1.8 to the fibration

$$U(R) \to (S\times I\times U(R))_{\theta_t(B)} \to S\times I/G.$$

Let s be a section with $s = \phi(\theta(B',B))$ over $S\times 1/G$. Then $s = \phi(\theta_t)$ where $\theta_t \in (S,U(R))^G_{\theta_t(B)}$ and $\theta_1 = \theta(B',B)$ while $\theta_0 \in (S,U(R))^G$ because $\theta_0(B) = *$. By hypothesis $[S,U(R)]^G$ consists of one element the class of $*$, so θ_0 is G homotopic to $*$ by some homotopy ψ_t. Then ψ_t and θ_t combine to give a trivialization $\theta_t(B',B)$ of B' relative to B.

In order for 1.14 ii) to be useful we need to know when $[S,U(R)]^G$ consists of one element. When R is $R'\oplus \mathbb{F}^\infty$ for some finite dimensional representation R' of G, the following reduction is useful.

LEMMA 1.15. If G acts freely on S, then $[S/G, U(\mathbb{F}^\infty)] = [S, U(\mathbb{F}^\infty)]^G \longrightarrow [S, U(R'\oplus \mathbb{F}^\infty)]^G$ is bijective.

Proof: The inclusion $U(\mathbb{F}^\infty) \to U(R'\oplus \mathbb{F}^\infty)$ given by $a \to 1_{R'}\oplus a$ is a G map which is a homotopy equivalence. (This is just the inclusion of U_∞ in U_∞ or O_∞ in O_∞.) The result follows from this and the interpretation of G maps as sections of a fiber bundle.

COROLLARY 1.16. Suppose G acts freely on S and $K^1(S/G) = 0$. If η is a G vector bundle over S, $B:s_\infty(\eta) \to s_\infty(\underline{R})$ is a complex P vector bundle with trivialization $\theta_t(B)$ and $B':s_\infty(\eta) \to s_\infty(\underline{R})$ is a G vector bundle isomorphism, then there is a trivialization of B' relative to $\theta_t(B)$.

Proof: $K^1(S/G) = [S/G, U_\infty]$ by definition. See [A_2]. Since $U_\infty = U(\mathbb{C}^\infty)$, 1.15 and the hypothesis here imply the hypothesis and hence conclusion of 1.14.

2. HOMOTOPY SURGERY LEMMA FOR FIXED H

In this section we treat question D) of the introduction by proving the Homotopy Surgery Lemma 0.4 in the case <u>H is a fixed p group for some $p \in A$</u>. This is Lemma 2.19. It requires less bundle data than the general case 3.12. The precise bundle data $\underline{\underline{B}} = (\underline{B}, \theta_t(\underline{B}))$ for 2.19 is given in 2.7 and 2.8.

To motivate the selection of data in 2.7 we note these points: The isomorphisms 2.7i-iii reflect the remarks in the introduction derived from the theorems of Smith and Atiyah. They reflect the geometric requirements for 0.4 in the smooth case. More precisely if $\eta = TW$, then η_H is the normal bundle of W^H in W and 2.7i)-ii) provide the necessary hypothesis to extend $\iota' = \partial\mu: S^k \to X^H$ (2.2) to an equivariant imbedding (immersion) of $G\times_H(S^k \times D^{n-k} \times D(V))$ into X ($V = (\eta_H)_x$ $x \in W^H$). This is used to attach $G\times_H(D^{k+1}\times D^{n-k}\times D(V))$ to W giving W' with 0 as G deformation retract. (See 0.4 and [D-P$_1$].) This is the case for $H \neq 1$. If $H = 1$ 2.7iii) gives rise to the relevant imbedding.

Let K be a subgroup of G and let S be a K space. Then $\mathrm{ind}_K^G S$ is the G space $G\times S/K$ where K acts on $G\times S$ via $k(g,s) = (gk^{-1}, ks)$ for $g \in G$ and $s \in S$. The important property of this construction is this: If S' is a G space and $f: S \to S'$ is a K map, there is a unique G map $\mathrm{ind}_K^G f: \mathrm{ind}_K^G S \to S'$ such that the composition $G\times S \to \mathrm{ind}_K^G S \to S'$ is given by $(g,s) \to gf(s)$.

We fix this notation through the paper:

2.1 W resp. Z are G spaces which contain X resp. Y as G invariant subspaces; $F: W \to Z$ is a G map whose restriction f to X carries X to Y.

Throughout this section H is a fixed subgroup of P (1.3) for some fixed $p \in A$. If C_{f^H} is the mapping cylinder of f^H, then by definition $\pi_{k+1}(f^H) = \pi_{k+1}(C_{f^H}, X^H)$. An element $\mu \in \pi_{k+1}(f^H)$ is represented by a homotopy class of diagrams

2.2
$$\mu: \quad \begin{array}{ccc} S^k & \xrightarrow{\iota'} & X^H \\ \downarrow{\scriptstyle j'} & & \downarrow{\scriptstyle f^H} \\ D^{k+1} & \xrightarrow{\kappa'} & Y^H \end{array}$$

$\partial\mu \in \pi_k(X^H)$ is represented by ι'

Let n be an integer larger than k and let

2.3
$$\mu':\quad \begin{array}{ccc} S & \xrightarrow{\ i\ } & X^H \\ \downarrow j & & \downarrow f^H \\ D & \xrightarrow{\ h\ } & Y^H \end{array} \qquad \begin{array}{l} D = D^{k+1}\times D^{n-k} \\ S = S^k\times D^{n-k} \\ \text{Define } \partial\mu' = i. \end{array}$$

be a diagram which gives μ by restriction to $(D^{k+1},S^k)\subset(D,S)$. Set $D' = \operatorname{ind}_H^G D$ and

2.4 $\quad O = W \bigcup_{\operatorname{ind} i} D'$ where $\operatorname{ind} = \operatorname{ind}_H^G$

Here H acts trivially on D. Extend F to $F'':O \to Z$ by setting $F''_{|D'} = \operatorname{ind} h$.

For each $q \in A$ we suppose given a G invariant subspace W_q of W where

2.5 $\quad W_q \subset \{x \in W | G_x \subset Q,\ G_x \neq 1\}$. Set $X_q = W_q \cap X$ and $f_q = f_{|X_q} : X_q \to Y$. (For definition of Q note 1.13.)

Then a class $\mu \in \pi_{k+1}(f_p^H)$ is represented by diagram 2.2 with f^H replaced by f_p^H. Similiarly 2.3 has f^H replaced by f_p^H.

2.6 $\quad$ For $q \in A$, define $O_q = W_q$ and $F''_q = F_{|W_q}$ if $q \neq p$. For $q = p$, set $O_p = W_p \bigcup_{\operatorname{ind} i} D' \subset O$ where $i = \partial\mu'$. Extend F_p to $F''_p:O_p \to Z$ by setting $F''_{p|D'} = \operatorname{ind} h$.

If η is a G vector bundle over W, set $\eta_q = \eta_{|W_q}$ for $q \in A$.

Let η and η' be a G vector bundle over W, $H \subset P$ and let

2.7 $\quad$ i) $B:s(\eta_p) \to s(\eta'_p)$ be a P vector bundle isomorphism

ii) $C:(\eta_p)_H \to (\eta'_p)_H$ be a P vector bundle isomorphism such that (+) $B_H = s(C)$.

iii) $B_\infty:s_\infty(E\times\eta) \to s_\infty(E\times\eta')$ be a G vector bundle isomorphism.

REMARK: By definition B is realized as a P vector bundle isomorphism $B:\eta_p\oplus\underline{M} \to \eta'_p\oplus\underline{M}$ for some representation M of G. The condition $B_H = s(C)$ in 2.7ii) means that $B_H = s_{M_H}(C) = C\oplus 1_{M_H}$.

Let us abbreviate the data in 2.7 by $\underline{B}:\eta \to \eta'$ and write $\underline{B} = (B,C,B_\infty)$. In the case $\eta' = \underline{R}$, $\theta_t(\underline{B}) = (\theta_t(B),\theta_t(C),\theta_t(B_\infty,B))$ denotes the following:

2.8 $\quad$ i) $\theta_t(B):W_p \to \Delta_p(R\oplus M)$ is a trivialization of B. See 1.13.

ii) $\theta_t(C):W_p^H \to \Delta_p(R_H)$ is a trivialization of C such that

(+) $(\theta_t(B)_H)|_{W_p^H} = s\theta_t(C)$. See 1.6 and subsequent caution.

Here $s = s_{M_H}$.

iii) $\theta_t(B_\infty, B)$: $E\times W_p \to U(R\oplus M\oplus \mathbb{F}^\infty)$ is a trivialization of $B_\infty|_{E\times W_p}$ relative to $\theta_t(1_E\times B)$. Call this a p trivialization of B_∞. See 1.13.

REMARKS: Note that for $t = 1$, 2.8ii (+) is implied by 2.7ii (+) because $\theta(B)_H|_{W_p^H} = \theta(B_H)$ by 1.10. The trivialization $\theta_t(1_E\times B)$ in iii) is the composition $E\times W_p \to W_p \to \Delta_p(R\oplus M\oplus \mathbb{F}^\infty)$ where the last map is $s_\infty\theta_t(B)$. We emphasize all maps in 2.8 are G maps with respect to actions defined in the previous section. If such a $\theta_t(\underline{B})$ is given, we call it a trivialization of $\underline{B}$ and denote $(\underline{B}, \theta_t(\underline{B}))$ by $\underline{\underline{B}}$ and write $\underline{\underline{B}}: \eta \to \underline{R}$.

Let η' be a G vector bundle over Z $F: W \to Z$, $\underline{\underline{B}}: \eta \to F^*\eta'$ and let $\mu \in \pi_{k+1}(f_p^H)$ if $H \neq 1$ and $\mu \in \pi_{k+1}(f)$ if $H = 1$. We define H vector bundle maps (covering $j: S \to D$ (2.3)).

2.9 i) $B_0: i^*s(\eta|_{X_p^H}) \to h^*s(\eta'|_{Y^H})$ $H \neq 1$

ii) $C_0: i^*(\eta|_{X_p})_H \to h^*(\eta'|_Y)_H$ $H \neq 1$

iii) $B_{\infty 0}: i^*s_\infty(\eta|_X) \to h^*s_\infty(\eta'|_Y)$ $H = 1$.

Cases i and ii): Let $(\mathcal{E}, \xi, \xi')$ be $(B, s(\eta|_{X_p^H}), s(\eta'|_{Y^H}))$ resp. $(C, (\eta|_{X_p})_H, (\eta'|_Y)_H)$. Then

$\mathcal{E}_0$: $i^*\xi \to h^*\xi'$ is the composition $i^*\xi \to i^*f_p^{H*}\xi' = j^*h^*\xi' \to h^*\xi^*$ where the first map is $i^*B|_{X_p^H}$ resp. i^*C.

Case iii): Let $(\mathcal{E}, \xi, \xi')$ be $(B_\infty, s_\infty(\eta|_X), s_\infty(\eta'|_Y))$. Denote by

$$\mathcal{E}_0': E\times i^*\xi \to E\times h^*\xi'$$

the composition

$$E\times i^*\xi \xrightarrow{\omega} E\times i^*f^*\xi = E\times j^*h^*\xi' \to E\times h^*\xi'$$

where $\omega = (1_E\times i)^*B_\infty|_X$. Then $\mathcal{E}_0: i^*\xi \to h^*\xi'$ is the composition $i^*\xi \to E\times i^*\xi \xrightarrow{\mathcal{E}_0'} E\times h^*\xi' \to h^*\xi'$. Note in this case $\mathcal{E}_0$ is equivariant only with respect to the trivial group.

Let ξ be a G vector bundle defined over some invariant subspace of W

containing X_p^H. For $x \in iS$ (2.3) let V denote the representation ξ_x of H on the fiber over x. If iS is not connected, we must suppose this is independent of x. Set

2.10 $\quad \omega(\xi) = D\times V$

This is an H vector bundle over D. (D has trivial H action.) Observe that $\omega(\xi) = \omega(\xi')$ if ξ and ξ' are two G vector bundles which agree on iS.

Let $(\mathcal{E},\xi,\xi')$ be one of the three choices in 2.9 cases i-iii). Choose any H vector bundle isomorphism $L(\mathcal{E}):\omega(\xi) \to h^*(\xi')$ (2.3). (This is possible because any two H vector bundles over D which have isomorphic representations of H on some fiber are isomorphic.) Then

2.11 $\quad \ell(\mathcal{E}):\omega(\xi)_{|S} \to i^*\xi$

is the unique H vector bundle isomorphism which makes the following diagram commutative

2.12
$$\begin{array}{ccc} \omega(\xi)_{|S} & \xrightarrow{\ \ell(\mathcal{E})\ } & i^*\xi \\ \downarrow & & \downarrow \mathcal{E}_0 \\ \omega(\xi) & \xrightarrow{\ L(\mathcal{E})\ } & h^*\xi' \end{array}$$

By definition two H vector bundle isomorphisms are regularly H homotopic if there is a homotopy of H vector bundle isomorphisms connecting them.

LEMMA 2.13. Suppose $\mu \in \pi_{k+1}(f_p^H)$ if $H \neq 1$ or $\mu \in \pi_{k+1}(f)$ if $H = 1$. Let $k < \dim {}^H = m$. Then there is an H vector bundle isomorphism $\ell(\underline{\underline{B}}):\omega(\eta)_{|S} \to i^*\eta_{|X_p^H}$ which is unique up to regular H homotopy such that $\ell(\underline{\underline{B}})_H = \ell(C)$, $s\ell(\underline{\underline{B}}) = \ell(B) = \ell(B)^H \oplus \ell s(C)$ for $H \neq 1$ and $s_\infty \ell(\underline{\underline{B}}) = \ell(B_\infty)$ for $H = 1$. (See 2.7ii) and subsequent remark.)

Proof: Case $H \neq 1$. First note that $B_{|W^H} = B^H \oplus B_H = B^H \oplus s(C)$ (2.7ii). This means $\ell(B) = \ell(B)^H \oplus s\ell(C)$. Now $\omega(\eta)^H{}_{|S}$ and $i^*(\eta_{|X_p})^H$ are stably trivial vector bundles over $S = S^k \times D^{n-k}$. The first is obvious by definition and the second is stably isomorphic to the first via $\ell(B)^H$. Since the fiber dimension exceeds k, they are isomorphic as vector bundles hence as H vector bundles as H acts trivially. Choose any vector bundle isomorphism b from $\omega(\eta)^H{}_{|S}$ to $i^*(\eta_{|X_p})^H$. Any other isomorphism differs from b by a map α of S to O_m. Since the inclusion of O_m in O_∞ induces an isomorphism in homotopy in dimensions less than m, we may choose α so that $s(\alpha b) = \ell(B)^H$. Set $\ell(\underline{\underline{B}}) = \alpha b \oplus \ell(C)$. The case $H = 1$ uses the same argument.

Using Lemma 2.13, we define a G vector bundle $\Gamma = \Gamma(\eta,\mu,\underline{\underline{B}})$ over O

and Γ_p over O_p by $\Gamma_p = \Gamma|_{O_p}$ with

2.14 $\Gamma(\eta,\mu,\underline{\underline{B}}) = \eta \bigcup_{\ell} \mathrm{ind}_H^G \omega(\eta)$ where $\ell = \mathrm{ind}_H^G \ell(\underline{\underline{B}})$.

Observe that Γ extends η and by 2.13

2.15 $s\Gamma = s(\eta) \bigcup_{\ell'} \mathrm{ind}_H^G \omega(s\eta)$

$\ell' = \mathrm{ind}_H^G \ell(B)$.

If η' is a G vector bundle over Z, then

2.16 $F''^*\eta' = F^*\eta' \bigcup \mathrm{ind}_H^G h^*(\eta'|_{Y^H})$

because $F''|_{D'} = \mathrm{ind}_H^G h$.

2.17 Note in the special case $\eta' = \underline{R}$ over Z, $F^*\eta' = \underline{R}$ over W. It is convenient to use this alternate view of $\underline{R}$ in the subsequent constructions.

LEMMA 2.18. Let $\eta' = \underline{R}$ over Z and let $B: s(\eta_p) \to s(F_p^*\eta')$ be a P vector bundle isomorphism. If $\theta: O_p \to \Delta_p(R \oplus M)$ is a G map extending $\theta(B)$, then there is a unique P vector bundle isomorphism $B'': s(\Gamma_p) \to s(F_p''^*\eta')$ extending B with $B''|_D = L(B)$ and $\theta(B'') = \theta$. (A similar statement holds for $C:(\eta_p)_H \to (F_p^*\eta')_H$.)

Proof: Note the decompositions 2.15 and 2.16 over $O_p = W_p \bigcup \mathrm{ind}_H^G D$. For $x \in D$, $g \in G$ set $B''(gx) = \theta(x)(g)gL(B)(x)g^{-1}$. This defines B'' over $D' = \mathrm{ind}_H^G D$ while B'' is B over W_p. To see that B'' is well defined use 2.12 and 2.15.

LEMMA 2.19. Let $|H| = p^n$, $\mu \in \pi_{k+1}(f_p^H)$ resp. $\mu \in \pi_{k+1}(f)$ if $H \neq 1$ resp. $H = 1$ and let $\underline{\underline{B}}: \eta \to \underline{R}$. If $k < \dim \eta^H$, there is an extension $\underline{\underline{B}}'': \Gamma \to \underline{R}$.

Proof: Case 1 $H \neq 1$. Note especially 2.17. By 2.8, 1.6 and 1.10 $\theta_t(B) = \theta_t(B)^H \oplus \theta_t(B)_H$. Use the G homotopy extension theorem [W] to extend $\theta_t(B)^H: W_p \to \Delta_p(R^H \oplus M^H) = \Delta$ to $\psi_t: O_p \to \Delta$ such that $\psi_0 = *$. Similiarly extend $\theta_t(C)$ to a G map $\gamma_t: O_p{}^H \to \Delta_p(R_H)$ such that $\gamma_0 = *$ (2.8). Let $\tilde{\theta}_t: O_p \to \Delta_p(R \oplus M)$ be the G map defined by: $\tilde{\theta}_t{}^H$ is ψ_t and $(\tilde{\theta}_t)_H$ is $\theta_t(B)_H$ on W_p and is $s(\gamma_t)$ on $O_p{}^H$. Since O_p is the union of W_p and $O_p{}^H$, this defines $\tilde{\theta}_t$. Use Lemma 2.18 to produce B'' and C'' with $\theta(B'') = \tilde{\theta}_1$ and $\theta(C'') = \gamma_1$. Then $\tilde{\theta}_t$ is a trivialization $\theta_t(B'')$ of B'' and γ_t is a trivialization $\theta_t(C'')$ such that $\theta_t(B'')_H = s_{M_H}\theta_t(C)$. (See 2.7ii.)

Use the G homotopy extension theorem to extend $\theta_t(B_\infty, B)$ (2.8) to

$\theta''_t: E\times O_p \to U(R\oplus M\oplus \mathbb{F}^\infty)$ with $\theta''_0 = *$. Then $\theta''_1 \circ s_\infty(1_E\times B''): s_\infty(E\times\Gamma_p) \to s_\infty(E\times\underline{R})$ is a G vector bundle isomorphism by 1.12 which together with B_∞ defines a G vector bundle isomorphism $B''_\infty: s_\infty(E\times\Gamma) \to s_\infty(E\times\underline{R})$ with p trivialization $\theta_t(B''_\infty) = \theta''_t$ because $s_\infty(E\times\Gamma) = s_\infty(E\times\eta) \cup s_\infty(E\times\Gamma_p)$.

Case 2. $H = 1$. We only need to extend B_∞: to $B''_\infty: s_\infty(E\times\Gamma) \to s_\infty(E\times\underline{R})$. By construction

$$s_\infty(E\times\Gamma) = s_\infty(E\times\eta) \underset{\ell'}{\cup} s_\infty(E\times \mathrm{ind}_1^G \omega(\eta))$$

where $\ell' = 1_E\times \mathrm{ind}_1^G \ell(B_\infty)$. Let $\overline{\ell}(B_\infty): s_\infty(E\times\omega(\eta)_{|S}) \to (1_E\times i)^* s_\infty(E\times\eta_{|X})$ be defined so as to make the following diagram commutative (see case 2.9iii):

$$\begin{array}{ccc}
s_\infty(E\times\omega(\eta)_{|S}) & \xrightarrow{\ \overline{\ell}(B_\infty)\ } & (1_E\times i)^* s_\infty(E\times\eta_{|X}) \\
\big\downarrow & & \big\downarrow B'_{\infty 0} \\
s_\infty(E\times\omega(\eta)) & \xrightarrow{\ 1_E\times L(B_\infty)\ } & s_\infty(E\times h^*\eta')
\end{array}$$

Then $\overline{\ell}(B_\infty)$ and $1_E\times\ell(B_\infty)$ are homotopic bundle isomorphisms as E is contractible. Thus we can replace ℓ' by $\ell'' = \mathrm{ind}_1^G \overline{\ell}(B_\infty)$. This gives a bundle Γ'' which is isomorphic to $s_\infty(E\times\Gamma)$ by an isomorphism which is the identity on $s_\infty(E\times\eta)$. It suffices to extend B_∞ to $B''_\infty: \Gamma'' \to s_\infty(E\times\underline{R})$. Take $B''_\infty = B_\infty$ on $s_\infty(E\times\eta)$ and $B''_\infty | \mathrm{ind}_1^G E\times D = \mathrm{ind}_1^G(1_E\times L(B_\infty))$.

3. THE HOMOTOPY SURGERY LEMMA

In this section we treat question D of the introduction by proving (0.4) for variable H and p. This is Theorem 3.12. It is basically a formal consequence of 2.19. The point is that to vary H within P, we need 2.7ii) for each $H \subset P$. This leads to the notion of a P-Π vector bundle isomorphism c_p. If we vary over P as well, we need isomorphisms β_p as in 2.7i), together with P-Π bundle isomorphisms c_p for all $p \in A$. This leads formally and naturally to the definition of $\underline{\underline{\beta}}$ in 3.9.

Let ξ be a G vector bundle over X and let $\Pi(\xi) = \{\xi_K | K \subset G\}$. See (1.1). Since G is abelian, each ξ_K is a G vector bundle which satisfies

3.1 $\quad (\xi_K)_L = (\xi_L)_{|X^K}$ whenever $L \le K$.

If ξ and ξ' are two G vector bundles over X, a Π bundle isomorphism $c: \Pi(\xi) \to \Pi(\xi')$ is a collection $c = \{c(K): \xi_K \to \xi'_K | K \subset G\}$ where each $c(K)$ is a G vector bundle isomorphism which respects 3.1 in the sense that

3.2 $\quad c(K)_L = c(L)_{|X^K}$ whenever $L \le K$.

If $P \subset G$ and we only want to consider ξ and ξ' as P vector bundles, we emphasize this by calling c a P-Π bundle isomorphism and then $c(K)$ is defined only for $K \subset P$. If $b:\xi \to \xi'$ is a G vector bundle isomorphism, it defines a Π vector bundle isomorphism $\Pi(b):\Pi(\xi) \to \Pi(\xi')$ via

3.3 $\quad \Pi(b)(K) = b_K$ for $K \subset G$.

Let $c:\Pi(\xi) \to \Pi(\underline{R})$ be a P-Π bundle isomorphism where ξ is a G vector bundle over X and R is a representation of G. Define

3.4 $\quad \theta(c) = \{\theta(c(K)) | K \subset P\}$ where $\theta(c(K)):X^K \to \Delta_p(R_K)$. (See 1.10.)

Because of 3.2, $\theta(c)$ gives the commutative diagram:

3.5
$$\begin{array}{ccc} X^L & \xrightarrow{\theta(c(L))} & \Delta_p(R_L) \\ \uparrow & & \uparrow \psi_{KL} \\ X^K & \xrightarrow{\theta(c(K))} & \Delta_p(R_K) \end{array} \qquad L \le K$$

where ψ_{KL} is induced by the projection $R_K \to R_L$.

3.6 $\quad$ A homotopy $\theta_t = \{\theta_t(K):X^K \to \Delta_p(R_K) | K \subset P\}$ which satisfies 3.5 and $\theta_0(K) = *$ and $\theta_1(K) = \theta(c(K))$ for all $K \subset P$ is called <u>a trivialization of c and denoted</u> $\theta_t(c)$. Here $0 \le t \le 1$. Compare 1.13.

Let η and η' be G vector bundles over W. By definition an <u>s-bundle isomorphism</u> $\underline{\beta}:\eta \to \eta'$ and written $\underline{\beta} = (\beta_\infty, \beta_p, c_p | p \quad A)$ consists of

3.7 $\quad$ i) A G vector bundle isomorphism $\beta_\infty:s_\infty(E\times\eta) \to s_\infty(E\times\eta')$

ii) A P vector bundle isomorphism $\beta_p:s(\eta_p) \to s(\eta_p')$ for each $p \in A$

iii) A P-Π vector bundle isomorphism $c_p:\Pi(\eta_p) \to \Pi(\eta_p')$ such that $\Pi(\beta_p) = s(c_p)$ for each $p \in A$.

A <u>bundle isomorphism</u> $\underline{\beta}:\eta \to \eta'$ and written $\underline{\beta} = (\beta_\infty, \beta_p | p \in A)$ consists of a β_∞ as in 3.7i) and P vector bundle isomorphisms $\beta_p:\eta_p \to \eta_p'$ $p \in A$. Clearly a bundle isomorphism determines an s bundle isomorphism with $c_p = \Pi(b_p)$.

In the special case $\eta' = \underline{R}$, we define a <u>trivialization</u> $\theta_t(\underline{\beta})$ of $\underline{\beta}$ (written $\theta_t(\underline{\beta}) = (\theta_t(\beta_\infty,\beta_p),\theta_t(\beta_p),\theta_t(c_p) | p \in A))$ to consist of

3.8 i) $\theta_t(\beta_\infty,\beta_p)$ - a p trivialization of β_∞

ii) $\theta_t(\beta_p)$ - a trivialization of β_p

iii) $\theta_t(c_p)$ - a trivialization of c_p with

$$(\theta_t(\beta_p)_H)_{|W_p^H} = s\theta_t(c_p(H)) \quad \text{for all } H \subset P.$$

If $\underline{\beta}$ is a bundle isomorphism only 3.8i) and 3.8ii) are required and $\theta_t(c_p)$ is deleted.

3.9 DEFINITION: a ts bundle isomorphism resp. t bundle isomorphism $\underline{\underline{\beta}}:\eta \to \underline{R}$ consists of an s bundle resp. bundle isomorphism $\underline{\beta}:\eta \to \underline{R}$ together with a trivialization $\theta_t(\beta)$ of $\underline{\beta}$. Write $\underline{\underline{\beta}} = (\underline{\beta},\theta_t(\underline{\beta}))$.

Let η be a G vector bundle over W, R a representation of G, $|H| = p^n$ $p \in A$ and let $\underline{\underline{\beta}}:\eta \to \underline{R}$ be a ts bundle isomorphism with $\underline{\underline{\beta}} = (\underline{\beta},\theta_t(\underline{\beta}))$ where $\underline{\beta} = (\beta_\infty,\beta_q,c_q|q \in A)$. Set $\underline{B} = (\beta_p,c_p(H),\beta_\infty)$, $\theta_t(\underline{B}) = (\theta_t(\beta_p),\theta_t(c_p(H)),\theta_t(\beta_\infty,\beta_p))$ and $\underline{\underline{B}} = (\underline{B},\theta_t(\underline{B}))$. For $\mu \in \pi_{k+1}(f_p^H)$ define

3.10 $\Gamma(\eta,\mu,\underline{\underline{\beta}}) = \Gamma(\eta,\mu,\underline{\underline{B}}) = \Gamma$ (2.14) and $\ell(\underline{\underline{\beta}}) = \ell(\underline{B}):\omega(\eta)_{|S} \to i^*(\eta_{|X_p^H})$

(2.13); so Γ is a G vector bundle over 0 and $\Gamma_q = \Gamma_{|0_q}$ is a G vector bundle over 0_q. See 2.6.

LEMMA 3.11. Suppose $d(H):(\Gamma_p)_H \to \underline{R}_H$ is a P vector bundle isomorphism which extends $c_p(H)$. Let $\theta_t(d(H))$ be a trivialization of $d(H)$ which extends $\theta_t(c_p(H))$. Then there is a unique extension $c_p'':\Pi(\Gamma_p) \to \Pi(\underline{R})$ of c_p with $c_p''(H) = d(H)$. There is a unique trivialization $\theta_t(c'')$ of c with $\theta_t(c''(H)) = \theta_t(d(H))$.

Proof: Let $L \subset P$. If L is not a subgroup of H, then $0^L = W^L$; so $c''(L) = c(L)$. Suppose $H \supset L$. Then $0^L - W^L = 0^H - W^H$; so $C''(L)$ is uniquely defined by $c''(L)_{|D'} = d(H)_{L|D'}$ $(D' = \mathrm{ind}_H^G D)$ because of 3.2. Similiarly $\theta_t(c''(L))$ is uniquely defined its restriction to $\mathrm{ind}_H^G D$ which must be $\psi_{HL}\theta_t(d(H))$ there by 3.5.

THEOREM 3.12. Suppose 2.1. Let $\underline{\underline{\beta}}:\eta \to \underline{R}$ be a ts bundle isomorphism (3.9) over W, let $H \subset G$ be a p group for some $p \in A$ (1.3), let $k < \dim \eta^H$ and $\mu \in \pi_{k+1}(f_p^H)$ if $H \neq 1$ or $\mu \in \pi_{k+1}(f)$ if $H = 1$. Then there is a G vector bundle $\Gamma = \Gamma(\eta,\mu,\underline{\underline{\beta}})$ over 0 (3.10) and a ts bundle isomorphism $\underline{\underline{\beta}}'':\Gamma \to \underline{R}$ over 0 extending $\underline{\underline{\beta}}$. (Compare [$DP_1$, 4.22] and [$P_2$, 3.4].)

Proof: Let $\underline{\underline{B}} = (\underline{B},\theta_t(\underline{B}))$ where $\underline{B} = (\beta_p, c_p(H), \beta_\infty)$. By Lemma 2.19 there is an extension $\underline{\underline{B}}'':\Gamma(\eta,\mu,\underline{\underline{\beta}}) \to \underline{R}$ extending $\underline{\underline{B}}$. This means $\underline{\underline{B}}'' = (\underline{B}'',\theta_t(\underline{B}''))$ where $\underline{B}'' = (\beta_p'', c_p''(H), \beta_\infty'')$. By Lemma 3.11, there is an extension $c_p'':\Pi(\Gamma_p) \to \Pi(\underline{R})$ of c_p and a trivialization $\theta_t(c_p'')$ of c_p'' which extends $\theta_t(c_p)$. To give $\underline{\underline{\beta}}''$ we need to specify $\underline{\beta}'' = (\beta_\infty'', \beta_q'', c_q'' | q \in A)$ together with trivialization $\theta_t(\underline{\beta}'')$. For $q \neq p$, $W_q = 0_q$ (2.6) so $\beta_q'' = \beta_q$ etc. For $q = p$ the required extensions were just given. This completes the proof.

4. FUNCTORIALITY AND EXISTENCE OF NON-TRIVIAL t BUNDLE ISOMORPHISMS

The definition of $\underline{\underline{\beta}}:\eta \to \underline{R}$ over W in the preceeding section (3.9) involved a collection of subspaces $\{W_p | p \in A\}$ as $\eta_p = \eta_{|W_p}$. It is useful to generalize this.

4.1 A G system $W_\star$ consists of G spaces W and W_p for $p \in A$ together with G maps $i_p:W_p \to W_\infty$. It is required that G/P act freely on W_p/P and $G_x \neq 1$ for all $x \in W_p$.

A G space W with subspaces W_p as in 2.5 give rise to a G system written $W_\star$ with $W_\infty = W$. A G map $f:W_\star \to W_\star'$ is a collection $f_\infty:W_\infty \to W_\infty'$ and $f_p:W_p \to W_p'$ of G maps which satisfy $f_\infty i_p = i_p f_p$. A G vector bundle η over a system $W_\star$ is by definition a G vector bundle η over W_∞. Set $\eta_p = i_p^*\eta$. With this setup, the meaning of $\underline{\underline{\beta}}:\eta \to \underline{R}$ is clear from the definition in the case of G spaces. In particular we have the notions of a t bundle isomorphism and a ts bundle isomorphism over a G system. See 3.9.

There is a universal G system $E_\star$ which we now describe.

4.2 $E_\star = \{E_\infty, E_p | p \in A\}$ where E_∞ is a point, $E_p = E/P$ where E is an acyclic G space with free action and $i_p:E_p \to E_\infty$ is the unique map.

LEMMA 4.3. Let $W_\star$ be any G system. Then there is a G map $\omega:W_\star \to E_\star$.

Proof: There is no ambiguity for ω_∞. Since G/P acts freely on W_p/P and $G = P \times G/P$, there is a G map $\omega_p:W_p \to E/P = B_P$ whose orbit map factors as $W_p/G \to B_{G/P} \to B_G$. Here $B_G = E/G$ is the classifying space of G. This produces ω_p for $p \in A$.

Let R and R' be two complex representations of G and suppose

4.4 $$R'-R \in \operatorname{Ker}(R(G) \to \prod_{p \in A} R(P)) = I$$

where $R(G)$ is the complex representation ring of G. Atiyah [A_2] has shown

4.5 $$I = \operatorname{Ker}(R(G) \to \hat{R}(G))$$

where $\hat{R}(G)$ is the completed character ring of G. Also he has identified $\hat{R}(G)$ with $K(B_G) = K_G(E)$ where E is our acyclic space with free action. If we interpret this geometrically, it gives an isomorphism

$$4.6 \qquad \beta_\infty : s_\infty(E\times R') \to s_\infty(E\times R)$$

of complex G vector bundles over E. The fact that $R'-R$ is in I, gives for each $p \in A$ an isomorphism of P representations

$$\varepsilon_p : R' \to R.$$

Since G is abelian, we may suppose both R' and R are represented by matrices in $\Omega(R)$ and that ε_p normalizes $\Omega(R)$. We utilize the notation $R(g)$ and $R'(g) \in \Omega(R)$ for the matrices representing $g \in G$. View R and R' as G vector bundles over a point E_∞. Using definition 1.9, we find that

$$4.7 \qquad \theta(\varepsilon_p)(x)(g) = \varepsilon_p R(g)\varepsilon_p^{-1} R'(g)^{-1} \in \Omega(R)$$

where $x \in E_\infty$ is the unique point. Thus the supposition in 1.9 is satisfied; so

$$4.8 \qquad \theta(\varepsilon_p) \in (E_\infty, \Delta_p(R))^G \text{ is defined.}$$

The isomorphism ε_p induces a P vector bundle isomorphism

$$4.9 \qquad \beta_p = 1\times\varepsilon_p : E_p\times R' \to E_p\times R \text{ over } E_p \text{ with } \theta(\beta_p) = \theta(\varepsilon_p)\circ\pi_p$$

$$\pi_p : E_p \to E_\infty .$$

Then $\underline{\beta} = (\beta_\infty, \beta_p | p \in A)$ is a bundle isomorphism $\underline{\beta}:\underline{R}' \to \underline{R}$. (See definition after 3.7) over the system E_∞. We seek a trivialization $\theta_t(\underline{\beta})$ (see definition after 3.8). Since $\underline{R}_p$ is a bundle over $S = E_p$ and G/P acts freely on S, we may apply 1.14 noting $H^1(E_p, \mathbb{Z}) = 0$. This gives a trivialization of β_p. To get a p trivialization of β_∞, we note this means a trivialization over $S = E\times E_p$ of $(1_E\times i_p)^*\beta_\infty$ relative to $\theta_t(1_E\times\beta_p)$ (see 2.8iii and subsequent remark). Note that G acts freely on S and S/G is a $K(\pi,1)$ where π is a finite group. By Atiyah $K^1(S/G) = 0$. See [A_2]. Now apply 1.16 to get the desired result. Thus we have proved this

<u>THEOREM 4.10</u>. Let R and R' be complex representations of G such that $R'-R \in I$ (4.4). Then there is a t bundle isomorphism $\underline{\underline{\beta}}_0 : \underline{R}' \to \underline{\underline{R}}$ over E_∞.

<u>REMARK:</u> $\underline{\underline{\beta}}_0$ of course determines a ts bundle isomorphism with $c_p = \Pi(\beta_p)$ for $p \in A$. See preceeding section.

We leave to the reader the simple project of verifying these notions are functorial for G maps of systems. Specifically if η is a G vector bundle over a G system W_*, $f:W_*' \to W_*$ is a G map and $\underline{\beta}:\eta \to \underline{R}$ is a ts bundle isomorphism over W_* then there is a ts bundle isomorphism

4.11 $f^*\underline{\underline{\beta}}: f^*\eta \to \underline{R}$ over W'.

COROLLARY 4.12. Let W be a G space with subspaces W_p $p \in A$ satisfying 2.5. Let R' and R be as in 4.10. Then there is a t bundle isomorphism $\underline{\underline{\beta}}: \underline{R}' \to \underline{R}$ over W.

Proof: Let $\underline{\underline{\beta}}_0$ be given by 4.10. Let $\omega: W_* \to E_*$ be given by 4.3. Then $\underline{\underline{\beta}} = \omega^*\underline{\underline{\beta}}_0$.

REMARK 4.13. Let $B: s(\eta) \to s(\underline{R}')$ be a G-vector bundle isomorphism over W and $C: \Pi(\eta) \to \Pi(\underline{R}')$ be a G-Π vector bundle isomorphism over W. This gives rise to a ts bundle isomorphism $\underline{\underline{\beta}}: \eta \to \underline{R}'$ with $\underline{\beta} = (\beta_\infty, \beta_p, c_p | p \in A)$ where $\beta_\infty = s_\infty(1_E \times B)$, $\beta_p = B_{|W_p}$ and $c_p = C_{|W_p}$ and all the trivializations in $\theta_t(\underline{\beta})$ are $*$. We note that this ts bundle isomorphism may be composed with a ts bundle isomorphism $\underline{\underline{\beta}}': \underline{R}' \to \underline{R}$ giving a new ts bundle isomorphism $\underline{\underline{\beta}}' \circ \underline{\underline{\beta}}: \eta \to \underline{R}$.

For further reference we summarize the data required of a $t(ts)$ bundle isomorphism. First we note that a t bundle isomorphism defines a unique ts bundle isomorphism. The only difference between these notions is that the latter is more flexible because some of its data only involves stable isomorphisms rather than unstable isomorphisms. The former is a little easier to describe; so we summarize the data required in its definition.

4.14 A t bundle isomorphism $\underline{\underline{\beta}}: \eta \to \underline{R}$ over W written $\underline{\underline{\beta}} = (\underline{\beta}, \theta_t(\underline{\beta}))$ consists of:

i) a G space W together with G subspaces W_p for $p \in A$ such that G/P acts freely on W_p/P and $G_x \neq 1$ for all $x \in W_p$.

ii) a G vector bundle η over W, a representation R of G. Set $\eta_p = \eta_{|W_p}$ for $p \in A$.

iii) $\underline{\beta} = (\beta_\infty, \beta_p | p \in A)$ where $\beta_\infty: s_\infty(E \times \eta) \to s_\infty(E \times R)$ is a G vector bundle isomorphism over $E \times W$ and $\beta_p: \eta_p \to \underline{R}$ is a P vector bundle isomorphism over W_p for $p \in A$.

iv) $\theta_t(\underline{\beta}) = (\theta_t(\beta_p), \theta_t(\beta_\infty, \beta_p) | p \in A)$ where $\theta_t(\beta_p)$ is a trivialization of β_p (1.13i) and $\theta_t(\beta_\infty, \beta_p)$ is a p trivialization of β_∞ (1.13ii and 2.8iii).

We note that $t(ts)$ bundle isomorphisms are functorial with respect to G maps $f: W \to W'$ such that $fW_p \subset W'_p$; so if $\underline{\underline{\beta}}: \eta' \to \underline{R}$ is a t bundle isomorphism over W', then $f^*\underline{\underline{\beta}}: f^*\eta' \to \underline{R}$ is a t bundle isomorphism over W.

Finally we specify the choice of the collection of subspaces $\{W_p | p \in A\}$ of W occurring in 4.14i) which is used [P_3] and [P_4]. There W is a smooth

G manifold and N is an open G regular neighborhood of $\{x \in W | G_x$ is not a p group for any p$\}$. Then for each $p \in A$

4.15 $W_p = \{x \in W-N \,|\, |G_x| = p^{\ell}$ for some $\ell > 0\}$.

Very briefly here is one of the points involved in this choice: If $F:W \to Z$ is a G map which is a homotopy equivalence, F^P is a mod p homology equivalence if $|P| = p^{\ell}$. Not much is required of W^P if P is not a p group for some $p \in A$. For this reason we remove the points of W whose isotropy groups are not p groups. To work with a compact G space we remove more namely N.

BIBLIOGRAPHY

[A_1] Atiyah, M.F., K theory, Benjamin, New York (1967).

[A_2] __________, Characters and cohomology, Inst. Hautes E'tudes Sci, Publ. Math. No. 9 (1961).

[A-S] Atiyah, M.F. and Segal, G., Equivariant K theory and completion, J. Diff. Geometry, Vol. 3, No. 1 (1969).

[B] Bredon, G., Introduction to compact transformation groups, New York (1972).

[DP_1] Dovermann, K.H. and Petrie, T., G surgery II, Mem AMS, No. 260, Vol. 37, May (1982).

[DP_2] __________, An Induction Theorem for Equivariant Surgery, Am. J. Math., to appear.

[H] Hirzebruch, F., Topological methods in algebraic geometry, Springer Vol. 131, New York (1966).

[O] Oliver, R., Group actions on disks, integral permutation representations and the Burnside ring, Proc. Sym. Pure

[P_1] Petrie, T., One fixed point actions on spheres I, Adv. in Math. (Sept. 1982).

[P_2] __________, One fixed point actions on spheres II, Adv. in Math. (Sept. 1982).

[P_3] __________, Isotropy representations on disks, In preparation.

[P_4] __________, Isotropy representations on spheres, In preparation.

[Sc] Schultz, R., Spherelike G manifolds with exotic equivariant tangent bundles, Adv. in Math.

[Sm] Smith, P.A., New results and old problems in finite transformation groups, Bull. A.M.S. 66 (1960).

[W] Wassermann, A., Equivariant differential topology, Topology 8 (1969).

DEPARTMENT OF MATHEMATICS
RUTGERS UNIVERSITY
NEW BRUNSWICK, NEW JERSEY 08903

Contemporary Mathematics
Volume 19, 1983

EMBEDDING CERTAIN COMPLEXES VIA UNSTABLE HOMOTOPY THEORY

Nigel Ray & Lionel Schwartz

In this note it is our aim to define embeddings of certain complexes in spheres. They are of low codimension, and the cohomology of the complexes satisfies various algebraic conditions. A central theme of these embeddings is their relationship with the problem of representing Toda's generators $\beta_1 \in {}_p\pi^S_{2p^2-2p-2}$ (p an odd prime) on suitably framed hypersurfaces.

First we give a general construction.

1. A CONSTRUCTION.

Suppose we are given a map

$$d : S^{n+1} \longrightarrow X * Y,$$

where the join is represented as $C_+X \cup_{X\times Y} C_-Y$ (and we shall label the cone points c_+, c_- respectively). If X and Y are finite CW complexes, we may assume they are given as open subsets of euclidean space, so that $X \times Y$ is a submanifold of $X * Y - \{c_+, c_-\}$ with trivial normal line bundle.

Thus d can be adjusted by a homotopy until it is transverse to $X \times Y$, whose inverse image will be a framed, codimension one submanifold $M^n \subset S^{n+1}$, dividing S^{n+1} into two pieces, Z and its n dual D_nZ.

Moreover, d will map Z into one side, say $X \times C_-Y$, of $X * Y$, and will map D_nZ into the other side $C_+X \times Y$. In other words, out of d we have constructed a diagram

(1.1)

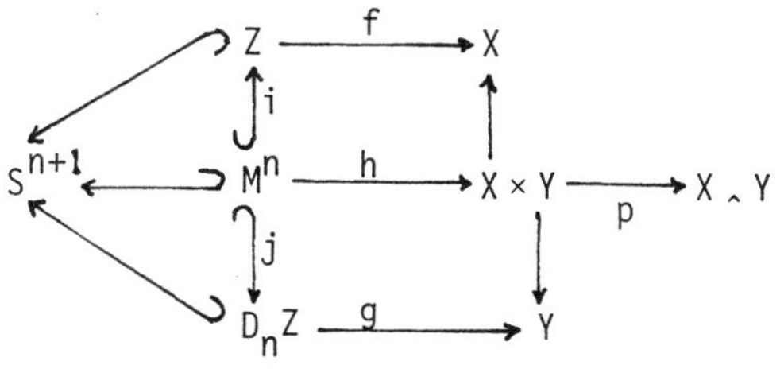

□

It is simple to check that the square

0271-4132/83 $1.00 + $.25 per page

$$\begin{array}{ccc} S^{n+1} & \xrightarrow{c} & \Sigma M^n \\ d\downarrow & & \downarrow \Sigma(p\circ h) \\ X * Y & \xrightarrow[k]{} & \Sigma X \wedge Y \end{array}$$

homotopy commutes, where c is the Pontrjagin-Thom collapse, and k pinches $\{c_+\} \times Y \cup X \times \{c_-\}$ to a point.

Of course, different choices of homotopy on d will yield different (but bordant) choices of M, and therefore different choices of Z and $D_n Z$.

(1.2) LEMMA. Let $g_n \in E_n(S^n)$ be a generator for any multiplicative (co)homology theory $E_*(\ )$. Then if $\sigma_n = c_* g_n \in E_n(M_+^n)$ is the corresponding orientation class, the composition

$$E^*(X) \xrightarrow[i^*f^*]{} E^*(M_+^n) \xrightarrow[\cap\sigma_n]{\cong} E_{n-*}(M_+^n) \xrightarrow[g_*j_*]{} E_{n-*}(Y)$$

is given by $\backslash h_*\sigma_n$.

PROOF. This follows from the definition of $\backslash$ (e.g. see [5]), or from the fact that

$$E^*(Z) \xrightarrow[j_*\circ(\cap\sigma_n)\circ i^*]{} E_{n-*}(D_n Z)$$

is the duality isomorphism $D_Z = \backslash (i \wedge j)_*\sigma_n$. □

We can apply these ideas to prove

(1.3) PROPOSITION. Suppose

$$\delta : S^{n+1} \longrightarrow \Sigma X \wedge Y$$

is an S-duality map (e.g. see [5]). Then there is a subcomplex $W \subset S^{n+1}$ admitting stable homotopy equivalances

$$W \underset{S}{\simeq} X \vee X^{\perp} \quad \text{and} \quad D_n W \underset{S}{\simeq} Y \vee Y^{\perp}$$

for suitable, and S-dual, complexes $X^{\perp}$, $Y^{\perp}$.

PROOF. Apply the construction to δ, in order to obtain a diagram of the form (1.1). This induces a commutative diagram of duality isomorphisms in integral (co)homology:

$$\begin{array}{ccc} H_*(Z) & \xrightarrow{f_*} & H_*(X) \\ D_{D_nZ}\uparrow\cong & & \cong\uparrow D_Y \\ H^{n-*}(D_n Z) & \xleftarrow[g^*]{} & H^{n-*}(Y) \end{array}$$

There is some map $\tilde{g}:\Sigma^t X \longrightarrow \Sigma^t Z$, t large enough, representing the dual of $g: D_n Z \longrightarrow Y$. Furthermore, the induced map $\tilde{g}_*: H_*(X) \longrightarrow H_*(Z)$ is defined by $\tilde{g}_* = D_{D_n Z} \circ g^* \circ D_Y^{-1}$.

Consider the composite

$$\Sigma^t X \xrightarrow[\tilde{g}]{} \Sigma^t Z \xrightarrow[\Sigma^t f]{} \Sigma^t X.$$

This induces $f_* \circ D_{D_n Z} \circ g^* \circ D_Y^{-1} = D_Y \circ D_Y^{-1} = 1$ in $H_*(\ )$, and so is a homotopy equivalence. Thus if we write $X^\perp$ for the cofibre of $\tilde{g}$, then

$$\Sigma^t Z \xrightarrow[\Sigma^t f + \text{proj}]{} \Sigma^t X \vee X^\perp$$

is a homotopy equivalence.

The proof for $D_n W$ is dual, so $X^\perp \simeq DY^\perp$ □

2. AN UNSTABLE HOMOTOPY ELEMENT.

In order to apply our construction to the elements β_1, we first describe an unstable homotopy class.

Recall that $H_*(CP^\infty) \cong DP(x)$, the divided power algebra on a single generator x, which is dual to the first Chern class $c \in H^2(CP^\infty)$. Thus

$$H_*(CP^\infty \times CP^\infty) \cong DP(x) \otimes DP(x).$$

Knapp has shown [2] that the element

$$x_0 = \frac{1}{p}(x^p \otimes x - x \otimes x^p) \in H_{2p+2}(CP^\infty \wedge CP^\infty; \mathbb{Z}_p)$$

is stably spherical for each prime p. We shall show that, in fact, x_0 arises unstably in $\pi_{2p+2}(\Omega(\Sigma CP^\infty \wedge CP^\infty))$. The proof involves two lemmas, for which we shall need some notation.

Let p be an odd prime, and I the set $\{1,2,\ldots, p-1\}$. For each $(i,j) \in I^2$, let $H_{i,j}$ be the subgroup of $H_*(CP^\infty \times CP^\infty; \mathbb{Z}_p)$ generated by classes $u \otimes v$, where $\deg(u) \equiv 2i \bmod 2(p-1)$ and $\deg(v) \equiv 2j \bmod 2(p-1)$. Let q be a primitive root of $1 \bmod p$, and $f_q: CP^\infty \longrightarrow CP^\infty$ a representative for $qc \in H^2(CP^\infty)$.

(2.1) LEMMA. There are spaces $X_{i,j}$ such that there is a p equivalence

$$\Sigma(CP^\infty \times CP^\infty) \xrightarrow[(p)]{\simeq} \bigvee_{(i,j)\in I^2} X_{i,j}$$

and $H_*(X_{i,j}; \mathbb{Z}_p) \cong \Sigma H_{i,j}$.

PROOF. There is a $\bmod p$ homology action of $\mathbb{Z}_p^* \times \mathbb{Z}_p^*$ on $\Sigma(CP^\infty \times CP^\infty)$ in the sense of [1]. This is given by $(q^s, q^t) \longmapsto \Sigma f_{q^s} \times f_{q^t}$.

Thus [1, (1.4)] applies to decompose $\Sigma(CP^\infty \times CP^\infty)$ into a wedge of components, one for each indecomposable ideal in the group ring $\mathbb{Z}_p(\mathbb{Z}_p^* \times \mathbb{Z}_p^*)$.

These ideals $J_{i,j}$ are specified by the idempotents $e_i \otimes e_j$, $1 \leq i,j \leq p-1$, where

$$e_i = -(1 + \omega^{-i}q + \omega^{-2i}q^2 + \ldots + \omega^{-(p-2)i}q^{p-2}) \in \mathbb{Z}_p(\mathbb{Z}_p^*)$$

and we write ω for q in the ground field. The $J_{i,j}$ primary part of $\Sigma(CP^\infty \times CP^\infty)$ is our $X_{i,j}$, and the corresponding summand of the $\mathbb{Z}_p(\mathbb{Z}_p^* \times \mathbb{Z}_p^*)$ module $H_*(\Sigma(CP^\infty \times CP^\infty); \mathbb{Z}_p)$ is $\Sigma H_{i,j}$. □

Now consider the space $X_{1,1}$, which is defined as the direct limit of the system

$$\cdots \xrightarrow[\varepsilon]{} \Sigma(CP^\infty \times CP^\infty) \xrightarrow[\varepsilon]{} \Sigma(CP^\infty \times CP^\infty) \xrightarrow[\varepsilon]{} \cdots$$

where $\varepsilon = \sum_{s,t} q^{-(s+t)} f_{q^s} \times f_{q^t}$, summing over the suspension cogroup structure. Thus $X_{1,1}$ is also a cogroup, and its first few mod p homology groups are given by

$$\begin{cases} \mathbb{Z}_p & \text{on } \sigma x \otimes x \qquad \text{dim } 5 \\ \mathbb{Z}_p \oplus \mathbb{Z}_p & \text{on } \frac{1}{p}\,\sigma x^p \otimes x,\ \frac{1}{p}\,\sigma x \otimes x^p \qquad \text{dim } 2p+3 \end{cases}$$

<u>(2.2) LEMMA.</u> There is an involution $\underline{\tau}: X_{1,1} \longrightarrow X_{1,1}$ such that $\underline{\tau}_*(\sigma u \otimes v) = -\sigma v \otimes u$ for each $\sigma u \otimes v \in H_*(X_{1,1}; \mathbb{Z}_p)$.

<u>PROOF</u>. Consider the involution

$$\tau: \Sigma(CP^\infty \times CP^\infty) \longrightarrow \Sigma(CP^\infty \times CP^\infty);\ \tau(t;x,y) = (1-t;y,x).$$

Then τ commutes with ε (so long as the terms in ε are suitably ordered). Thus, in the limit, we obtain $\underline{\tau}: X_{1,1} \longrightarrow X_{1,1}$ with the desired properties. □

We can now apply [1] again, and deduce a p equivalence $X_{1,1} \simeq X_+ \vee X_-$, where $X_\pm$ is the ± 1 eigenspace of τ. Thus (2.1) extends to a p equivalence.

$$\Sigma(CP^\infty \times CP^\infty) \xrightarrow[(p)]{\simeq} X_+ \vee X_- \vee \bigvee_{(i,j) \neq (1,1)} X_{i,j}$$

<u>(2.3) PROPOSITION</u>. The class $\sigma x_0 \in H_{2p+3}(\Sigma CP^\infty \wedge CP^\infty; \mathbb{Z}_p)$ is spherical.

<u>PROOF</u>. In the above decomposition, $H_*(X_+; \mathbb{Z}_p)$ is the $+1$ eigenspace of $\underline{\tau}_*$ on $H_*(X_{1,1}; \mathbb{Z}_p)$. So σx_0 is the bottom homology class in this space, recalling (2.2). Thus σx_0 is spherical in $H_{2p+3}(X_+; \mathbb{Z}_p)$, and the result follows. □

We therefore have constructed a map

$$d : S^{2p+3} \longrightarrow CP^{\infty} * CP^{\infty}$$

such that $d_* g_{2p+2} = x_0$ in $H_{2p+2}(CP^{\infty} \wedge CP^{\infty}; \mathbb{Z}_p)$.

3. AN APPLICATION.

We now apply (1.1) to (2.3), and deduce the existence of a diagram

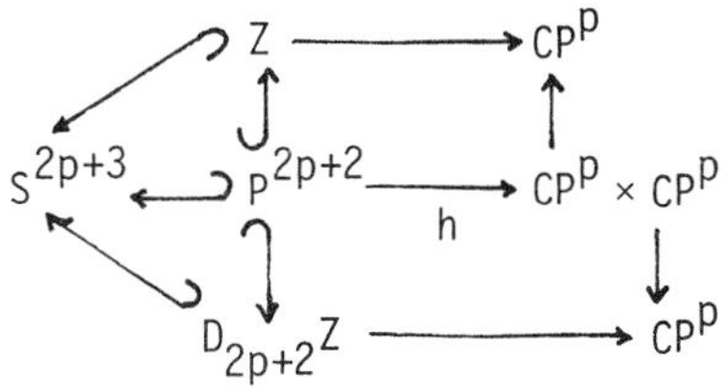

with $h_* \sigma_{2p+2} = x_0 \bmod p$. Thus we have

(3.1) THEOREM. For each prime p, there is a subcomplex $Z \subset S^{2p+3}$ with $w \in H^2(Z)$ such that $w^p \in H^{2p}(Z)$ does not vanish mod p. Moreover, $D_{2p+2}Z$ has the same property.

PROOF. Consider Z defined as above, and choose $w = f^*c$. Then by (1.2), the composition

$$H^{2p}(CP^p) \xrightarrow[f^*]{} H^{2p}(Z) \xrightarrow[j_* \circ \cap \sigma_{2p+2} \circ i^*]{} H_2(D_{2p+2}Z) \xrightarrow[g_*]{} H_2(CP^p)$$

is given by $\backslash x_0$, and so maps c^p to $(p-1)!x$. Hence $f^*c^p \not\equiv 0 (\bmod p)$.

The proof for $D_{2p+2}Z$ is identical. □

Exhibiting such a Z explicitly is a fascinating problem (equivalent to (2.3)), whose solution will give a specific model for P^{2p+2}. Note that $\times S^2$ fails precisely because any 2 dimension cohomology class y satisfies $y^{p^p} \equiv 0 (\bmod p!)$. Of course, P^{2p+2} carries classes w_1, w_2 such that w_1^p, $w_2^p \not\equiv 0 (\bmod p)$ and $w_1^p w_2 \neq 0 \neq w_1 w_2^p$.

If $p = 2$, we may choose $Z = CP^2 \subset S^7$. Then P^6 is the normal sphere bundle of the embedding.

Next let p be odd, and consider the hypersurface $N = T^2 \times (P^{2p+2})^{p-2}$, where $T^2 = S^1 \times S^1$. This bounds

$$T^2 \times (P^{2p+2})^{p-3} \times Z \subset S^{2p^2-2p-1}.$$

Then we can define a map $\alpha : N \longrightarrow SO$ by means of the following diagram (which also defines λ and μ):

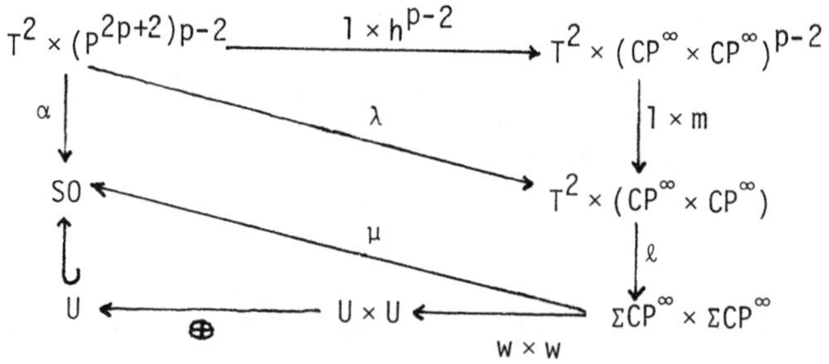

where m is multiplication in $K(\mathbb{Z}\oplus\mathbb{Z}, 2)$, ℓ the collapse and w the complex reflection map.

(3.2) PROPOSITION. The hypersurface N reframed by α represents a non-zero element $[N, \alpha] \in {}_p\pi^S_{2p^2-2p-2}$: thus $[N, \alpha] = n\beta_1$ for some $n \not\equiv 0 \bmod p$.

PROOF. Let the fundamental class $\tilde{\sigma} \in H_{2p^2-2p-2}(N)$ be $g_2 \otimes \sigma \otimes \ldots \otimes \sigma$. Then in $H_{2p^2-2p-2}(T^2 \times CP^\infty \times CP^\infty; \mathbb{Z}_p)$, we have

$$\lambda_*\tilde{\sigma} = g_2 \otimes \sum_{i=0}^{p-2} \binom{p-2}{i}(-1)^i p^{2+i-p} x^{p(p-2)-i(p-1)} \otimes p^{-i} x^{p-2+i(p-1)}$$

$$= -g_2 \otimes \sum_{i=0}^{p-2} (i+1)^{-1} b_{p(p-2)-i(p-1)} \otimes b_{p-2+i(p-1)}$$

where $j!b_j = x^j \ \forall j$.

Also, w represents the generator $\eta - 1$ of $KU^0(CP^\infty)$.
So writing $H_*(SO; \mathbb{Z}_p) \cong E(a_1, a_2, \ldots)$ (see [3]), a simple characteristic class calculation shows

$$\mu_*\{(g_1 \otimes b_{2j-1}) \otimes (g_1 \otimes b_{2k-1})\} = 4a_j a_k.$$

So

$$\alpha_*\tilde{\sigma} = 4 \sum_{i=0}^{p-2} (i+1)^{-1} a_{m(p-i-1)} a_{m(i+1)} \qquad (m = \tfrac{1}{2}(p-1))$$

$$= 4 \sum_{s=1}^{p-1} s^{-1} a_{ms} a_{m(p-s)}.$$

But $[N, \alpha] \in \pi_{2p^2-2p-2}$ has hurewicz image $J_*\alpha_*\tilde{\sigma}$ in $H_*(SF; \mathbb{Z}_p)$, where $J : SO \longrightarrow SF$. Thus by [3] we deduce that $J_*\alpha_*\tilde{\sigma} \neq 0$. □

This calculation is closely connected with Knapp's work [2] on the transfer. An alternative formulation can be given with the aid of the following construction.

Let Y be an H space with multiplication m, and $h(m) : \Sigma Y \wedge Y \longrightarrow \Sigma Y$ the associated hopf map. Then there is a product map

$$n : \Omega\Sigma Y \wedge \Omega\Sigma Y \longrightarrow \Omega\Sigma Y$$

defined on pairs of loops $\ell_1, \ell_2 : S^1 \longrightarrow \Sigma Y$ by the diagram

$$S^1 \xrightarrow{\ell_1} \Sigma Y = S' \wedge Y \xrightarrow{\ell_2 \wedge 1} \Sigma Y \wedge Y \xrightarrow{h(m)} \Sigma Y, \qquad n(\ell_1, \ell_2) : S^1 \longrightarrow \Sigma Y$$

This product may be iterated to give

$$n : (\Omega\Sigma Y)^{\wedge k} \longrightarrow \Omega\Sigma Y.$$

Now choose $Y = CP^\infty \times CP^\infty$, and $d_1 : S^{2(p+1)} \longrightarrow \Omega\Sigma Y$ the adjoint of the map representing x_0. Let d_2 be the composite

$$S^{2(p+1)(p-2)} \xrightarrow[d_1^{p-2}]{} (\Omega\Sigma Y)^{\wedge(p-2)} \xrightarrow[n]{} \Omega\Sigma Y,$$

and $d_3 : S^{2(p+1)(p-2)+1} \longrightarrow \Sigma CP^\infty \wedge CP^\infty$ be the projection of the adjoint of d_2.

<u>(3.3) LEMMA</u>. $(d_3)_*(g_{2(p+1)(p-2)}) = \sum_{i=0}^{p-2} (i+1)^{-1} b_{p(p-2)-i(p-1)} \otimes b_{p-2+i(p-1)}$

<u>PROOF</u>. Applying the definition of n shows that $(d_3)_*(g_{2(p+1)(p-2)} = x_0^{p-2}$. Then use the proof of (3.2). $\square$

Thus $(d_3)_*(g_{2(p+1)(p-2)})$ lies in $H_{p-2,p-2}$, and hence is given by the same formula when followed by the projection $\Sigma CP^\infty \wedge CP^\infty \longrightarrow X_{p-2,p-2}$

Moreover, by (easier!) analogy with (2.1), there is a p equivalence

$$\Sigma CP^\infty \xrightarrow{\simeq} \bigvee_{i=1}^{p-1} X_i \qquad \text{(e.g. see [4])}$$

given by the action of $\mathbb{Z}_p^*$. X_i involves all the cells of ΣCP^∞ in dimensions $\equiv (2i+1) \bmod 2(p-1)$, so there is a further p equivalence $\Sigma X_{p-2,p-2} \simeq X_{p-2} \wedge X_{p-2}$, where

$$X_{p-2} = S^{2p-3} \cup e^{4p-5} \cup \ldots \cup e^{2t(p-1)+2p-3} \cup \ldots \quad .$$

Thus Σd_3 gives rise to a map

$$d_4 : S^{2(p+1)(p-2)+2} \longrightarrow X_{p-2} \wedge X_{p-2}$$

which by (3.3) we can describe in mod p homology by

$$(d_4)_*(g_{2(p+1)(p-2)+2}) = \sum_{i=0}^{p-2} (i+1)^{-1} \sigma b_{p(p-2)-i(p-1)} \otimes \sigma b_{p-2+i(p-1)} \ .$$

We deduce our alternative embedding result:

<u>(3.4) THEOREM</u>. There is a self $(2p^2-2p-2)$-dual complex W embedded in S^{2p^2-2p-1} such that

$$P^{p-2} : H^{2p-3}(W; \mathbb{Z}_p) \longrightarrow H^{2p^2-4p-3}(W; \mathbb{Z}_p)$$

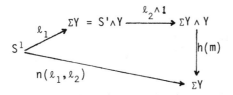

is non-zero.

PROOF. By virtue of its action in homology, $\Sigma d_4 : S^{2p^2-2p-2} \longrightarrow X'_{p-2} * X'_{p-2}$ is an S duality, where X'_{p-2} is the $2p^2-4p+1$ skeleton of X_{p-2}. Furthermore, P^{p-2} is non-zero on $H^{2p-3}(X'_{p-2}; \mathbb{Z}_p)$.

But (1.3) now gives $W \underset{S}{\simeq} X'_{p-2} \vee (X'_{p-2})^{\perp}$, where X'_{p-2} is self dual. Hence so are W and $(X'_{p-2})^{\perp}$, and P^{p-2} is non-zero on $H^{2p-3}(W; \mathbb{Z}_p)$. □

The bounding $2p^2-2p-2$ dimensional hypersurface Q of W comes equipped with a map $h : Q \longrightarrow X'_{p-2} \times X'_{p-2}$, and hence a map $\beta : Q \longrightarrow SO$. The fact that P^{p-2} is non-zero on W is sufficient to deduce that $[Q,\beta]$ is a non-zero multiple of β_1; the description above makes this transparent since we already have the correct hurewicz image.

Thus the simplest of all constructions of β_1 seems to be by direct exhibition of a complex W satisfying the conditions of (3.4). This too is a fascinating, but apparently unsolved, geometric problem.

REFERENCES.

1. Cooke, G.E. & Smith, Larry, 'Mod p decompositions of co H-spaces and applications', Math. Z. 157, 155-77 (1977).

2. Knapp, K., 'Some applications of K-theory to framed bordism : e invariant and transfer', Habilitationsschrift, Bonn University, (1979).

3. May, J.P., 'The homology of iterated loop spaces', Springer Lecture Notes in Math. 533 (1976).

4. Mimura, M., Nishida, G. & Toda, H., 'Localization of CW-complexes and its applications', J. Math. Soc. Japan, 23, 593-624 (1971).

5. Switzer, R.M., 'Algebraic Topology - Homotopy and Homology', Springer-Verlag 1975.

6. Toda, H., 'Composition methods in homotopy groups of spheres', Ann. of Math. Studies no.49 (1962).

Mathematics Department
The University
Manchester
M13 9PL.

Université de Paris-Sud
Centre D'Orsay
91405 ORSAY
Mathématique.

Contemporary Mathematics
Volume 19, 1983

EXOTIC SPHERES ADMITTING CIRCLE ACTIONS WITH CODIMENSION FOUR STATIONARY SETS

Reinhard Schultz[1]

ABSTRACT. The only homotopy n-sphere ($n \geq 5$) admitting a smooth circle action with a codimension 2 fixed point set is the standard sphere; on the other hand, it is known that every exotic sphere bounding a π-manifold admits a semifree circle action with a codimension 4 fixed point set. This paper studies the question of which exotic spheres admit smooth circle actions with codimension 4 fixed point sets. Complete information is obtained on the Pontrjagin-Thom invariants, and partial information is obtained regarding boundaries of parellelizable manifolds. Previously the semifree case had been settled; the nonsemifree cases appear very similar to the semifree case in some ways but very different in others.

If the circle group S^1 acts linearly on an n-sphere, it is well-known that the stationary (or fixed point) set is a k-sphere and the difference $n-k$ is even. Thanks to P. A. Smith theory we know that the analogous statement is true for (say) continuous actions on compact $\mathbb{Z}$-homology spheres (compare [3, Ch.III]). Therefore, if S^1 acts smoothly on a closed homotopy n-sphere Σ^n, the minimum nontrivial value for $n-k$ is 2. In this extremal case we have the following complete description of all possible actions from the work of W.-Y. Hsiang [14].

THEOREM. *Let* S^1 *act smoothly on the closed homotopy sphere* Σ^n, *and suppose the fixed point set* F *has dimension* $n-2$.
(i) *The orbit space* Σ^n/S^1 *is a contractible manifold* K *with boundary* F.
(ii) *There is a canonical smooth structure on* K *such that* Σ^n *is diffeomorphic to* $\partial(K\times D^2)$ [*with rounded corners*]; *in fact, if* S^1 *acts trivially on* K *and by ordinary complex multiplication on* D^2, *then the diffeomorphism may be chosen to be equivariant*.
(iii) *If* $n \geq 5$, *then* Σ^n *is (nonequivariantly) diffeomorphic to* S^n.

At this point we shall merely note that (ii) implies (iii) by contractibility of $K \times D^2$ and the h-cobordism theorem.

AMS Subject Classifications 57S17, 57S25.

[1] Partially supported by National Science Foundation Grant MSC 78-02913 and MCS 81-04852.

During the past two decades there have been many studies of the case $n - k = 4$ (the next lowest codimension), by far the most attention has been placed upon the semifree case (i.e., the action is free off the fixed point set). One feature not present if $n - k \geq 6$ makes these actions extremely attractive; namely, if $n - k = 4$ the orbit space has a canonical smooth structure such that the fixed point set is a smooth submanifold. It turns out that the study of semifree actions with codimension 4 fixed point sets reduces to the study of knotted homotopy (n-4)-spheres in homotopy (n-1)-spheres and the latter can be attacked by methods such as Levine's [19,20]. A good overall picture of this subject is presented in [20].

If $n - k = 4$ but the action is not semifree, it is still possible to view the orbit space as a smooth manifold; we shall formulate this in a manner suitable for our purposes in Section 1. This description has been exploited very effectively by E. V. Stein [31] and R. Fintushel and P. Pao [12]. We shall use the results of [12] heavily in this paper, to study the following question:

Let V *be a* 4-*dimensional real representation of* S^1 *that is faithful and has fixed point set* $\{0\}$. *Which homotopy* (n+4)-*spheres* $\Sigma (n \geq 1)$ *admit smooth* S^1 *actions with* n-*dimensional fixed point sets and local representation* $n + V^2$?

If the action on Σ (equivalently, on V) is semifree, separate answers to this question are known from [19] and [28; details will appear in 29:IV] (it would be interesting to understand the relationship between these descriptions better). Since [28] does not give the answer explicitly, we shall state it here in the interests of clarity (however, we shall not use this statement in our arguments except to say that $SF_{n,4}/b\,P_{n+1}$ has exponent 2):

THEOREM @. *Let* $SF_{n,4} \subseteq \Theta_{n+4}$ *be the set of all homotopy* (n+4)-*spheres admitting smooth semifree circle actions with* n-*dimensional fixed point sets.*
(i) $SF_{n,4}$ *is a subgroup containing* bP_{n+5}.
(ii) *The Pontrjagin-Thom construction* $\mathfrak{P}$ *maps* $SF_{n,4}$ *into the set of all* $\alpha \in \pi_{n+4}/\mathrm{Im}\, J$ *that are expressible as Toda brackets* $\langle x,\nu,\eta\rangle$ *for some* $x \in \pi_{n-1}$ *that desuspends to* $\pi_{n+2}(S^3)$ ■

EXAMPLES 1. Take $x = 0$, n arbitrary. Then $\langle 0,\nu,\eta\rangle = \pi_{n+3}\,\eta$ and hence $\pi_{n+3}\,\eta \subseteq \mathfrak{P}(SF_{n,4})$. This was originally proved by Bredon [1].

2. Take $n = 4$, $x = 2\nu \in \pi_3$. Then $\langle 2\nu,\nu,\eta\rangle$ is nonzero in $\pi_8/\mathrm{Im}\,\zeta = \mathbb{Z}_2$, and this reproves that the exotic 8-sphere admits such a circle action [20,26].

The main result of this paper is that $SF_{n,4}$ gives ALL homotopy (n+4)-spheres admitting (possibly nonsemifree) circle actions with n-dimensional fixed point sets. In order to state our results precisely, we must first develop some notation.

Let W denote an n-dimensional representation of S^1 with positive-dimensional fixed point set, and define

$$\Theta_n(W)$$

to be the subgroup of all $\Sigma^n \in \Theta_n$ admitting S^1 actions with local representation W at the fixed points (existence of equivariant connected sums along the fixed point sets implies that one gets a subgroup). There is a natural subgroup

$$\hat{\Theta}_n(W) \subseteq \Theta_n(W)$$

consisting of all Σ^n admitting smooth circle actions with the same set of orbit types as W; local linearity implies that every Σ^n has every orbit type present in W, but there are many examples (e.g.,[31]) where Σ^n has more.

Let V_k be the standard 1-dimensional unitary representation of S^1 given by $(z,v) \mapsto z^k v$. In the notation developed thus far, Theorem @ gives a computation of $\Theta_n(V_1+V_1+(n-4))$, and it is known that $\Theta_n = \hat{\Theta}_n$ in this case.

We need one more subgroup. Assume $n \not\equiv 1 \bmod 4$. Then by [17] and [6] we have a canonical splitting

$$\Theta_n = bP_{n+1} \oplus \Phi_n$$
$$(n \text{ even implies } bP_{n+1}=0),$$

and we define

$$\Theta_n'(V_1+V_1+(n-4)) = K_n + \Theta_{n-1}\,\eta,$$

where K_n is the kernel of projection onto bP_{n+1} and $\Theta_{n-1}\,\eta$ denotes the image of the Bredon pairing

$$\pi_1 \times \Theta_{n-1} \longrightarrow \Theta_n$$

(see [1]).

If W^n is a faithful representation of S^1 with (n-4)-dimensional fixed point set, then

$$W^n = V_a + V_b + (n-4)$$

where $a, b \geq 1$ are relatively prime; the case $a = b = 1$ is the semifree case.

Finally, here are our results:

THEOREM A. *If* $a, b \geq 1$ *are relatively prime, then*

$$\mathfrak{p}(\Theta_n(V_a \oplus V_b \oplus (n-4))) = ab\,\mathfrak{p}(\Theta_n(2V_1 \oplus (n-4))).$$

THEOREM B. *If* $a > b = 1$ *and* $n \not\equiv 1 \bmod 4$, *then*

$$\hat{\Theta}_n(V_a \oplus V_b \oplus (n-4))) = \Theta_n(V_a \oplus V_b \oplus (n-4)) = a\Theta_n'(2V_1 \oplus (n-4)).$$

THEOREM C. *If* $a, b > 1$ *are relatively prime and* $n \not\equiv 1 \bmod 4$, *then*

$$\hat{\Theta}_n(V_a \oplus V_b \oplus (n-4)) = ab\Theta_{n-1}\,\eta.$$

In order to illustrate the scope of these results, we shall discuss some special cases. If $n = 8k + 1 \geq 9$, let $b\,\mathrm{Spin}_{8k+2} \subseteq \Theta_{8k+1}$ be the index 2 subgroup of exotic spheres bounding spin manifolds [7]. By [20] we know that $\Theta_{8k+1}(2V_1 \oplus (8k-3)) \subseteq b\,\mathrm{Spin}_{8k+2}$, and therefore by Theorem A we also know $\Theta_{8k+1}(V_a \oplus V_b(8k-3)) \subseteq b\,\mathrm{Spin}_{8k+2}$. This yields the following conclusion.

COROLLARY D. *If* Σ^{8k+1} *admits a smooth circle action with a fixed point set of codimension* 4, *then* Σ *bounds a spin manifold* ■

The analogous result for codimension 6 and higher is systematically false [15,27]. Next, let us consider what Theorems A-C mean if $n = 8$, where we know $\Theta_8(2V_1 \oplus 4) = \Theta_8 = \mathbb{Z}_2$. In this case we get

$$\Theta_8(V_a \oplus V_b \oplus 4) = \begin{cases} \Theta_8 & a \text{ and } b \text{ odd} \\ 0 & a \text{ or } b \text{ even.} \end{cases}$$

On the other hand, $\Theta_7\,\eta = 0$ implies $\hat{\Theta}_8(V_a \oplus V_b \oplus 4) = 0$ if $a, b > 1$. Therefore we obtain the following surprising conclusion.

COROLLARY E. *Let* $a, b > 1$ *be relatively prime, and let* Σ^8 *be the exotic 8-sphere. Then* Σ^8 *admits smooth circle actions with local representation* $V_a \oplus V_b \oplus 4$, BUT *it admits no such actions with only four orbit types. In particular, as with Stein's action the fixed point set of* $\mathbb{Z}_{ab}$ *is the disjoint union of the fixed point set of* S^1 *and a second nonempty submanifold* ■

This result demolishes an apparently reasonable "regularity conjecture": If an exotic sphere Σ^n admits a smooth circle action, it admits one with the same local representation but with connected fixed point sets for all isotropy subgroups. This sobering observation balances the optimistic-looking reduction to the semifree case in Theorem A.

The restriction $n \not\equiv 1 \bmod 4$ is made to avoid confronting Kervaire invariant problems; we hope to take them up in a later paper. Suffice it to say that one can at least prove an analog of Theorem B if $n \equiv 1 \bmod 8$.

In subsequent work we shall also consider the following related question: Which exotic n-spheres admit smooth r-torus actions with fixed point sets of codimension $(n-2r-2)$? It is known from cohomological considerations that the minimum codimension is $(n-2r)$, and by Hsiang's result [14] and weight considerations the exotic spheres in the latter case must be standard (also see [2]).

Acknowledgments. This work was mainly inspired by the work of Ron Fintushel and Peter Pao, and Elliot Stein's example. I am grateful to them for sharing their knowledge with me and being patient with my earlier mistakes in formulating the theorems and their proofs. I am also grateful to Glen Bredon for sending me his unpublished manuscript [4] and to Julius Shaneson for helpful conversations on the material in the addendum to Section 3.

1. Smooth orbit space data

It is well-understood that the orbit space of a smooth action generally is not a topological manifold - much less a smooth one, but in any case the orbit space can be regarded as a "differentiable space with singularities" in an appropriate sense. Specifically, a map $M/G \longrightarrow \mathbb{R}$ is "smooth" if and only if the composite $M \longrightarrow M/G \longrightarrow \mathbb{R}$ is smooth in the usual sense. If G acts with only one orbit type, then this differentiable space is differentiably isomorphic to a smooth manifold and the projection $M \longrightarrow M/G$ is smooth. One cannot expect similar behavior if there are two or more orbit types, even if M/G is a smoothable manifold. However, one frequently can say something close to this that is true; we shall develop such notions precisely in this section.

Although the orbit space of a smooth - more generally, locally smoothable [3] - action on a manifold is usually not a topological manifold, there are some cases where one does get a manifold as the orbit space. For the sake of simplicity we assume M is unbounded (WARNING - M/G may have a boundary even if M does not).

In some cases a smooth structure can be placed M/G in an obvious way if G acts smoothly on M^n. For example, if $G = \mathbb{Z}_2$ and $\dim M^G = n - 1$, then there is no problem putting a smooth structure on $M - M^G/G$ and M^G separately. If one looks at an invariant tubular neighborhood E of M^G, then $E/G \cong M^G \times [0,\infty)$ by the map $v \mapsto (\text{proj.}(v), |v|)$ and the induced smooth structure on $E - M^G/G$ coincides with the smooth structure it inherits as an open set in $M - M^G/G$. Results of this sort are essentially folklore to workers in transformation groups (compare [2,4,9-11,14,18,20,22,23,31]); we shall summarize what is essentially known in this direction, both for the sake of completeness and to provide us with some further observations that we need.

There are five basic cases in which a G-action on an unbounded manifold has a topological manifold as its orbit space:

I. Involutions with codimension one fixed point sets.

II. Semifree S^1 and S^3 actions with codimension 2 and 4 (resp.) fixed point sets.

III. Semifree $\mathbb{Z}_k$ actions with codimension 2 fixed point sets.

IV. Semifree S^1 and S^3 actions with codimension 4 and 8 (resp.) fixed point sets.

V. Special S^1 actions with codimension 4 fixed point sets - Specifically, the local representation at each fixed point is $V_a \oplus V_b \oplus k$ with $a, b \geq 1$ relatively prime, $ab > 1$, and the isotropy groups are $S^1, 1, \mathbb{Z}_a, \mathbb{Z}_b$, and perhaps $\mathbb{Z}_{ab}$.

If M is a $\mathbb{Z}$-homology sphere, then every S^1 action with codimension 4 fixed point set is special in the sense of (V) [12]. As in [12], (V) is really a consequence of (III) and (IV). Suppose $a > 1$ without less of generality. If $M_1 = M/\mathbb{Z}_a$, then by (IV) M_1 has a (locally smoothable) circle action with codimension 4 fixed point set and local representation $V_1 \oplus V_b \oplus k$. If $b = 1$ then the action is in fact semifree, so (III) applies. If $b > 1$, let $M_2 = M_1/\mathbb{Z}_b$ and notice that another application of (IV) implies that M_2 has a (locally smoothable) semifree circle action with codimension 4 fixed point set.

The known facts on the existence of smooth structures can be summarized as follows:

THEOREM 1.1. *Let G^g act smoothly on the unbounded manifold M^m satisfying one of I-V above. Then there is a smooth structure on the orbit space "M/G" such that the following hold:*

(a) *In cases I and II, ∂ "M/G" $= F^f$, the fixed point set. In cases III and IV, ∂ "M/G" $= \phi$.*

(b) *The orbit space projection*

$$p: M \longrightarrow \text{"M/G"}$$

is smooth (as smooth as the action).

(c) *The orbit map sends F diffeomorphically to $p(F)$ and the composite*

$$F \longrightarrow p(F) \overset{\subseteq}{\longrightarrow} "M/G"$$

is a smooth embedding.

(d) The induced smooth structure on the set of free orbits "M/G" - Sing. Set is the usual one.

(e) In each case there is a standard C^∞ map

$$q\colon \mathbb{R}^{m-f} \longrightarrow \mathbb{R}^{m-f-g}$$

such that at $x \in F$ the orbit map is equivariatly smoothly equivalent to

$$Q\colon \mathbb{R}^m \longrightarrow \mathbb{R}^{m-q},\ Q(x,y) = (x,q(y)) \quad \text{for } x \in \mathbb{R}^f \text{ and } y \in \mathbb{R}^{m-f}.$$

In Case V a similar statement also holds if $x \in$ Sing. Set but $x \notin F$ ∎

The choices for q in the various cases are summarized in the table below.

CASE DATA	FUNCTION: VARIABLES	FUNCTION: FORMULA
I & II	$\mathbb{R}$ & $\mathbb{C}$, $\mathbb{K}$	$q(x) = \lvert x\rvert^2$
III	$\mathbb{C}$	$q(z) = z^k$
IV	$\mathbb{C}^2$, $\mathbb{K}^2$	$q(x,y) = (\lvert x\rvert^2 - \lvert y\rvert^2,\ \bar{x}y)$
V, fixed pt.	$\mathbb{C}^2 = V_a \oplus V_b$	$q(x,y) = (\lvert x\rvert^{2b} - \lvert y\rvert^{2a},\ \bar{x}^b y^a)$
V, $\mathbb{Z}_a$ or $\mathbb{Z}_b$ orbit	Same as III with $k = a$ or b	
V, $\mathbb{Z}_{ab}$ orbit	$\mathbb{C}^2 = V_a \oplus V_b$	$q(x,y) = (x^b, y^a)$

($\mathbb{K}$ denotes the quaternions)

TABLE I. LOCAL MODELS FOR SMOOTHED ORBIT SPACE PROJECTIONS

Suppose now that M is given a G-invariant Riemannian metric; we shall say that the metric is *good* if, on each tubular neighborhood of a fixed point set $M^H (H \subseteq G)$, the metric is locally the product of a metric on M^H with a Euclidean metric on the fibers. We say that the equivalence in (1.1(e)) is *compatible with a good G-invariant metric* if the charts h, k with $k^{-1} ph|\varepsilon\text{-disk} = Q|\varepsilon\text{-disk}$ may be chosen so that h^{-1} maps the metric on M isometrically to the product of a metric on $\mathbb{R}^f$ and a Euclidean metric, sending F to $\mathbb{R}^f$ and the linear fibers linearly to the sets $\{y\} \times \mathbb{R}^{m-g-f}$. An elementary argument implies that the M's under consideration all admit good G-invariant Riemannian metrics, and the following extra conclusion for (1.1) is valid:

(f) *The charts in* (e) *may be chosen to be compatible with a given good G-invariant Riemannian metric.*

The constructions for the smooth structures on "M/G" as outlined in [2-4,9-12,14,18,22,23,31] involve several choices, and it probably is not obvious if there is a uniqueness statement. Therefore we shall show that the properties in (1.1) characterize the smooth structure uniquely up to C^∞ diffeomorphism. The main step is contained in the following result:

THEOREM 1.2. *Let* $\mathcal{A}$ *and* $\mathcal{B}$ *be two smooth alases on* M/G *such that conditions* (1.1a) - (1.1f) *hold (assume one uses the same good G-invariant metric). Then the identity map* $(M/G,\mathcal{A}) \longrightarrow (M/G,\mathcal{B})$ *is a diffeomorphism.*

Thanks to this result we know that the induced smooth structure depends only on the metric.

PROOF. Let us begin by considering Cases I-IV, where G acts semifreely. By (1.1d) the identity is a C^∞ diffeomorphism on $M/G - F$, so it remains to check its behavior at fixed points. Because of (1.1e) and (1.1f), we can write out the identity map $(M/G,\mathcal{A}) \longrightarrow (M/G,\mathcal{B})$ in local coordinates at a fixed point as follows

$$(1.3)\qquad \begin{array}{ccc} \mathbb{R}^f \times \mathbb{R}^{m-f-g} & \xrightarrow{\quad\varphi\quad} & \mathbb{R}^f \times \mathbb{R}^{m-f-g} \\ \uparrow{\scriptstyle 1\times q} & & \uparrow{\scriptstyle 1\times q} \\ \mathbb{R}^f \times \mathbb{R}^{m-f} & \xrightarrow{\quad\psi\quad} & \mathbb{R}^f \times \mathbb{R}^{m-f} \end{array}$$

The map φ gives the identity on M/G in local coordinates, and the map ψ is a lifting of φ to a map giving the identity on M in local coordinates. By (1.1e) and (1.1f) we can make our choices so that $\psi(x,y) = (x,\theta(x)y)$, where

$$\theta\colon \mathbb{R}^f \longrightarrow \Gamma = \mathrm{Centralizer}_G(\mathbb{R}^{m-f})$$

is smooth. We now apply the following auxiliary result, whose proof is left as an exercise:

LEMMA 1.4. <u>In the notation above, there exists a smooth homomorphism</u> $\sigma: \Gamma \to O_{m-f-g}$ <u>such that</u> q <u>is</u> σ-<u>covariant</u> $(q(ax) = \sigma(a)q(x))$ ∎

(Of course, one must verify each case separately).

Returning to (1.2), we now conclude that the identity map from $(M/G, \mathscr{A})$ to $(M/G, \mathscr{B})$ is smoothly equivalent at a fixed point to a map satisfying

$$\varphi(x, q(y)) = (x, \sigma\Theta(x)\cdot q(y)).$$

But this map is obviously smooth, so the identity from $(M/G, \mathscr{A})$ to $(M/G, \mathscr{B})$ must be smooth at fixed points. This concludes the proof in Cases I - IV because those actions have fixed points and free orbits.

In Case V, it remains to show smoothness at exceptional orbits. There are essentially two cases; namely, the isotropy subgroup may be $\mathbb{Z}_a$, $a > 1$, or $\mathbb{Z}_{ab}$ (the $\mathbb{Z}_b$ case, $b > 1$, is parallel to the $\mathbb{Z}_a$ case). In the first case a local model for the orbit space projection is given by the map $Q_0 = \mathrm{id}(\mathbb{R}^{n-3}) \times q_0$, where

$$(1.5) \qquad q_0: S^1 \times_{\mathbb{Z}_a} W^2 \longrightarrow (\text{same})/S^1 = W^2/\mathbb{Z}_a \longrightarrow \mathbb{R}^2 = \mathbb{C}$$

is the map $q_0[z,v] = q(v) = v^a$ with q as in line III of Table I and W^2 a free representation of $\mathbb{Z}_a$. One also has the following diagram in analogy with (1.3):

$$(1.5a) \qquad \begin{array}{ccc} \mathbb{R}^{n-3} \times S^1 \times_{\mathbb{Z}_a} \mathbb{R}^2 & \xrightarrow{\psi_0 = S^1 \times_{\mathbb{Z}_a} "\psi"} & \mathbb{R}^{n-3} \times S^1 \times_{\mathbb{Z}_a} \mathbb{R}^2 \\ \downarrow & & \downarrow \\ \mathbb{R}^{n-3} \times \mathbb{R}^2 & \xrightarrow{\quad\varphi\quad} & \mathbb{R}^{n-3} \times \mathbb{R}^2 . \end{array}$$

Using this diagram one can proceed as Case III fixed points to prove smoothness. Finally, we must consider $\mathbb{Z}_{ab}$ isotropy. In this case the orbit space projection locally corresponds to the map $\mathrm{id}(\mathbb{R}^{n-5}) \times q_1$, where

$$(1.5b) \qquad q_1: S^1 \times_{\mathbb{Z}_{ab}} (V_a \oplus V_b) \longrightarrow \mathbb{C} \oplus \mathbb{C}$$

is the map $q_1[z;x,y] = (x^b, y^a)$. This description allows one to proceed along previous lines; the only new element is an extension of the covariance lemma (1.4) to cover the map q_1 ∎

COROLLARY 1.6. <u>The equivalence class of the smoothing of</u> M/G[32 ,p.154] <u>is uniquely determined by (1.1). In particular, it does not depend upon the choice of metric.</u>

PROOF. Consider $M \times (-\varepsilon, 1+\varepsilon)$ with a good G-invariant metric satisfying the following conditions.

(*) On $(-\varepsilon,\varepsilon)$ and $(1-\varepsilon,1+\varepsilon)$ the metric is the product of good metrics on M with the Euclidean metric on the interval.

Standard considerations involving invariant smooth partitions of unity imply that such metrics exist. Let $J = (-\varepsilon,1+\varepsilon)$ to simplify notation. Then the construction of smooth structures shows that we obtain a smooth structure on $M/G \times J$ such that on $M/G \times \{(-\varepsilon,\varepsilon)$ or $(1-\varepsilon,1+\varepsilon)\}$ one gets the product of the two smooth structures on M/G (associated to the good metrics on M) with the appropriate intervals. In addition, everything in sight is compatible with the obvious submersions into J, and these submersions are always smooth. A routine analysis now shows that "$M/G \times \{0\}$" and "$M/G \times \{1\}$" are diffeomorphic; in fact, the diffeomorphism between the latter is topologically isotopic to the map $(y,0) \longrightarrow (y,1)$ ■

The smooth orbit space structures fit together nicely in many ways. In this paper we shall need the following extra information regarding Case V.

THEOREM 1.7. <u>Assume</u> S^1 <u>acts smoothly on</u> M <u>and the conditions of Case V hold; let</u> $a > b \geq 1$, <u>let</u> $A =$ <u>fixed point set of</u> $\mathbb{Z}_a$, <u>and let</u> $B =$ <u>fixed point set of</u> $\mathbb{Z}_b$ <u>if</u> $b > 1$.

(i) <u>If one defines smooth structures</u> "A/S^1" <u>and (if meaningful)</u> "B/S^1" <u>as in Case II, then</u> "A/S^1" <u>and</u> "B/S^1" <u>are smoothly embedded in</u> "M/S^1" <u>with boundary</u> F.

(ii) <u>If</u> $a, b > 1$, <u>write</u> $A/S^1 \cap B/S^1 = F^{n-4} \amalg E^{n-5}$. <u>Then</u>

$$\{"A/S^1" \cup "B/S^1"\} - E$$

<u>is a smooth submanifold without boundary, and</u> "A/S^1" $- F$ <u>meets</u> "B/S^1" $- F$ <u>transversely in</u> E.

PROOF. (i) Away from the fixed point set F, this result follows from the local description of the orbit space projection in (1.5). At F the result follows from the local description of orbit space projections in Cases II and V of Table I.

(ii) The transversality statement is immediate from the local description in (1.5b). The first statement is however rather surprising at first glance. It says that "A/S^1" $- E$ and "B/S^1" $- E$ meet very nicely at F - Specifically they meet tangentially and from opposite directions. This requires the explicit definition of $q\colon V_a \oplus V_b \longrightarrow \mathbb{C} \oplus \mathbb{R} = \mathbb{R}^3$. It is elementary to check that near F the manifold "A/S^1" corresponds to all points $(w;q(x,0)) = (w;|x|^{2b},0)$ while "B/S^1" corresponds to all points $(w;q(0,y)) = (w;-|y|^{2a},0)$[here $w \in \mathbb{R}^{m-4}$, x & $y \in \mathbb{C}$] ■

The real importance of these results on orbit spaces is the existence of a converse theory. Any candidate for orbit space data that is consistent with basic necessary conditions is realizable as the smooth orbit space data of an action. In Cases I - IV this is folklore, and in Case V this is contained in the work of R. Fintushel and P. Pao [12] (also see [9-11,23]). Of course, their main interest and ours lie with actions on homotopy spheres. In this case one has the following additional necessary conditions:

(1.8a) M/S^1 is a homotopy sphere.

(1.8b) F is a $\mathbb{Z}$-homology sphere.

(1.8c) A/S^1 and B/S^1 are $\mathbb{Z}_a$ and $\mathbb{Z}_b$ (resp.) homology disks.

An important feature of their work is that if (1.8a) - (1.8c) hold, the Case V S^1-manifold constructed is a homotopy sphere.

Analogous results hold for manifolds with boundary, and we shall need these later. First of all, if M is a bounded manifold with a Case V circle action, then M/S^1 is a manifold with $\partial(M/S^1) = (\partial M)/S^1$, and the fixed set F is a smooth submanifold with $(\partial M)^G = \partial F$. The exceptional set E/S^1 is likewise a properly embedded submanifold. The analogs of A/S^1 and B/S^1 are a little more complicated; to avoid duplication, we only discuss the first case. Topologically $(\mathrm{Fix}\,\mathbb{Z}_a)/S^1$ is a manifold with boundary, and $W = (\mathrm{Fix}\,\mathbb{Z}_a)/S^1 - \partial F$ is in fact a smooth submanifold with $\partial W = (F-\partial F) \amalg ((\partial \mathrm{Fix}\,\mathbb{Z}_a - \partial F)/S^1)$. The problem is that the two boundary pieces meet at a 90 degree angle at F (see Figure 1).

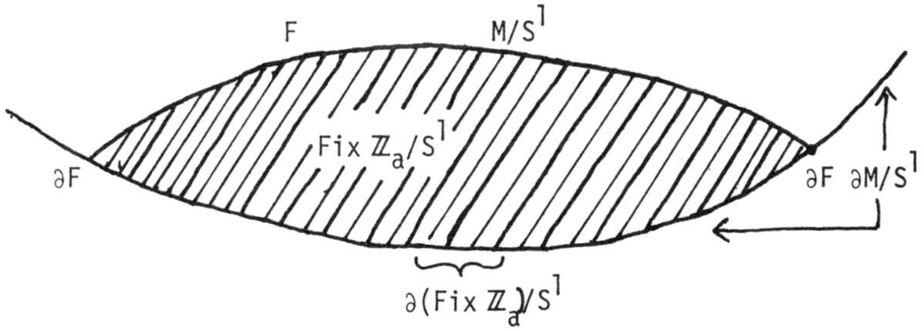

FIGURE 1

With this notation we can state some extensions of [12] that we need; the proofs are all direct adaptations of their homological arguments.

THEOREM 1.9. Suppose we have a candidate for Case V orbit data with the following homological and homotopical restrictions:

(i) "M/S^1" is an R-homological h-cobordism of S-homology spheres, where $\mathbb{Z} \subseteq S \subseteq R \subseteq \mathbb{Q}$.

(ii) Similarly for "F".

(iii) "A/S^1" *is a smooth* $R \otimes \mathbb{Z}_{(a)}$-*homology disk with a cornered boundary*:

(1.9a) ∂"A/S^1" = [∂"A/S^1" $\cap\ \partial_-$ "M/G"] $\cup$ [∂"A/S^1" $\cap\ \partial_+$ "M/G"] $\cup$ F.

Furthermore, the first two pieces in decomposition (1.9a) *are* $S \otimes \mathbb{Z}_{(a)}$-*homology disks*.

(iv) *Likewise for* "B/S^1" *if* $b > 1$.

(v) *The set* "E/S^1" *of "fifth orbit types" is empty*.

(vi) *The normal bundles of all relevant embeddings are trivial*.

THEN the realization M *of the action is an* R-*homology* h-*cobordism between two* S-*homology spheres* ■

Finally, we need two minor variants of (1.9).

COMPLEMENT 1.10. *One can also assume that different subrings* S_0, S_1 *are used on the two ends. In this case the respective ends are* S_i-*homology spheres* ■

COMPLEMENT 1.11. *Suppose in the situation of* 1.9 *we change the conditions as follows*:

(i) "M/S^1" *is an* R-*homology disk and* ∂"M/S^1" *is an* S-*homology sphere*.

(ii) *Similarly for* "F".

(iii) "A/S^1" *is an* $R \otimes \mathbb{Z}_{(a)}$-*homology disk with cornered boundary* F $\cup$ [∂"A/S^1" $\cap\ \partial$"M/G"], *and the second piece is an* $S \otimes \mathbb{Z}_{(a)}$- *homology disk*.

(iv) *Similarly for* B.

(v) *There are no orbits with istropy group* $\mathbb{Z}_{ab}$.

(vi) *All relevant normal bundles are trivial*.

THEN M *is an* R-*homology disk and* ∂M *is an* S-*homology sphere* ■

2. The Pontrjagin-Thom Construction

In this section we shall prove Theorem A. The first step is to prove

(2.0) $$\mathfrak{b}(\Theta_n(V_a \oplus V_b \oplus (n-4))) \subseteq ab\, \mathfrak{b}(\Theta_n(2V_1 \oplus (n-4))).$$

The crucial information is contained in the following result:

LEMMA 2.1. *Let* Σ^n *be a homotopy sphere admitting a smooth circle action with local representation* $V_a \oplus V_b \oplus n-4$ *at fixed points. Assume* $a, b \geq 1$ *and* $a > 1$.

(i) *The orbit manifold* $\Sigma^n/\mathbb{Z}_a$ *is a smooth homotopy sphere admitting a smooth circle action with local representation* $V_1 \oplus V_b \oplus (n-4)$ *at fixed points, and the orbit map* $\Sigma \longrightarrow \Sigma/\mathbb{Z}_a$ *is a cyclic branched covering along* A = *fixed point set of* A. (*If* $b = 1$, *the action is semifree*.)

(ii) *If in addition* $b > 1$, *then* $\Sigma^n/\mathbb{Z}_{ab}$ *is a smooth homotopy sphere admitting a smooth SEMIFREE circle action with* $(n-4)$-*dimensional fixed point set, and the map* $\Sigma/\mathbb{Z}_a \longrightarrow \Sigma/\mathbb{Z}_{ab}$ *is a cyclic branched covering along* $B/S^1 = \text{Fix}\,(\mathbb{Z}_b,\Sigma)/S^1$.

PROOF. The branched covering and local representation statements follow from the results of Section 1, so the only assertion remaining to be proved is the one about the orbit space being a homotopy sphere. In each case one is taking the orbit space of a $\mathbb{Z}_k$ action on some U^n with $(n-2)$-dimensional set; but it is well-known that the orbit space is always a homotopy sphere for such actions [3, II.6 and Exc.3] ■

The next step in proving (2.0) is the following result of Bredon [4, Ch.III, Sect.12, pp.84-86]; since [4] was (regrettably) never published, we shall reproduce Bredon's argument for the sake of completeness:

LEMMA 2.2 (Bredon). *Let* Σ^n *and* V^n *be two homotopy spheres, and let* $f\colon \Sigma \longrightarrow V$ *be a* k-*fold cyclic branched covering along a codimension* 2 *submanifold. Then* $\Sigma^n = kV^n \# W^n$ *for some* $W^n \in bP_{n+2}$.

PROOF. We first reduce to the case $V^n = S^n$; specifically, take a connected sum of V^n with $-V^n$ away from the branch set, and notice that this gives a new branched covering $\Sigma^n \# -kV^n \longrightarrow V^n \# -V^n = S^n$. Thus it suffices to show that $V^n = S^n$ implies $\Sigma^n \in bP_{n+1}$.

By a discussion as in [3,p.335] one can construct a smooth map

$$\tau\colon S^n \longrightarrow D^2 \subseteq \mathbb{C}$$

which is the projection to the fiber for an appropriate trivialization $D^2 \times B^{n-2}$ of the branch set's nomal bundle. Now τ can be extended to a smooth map $D^{n+1} \longrightarrow \mathbb{C}$ with $0 \in \mathbb{C}$ as a regular value (this is essentially an elementary application of Sard's Theorem). If we let $\psi^k\colon \mathbb{C} \longrightarrow \mathbb{C}$ denote the k-th power map, then we have the following pullback diagram:

$$\begin{array}{ccc} \Sigma^n & \longrightarrow & \mathbb{C} \\ {\scriptstyle f}\downarrow & & \downarrow{\scriptstyle \psi^k} \\ S^n & \xrightarrow{\ \tau\ } & \mathbb{C} \end{array}$$

However, Σ is clearly the boundary of the following pullback:

$$\begin{array}{ccc} W^{n+1} & \longrightarrow & \mathbb{C} \\ \Big\downarrow & & \Big\downarrow \psi^k \\ D^{n+1} & \xrightarrow{\ \tau'\ } & \mathbb{C} \end{array}$$

The manifold W with $\partial W = \Sigma$ is a smooth submanifold of $D^{n+1} \times \mathbb{C}$. Furthermore, it has a trivial normal bundle because it is the inverse image of a regular value of $\tau' - \psi^k\colon D^{n+1} \times \mathbb{C} \longrightarrow \mathbb{C}$. Thus W^{n+1} is a π-manifold with Σ as its boundary ∎

REMARK. There are many exotic $\mathbb{Z}_k$ actions on Brieskorn spheres with codimension two fixed point sets; it follows that the conclusion of 2.2 is best possible.

PROOF OF THEOREM A. Inclusion (2.0) is immediate from (2.1) and (2.2). In order to prove the reverse inclusion, it will simplify matters greatly to use the following consequence of Theorem @ in the introduction:

(2.4) <u>The group</u> $\Theta_n(2V_1 \oplus (n-4))/bP_{n+1}$ <u>has exponent at most</u> 2.

[To see this, notice that $0 \in 2\langle ?,\nu,\eta\rangle \subseteq \langle ?,\nu,2\eta=0\rangle$, and the indeterminacy of the right hand side is $?\circ\pi_5$; but $\pi_5 = 0$] ∎

Thanks to 2.4 we may restrict ourselves to the case where a and b are both odd henceforth; if either is even, then (2.0) and (2.4) imply $\mathfrak{F}(\Theta_n(V_a \oplus V_b \oplus (n-4))) = 0$.

We shall begin with a very useful reduction principle that allows us to assume $F^{n-4} = \mathrm{Fix}\,(S^1,\Sigma)$ is a homotopy sphere.

LEMMA 2.5. <u>Suppose</u> Σ^n <u>admits a smooth circle action with an</u> $(n-2r)$-<u>dimensional fixed point set</u> F, <u>where</u> $2r \geq 4$ <u>and</u> $n-2r \geq 5$. <u>Assume the equivariant normal bundle of</u> F <u>in</u> Σ <u>is trivial. Then</u> Σ <u>admits a smooth circle action with fixed point set a</u> HOMOTOPY $(n-2r)$-<u>sphere</u>.

PROOF. As in [13] there is a simply connected homology h-cobordism W from F to a homotopy sphere H (i.e., W and H are simply connected, but F might not be). Let Ω^{2r} be the nontrivial part of the local representation at fixed points. Construct the S^1-manifold

$$X^{n+1} = \Sigma \times I \cup_{F\times D(\Omega)\times\{1\}} W \times D(\Omega),$$

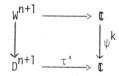

and round the corners equivariently to obtain a smooth S^1-manifold. Standard computations imply that X is a (homotopy) h-cobordism from Σ to Σ', where Σ' has an action of the sort desired. Since Σ and Σ' are (nonequivariantly) diffeomorphic by the h-cobordism Theorem, the result is proved ■

To conclude the proof of Theorem A, we must construct some actions, and the reader may well guess that we shall be using the realization Theorem of Fintushel and Pao. The following result gives us the data that we need in order to apply their work; we shall postpone the proof until after we show how the result implies Theorem A.

PROPOSITION 2.6. Let $K^{m-3} \subseteq S^m$ be a smooth knotted homotopy sphere (say $m \geq 8$ for convenience). Then for some odd integer q we have

$$\#^q K = K_1 \# K_2 \qquad \text{(knot connected sum)},$$

where K_1 and K_2 bound parallelizable and $\mathbb{Z}[1/2]$-acyclic submanifolds of S^{n-1} respectively.

PROOF OF THEOREM A CONCLUDED. Recall that ab is odd.

We are given a smooth semifree circle action on an exotic sphere Σ. Since $\mathfrak{p}(\Theta_n(2V_1 \oplus (n-4))$ is 2-torsion, it suffices to show that $q'\Sigma \in \Theta_n(V_a \oplus V_b \oplus (n-4)) + bP_{n+1}$ for some suitable odd integer q'.

The following assertion disposes of some easy cases

(2.8) We have $\Theta_{n-1}\,\eta \subseteq \hat{\Theta}_n(V_a \oplus V_b \oplus (n-4)) \subseteq \Theta_n(V_a \oplus V_b \oplus (n-4))$.

The fastest proof of (2.8) involves the Browder-Quinn theory of stratified surgery [5]; if $b = 1$ this construction is essentially identical to one of Bredon [1]. Let M^* denote the orbit space of $S(V_a \oplus V_b \oplus (n-5))$ with its standard smooth stratification. Given $V \in \Theta_{n-1}$, let $N^* = M^* \# V$, where connected sum is taken along the nonsingular stratum. The homomorphism $fN^* \longrightarrow M^*$ is a smooth stratified isomorphism away from a disk in the nonsingular set, and consequently it defines a stratified normal map with normal invariant in $[M^*, G/O] = \pi_{n-1}(G/O)$ (since $M^* = S^{n-1}$). It is almost immediate that $\mathfrak{p}(V) \in \pi_{n-1}(G/O)$ represents the normal invariant of f. A standard pullback construction (compare [3] for the topological case; also see work of M. Davis, A. Durfee, and L. Kauffman) implies that N^* is the smoothly stratified orbit space of some action on a homotopy sphere Σ. The Pontrjagin-Thom construction is then given by the composite of (V) with the orbit map

$$h: S^n = S(V_a \oplus V_b \oplus (n-4)) \longrightarrow S(\cdots)/S^1 = S^{n-1}.$$

It is a routine exercise to check that the class of h in $\pi_n(S^{n-1}) = \mathbb{Z}_2$ is $ab\eta = \eta$ (since ab is odd). Therefore $\mathfrak{p}(\Sigma) = \mathfrak{p}(V)\eta$. In order to get $\Sigma = V\eta$ exactly, one must reformulate Browder-Quinn to deal with transverse isovariant homeomorphisms. In this case the transverse isovariant topological smoothings of M are classified by $[M/G, Top/O]$. The argument outlined above goes through in this setting with only obvious and minor changes of terminology ∎

The virtue of (2.8) is that it allows us to alter the differential structure on the orbit space without losing our hold on Σ.

Returning to the main line of argument, we may as well assume $\Sigma/S^1 = S^{n-1}$ by (2.8) and its proof (if $\Sigma/S^1 = V$, take connected sums with $M^* \# - V$ (M^* = standard orbit space) to get an action on $\Sigma \# - V\eta$). Thus we may let (S^{n-1}, K^{n-3}) denote the action's orbit data.

If K has the special form of K_1 or K_2 in (2.6) then the argument is fairly direct:

Case 1. K bounds a parallelizable manifold in S^{n-1}. We claim $\Sigma^n \in bP_{n+1} + \Theta_{n-1}\eta$. This will follow from [29:I, Thm.3.3] if we can show that the action's knot invariant vanishes. One encouraging sign in this direction is the vanishing of the original Levine knot invariant for (S^{n-1}, K) [18] in $\pi_{n-4}(G_3)$ (recall that K bounds a π-manifold). Furthermore the map

$$F_{S^1}(S^3) \longrightarrow G_3 = \text{self maps of } S^2$$

given by passage to orbit spaces induces an isomorphism in homotopy for $n - 4 > 3$. Therefore it suffices to check that passage to orbit spaces sends the S^1-knot invariant of K in Σ^n to the Levine knot invariant of K in S^{n-1}. However, this is immediate from the diagram below and the fact that the knot invariants are αi and βj, where α and β are inverses to the horizontal composites $i|S^3$ and $j|S^2$:

$$\begin{array}{ccccc} S^3 & \subseteq & K \times S^3 & \xrightarrow{\ i\ } & \Sigma - K \\ \downarrow & & \downarrow & & \downarrow \\ S^2 & \subseteq & K \times S^2 & \xrightarrow{\ j\ } & (\Sigma/S^1) - K \end{array}$$

(The vertical arrows are orbit maps).

Case 2. K bounds a $\mathbb{Z}[1/2]$-acyclic manifold W. Suppose $b = 1$. In this case (S^{n-1}, K, W) is a good choice for orbit space data and it is realized by the work of Fintushel and Pao (K = fixed point set, W = orbit space of Fix $\mathbb{Z}_a$). To find the Pontrjagin-Thom invariant, look first at the semifree action with data (S^{n-1}, K) and ambient sphere Σ. If Σ' is the exotic sphere with data (S^{n-1}, K, W), then as in 2.3 we have $\Sigma' = a\Sigma$ + elt. of bP_{n+1} = (using (2.4)) Σ + elt. of bP_{n+1}.

If $b > 1$, we want an orbit space candidate with copies of W as the fixed orbits for $\mathbb{Z}_a$ and $\mathbb{Z}_b$. Let W' be a copy of W that is parallel to W near the boundary and so that W and W' meet transversely in their interiors.

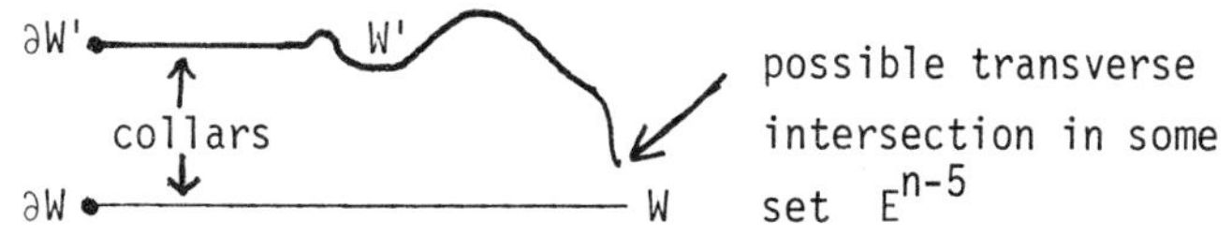

FIGURE 2

Construct a new manifold W'' such that $W'' = W'$ off a collar but W'' meets W' at their common boundary, tangentially in opposite directions. The figure below suggests how this may be done. An explicit description involving bump functions is left to the reader.

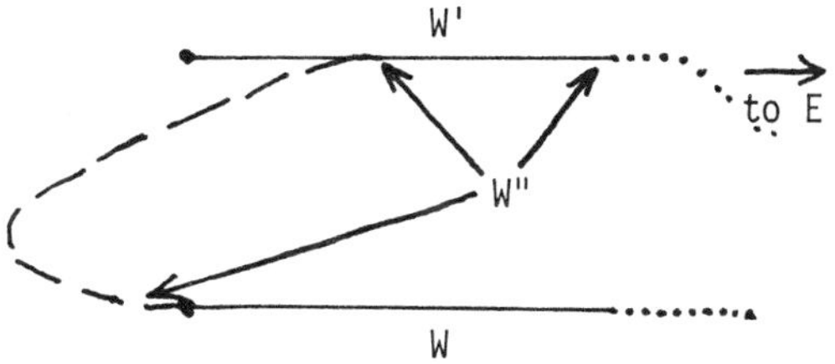

FIGURE 3

It follows that (S^{n-1},K,W,W'') give data for realizing an action, and once again we have $\Sigma' \equiv ab\Sigma \bmod bP_{n+1}$ or equivalently $\Sigma' \equiv \Sigma \bmod bP_{n+1}$ if Σ' is the ambient homotopy sphere.

The conclusions of cases 1 and 2 may be summarized with (2.8) as follows:

(2.9) <u>If Σ admits a semifree circle action with orbit data</u> $(\Sigma/S^1, K_1\#K_2)$, <u>then</u> $\Sigma \in \Theta_n(V_a \oplus V_b \oplus (n-4)) + bP_{n+1}$ ■

We are now ready to apply 2.6. Notice first that if (S^{n-1},K) is the data for a semifree action on Σ, then $(S^{n-1},\#^q K)$ is the data for a semifree action on $q\Sigma$; this is more or less immediate from the fact that passage to orbit spaces preserves connected sum. Then (2.6) implies that if Σ admits an appropriate semifree circle action, then for some odd q one has an action on Σ satisfying the hypotheses of (2.9). Hence $q\Sigma \in \Theta_n(V_a \oplus V_b \oplus (n-4)) + bP_{n+1}$; since $q\Sigma \equiv \Sigma \bmod bP_{n+1}$, one sees that the conclusion of (2.9) is still valid in the general case. Therefore we have shown

$$\Theta_n(2V_1 \oplus (n-4)) \subseteq \Theta_n(V_a \oplus V_b \oplus (n-4)) + bP_{n+1}.$$

that is just the second half of Theorem A when a and b are odd (as they have been above) ■

Now we must turn to proving (2.6). This will require a large amount of auxiliary information. In order to present the line of argument more clearly, we shall first digress to establish what we need.

We begin with a complete statement of what is needed about the coexact sequence $S^p \longrightarrow S^p \times S^q \longrightarrow (S^p \times S^q)/S^p = S^{p+q} \vee S^q$:

(2.10) In the Puppe sequence

$$[S^{p+q} \vee S^q, X] \xrightarrow{\alpha} [S^p \times S^q, X] \xrightarrow{\beta} \pi_p(X)$$

one has the following exactness properties: There is a natural (in X) group structure on the left hand set and a natural action of $[S^{p+q} \vee S^q, X]$ on $[S^p \times S^q, X]$ so that $\beta u = \beta v$ implies $v = \alpha(t)\cdot w$ for some $t \in [S^{p+q} \vee S^q, X]$ ∎

We also need the following statement about localizations and Thom spaces; this is an unstable analog of [30,Thm.2.1]:

(2.11) Let ℓ be a set of primes, and let Γ_k be the fiber of the localization map $BSO_k \longrightarrow (BSO_k)_\ell$. Let $M\Gamma_k$ be the Thom complex of the obvious k-plane bundle over $\Gamma_{k\ell}$. Then the map of pairs from $(MSO_k, M\Gamma_k)$ to $((MSO_k)_\ell, (M\Gamma_k)_\ell = S^k_\ell)$ induces isomorphisms in homology and homotopy.

PROOF (SKETCH). The homological statement follows from the same method used in [30,§2]. Since both pairs are $k \geq 2$-connected the homotopy statement is an elementary consequence of the Hurcwicz and Whitehead Theorems ∎

PROOF OF 2.6. The first order of business is to prove that $\#^q K$ bounds a sufficiently nice manifold in S^{m+3}. Our first step is the following well-known fact, which follows from (i) the fact that any embedding may be isotoped so that a coordinate disk maps linearly, (ii) the existence of an h-cobordism from the normal sphere bundle of K to $S^{m-3} \times S^2$, (iii) the triviality of all normal bundles for knotted codimension 3 homotopy spheres [$O_3 \longrightarrow G_3 \longrightarrow S^2$ is bijective in homotopy above dimension 2, so fiber homotopy triviality ⇔ triviality]:

(2.12) Let $D_0 \subseteq K$ be a coordinate disk. Then there is a diffeomorphism

$$f: S^2 \times (K; D_0, K\text{-In} + D_0) \longrightarrow S^2 \times (S^{m-3}; D_+^{m-3}, D_-^{m-3})$$

such that

(i) $K \subseteq S^{m-3}$ is equivalent to the knot

$$K \longrightarrow \{e\} \times K \subseteq S^2 \times K \cong S^2 \times S^{m-3} \xrightarrow[\text{inclusion}]{\text{std.}} S^m;$$

in fact, f extends to a diffeomorphism $F: D^3 \times K \cong D^3 \times S^{m-3}$ such that K is embedded as the composite

$$\{0\} \times K \subseteq D^3 \times K \xrightarrow{\cong} D^3 \times S^{m-3} \xrightarrow[\text{inclusion}]{\text{std.}} S^{m-3}.$$

(ii) f restricted to $S^2 \times D_0 \longrightarrow S^2 \times D_+^{m-3}$ is the identity on S^2 crossed with a standard identification of disks ∎

The significance of (2.12) is that it reduces the bounding question to a bordism question of a standard type. Namely, we wish to show that $K \subseteq S^2 \times S^{m-3}$ and $\{e\} \times S^{m-3} \subseteq S^{m-3}$ are cobordant; i.e. $\{e\} \times S^{m-3} \times \{1\} \cup -K \times \{0\} = \partial W$, $W \subseteq S^2 \times S^{m-3} \times I$. If we cap off the top, we then get a coboundary for K.

It is almost trivial to show that K bounds something. As a framed submanifold of $S^{m-3} \times S^2$ it determines a Pontrjagin-Thom homotopy class ξ in

$$[S^{m-3} \times S^2, S^2].$$

If ξ = projection on S^2 (= P.-T. class of $\{e\} \times S^{m-3} \subseteq S^2 \times S^{m-3}$), then $S^{m-3} \amalg -K$ bounds a framed submanifold. To show that K bounds something, it is enough to show that the images of ξ and proj (S^2) under the map

$$[S^{m-3} \times S^2, S^2] \longrightarrow [S^{m-3} \times S^2, MSO_2 = K(\mathbb{Z},2)]$$

are the same. But this is immediate because we are dealing with a $K(\mathbb{Z},2)$ in the codomain and the restriction of ξ to S^2 is homotopic to the identity by construction.

The discussion above is a little too crude for our purposes. The first step towards sharpening the result is to consider the difference class $\Delta(\xi) \in \pi_{m-1} \times \pi_{m-3}(S^2) = [S^{n-3} \vee S^{m-1}, S^2]$ given by (2.10) such that $\xi = \Delta(\xi)\cdot\pi$. The additivity property $\Delta(\xi_{K\#L}) = \Delta(\xi_K) + \Delta(\xi_L)$ is a straightforward consequence of the definitions.

Since $\pi_k(S^2)$ is finite if $k > 3$, the additivity property tells us that $q\Delta(\xi_K) = \Delta(\xi_{q\#K})$ is 2-primary for some odd q. This is the value of q that we want. For simplicity, write $M = \#^q K$ henceforth.

Let ℓ denote localization away from 2. In order to prove (2.6) we must show that $\Delta(\xi_M)$ goes to zero in $[S^{m-1} \vee S^{m-3}, M\Gamma_2]$.

To do this, consider the following large commutative diagram (Λ = fiber of $M\Gamma_2 \longrightarrow MSO_2$ = fiber of $S^2_\ell \longrightarrow (MSO_2)_\ell$; recall $MSO_2 = K(\mathbb{Z},2)$).

$$\begin{array}{ccccc}
 & & [S^{m-1}\vee S^{m-3},\Omega MSO_2] & \xrightarrow{\text{loc.}} & [S^{m-1}\vee S^{m-3},(\Omega MSO_2)_\ell] \\
 & & \downarrow & & \downarrow h \\
 & & [S^{m-1}\vee S^{m-3},\Lambda] & \xrightarrow{=} & [S^{m-1}\vee S^{m-3},\Lambda] \\
\{\Delta(\xi)\epsilon\} & & \downarrow g & & \downarrow \\
[S^{m-1}\vee S^{m-3},S^2] & \xrightarrow{i} & [S^{m-1}\vee S^{m-3},M\Gamma_2] & \xrightarrow{j} & [S^{m-1}\vee S^{m-3},S^2_\ell] \\
 & & \downarrow f & & \downarrow \\
 & & [S^{m-1}\vee S^{m-3},MSO_2] & \xrightarrow{\text{loc.}} & [S^{m-1}\vee S^{m-3},(MSO_2)_\ell].
\end{array}$$

(The columns are exact, and ji is induced by localization at ℓ).

We already noted that $fi\,\Delta(\xi) = 0$, and $ji\,\Delta(\xi) = 0$ since $\Delta(\xi)$ is 2-primary. Hence $i\Delta(\xi) = g(x)$ for some x, and $jg(x) = 0$. Therefore $x = h(y)$ for some y. But $(\Omega\, MSO_2)_\ell = K(\mathbb{Z}_\ell,1)$, so the group in the upper right corner is zero $(m-3\geq 4)$. Hence $y = 0$, so that $x = 0$, and consequently $g(x)=0$ as well. This proves $i\Delta(\xi) = 0$, which gives the desired bounding condition on M.

We may now let $M = \partial W$ for $W \subseteq S^m$ a Γ_2-structured manifold. We wish to perform surgery on W as far as possible, and to do so within S^m. As in Levine's work [19], there is no additional problem in doing the surgery ambiently since M has codimension 3, and the ultimate obstruction lies in the group $L_{m-2}(\mathbb{Z}[1/2])$. (<u>Note</u>: Unlike Levine's case, one cannot always kill the entire fundamental group. However, the techniques of [30] may be extended to show that fundamental group problems do not introduce any complications into the conclusions). Since $L_{m-2}(\mathbb{Z}[1/2]) = 0$ unless $(m-2) \equiv 0 \bmod 4$, this suffices to prove (2.6) in those cases. In fact, in these cases one can let K_1 be trivial.

If $m - 2 \equiv 0 \bmod 4$ the considerations of Levine's work show that surgery cannot always be completed even if M bounds a parallelizable manifold W; the signature of W interferes with this. The obvious thing to do is to see if one can kill off the signature of W by adding on an almost closed parallelizable manifold Q^{m-2}, which we can assume lies inside a very small m-disk inside S^m. We could then add $-Q^{m-2}$ on to recover M, and our candidates for K_1 and K_2 would be $-\partial Q$ and $\partial Q \# M$. Since we may choose Q to have signature an arbitrary multiple of 8 ($m - 2 \equiv 0 \bmod 4$ and $m - 3 \geq 4$ imply $m \geq 10$), we can kill off the signature of W if it is divisible by 8. Fortunately, this is the case; by construction the stable normal bundle of W has a unitary structure (since $SO_2 = U_1$!), and W has no rational characteristic classes. Therefore a result of G. Brumfiel [6,1.6] implies $\operatorname{sgn} W \equiv 0 \bmod 8$. Therefore we may as well assume $\operatorname{sgn} W = 0$.

But the obstruction to completing surgery, which is in $L_0(\mathbb{Z}[1/2])$, is just the Witt class of the middle-dimensional cup product form, and $L_0(\mathbb{Z}[1/2])$ is completely specified by the signature and discriminant invariants [21]. We know the signature is zero, so we must consider the discriminant. But let $\hat{W} = W \cup \text{Cone}(\partial W)$ (recall ∂W is a homotopy sphere). This is a topological manifold with boundary, and hence its middle dimensional cup product form is integrally unimodular. Since $\hat{W}$ and W have the same middle dimensional homology, this implies the form on W has no discriminant. Therefore there is no obstruction to completing homology surgery on W, and accordingly K_2 bounds a $\mathbb{Z}[1/2]$-acyclic manifold ■

3. Linearly stratified actions

The results of Section 2 are formulated to sidestep questions about boundaries of π-manifolds. In particular, they shed almost no light at all on the following simple question. If an exotic sphere Σ^n bounds a π-manifold, is $\Sigma \in \Theta_n(V_a \oplus V_b \oplus (n-4))$? (we know that $\Sigma \in \Theta_n(V_1 \oplus (n-4))$). The results of this section answer this question for $n \equiv 3 \bmod 4$ and actions with three or four orbit types (in particular, this disposes of the case $\Theta_n(V_a \oplus V_1 \oplus (n-4))$, where $a > 1$ and $n \equiv 3 \bmod 4$). In contrast to the semifree case, the conclusions of Theorems B and C imply that elements of bP_{4k} almost never belong to $\Theta_{4k-1}(V_a \oplus V_1 \oplus (n-4))$ or $\hat{\Theta}_{4k-1}(V_a \oplus V_b \oplus (n-4))$ [here $a, b > 1$]. A precise statement is given in Proposition 3.5 below.

<u>Definition</u>. Let Φ be a smooth circle action on a manifold, and assume that the fixed point set of Φ is connected. Let $V(\Phi)$ be the local representation at a fixed point. We say that Φ is <u>linearly stratified</u> if every isotropy subgroup of the action on M is an isotropy subgroup of the action on Φ.

If M^n is a homotopy sphere and $\dim \text{Fix}(\Phi) = n-4$, then an action is linearly stratified if and only if it has two, three, or four orbit types. In particular, the only exceptions have $V(\Phi) = V_a \oplus V_a \oplus (n-4)$ with $a, b > 1$ and five isotropy subgroups; namely, $1, \mathbb{Z}_{ab}, \mathbb{Z}_a, \mathbb{Z}_b, S^1$ (e.g., Stein's action [31]).

The cases of three and four orbit types are treated separately in Theorems B and C respectively.

PROOF OF THEOREM B. The equality $\hat{\Theta}_n(V_a \oplus V_1 \oplus (n-4)) = \Theta_n(V_a \oplus V_1 \oplus (n-4))$ is known (compare [12]), and we know $\mathfrak{P}(\Theta_n(V_a \oplus V_1 \oplus (n-4))) = a\,\mathfrak{P}(\Theta_n(2V_1 \oplus (n-4)))$ by Theorem A. The latter immediately settles the case where n is even.

Therefore the main problem left is to determine the image of $\Theta_n(V_a \oplus V_1 \oplus (n-4))$ under the Brumfiel splitting map from Θ_n to bP_{n+1}.

This question splits naturally into two parts, one involving actions whose orbit spaces are standard spheres and the other involving the possibility of an exotic differential structure on the orbit space. The second part may be handled using 2.8; in particular, it yields $a\,\Theta_{n-1}\,\eta \subseteq \Theta_n(V_a \oplus V_1 \oplus (n-4))$. Because of this, it will suffice to prove the following assertion.

(3.1) <u>Let</u> $\Sigma \in \Theta_n(V_a \oplus V_1 \oplus (n-4))$, $n \equiv 3 \bmod 4$, <u>and assume that</u> Σ/S^1 <u>is diffeomorphic to</u> S^{n-1}. <u>Then the Brumfiel splitting invariant of</u> Σ <u>is zero in</u> bP_{n+1}.

The idea is very simple. Start with the orbit space data over S^{n-1} and find cobounding orbit space data in D^n Try to realize this data by an action on some Δ^{n+1} using (1.10). Then (1.10) will imply that Δ is acyclic and hence has no signature. But the latter implies that the splitting invariant must vanish.

Unfortunately, this discussion has some gaps that must be filled, so we must go through the construction of the cobounding action in greater detail. In particular it is not immediate from (1.10) that one can find the action desired.

To begin, we describe the orbit space data more precisely. Take $W \times [1/2,1] \subseteq S^{n-1} \times [0,1] \cong D^n\ \{0\}$, and round off the 90 degree angle at which $\partial W \times [1/2,1]$ and $W \times \{1/2\}$ meet. Let Z denote

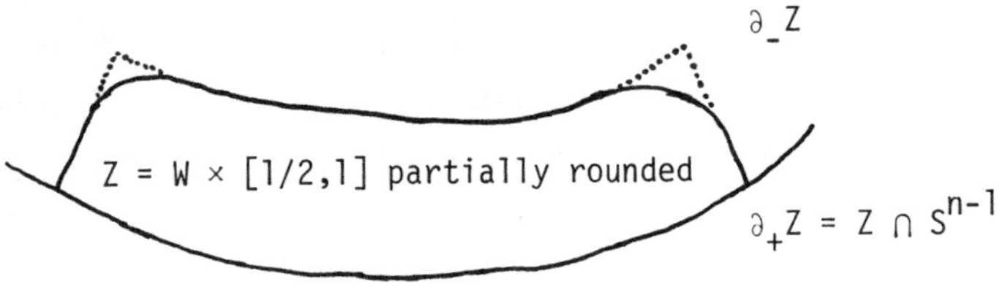

FIGURE 4

the resulting smooth manifold with 90 degree angle at the intersection of $\partial W \times [1/2],1]$ and $W \times \{1\}$.

We claim that $(D^n, Z, \partial_- Z)$ can be realized by an action of the desired type. This will be done in two steps, the first being to construct a semifree action with orbit data $(D^n, \partial_- Z)$ and the second being to take an a-fold cyclic branched covering along the inverse image of Z. The latter is fairly standard, so we shall only consider the semifree problem here. There are two conditions that must hold in order to construct the action we want:

(i) The classifying map $f\colon \partial_- Z \longrightarrow BSO_3$ for the normal bundle of $\partial_- Z$ must lift to BU_2. Here $BU_2 \longrightarrow BSO_3$ is induced by the homomorphism $U_2 \longrightarrow SO_3$ mentioned in Section 1.

(ii) Let ν denote the bundle classified by f in (i), and let ξ be the lifting to a complex 2-plane bundle. If $S(\xi) \longrightarrow S(\nu)$ is the principal S^1 bundle obtained by taking the unit sphere bundle of ξ and factoring out the natural free S^1 action to obtain $S(\nu)$, then the classifying map $c\colon S(\nu) \longrightarrow \mathbb{C}P^\infty$ for this principal bundle must extend to $D^n - \partial_- Z$. This extension must also coincide with the canonical extension of $c|\partial S(\nu)$ to $S^{n-1} - \partial W$.

Point (i) causes no real difficulty because the classifying map f has a canonical lifting to BSO_2. Thus it is merely necessary to find a lifting homomorphism to make the diagram below commute:

$$\begin{array}{ccc} & \overset{?}{\dashrightarrow} & U_2 \\ & & \downarrow \\ SO_2 & \longrightarrow & SO_3 \end{array}$$

An unbeatable choice is given by using the identity $SO_2 = U_1$ and mapping U_1 into U_2 by a standard stabilization homomorphism. In particular, if $\nu = (\mathcal{R}e(\eta)) \oplus \mathbb{R}$, then $\xi = \eta \oplus \mathbb{C}$ with this choice.

The second point is basically an obstruction theoretic problem that can be solved affirmatively. Here is one fast if inelegant solution. Look at the double $\mathcal{D}$ of $\partial_- Z$ in S^n = double of D^n. Since $S(\nu(\mathcal{D}))$ is the double of $S(\nu)$, the classifying map $S(\nu) \longrightarrow \mathbb{C}P^\infty$ extends to $S(\nu(\mathcal{D}))$. The associated class in $H^2(S(\nu(\mathcal{D})); \mathbb{Z})$ has dual submanifold given by the standard cross section of $\nu(\mathcal{D}) = \mathcal{R}e(\eta) \oplus \mathbb{R}$. We must show that this cross section bounds a nice submanifold of S^n - Int $E(\nu(\mathcal{D}))$. However, this is no problem because the double $Z \cup_{\partial_+ Z} Z$ is an appropriate coboundary. Finally, we have to check that this extension restricts to a standard map on $S^{n-1} - F$. But this is trivial because the extension restricts to the standard map $S^2 \longrightarrow \mathbb{C}P^\infty$ on $S^2 \subseteq S(\nu(F, S^{n-1})) \subseteq S^{n-1} - F$ and the inclusion of S^2 in $S^{n-1} - F$ is a homotopy equivalence.

Having constructed the action, we must prove that the ambient bounded manifold is a unitary manifold in order to apply Brumfiel's results [6]. We again break this into two pieces, one for the semifree action and one for the a-fold cyclic branched covering.

Let Γ be the semifree S^1-manifold with $(D^n, \partial_- Z)$ as orbit data. Since $D^n - \partial_- Z$ has a framing induced by being a codimension zero submanifold of D^n and S^1 acts freely on $\Gamma - \partial_- Z$ we can frame $\Gamma - \partial_- Z$ by pulling back this framing and adding the standard trivialization for the 1-dimensional bundle of tangents along the S^1 fibers. On the other hand, the tubular neighborhood

$E(\xi + \partial_- Z)$ also admits a unitary structure since $\partial_- Z$ does (its normal bundle in D^n reduces to $SO_2 = U_1$) and ξ is a complex vector bundle. The following assertion is then a consequence of the definitions:

(3.2) Let $\mathcal{S}_1$ be the canonical unitary structure on $E(\xi)$ given above, and let $\mathcal{S}_2$ be the canonical framing of $\Gamma - \partial_- Z$. Then the unitary structures given by restricting $\mathcal{S}_1$ and $\mathcal{S}_2$ to $E - \partial_- Z$ are equivalent.

This allows us to piece together a unitary structure on Γ. Let Δ be the a-fold cyclic branched covering of Γ along the inverse image of Z. The homological calculations of [12] imply that Δ is rationally acyclic, so it remains to check that Δ is a unitary manifold. But a cutting and pasting argument similar to the previous one shows the following:

(3.3) Let $X \longrightarrow Y$ be a smooth cyclic branched covering along a codimension 2 submanifold; denote the branch set by B. Suppose that Y admits a unitary structure such that its restriction to a tubular neighborhood $E(\eta)$ of B is the sum of the unitary structure on η with a unitary structure on B. Then X admits a unitary structure whose restriction to $X - B$ is the covering space pullback of the unitary structure on $Y - B$ ■

This concludes the proof of Theorem B ■

PROOF OF THEOREM C. This case is much different from the others because $(\mathrm{Fix}\,\mathbb{Z}_a/S^1) \cup (\mathrm{Fix}\,\mathbb{Z}_b/S^1)$ is a smooth unbounded, oriented, codimension 2 submanifold of Σ^{n-1}. Since the normal bundle of such a manifold is always trivial, we see immediately that F^{n-4} bounds a parallelizable submanifold of $\mathbb{R}^{n-1}$. Since it also bounds an acyclic manifold and $n \not\equiv 1 \bmod 4$, by the results of Levine [19] we in fact know that F is an unknotted standard sphere. (Incidentally, the second conclusion is systematically false if $n \equiv 1 \bmod 4$).

The first part of the proof is to determine $(\hat{\Theta}_n(V_a \oplus V_b \oplus (n-4))$. We already know it contains $(ab\Theta_{n-1}\eta)$ by (2.8), and we wish to show that nothing else is possible. The key idea is to construct a transverse linear isovariant normal map f (in the sense of [5]) from Σ to the linear model $S(V_a \oplus V_b \oplus (n-3))$. It will then follow from [5] that $\mathfrak{p}(\Sigma)$ lies in the image of the map

$$q^*\colon [S(V_a \oplus V_b \oplus (n-3))/S^1, G/O] \longrightarrow \pi_n(G/O)$$

where $q\colon S^n \longrightarrow S(V_a \oplus V_b \oplus (n-3))/S^1 = S^{n-1}$ is ab times the Hopf map. In fact, if $N(f/S^1)$ denotes the normal invariant for the map f/S^1, then $\mathfrak{p}(\Sigma)$ is given by $q^*N(f/S^1)$. Since Σ/S^1 is a homotopy sphere, f/S^1 is a homotopy equivalence and $N(f/S^1) = \mathfrak{p}(\Sigma/S^1)$. Hence $\mathfrak{p}(\Sigma) = ab\,\mathfrak{p}(\Sigma/S^1)\eta$.

Construction of f proceeds stratum by stratum. If F^{n-4} is the fixed point set, begin by defining the map to be the identity on $D(V_a \oplus V_b)$ crossed with the standard degree one map from F to S^{n-4}. Extend this map to tubular neighborhoods of $\mathrm{Fix}(\mathbb{Z}_a)$ and $\mathrm{Fiz}(\mathbb{Z}_b)$ using the triviality of their normal bundles of $\mathrm{Fix}(\mathbb{Z}_a)/S^1$ and $\mathrm{Fix}(\mathbb{Z}_b)/S^1$). It remains to extend the map over the set of free orbits. If we let $\mathcal{S}$ denote an S^1 invariant regular neighborhood of the singular set (plumb together the equivariant normal bundles of the fixed point sets and round out the corners) and P_0 is the set of principal orbits with $\mathrm{Int}\,\mathcal{S}$ removed, the entire question reduces to finding a map from P_0 to S^1 that will make the following diagram commute:

$$\begin{array}{ccc} P_0/_{S^1} & \xrightarrow{\ ?\ } & P_0(\text{linear model})/_{S^1} = S^1 \\ \uparrow & & \uparrow \\ \partial\mathcal{S}/_{S^1} & \xrightarrow[f/S^1]{\text{partial}} & \partial\mathcal{S}(\text{linear model})/_{S^1} = S^{n-3} \times S^1. \end{array}$$

In other words, we must extend the associated class in

$$H^1(\partial\mathcal{S}/S^1;\mathbb{Z}) \text{ to } H^1(P_0/S^1;\mathbb{Z}).$$

Under Alexander duality the map

$$j^*\colon H^1(P_0/S^1) \longrightarrow H^1(\partial\mathcal{S}/S^1)$$

corresponds to the map

$$i_*\colon H_{n-3}(\text{Sing set}/S^1) \longrightarrow H_{n-3}(\text{Sing set}/S^1 \amalg P_0/S^1).$$

Since i_* is split injective, the same is true of j^*. On the other hand, rational calculations and universal coefficient considerations imply that the domain and codomain of j^* are infinite cyclic. Therefore j^* must also be onto. This allows us to complete construction of f, and as noted before it implies the assertion on Pontrjagin-Thom invariants.

Now assume $n \equiv 3 \mod 4$. As in the proof of Theorem B, if we have an action Σ with four orbit types and $\Sigma/S^1 = S^{n-1}$, we wish to find orbit space data in D^n that bound the data in S^{n-1}.

We do this in two steps. The first is to construct a cobordism U from the original data (S^{n-1},A,B,F) to new data (S^{n-1},A',B',F') with the following properties:

(i) U <u>is diffeomorphic to</u> $S^{n-1} \times I$

(ii) <u>The cobordism between</u> F <u>and</u> $F' = S^{n-4}$ <u>is diffeomorphic to</u> $B - \mathrm{Int}\,h(D^{n-3})$, <u>where</u> $h\colon D^{n-3} \longrightarrow \mathrm{Int}\,B$ <u>is a smooth embedding</u>.

(iii) B' <u>is diffeomorphic to</u> D^{n-3}.

(iv) <u>The cobordism between</u> $A \cup B$ <u>and</u> $A' \cup B'$ <u>is diffeomorphic to</u>

$(A \cup B) \times I$.

The construction is simple. Let f be a smooth function on $B - \mathrm{Int}(D^{n-3})$ which is zero on ∂B and one on S^{n-4}, and let $\mathcal{F} \subseteq (A \cup B) \times I$ be the graph of f. Then $(A \cup B) \times I - \mathcal{F}$ is separated into two components, and the component $\mathcal{B}_0$ containing $\mathrm{Int}\, B \times \{0\}$ is a cobordism from that manifold to $\mathrm{Int}\, D^{n-3} \times \{1\}$. Let $\mathcal{A}_0$ be the other component. Then the candidate for orbit space data is given by $(S^{n-1}, \mathcal{A}, \mathcal{B}, \mathcal{F})$, where $\mathcal{A}$ and $\mathcal{B}$ are the closures of $\mathcal{A}_0$ and $\mathcal{B}_0$ respectively. Observe that $\mathcal{F}$ has the rational homology of S^{n-4}, while each of $\mathcal{A}, \mathcal{B}$, $\mathcal{A} \cap S^{n-1} \times \{0\}$, $\mathcal{A} \cap S^{n-1} \times \{1\}$, $\mathcal{B} \cap S^{n-1} \times \{0\}$, $\mathcal{B} \cap S^{n-1} \times \{1\}$ is rationally acyclic. Furthermore, all relevant normal bundles are trivial. It follows from the realization results (1.9) that $(S^{n-1} \times I, \mathcal{A}, \mathcal{B}, \mathcal{F})$ is the orbit space data of an S^1 action on some cobordism W with $\partial_- W = \Sigma$. Furthermore $\partial_+ W$ and W are both rational homology spheres.

The next step is to show that the orbit space data for $\partial_+ W$ bounds rationally acyclic orbit space data. We may construct cobounding data as follows: Take a Seifert hypersurface H^{n-2} in S^{n-1} that bounds $A \cup B$, and using an C^∞ function $\varphi: H \longrightarrow [0,1/2)$ with $\varphi^{-1}(0) = \partial H$, construct a proper embedding in $S^{n-1} \times (1/2,1]$ by sending x to $(x, 1-\varphi(x))$. Actually, <u>any</u> properly embedded coboundary H^{n-2} can be used for the subsequent discussion up to 3. (This is important). Let $A_1 = A \cup B - \mathrm{Int}\, h(D^{n-3}))$ and let $B_1 = h(D^{n-3})$. Let $\mathcal{F}' \subseteq (A \cup B) \times [0,2) = \partial H \times [0,2) \subseteq H$ be the embedded disk $\{(x,t) \mid x = h(y),\ t = \sqrt{1-|y|^2}\}$, let $\mathcal{B}' = \{(x,t) \mid x = h(y),$ $0 \leq t \leq \sqrt{1-y^2}\}$ and let $\mathcal{A}' = \mathrm{Closure}\,(H - \mathcal{B}')$.

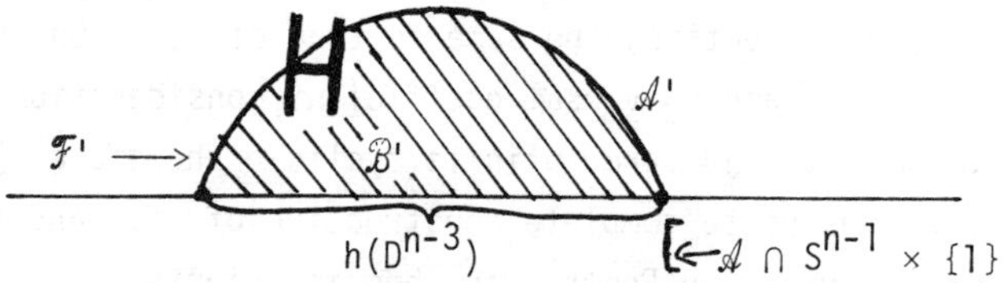

FIGURE 5

Then $(D^n, \mathcal{A}', \mathcal{B}', \mathcal{F}')$ give a candidate for orbit space data, and the triviality of all normal bundles guarantees that one can realize this data by an action, say on X.

We now need the following result, which is a variation of Kervaire's theorem on the triviality of even-dimensional knot cobordism [16]. A proof is given in the addendum to this section.

THEOREM 3.4. <u>Let</u> $M^k \subseteq S^{k+2}$ <u>be a smooth oriented rational homology sphere</u>; <u>assume</u> $k \geq 6$ <u>is even. Then</u> $M^k = \partial H^{k+1}$, <u>where</u> H^{k+1} <u>is an oriented and rationally acyclic submanifold of</u> D^{k+3} ■

Suppose that we use the choice of H given by 3.4 to construct X (recall that $n \equiv 3 \bmod 4$, so $n - 3$ is indeed even). Then the standard homological calculations imply that X is a rationally acyclic. If we glue together W and X, we get a rationally acyclic S^1-manifold with boundary Σ. If we can show that $W \cup X$ is a unitary manifold, then it will follow that the Brumfiel splitting invariant of Σ is zero and we shall be finished.

We do this in two steps. First we prove it for the associated semifree action with orbit space data $(D^n, \mathcal{F} \cup \mathcal{F}')$ and then we prove it for the action of primary interest by means of two branched coverings. Since the normal bundle of $\mathcal{F} \cup \mathcal{F}'$ is trivial, its equivariant normal in the associated semifree S^1 action is also trivial. This trivialization gives us a framing of $(\mathcal{F} \cup \mathcal{F}') \times S^3 \times (0,1)$. On the other hand, the framing that the orbit manifold $(\mathcal{F} \cup \mathcal{F}') \times S^2 \times (0,1)$ inherits as a submanifold of D^n and the fundamental vector field of the free S^1 action on $(\mathcal{F} \cup \mathcal{F}') \times S^3 \times (0,1)$ combine to yield a second framing on $\mathcal{F} \cup \mathcal{F}' \times S^3 \times (0,1)$. However, the argument of (3.2) implies that these framings are at least equivalent as unitary structures. This is enough for our purposes, for it guarantees that the ambient manifold Q^{n+2} for the semifree action is a unitary manifold.

The manifold $W \cup X$ is obtained from Q by taking two cyclic branched coverings. Therefore two applications of the principle given in (3.3) will guarantee that $W \cup X$ is also a unitary manifold. By what was said before, this concludes the proof of Theorem C ■

Finally, we have the following result:

PROPOSITION 3.5. Let $a > b \geq 1$, $k \geq 2$. Then $\hat{\Theta}_{4k-1}(V_a \oplus V_b \oplus (n-4)) \cap bP_{4k} = ab\, \Theta_{4k}\eta \cap bP_{4k}$.

NOTE. Results of [6] imply that this group is 0 if k is even and $\mathbb{Z}_2$ if k is odd (the latter is related to the stable stem identity $\mu_k \eta^2 = 2$-torsion in Image J).

PROOF. The first contains the second, and by (2.8) we know that any remaining elements must be classes in bP with S^1 actions for which $\Sigma/S^1 = S^{4k-2}$. But the proofs of Theorems B and C show that any such Σ has trivial Brumfiel splitting invariant. Since $\Sigma \in bP_{4k}$, this forces Σ to be a standard sphere ■

Addendum to Section 3: Even-dimensional knot cobordism with coefficients

In this addendum we shall prove Theorem 3 using the methods of Cappell and Shaneson (see [8, Cor. 6.5 and Thm. 6.6, pp. 314-315]).

Let φ_0 be the following diagram:

$$\begin{array}{ccc} \mathbb{Q}[\mathbb{Z}] & \xrightarrow{1} & \mathbb{Q}[\mathbb{Z}] \\ {\scriptstyle 1}\downarrow & & \downarrow{\scriptstyle \mathcal{F}_0} \\ \mathbb{Q}[\mathbb{Z}] & \xrightarrow{\mathcal{F}_0} & \mathbb{Q} \end{array}$$

$(\mathcal{F}_0$ = augmentation)

Cappell and Shaneson [8] define homology surgery obstruction groups $\Gamma_*(\mathcal{F}_0)$, $\Gamma_*(\varphi_0)$ as follows[1]: Elements of $\Gamma_{\text{even}}(\mathcal{F}_0)$ are given by suitable Hermitian forms over $\mathbb{Q}(\mathbb{Z})$ which become nondegenerate when one tensors with $\mathbb{Q}$ over the map $\mathcal{F}_0$, and a form represents zero if it admits a self-orthogonal submodule which becomes a subkernel when one tensors with $\mathbb{Q}$. Elements of $\Gamma_{\text{odd}}(\mathcal{F}_0)$ are essentially elements of $L_{\text{odd}}(\mathbb{Q})$ that have representatives of a special type with respect to the map $\mathcal{F}_0$; since $L_{\text{odd}}(\mathbb{Q}) = 0$ [24] we shall not elaborate. In both cases one has natural maps $L_*(\mathbb{Q}[\mathbb{Z}]) \longrightarrow \Gamma_*(\mathcal{F})$, $\Gamma_*(\mathcal{F}) \longrightarrow L_*(\mathbb{Q})$. These Γ groups represent the obstructions to performing rational homology surgery on a normal map into a closed manifold with fundamental group $\mathbb{Z}$ (i.e., the obstruction to finding a cobordant representative with $f_*: H_*(X;\mathbb{Q}) \longrightarrow H_*(M;\mathbb{Q}))$ an isomorphism. The groups $\Gamma_*(\varphi_0)$ are defined to be analogous obstruction groups for surgery problems involving manifolds with boundary.

The arguments of [8, Section 6] generalize with only minor changes to show that the knot cobordism groups for rational homology n-spheres in S^{n+2} are isomorphic to the groups $\Gamma_{n+3}(\varphi_0)$. We claim this group vanishes if n is even; the proof follows the discussion of [8, pp.314-5] very closely. There is an exact sequence

$$\Gamma_{n+3}(\mathcal{F}_0) \xrightarrow{i} \Gamma_{n+3}(\varphi_0) \xrightarrow{j} L_{n+2}(\mathbb{Q}\,[\mathbb{Z}]) \xrightarrow{k} \Gamma_{n+2}(\mathcal{F}_0).$$

Since n is even, we have $\Gamma_{n+3}(\mathcal{F}_0) \subseteq L_{n+3}(\mathbb{Q}) = 0$ and therefore j is monic. It suffices to prove that k is also monic. But by construction the composite of k with the forgetful map $\Gamma_{n+2}(\mathcal{F}_0) \longrightarrow L_{n+2}(\mathbb{Q})$ is just the usual coefficient map $L_{n+2}(\mathbb{Q}[\mathbb{Z}]) \longrightarrow L_{n+2}(\mathbb{Q})$. Since n is even, the latter map is an isomorphism by a result of Ranicki [25, Thm.4.2, p.155] (an algebraic analog of Shaneson's result for $L_*(\mathbb{Z}[\mathbb{Z}])$) and the vanishing of $L_{\text{odd}}(\mathbb{Q})$. This implies that k is a split monomorphism, and therefore Image $j = 0$; consequently, $\Gamma_{n+3}(\varphi_0) = 0$ as claimed ■

[1] Strictly speaking, they only do this when the domains involve integral group rings, but the generalization to rational coefficients is immediate.

REFERENCES

1. G. Bredon, A π_*-module structure for Θ_* and applications to transformation groups, Ann. of Math. 86(1967), 434-448.
2. ________, Exotic actions on spheres, Proc. Conf. on Transformation Groups (New Orleans, 1967), 47-76. Springer, New York, 1968.
3. ________, Introduction to Compact Transformation groups, Pure and Applied Mathematics Vol. 46. Academic Press, New York, 1972.
4. ________, Biaxial actions of classical groups, unpublished manuscript (166 pp.), Rutgers University, 1973.
5. W. Browder and F. Quinn, A surgery theory for G-manifolds and stratified sets, Manifolds - Tokyo 1973 (Conf. Proc.), 27-36. Univ. of Tokyo 1973 (Conf. Proc.), 27-36. Univ. of Tokyo Press, Tokyo, 1975.
6. G. Brumfiel, On the homotopy groups of BPL and PL/O-I, Ann. of Math. 88 (1968), 291-311.
7. ________, ibid. II, Topology 8 (1969), 39-46.
8. S. Cappell and J. Shaneson, The codimension two placement problem and homology equivalent manifolds, Ann. of Math. 99(1974), 277-348.
9. R. Fintushel, Locally smooth actions on homotopy 4-spheres, Duke Math. J. 73(1976), 63-70.
10. ________, Circle actions on simply connected 4-manifolds, Trans. Amer. Math. Soc. 230(1977), 147-171.
11. ________, Classification of circle actions on 4-manifolds, Trans. Amer. Math. Soc. 242(1978), 377-390.
12. ________ and P. Pao, Circle actions on spheres with codimension 4 fixed point set, Pac. J. Math., to appear.
13. W.-C. Hsiang and W.-Y. Hsiang, Differentiable actions of compact connected classical groups: I, Amer. J. Math. 89(1967), 705-786.
14. W.-Y. Hsiang, On the unknottedness of the fixed point set of differentiable circle group actions on sphreres - P. A. Smith conjecture, Bull. Amer. Math. Soc. 70(1964), 678-680.
15. V. J. Joseph, Smooth actions of the circle group on exotic spheres, Pac. J. Math. 95(1981), 323-336.
16. M. Kervaire, Les noeuds de dimension supérieure, Bull. Soc. Math. France 93(1965), 225-271.
17. ________ and J. Milnor, Groups of homotopy spheres, Ann. of Math. 77 (1963), 504-537.
18. H.-T. Ku and M.-C. Ku, Semifree differentiable actions of S^1 on homotopy (4k + 3)-spheres, Michigan Math. J. 15(1968), 471-476.
19. J. Levine, A classification of differentiable knots, Ann. of Math. 82 (1965), 15-50.
20. ________, Semifree circle actions on spheres, Invent. Math. 22(1973), 161-186.
21. J. Milnor and D. Husemoller, Symmetric Bilinear Forms, Ergeb. d. Math. u. I. Grenzgebiete Bd. 73. Springer, New York, 1972.
22. D. Montgomery and C.-T. Yang, Differentiable group actions on homotopy seven spheres. I, Trans. Amer. Math. Soc. 133(1967), 480-498.

23. P. Pao, Nonlinear circle actions on the 4-sphere and twisting spun knots, Topology 17(1978), 291-296.

24. W. Pardon, Local surgery and the exact sequence of a localization for Wall groups, Memoirs A.M.S. Vol. 12, Issue 2, No. 196(1977).

25. A. A. Ranicki, Algebraic L-theory: II. Laurent extensions, Proc. London Math. Soc. (3) 27(1973), 126-158.

26. R. Schultz, Circle actions on homotopy spheres bounding plumbing manifolds, Proc. Amer. Math. Soc. 36(1972), 297-300.

27. _________, Circle actions on homotopy spheres not bounding spin manifolds, Trans. Amer. Math. Soc. 213(1975), 89-98.

28. _________, Smooth actions of small groups on exotic spheres, Proc. A.M.S. Sympos. Pure Math. 32, Pt. 1(1978), 155-160.

29. _________, Differentiable group actions on homotopy spheres:
I. Differential structure and the knot invariant, Invent. Math. 31(1975), 105-128.
IV. Normal invariant formulas and applications, in preparation.

30. _________, Differentiability and the P. A. Smith theorems for spheres: I. Actions of prime order groups, Current Trends in Algebraic Topology, Proc. of Conf., London, Ont., 1981), C.M.S. Monograph Series, to appear.

31. E. V. Stein, On the orbit types in a circle action, Proc. Amer. Math. Soc. 66(1977), 143-147.

32. R. Lashof, The immersion approach to triangulation and smoothing, Proc. A.M.S. Sympos. Pure Math. 22(1971), 131-164.

DEPARTMENT OF MATHEMATICS
PURDUE UNIVERSITY
WEST LAFAYETTE, IN 47907

Contemporary Mathematics
Volume **19**, 1983

NOTE ON SYMBOLS IN K_n

Nobuo Shimada

In this note we shall introduce some kind of symbols in higher algebraic K-groups $K_n(A)$ of a commutative ring with unit, A, which appear as a natural generalization of the known symbols by Steinberg [M], Dennis-Stein [D-S], Loday [L] and Keune [K].

1. SPECIFIC POLYNOMIALS. Since each of these symbols corresponds its own polynomial with integral coefficients, we begin by generalizing such polynomials.

Put

$$f_{n,0} = 1 - x_1 \cdots x_n,$$

and

$$f_{n,k} = (1-x_k) - f_{n,k-1} \cdot x_{n+k} \quad (\text{for } n > k \geqq 1)$$

as an element of the polynomial ring $\mathbb{Z}[x_1,\dots,x_{n+k}]$. Note that all non-zero coefficients of monomials in $f_{n,k}$ are either +1 or -1, $f_{n,k}(1,\dots,1) = 0$ and $f_{n,k}(0,\dots,0) = 1$.

Recall that a noetherian ring A is called *regular*, if its localization A_m is a regular local ring for any maximal ideal m. The proof of the following lemma I owe to H. Matsumura.

LEMMA 1.1. *Let f be an irreducible polynomial in $\mathbb{Z}[x_1,\dots,x_n]$. Assume that f together with its derivatives $\partial_i f = \frac{\partial f}{\partial x_i}$ $(j=1,\dots,n)$ generate the unit ideal in $\mathbb{Z}[x_1,\dots,x_n]$. Then the quotient ring $A = \mathbb{Z}[x_1,\dots,x_n]/(f)$ of $\mathbb{Z}[x_1,\dots,x_n]$ by the principal ideal (f), generated by f, is regular.*

PROOF. Let P denote $\mathbb{Z}[x_1,\dots,x_n]$ and $\phi\colon P \to A$ the projection. Take a maximal ideal $m \subset A$ and put $M = \phi^{-1}(m)$. Then $A_m \cong P_M/(f)\cdot P_M$ and $A/m \cong P/M$ is a finite field. We know that P is regular and dim P_M =

1980 Mathematics Subject Classification. 18F55, 18G12.

$\dim_{P/M}(M/M^2) = n+1$ (dim P_M means the Krull dimension). Since $\dim A_m = \dim(P_M/(f)\cdot P_M) = n$, it is sufficient to see that $\dim_{A/m}(m/m^2) = n$.

Note that $f \notin M^2$, because if otherwise all $\partial_i f \in M$ and this contradicts the assumption of Lemma. Therefore, we have $\dim_{A/m}(m/m^2) = \dim_{P/M}(M/M^2) - \dim_{P/M}(((f)+M^2)/M^2) = n$.

LEMMA 1.2. *The following equalities hold*

$$f_{n,0} - x_n\cdot\partial_n f_{n,0} = 1, \quad f_{n,k} - x_k\cdot\partial_k f_{n,k} + x_n\cdot\partial_n f_{n,k} - x_{n+k}\cdot\partial_{n+k} f_{n,k} = 1,$$

for $n > k \geqq 1$.

As a corollary of the lemmas, we have

PROPOSITION 1.3. *The quotient ring* $Q_{n,k} = \mathbb{Z}[x_1,\ldots,x_{n+k}]/(f_{n,k})$ *are regular for* $n > k$.

2. LOCALIZATION EXACT SEQUENCES AND SYMBOLS. Let A be a regular ring, $a \in A$ an element of A such that $A/(a)$ is also regular. Following [L], we consider the localization long exact sequence of K-groups (Quillen [Q]):

$$(*) \qquad \cdots \to K_i(A/(a)) \xrightarrow{\iota} K_i(A) \xrightarrow{\eta} K_i(A[a^{-1}]) \xrightarrow{\partial} K_{i-1}(A/(a)) \to \cdots.$$

CASE I. Applying this long exact sequence (*) to the case $A = \mathbb{Z}[x_1,\ldots,x_{n+k}]$ and $a = f_{n,k}$, $n > k$, we get a canonical isomorphism

$$\text{(i)} \qquad \partial_I : \tilde{K}_i(\mathbb{Z}[x_1,\ldots,x_{n+k},(f_{n+k})^{-1}]) \tilde{\to} K_{i-1}(\mathbb{Z}[x_1,\ldots,x_{n+k}]/(f_{n,k})),$$

where $\tilde{K}_i$ denotes the kernel of the augmentation $\varepsilon : K_i(A[a^{-1}]) \to K_i(\mathbb{Z})$, defined by sending all x_i to 0.

CASE II. Assuming $n > k$, we take $A = \mathbb{Z}[x_1,\ldots,x_{n+k}]/(f_{n,k})$ and $a = \overline{f_{n,k-1}} \in A$, where $\overline{f_{n,k-1}}$ means the residue class of $f_{n,k-1}$. We have then $A[a^{-1}] \cong \mathbb{Z}[x_1,\ldots,x_{n+k-1},(f_{n,k-1})^{-1}]$ and $A/(a) \cong \mathbb{Z}[x_1,\ldots,\check{x}_k,\ldots,x_{n+k}]/(f_{n,k-1}(x_k=1))$. In this case the ring $A/(a)$ becomes isomorphic to $\mathbb{Z}[x_1,\ldots,x_{n+k-1}]/(f_{n-1,k-1})$, which is regular by the above lemma. Thus we can apply the localization exact sequence (*) to this case. The boundary homomorphism ∂_{II} is proved to be surjective (see Lemma 2.3. (2) below) and $\iota_{II} = 0$, and we have a canonical isomorphism

$$\text{(ii)} \quad \eta_{II} : K_i(\mathbb{Z}[x_1,\ldots,x_{n+k}]/(f_{n,k})) \tilde{\to} \tilde{\tilde{K}}_i(\mathbb{Z}[x_1,\ldots,x_{n+k-1},(f_{n,k-1})^{-1}]),$$

where $\tilde{\tilde{K}}_i$ denotes $\operatorname{Ker} \partial_{II}$.

THEOREM 2.1. *For any pair* (n,k), $n \geq 2$, $n > k \geq 0$, *there exist universal symbols*

$$\{x_1,\dots,x_{n+k}\}_k \in K_{n+k-1}(\mathbb{Z}[x_1,\dots,x_{n+k}]/(f_{n,k}))$$

and

$$\langle x_1,\dots,x_{n+k}\rangle_k \in K_{n+k}(\mathbb{Z}[x_1,\dots,x_{n+k},(f_{n,k})^{-1}])$$

which are generators of infinite cyclic direct summands of the respective K-groups and related with each other as follows:

$$\langle x_1,\dots,x_{n+k}\rangle_k = \partial_I^{-1}\{x_1,\dots,x_{n+k}\}_k$$

by the isomorphism (i), and

$$\{x_1,\dots,x_{n+k}\}_k = \eta_{II}^{-1}\langle x_1,\dots,x_{n+k-1}\rangle_{k-1} \quad (k \geq 1)$$

by the isomorphism (ii).

For a sequence of elements $a_1,\dots,a_{n+k}$ of a commutative ring A such that $f_{n,k}(a_1,\dots,a_{n+k}) = 0$ (resp. $f_{n,k}(a_1,\dots,a_{n+k}) \in A^\times$, the set of units in A), we define $\{a_1,\dots,a_{n+k}\}_k \in K_{n+k-1}(A)$ (resp. $\langle a_1,\dots,a_{n+k}\rangle_k \in K_{n+k}(A)$) as the natural image of the universal symbol $\{x_1,\dots,x_{n+k}\}_k$ (resp. $\langle x_1,\dots, x_{n+k}\rangle_k$).

THEOREM 2.2. *Denote by* $Q_{n,k}$ *the quotient ring* $\mathbb{Z}[x_1,\dots,x_{n+k}]/(f_{n,k})$ *and by* $L_{n,k}$ *the localized ring* $\mathbb{Z}[x_1,\dots,x_{n+k},f_{n+k}^{-1}]$.

(1) The graded ring $K_*(Q_{n,k}) = \sum_{i\geq 0} K_i(Q_{n,k})$ *for* $n > k$, *is a free* $K_*(\mathbb{Z})$*-module with basis* $1 \in K_0(Q_{n,k})$ *and*

$$\{x_1,\dots,x_k,x_{i_{k+1}},\dots,x_{i_s}x_{j_1}\cdots x_{j_{n-s}},x_{n+1},\dots,x_{n+k}\}_k \in K_{s+k}(Q_{n,k}),$$

where the suffices $(i_{k+1},\dots,i_s,j_1,\dots,j_{n-s})$ *run over such shuffles of* $(k+1,\dots,n)$ *that*

$$k < i_{k+1} < \dots < i_s < n \quad \text{and} \quad k < j_1 < \dots < j_{n-s} = n,$$

(2) The graded ring $K_*(L_{n,k})$, *for* $n > k$, *is a free* $K_*(\mathbb{Z})$*-module with basis* $1 \in K_0(L_{n,k})$, $\{f_{n,k}\} \in K_1(L_{n,k})$ *and*

$$\langle x_1,\dots,x_k,x_{i_{k+1}},\dots,x_{i_s},x_{j_1}\cdots x_{j_{n-s}},x_{n+1},\dots,x_{n+k}\rangle_k \in K_{s+k+1}(L_{n,k}),$$

where the suffices $(i_{k+1},\dots,i_s,j_1,\dots,j_{n-s})$ *run over the same as above.*

PROOF OF THEOREMS 2.1 AND 2.2. Recall that [Q,L]

$$K_*(Q_{n,0}) = \bigoplus_{0<i_1<\cdots<i_s<n} K_*(\mathbb{Z})\cdot\{x_{i_1},\ldots,x_{i_s}\} \oplus K_*(\mathbb{Z})\cdot 1,$$

where $\{x_{i_1},\ldots,x_{i_s}\}$ means the Steinberg symbol and it is denoted by $\{x_{i_1},\ldots,x_{i_s},x_{j_1}\cdots x_{j_{n-s}}\}_0$ in Theorem 2.2 (1). Then the theorems can be proved by induction on k as follows.

Assume that the theorems hold for $K_*(Q_{n,j})$ and $K_*(L_{n,j-1})$ for $0 \leqq j < k < n$. By the isomorphism (i), we can determine $K_*(L_{n,k-1})$ as described in the theorems. To determine $K_*(Q_{n,k})$ we need the following

LEMMA 2.3. *(1) Denote by* $\tilde{\tilde{K}}_*(L_{n,k-1})$ *the kernel of the homomorphism* $\partial_{II}: K_*(L_{n,k-1}) \to K_{*-1}(Q_{n-1,k-1})$ *in the localization exact sequence (*). Then we have*

$$\tilde{\tilde{K}}_*(L_{n,k-1}) = \mathrm{Ker}\ \rho_k \oplus K_*(\mathbb{Z})\cdot 1,$$

where $\rho_k : K_*(L_{n,k-1}) \to K_*(L_{n-1,k-1})$ *is the natural* $K_*(\mathbb{Z})$*-module map induced by the correspondence* $x_i \longmapsto x_i$ $(i < k)$, $x_k \longmapsto 1$ *and* $x_i \to x_{i-1}$ $(i > k)$,

(2) the homomorphism ∂_{II} *is onto,*

(3) $\mathrm{Ker}\ \rho_k$ *is spanned over* $K_*(\mathbb{Z})$ *by such basis elements* $\langle x_1,\ldots,x_{k-1},x_{i_k},\ldots,x_{i_s},x_{j_1}\cdots x_{j_{n-s}},x_{n+1},\ldots,x_{n+k-1}\rangle_{k-1}$ *of* $K_{s+k}(L_{n,k-1})$ *that* $i_k = k$.

PROOF OF LEMMA 2.3. The first assertion (1) follows easily from the following commutative diagram

$$\begin{array}{ccc}
K_*(L_{n,k-1}) & \xrightarrow{\partial_I} & K_{*-1}(Q_{n,k-1}) \\
\rho_k \downarrow & \searrow^{\partial_{II}} & \downarrow \rho_k \\
K_*(L_{n-1,k-1}) & \xrightarrow{\partial_I} & K_{*-1}(Q_{n-1,k-1}).
\end{array}$$

To prove (2) and (3) it saffices to see

$$\rho_k\langle x_1,\ldots,x_{k-1},x_{i_k},\ldots,x_{i_s},x_{j_1}\cdots x_{j_{n-s}},x_{n+1},\ldots,x_{n+k-1}\rangle_{k-1}$$

$$= \begin{cases} \text{one of the basis element} & (i_k \neq k) \\ 0 & (i_k = k). \end{cases}$$

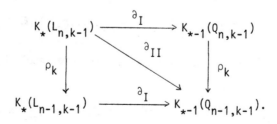

The case $i_k \neq k$ is obvious, because then x_k must be one of $x_{j_1},\ldots,x_{j_{n-s-1}}$. Thus ρ_k and therefore $\partial_{II} = \rho_k \circ \partial_I$ is onto. The case $i_k = k$ is verified as follows:

Consider the following ladder of commutative squares

$$\begin{array}{ccccccccc} \hat{K}_{s+k}(L_{n,k-1}) & \xrightarrow{\partial_I} & \tilde{K}_{s+k-1}(Q_{n,k-1}) & \xrightarrow{\eta_{II}} & \hat{K}_{s+k-1}(L_{n,k-2}) & \xrightarrow{\partial_I} & \cdots \to & \tilde{K}_s(Q_{n,0}) \\ \downarrow \rho_k & & \downarrow \rho_k & & \downarrow \rho_k & & & \downarrow \rho_k \\ \hat{K}_{s+k}(L_{n-1,k-1}) & \xrightarrow{\partial_I} & \tilde{K}_{s+k-1}(Q_{n-1,k-1}) & \xrightarrow{\eta_{II}} & \hat{K}_{s+k-1}(L_{n-1,k-2}) & \xrightarrow{\partial_I} & \cdots \to & \tilde{K}_s(Q_{n-1,0}) \end{array} \tag{2.4}$$

where $\hat{K}_*(L_{n,j}) = K_*(L_{n,j})/(K_*(\mathbb{Z})\cdot\{f_{n,j}\} \oplus K_*(\mathbb{Z})\cdot 1)$ and $\tilde{K}_*(Q_{n,j})$ means the kernel of the augmentation $\varepsilon\colon K_*(Q_{n,j}) \to K_*(\mathbb{Z})$ induced by the correspondence $x_i \longmapsto 1$ $(1 \leqq i \leqq n+j)$.

We note that each horizontal homomorphism is monomorphic. The element

$$\langle x_1,\ldots,x_{k-1},x_k,x_{i_{k+1}},\ldots,x_{i_s},x_{j_1}\cdots x_{j_{n-s}},x_{n+1},\ldots,x_{n+k-1}\rangle_{k-1} \in K_{s+k}(L_{n,k-1})$$

is mapped to

$$\{x_1,\ldots,x_k,x_{i_{k+1}},\ldots,x_{i_s},x_{j_1}\cdots x_{i_{n-s}}\}_0 \in K_s(Q_{n,0})$$

by the horizontal sequence of maps, and the latter maps to zero by ρ_k. Thus we have proved the case $i_k = k$.

Now $K_*(Q_{n,k})$ is determined by this lemma and the isomorphism (ii), so that the induction step and therefore the proof of Theorems 2.1 and 2.2 are completed.

REMARK. Note that $\langle x_1,\ldots,x_n\rangle_0 = (-1)^{n-1}\langle(-1)^n x_1,\ldots,(-1)^n x_{n-1},x_n\rangle$ by the Loday symbol [L] and $\{x_1,x_2,x_3\}_1 = \langle x_1,x_2,x_3\rangle_*$ by the Keune symbol [K].

3. PROPERTIES.

We shall list some of properties of the symbols.

PROPOSITION 3.1. *Let A be a commutative ring with unit. The following relations hold in $K_*(A)$ (provided that the symbols are defined):*

(1) $\{a_1,\ldots,a_{n+k}\}_k$ (resp. $\langle a_1,\ldots,a_{n+k}\rangle_k$) are skew-symmetric with respect to $a_{k+1},\ldots,a_{n-1}$ $(n > 2)$,

(2) $\{a_1,\ldots,a_{n+k}\}_k = 0$ (resp. $\langle a_1,\ldots,a_{n+k}\rangle_k = 0$) if $a_i = 1$ for some i, $k+1 \leqq i \leqq n-1$ $(n > 2)$,

(3) $\{a_1,a_2,\ldots,a_{n+1}\}_1 + \{a_1,a_2',a_3,\ldots,a_n,a_{n+1}'\}_1 = \{a_1,a_2'',a_3,\ldots,a_n,a_{n+1}''\}_1$

(resp. $\langle a_1,a_2,\ldots,a_{n+1}\rangle_1 + \langle a_1,a_2',a_3,\ldots,a_n,a_{n+1}'\rangle_1 = \langle a_1,a_2'',a_3,\ldots,a_n,a_{n+1}''\rangle_1)$ *if* $a_2'' = a_2 + a_2' - a_1a_2a_2'a_3\cdots a_n$, $a_{n+1} = (1-a_1a_2'a_3\cdots a_n)a_{n+1}''$ *and* $a_{n+1}' = (1-a_1a_2\cdots a_n)a_{n+1}''$ $(n \geqq 2)$

(4) $\{a_{n+1},a_2,\ldots,a_n,a_1\}_1 = -\{a_1,\ldots,a_{n+1}\}_1$ *(resp.* $\langle a_{n+1},a_2,\ldots,a_n,a_1\rangle_1 = -\langle a_1,\ldots,a_{n+1}\rangle_1)$ $(n \geqq 2)$.

PROOF. Considering the universal models and the diagram (2.4) with ρ_i $(k \leqq i \leqq n-1)$ in the place of ρ_k, we can reduce the relations (1), (2) and (3) to the corresponding known relations among the Steinberg and the Loday symbols ([L]).

To prove (4), first note that, in the universal model $K_{n+1}(L_{n,1})$, the relation

$$\langle x_{n+1},x_2,\ldots,x_n,x_1\rangle_1 = -\langle x_1,\ldots,x_{n+1}\rangle_1$$

is a direct consequence of the relation in $K_n(Q_{n,1})$:

$$\{x_{n+1},x_2,\ldots,x_n,x_1\}_1 = -\{x_1,\ldots,x_{n+1}\}_1$$

through the isomorphism (i) in §2, which we shall prove in the sequel. We consider an automorphism τ of $K_n(Q_{n,1})$ induced by the exchange of x_1 and x_{n+1}. We characterize the direct summand $\mathbb{Z}\{x_1,\ldots,x_{n+1}\}_1$ in $K_n(Q_{n,1})$ as follows:

$$\mathbb{Z}\cdot\{x_1,\ldots,x_{n+1}\}_1 = \begin{cases} \text{Ker } \varepsilon = \text{the augmentation kernel} & (n = 2) \\ \bigcap_{i=2}^{n-1} \text{Ker } \rho_i & (n > 2). \end{cases}$$

Thus the infinite cyclic subgroup $\mathbb{Z}\cdot\{x_1,\ldots,x_{n+1}\}_1$ is invariant under the automorphism τ, and hence $\tau = \pm$ *identity* on this group:

$$\tau\{x_1,\ldots,x_{n+1}\}_1 = \{x_{n+1},x_2,\ldots,x_n,x_1\}_1 = \pm\{x_1,\ldots,x_{n+1}\}_1.$$

To determin τ, we utilize the calculation of the Hodge-Chern classes (See Proposition 3.3. (3) below), and have only to verify

$$c_n\{x_{n+1},x_2,\ldots,x_n,x_1\}_1 = -c_n\{x_1,\ldots,x_{n+1}\}_1 \neq 0.$$

PROPOSITION 3.2. *Let* A *be a commutative ring with unit and* $A^\times$ *the set of units in* A*. The following equalities hold in* $K_*(A)$*:*

(1) $\{a_1,\ldots,a_{n+k}\}_k = \langle a_1,\ldots,a_{n+k-1}\rangle_{k-1}$ *if* $f_{n+k}(a) = 0$ *and* $f_{n,k-1}(a) \in A^\times$,

(2) $\langle a_1,\dots,a_{n+k}\rangle_k = \langle a_1,\dots,a_{n+k-1}\rangle_{k-1}\cdot\{f_{n,k}(a)\}$ *if* $f_{n,k}(a)\in A^{\times}$ *and* $f_{n,k-1}(a)\in A^{\times}$, *for* $n > k \geq 1$.

PROOF. Since $Q_{n,k}[f_{n,k-1}^{-1}] \cong L_{n,k-1}$ in the universal models, (1) follows immediately from the definition.

To prove (2), consider the following commutative diagram:

$$\begin{array}{ccc} K_{n+k}(L_{n,k}) & \xrightarrow{\partial_I} & K_{n+k-1}(Q_{n,k}) \\ \downarrow{\scriptstyle\eta} & & \downarrow{\scriptstyle\eta} \\ K_{n+k}(L_{n,k}[f_{n,k-1}^{-1}]) & \xrightarrow{\partial_I} & K_{n+k-1}(L_{n,k-1}). \end{array}$$

Then we have

$$\partial_I\eta\langle x_1,\dots,x_{n+k}\rangle_k = \langle x_1,\dots,x_{n+k-1}\rangle_{k-1}.$$

On the other hand, from the localization theorem, we have the following commutative diagram:

$$\begin{array}{ccccccc} & & 0 & & 0 & & \\ & & \downarrow & & \downarrow & & \\ & & \tilde{K}_*(L_{n,k}) & \xrightarrow[\cong]{\partial_1} & K_{*-1}(Q_{n,k}) & & \\ & & \downarrow{\scriptstyle\eta_2} & & \downarrow{\scriptstyle\eta_2} & & \\ 0 \to \tilde{K}_*(L_{n,k-1}[x_{n+k}]) & \xrightarrow{\eta_1} & \tilde{K}_*(L_{n,k}[f_{n,k-1}^{-1}]) & \xrightarrow{\partial_1} & K_{*-1}(L_{n,k-1}) & \to & 0 \\ \cong\downarrow{\scriptstyle\partial_2} & & \downarrow{\scriptstyle\partial_2} & & \downarrow{\scriptstyle\partial_2} & & \\ 0 \to K_{*-1}(Q_{n,k-1}[x_{n+k}]) & \xrightarrow{\eta_1} & K_{*-1}(Q_{n,k-1}[x_{n+k},(1-x_k)^{-1}]) & \xrightarrow{\partial_1} & K_{*-2}(Q_{n-1,k-1}) & \to & 0 \\ & & \downarrow & & \downarrow & & \\ & & 0 & & 0 & & \end{array}$$

By the fundamental theorem, (cf. [Q], [L^2]), the horizontal short exact sequences split. Thus we have

$$\tilde{K}_*(L_{n,k}[f_{n,k-1}^{-1}]) = \tilde{K}_*(L_{n,k-1}[x_{n+k}])\oplus K_{*-1}(L_{n,k-1})\cdot\{f_{n,k}\},$$

$$K_{*-1}(Q_{n,k-1}[x_{n+k},(1-x_k)^{-1}]) = K_{*-1}(Q_{n,k-1}[x_{n+k}])\oplus K_{*-2}(Q_{n-1,k-1})\cdot\{1-x_k\}.$$

In this direct sum decomposition, we have

$$\eta_2\langle x_1,\dots,x_{n+k}\rangle_k = \eta_1(\alpha)\oplus\beta,\quad \alpha\in K_{n+k}(L_{n,k-1}[x_{n+k}]),$$

$$\beta\in K_{n+k-1}(L_{n,k-1})\cdot\{f_{n,k}\}.$$

Applying ∂_2 to this, we have

$$0 = \partial_2\eta_2\langle x_1,\ldots,x_{n+k}\rangle_k = \partial_2\eta_1\alpha = \eta_1\partial_2(\alpha),$$

but $\eta_1\partial_2$ is injective, so we obtain $\alpha = 0$. On the other hand, we know that $\partial_1\eta_2\langle x_1,\ldots,x_{n+k}\rangle_k = \partial_1\beta = \langle x_1,\ldots,x_{n+k-1}\rangle_{k-1}$. Thus $\beta = \langle x_1,\ldots,x_{n+k}\rangle_{k-1}\times\{f_{n,k}\}$ and we have proved (2).

We shall calculate the Hodge-Chern classes $c_n : K_n(A) \to \Omega^n_{A/\mathbb{Z}}$ for the symbols, where $\Omega^n_{A/\mathbb{Z}} = \Lambda^n\Omega^1_{A/\mathbb{Z}}$, the exterior n-th power of the A-module of absolute Kähler differentials $\Omega^1_{A/\mathbb{Z}}$ (See [Ge], [Gr], [B].).

PROPOSITION 3.3. *The values of the Hodge-Chern classes on symbols are as follows:*

(1) $c_{n-1}\{a_1,\ldots,a_n\}_0 = (-1)^n(n-2)!\,a_n da_1\wedge\cdots\wedge da_{n-1} \quad (n \geqq 2),$

(2) $c_{n+k}\langle a_1,\ldots,a_{n+k}\rangle_k = (-1)^n(n+k-1)!\,f^{-1}_{n,k}(a)da_1\wedge\cdots\wedge da_{n+k} \quad (n > k \geqq 0),$

(3) $c_{n+k-1}\{a_1,\ldots,a_{n+k}\}_k = (-1)^n(n+k-2)!\,(a_{n+k}da_1\wedge\cdots\wedge da_{n+k-1}$
$+(-1)^n a_k da_1\wedge\cdots\wedge \check{da_k}\wedge\cdots\wedge da_{n+k}$
$+(-1)^{k-1}a_n da_1\wedge\cdots\wedge \check{da_n}\wedge\cdots\wedge da_{n+k})$
$(n > k \geqq 1).$

PROOF. (1) is well-known [B,L]. For (2), we use the formula for product (Theorem 2.3 in §3, Chap. I of [B]):

$$c_{m+n}(a\cdot b) = -\frac{(m+n-1)!}{(m-1)!(n-1)!}\,c_m(a)\wedge c_n(b)$$

for $a \in K_m(A)$, $b \in K_n(A)$, and apply this formula, in the case of the above Proposition 3.2. (2), to the universal model, that is,

$$\langle x_1,\ldots,x_{n+k-1}\rangle_{k-1}\cdot\{f_{n,k}\} \quad \text{in} \quad K_{n+k}(L_{n,k}[f^{-1}_{n,k-1}]).$$

Then, using the induction on k and the identification $\Omega^*_{A_S/\mathbb{Z}} = A_S \otimes_A \Omega^*_{A/\mathbb{Z}}$ for localization, we obtain (2).

For (3), similarly we consider the localized situation $Q_{n,k}[f^{-1}_{n,k-1}] \cong L_{n,k-1}$ (the above Proposition 3.2. (1)). In this case, we need to pull back

$$c_{n+k-1}\langle x_1,\ldots,x_{n+k-1}\rangle_{k-1} = (-1)^n(n+k-2)!\,f_{n,k-1}dx_1\wedge\cdots\wedge dx_{n+k-1} \quad (\text{in } \Omega^*_{L_{n,k-1}/\mathbb{Z}})$$

into $\Omega^*_{Q_{n,k}/\mathbb{Z}}$. Then we seek such an element ω of $\Omega^{n+k-1}_{Q_{n,k}/\mathbb{Z}}$ that

$$f_{n,k-1}\cdot\omega = dx_1 \wedge\cdots\wedge dx_{n+k-1}$$

and

$$\rho_i(\omega) = 0 \quad \text{for} \quad k+1 \leqq i \leqq n-1,$$

so as to be compatible with the Proposition 3.1. (2), where

$$\rho_i : \Omega^*_{Q_{n,k}/\mathbb{Z}} \to \Omega^*_{Q_{n-1,k}/\mathbb{Z}}$$

is the homomorphism induced from that of $Q_{n,k}$ to $Q_{n-1,k}$ by the mapping $x_j \longmapsto x_j$ $(j < i)$, $x_i \longmapsto 1$ and $x_{j+1} \longmapsto x_j$ $(j > i)$.

Now using the relation $f_{n,k} = 0$ in $Q_{n,k}$ and its derivation in $\Omega^1_{Q_{n,k}/\mathbb{Z}}$, we can choose ω uniquely under the above condition:

$$\omega = x_{n+k}dx_1 \wedge\cdots\wedge dx_{n+k-1} + (-1)^n x_k dx_1 \wedge\cdots\wedge \check{dx}_k \wedge\cdots\wedge dx_{n+k}$$
$$+ (-1)^{k-1} x_n dx_1 \wedge\cdots\wedge \check{dx}_n \wedge\cdots\wedge dx_{n+k}.$$

Thus we have proved (3).

4. EXAMPLES. (1) For example, we have a non-zero element

$$\{x_1,x_2,\ldots,x_n,1\}_1 \quad \text{in} \quad K_n(\mathbb{Z}[x_1,\ldots,x_n]/(x_1\cdot(1-x_2\cdots x_n)),$$

which is of infinite order, with the Chern class

$$c_n\{x_1,\ldots,x_n,1\} = (-1)^n(n-1)!dx_1\wedge\cdots\wedge dx_n.$$

Similarly, the element

$$\{x_1,\ldots,x_{k-1},1,x_k,\ldots,x_{2k}\}_k \in K_{2k}(\mathbb{Z}[x_1,\ldots,x_{2k}]/(x_{2k}\cdot f_{k,k-1}))$$

is non-zero and of infinite order.

(2) $\langle -1,-2,-3\rangle_1 = \langle -1,-2\rangle_0\cdot\{-1\} = \{-1,-1,-1\}$ in $K_3(\mathbb{Z})$,

and I could find no interesting example of symbol in $K_*(\mathbb{Z})$ that would not be reduced to the Steinberg symbol. On the other hand, in $K_3(\mathbb{Z}[\sqrt{6}])$, the element $\langle 2+\sqrt{6},\ 2+2\sqrt{6},\ -2+\sqrt{6}\rangle_1$ seems to be non-trivial, but I have not succeeded in proving it.

BIBLIOGRAPHY

[B]. S. Bloch, "Algebraic K-theory and the christalline cohomology", Publ. I.H.E.S. 47 (1977), 187-268.

[D-S]. R. K. Dennis and M. R. Stein, "K_2 of discrete valuation rings", Adv. in Math. 18 (1975), 182-238.

[Ge]. S. Gersten, "Higher K-theory of rings in Algebraic K-theory I", Lect. Notes in Math. 341, Springer-Verl. (1973), 211-243.

[Gr]. A. Grothendieck, Classes de Chern et représentations linéaires des groupes discrets, in Dix exposés sur le cohomologie des schémas, North Holland, 1968.

[K]. F. Keune, "Another presentation for the K_2 of a local domain", J. of Pure and Appl. Alg. 22 (1981), 131-141.

[L]. J.-L. Loday, "Symboles en K-théorie algébrique supérieure", C. R. Acad. Sc. Paris, 292 (1981), 863-866.

[L^2]. J.-L. Loday, "K-théorie algébrique et représentations de groupes", Ann. scient. Éc. Norm. Sup., t.9 (1976), 309-377.

[M]. J. Milnor, Introduction to Algebraic K-theory, Ann. of Math. Studies 72, Princeton U. Press, 1971.

[Q]. D. Quillen, "Higher algebraic K-theory, I", in Lect. Notes in Math. 341, Springer-Verl. 1973, 85-147.

RESEARCH INSTITUTE FOR MATHEMATICAL SCIENCES
KYOTO UNIVERSITY
SAKYO-KU, KYOTO 606, JAPAN

Contemporary Mathematics
Volume 19, 1983

THE COHOMOLOGY OF SUBGROUPS OF $GL_n(F_q)$

M.Tezuka and N.Yagita

§1. Introduction. Let $G=GL_n(k)$ be the finite group of invertible $n\times n$-matrices over a finite field k of p^d elements. We consider the cohomology ring $H^*(G;Z/\ell Z)$ where ℓ is a prime number. When ℓ is prime to p, Quillen determined it, namely, it is a tensor product of a polynomial algebra and an exterior algebra (except for $\ell=2$) [10]. However if $\ell=p$, the situation seems quite different. For example if n=3 $p\geq 3$ d=1, then $H^*(G)_{(p)}$ has p+2 generators, modulo nilpotent elements, but its Krull dimension is 2 [12]. It seems difficult to characterize such a ring by generators and relations.

For ease of arguments, in this paper we always treat $H^*(G;k)$ and simply write $H^*(G)=H^*(G;k)$. We consider the restriction maps of typical subgroups

$$H^*(G)\xrightarrow[i_1^*]{} H^*(B)\xrightarrow[i_2^*]{} H^*(U)\xrightarrow[j^*]{} H^*(\langle I+E_{ij}\rangle).$$

Here B is the Borel subgroup, U is the maximal unipotent subgroup which is a p-Sylow subgroup of G, and I is the identity matrix and E_{ij} is the elementary matrix with 1 in (i,j)-entry. By using Evens' norm [4], we define an element v_{ij} in $H^*(U)$ such that it generates Image j^* modulo nilpotent elements. The Image i_2^* is an invariant set under the action of diagonal matices T and Imagei_1^* is a set of stable elements under the Weyl group S_n actions. We study what polynomials of v_{ij} are in Image i_2^* and Image$(i_1i_2)^*$.

As an example, in the last section we compute $H^*(GL_4(F_2))$ and we can see how the cohomology theory and the invariant theory relate.

We gratefully thank Leonard Evens and Stewart Priddy who both give us many valuable suggestions and corrected our errors in the earlier versions. We also thank Nobuo Shimada and Tokushi Nakamura for many conversations and suggestions.

1980 Mathematic Subject Classification. 18H10.

§2. The Evens norm and the cohomology of the coefficient ring k.

We recall the Evens' norm, which is a map, for given subgroup K' of a finite group K with $\ell = [K;K']$,

$$N[K' \subset K] : H^*(K';A) \to H^{\ell *}(K;A) \qquad (* = \text{even if } ch(A) \neq 2)$$

defined by $N[K' \subset K](x) = \phi^*(1 \int x^\ell)$ where $\phi : K \subset K' \int S_\ell$ is the natural embedding and S_ℓ is the symmetric group. (For details see [4].) This operation satisfies the following.

2.1. Transitivity ; If $G \supset K \supset H$,

$$N[H \subset G](y) = N[K \subset G](N[H \subset K](y)).$$

2.2. Naturality ; If $f : G' \to G$ is a homomorphism with $[G';f^{-1}(H)] = [G;H]$,

$$f^*(N[H \subset G](y)) = N[f^{-1}(H) \subset G'](f_0^* y) \text{ where } f_0 = f|f^{-1}(H)$$

2.3. Multiplicative property ;

$$N[H \subset G](yy') = N[H \subset G](y) \cdot N[H \subset G](y').$$

2.4. Double coset formula ;

$$N[G' \subset G](y)|K = \prod_{G'gK} N[K \cap g^{-1}G'g \subset K](i_g^* y)$$

where $x|K = res_{K \subset G}(x)$ and g runs over a system of representatives for (G',G) double cosets, and i_g is the homomorphism $K \cap g^{-1}G'g$ to G' given by $i_g(x) = gxg^{-1}$.

Let $k = F_q$ be the finite field of elements $q = p^d$. We always consider the cohomology theory with the coefficient k. Hereafter we put $H^*(X) = H^*(X;k)$.

By Galois theory $Gal(k/F_p)$ is generated by the Frobenius operation $x^i : \lambda \mapsto \lambda^{p^i}$ for $\lambda \in k$ and

$$H^1(k) \simeq Hom(k;k) \simeq kx^0 \oplus kx^1 \oplus \ldots \oplus kx^{d-1}.$$

Then it is easily seen

$$(2.5) \quad H^*(k) = \begin{cases} k[y(0),\ldots,y(d-1)], & y(b) = x^b \text{ if } p=2 \\ k[y(0),\ldots,y(d-1)] \otimes \Lambda(x^0,\ldots,x^{d-1}), & y(b) = \beta x^b \text{ if } p \geq 3 \end{cases}$$

where the Bockstein β is defined by the exact sequence

$$0 \longrightarrow R_{\wp}/\pi \xrightarrow{\pi} R_{\wp}/\pi^2 \longrightarrow R_{\wp}/\pi \longrightarrow 0$$

Here R is the integral closure of Z in $Q(\sqrt[p^n-1]{1})$ and $\wp$ is the prime in R over p and $\mathfrak{P} = (\pi)$ in $R_{\wp}$.

Let $\xi : k \to k$ be the multiplication by ξ. This induces the automorphism of $H^*(k)$.

(2.6) $\xi^* x^j = \xi^{p^j} x^j$ and $\xi^* y(j) = \xi^{p^j} y(j)$.

Let $V = \bigoplus_{i=1}^{m} ke_i$ be a k-vector space with basis $\{e_i\}$. By the Künneth theorem

$$H^*(V) \simeq \begin{cases} \bigotimes_{i,b} k[y_i(b)] & \text{if } p=2 \\ \bigotimes_{i,b} k[y_i(b)] \otimes \Lambda(x_i^b) & \text{if } p \geq 3 \end{cases}$$

where x_i^b is the Frobenius map induced by the dual to e_i.

Theorem 2.7. <u>In the above notations</u>

$$N[ke_m \subsetneq V](y_m(b)) = \prod_{\lambda_i \in k} (y_m(b) + \lambda_1 y_1(b) + \ldots + \lambda_{m-1} y_{m-1}(b))$$

<u>modulo</u> <u>Ideal</u>$(x_i^j, y_i(h) \mid h \neq b)$. (2.8)

Proof. We prove only the case b=0. The other cases are also proved by the Frobenius operation. First we consider the restriction to $\oplus^{m-1} ke_i = V_{m-1}$.

$$\begin{aligned} N|V_{m-1} &= N[ke_m \subsetneq V](y_m)|V_{m-1} \\ &= \prod N[V_{m-1} \cap (-g + ke_m + g) \subsetneq V_{m-1}](i_g^* y_m) \\ &= N[0 \subsetneq V_{m-1}](y_m|0) = 0. \end{aligned}$$

This means

(2.9) $N \in \operatorname{Ker}(i^* : H^*(V) \to H^*(V_{m-1})) = \operatorname{Ideal}(x_m^h, y_m(h))$.

Next consider the map

$$f : \begin{cases} e_m \longmapsto e_m \\ e_i \longmapsto e_i + \lambda_i e_m, & 1 \leq i \leq m-1 \end{cases}$$

which induces $f^* y_i = y_i$ and

(2.10) $f^* y_m = \lambda_1 y_1 + \ldots + \lambda_{m-1} y_{m-1} + y_m$.

The norm N is invariant under this map

$$f^* N = N[f^{-1}(ke_m) \subsetneq V](f^* y_m) = N,$$

since $f^{-1}(ke_m) = ke_m$ and $f^* y_m = y_m$ in $H^*(ke_m)$. From (2.9), N contains as a factor y_m modulo (2.8) and so contains (2.10) modulo (2.8). Therefore

(2.11) $N \in \operatorname{Ideal}(\prod (y_m + \lambda_1 y_1 + \ldots + \lambda_{m-1} y_{m-1}), (2.8))$.

Lastly we consider the restiction map to ke_m.

$$N|ke_m = \prod_{ke_m+g+ke_m} N[ke_m\cap(-g+ke_m+g)\hookrightarrow ke_m](i_g^* y_m)$$
$$= \prod_{g\in V_{m-1}} N[ke_m\hookrightarrow ke_m](y_m) = y_m^{p(m-1)d}.$$

Hence from (2.11), we have the theorem. q.e.d.

Remark 2.12. When d=1 (Evens, Proposition 5 in [5]) It is not necessary to mod out by (2.8).

§3. The p-Sylow subgroup U.

Let $G=GL_n(k)$ and $U=U_n$ be the unipotent subgroup generated by upper triangular matrices with diagonal elements=1.

Since
$$|G|=(q^n-1)(q^n-q)\dots(q^n-q^{n-1}) \quad \text{and}$$
$$|U|= q^{n(n-1)/2},$$
U is a p-Sylow subgroup of G. Let $x_{ij}(\lambda) = I+\lambda E_{ij}$ for $\lambda\in k$ and d_δ be the diagonal matrix with 1 except for δ in 1-1 entry. Then it is well known

(3.1) $SL_n(k) = \langle x_{ij}(\lambda) \mid i\neq j\rangle$
$GL_n(k) = \langle x_{ij}(\lambda)\ \ d_\delta \mid i\neq j\rangle$

(3.2) $U_n = \langle x_{ij}(\lambda) \mid i<j\rangle$ and relations are given by
$$x_{ij}(\alpha)x_{ij}(\beta) = x_{ij}(\alpha+\beta)$$
$$[x_{ij}(\alpha),\ x_{km}(\beta)] = \begin{cases} x_{im}(\alpha\beta) & \text{if } j=k\\ 1 & \text{if } j\neq k.\end{cases}$$

Let I_{1n} be the upper row subgroup, i.e., the subgroup of U generated by $x_{1i}(\alpha)$, $2\leq i\leq n$. There is an additive isomorphism
$$I_{1n} = \langle x_{12}(\alpha),\dots,x_{1n}(\alpha)\rangle \simeq \oplus^{n-1}k \simeq (Z/pZ)^{(n-1)d}.$$
Consider the restriction map
$$\tilde{i}^*: H^*(U) \longrightarrow H^*(I_{1n})/(\sqrt{0},\ y_{ij}(b)\mid b\geq 1)$$
$$\cong k[y_{12},\dots,y_{1n}]$$
where $\sqrt{0}$ is the ideal of nilpotent elements ($\sqrt{0} = (x_{1i}{}^b)$) and
$$y_{1j} = y_{1j}(0) = \begin{cases} x_{1i}^{\#} & \text{if } p=2\\ \beta x_{1i}^{\#} & \text{if } p\geq 3\end{cases}$$
where $x_{1i}^{\#}$ is the Frobenius operation induced by the dual of x_{1i}.

Imbed U_{n-1} as the subgroup $\langle x_{ij}\mid i\neq 1\rangle$ of U_n Given $A\in U_{n-1}$ and $X\in I_{1n}$, the fact $AXA^{-1}\in I_{1n}$ induces the automorphism A^* of $H^*(I_{1n})$

by the conjugation by A. Let $(a_{ij})=A$. Then

$$A^* x_{1j}^{\#}(x_{1i}) = x_{1j}^{\#}(A x_{1i} A^{-1}) = x_{1j}^{\#}(\sum a_{ik}x_{ik}) = a_{ij}.$$

Hence $A^* x_{1j}^{\#} = \sum a_{ij} x_{1i}^{\#}$ and so $A^* y_{1j} = \sum a'_{ij} y_{1i}$, this shows

$$A^*(y_{12},\dots,y_{1n}) = (y_{12},\dots,y_{1n})A \quad .$$

Since inner automorphisms induce the identity, we see

$$\text{Image } \tilde{i}^* \subset k[y_{12},\dots,y_{1n}]^{U_{n-1}} .$$

Proposition 3.5.

$$k[y_{12},\dots,y_{1n}]^{U_{n-1}} = k[\tilde{v}_{12},\dots,\tilde{v}_{1n}]$$

where

$$\tilde{v}_{1i} = \prod_{\lambda \in k} (y_{1i}+\lambda_2 y_{12}+\dots+\lambda_{i-1}y_{1i-1}).$$

Proof. (Reference the proof of Theorem 3.4 in [7].) It is obvious $\tilde{v}_{1i}$ is invariant under U_{n-1}. Let $f=gy_{1n}+h$, $h \in k[y_{12},\dots,y_{1n-1}]$ be invariant. Then h is invariant under U_{n-2}. By induction it sufficient to prove f-h is in $k[\tilde{v}_{12},\dots,\tilde{v}_{1n}]$. Since f-h contains y_{1n} as a factor, f-h also contains

$$y_{1n}+\lambda_2 y_{12}+\dots+\lambda_{n-1}y_{1n-1}.$$

Hence f-h contains $\tilde{v}_{1n}$. Applying the same argument to $(f-h)/\tilde{v}_{1n}$, we have the proposition. q.e.d.

To simplify the notations, given a subgroup S of U, we denote by U-S the subgroup generated by the $x_{ij}(\alpha)$ which are not contained in S. Moreover we write $\langle x_{ij}\rangle = \langle x_{ij}(\alpha)\rangle$.

Now we use the Evens norm. Let $I_{1i-1}=\langle x_{12},\dots,x_{1i-1}\rangle$. Since there is the quotien map:

$$U\text{-}i_{1i-1} \longrightarrow U\text{-}I_{1i-1}/U\text{-}I_{1i} = \langle x_{1i}\rangle ,$$

we can take $y_{1i}(b)$ in $H^*(U\text{-}I_{1i-1})$. Define

$$(3.6) \qquad v_{1i}(b) = N[U\text{-}I_{1i-1} \hookrightarrow U](y_{1i}(b)) \quad .$$

Proposition 3.7. $v_{1i}(b)|I_{1n} = \tilde{v}_{1i}(b)$ <u>mod</u> (2.8).

Proof.
$$\begin{aligned} v_{1i}|I_{1n} &= \prod_{(U-I_{1i})gI_{1n}} N[I_{1n}\cap g^{-1}(U\text{-}I_{1i-1})g \hookrightarrow I_{1n}](y_{1i}) \\ &= N[I_{1n}-I_{1i-1} \hookrightarrow I_{1n}](y_{1i}) \\ &= N[\langle x_{1i}\rangle \hookrightarrow I_{1i}](y_{1i}) \quad \text{(Apply 2.2 to the map : } I_{1n}\to I_{1i}) \\ &= \tilde{v}_{1i} \text{ mod (2.8) (by Theorem 2.7, Prop. 3.5).} \end{aligned}$$
q.e.d.

Corollary 3.8. $\text{Image } \tilde{i}^* = k[y_{12},\dots,y_{1n}]^{U_{n-1}}$.

Remark 3.9. Let $j : U \to GL_n(F_p)$ be the natural inclusion. Then Priddy has proved [8]

$$\mathrm{Image}(ij)^*/\sqrt{0} \;=\; F_p[y_{12},\ldots,y_{1n}]^{GL_{n-1}(F_p)}.$$

Let y_{ij} be the obvious element in $H^*(U-I_{ij-1})$ and define

$$v_{ij} = N[U-I_{ij-1} \hookrightarrow U](y_{ij}) \quad \text{where } I_{ij-1}=\langle x_{ii+1},\ldots,x_{ij-1}\rangle.$$

Lemma 3.10. $v_{ij}|\langle x_{ij}\rangle = y_{ij}^{p^{j-i-1}}$ <u>mod</u> (2.8)

$$v_{ij}|\langle x_{\ell m} \mid \ell > i\rangle = v_{ij}|\langle x_{\ell m} \mid m<j\rangle = 0$$

<u>Moreover</u> $y_{ij}^{p^{j-i-2}} \notin$ <u>Image</u> $(i^*: H^*(U) \to H^*(\langle x_{ij}\rangle)$ <u>and hence</u> v_{ij} <u>is a ring generator</u>.

Proof. Put $L=\langle x_{\ell m} \mid \ell > i\rangle$. Then

$$v_{ij}|L = \prod_{U-I_{1j-1}gL} N[L\cap g^{-1}(U-I_{ij-1})g \hookrightarrow U](i_g^* y_{ij}).$$

That $y_{ij}|L=0$ implies $v_{ij}|L=0$. From Proposition 3.7

$$v_{ij} \mid x_{ij} = (v_{ij}|I_{ij}) \mid x_{ij} = y_{ij}^{p^{i-j-1}} \quad \mathrm{mod}\ (2.8).$$

From Corollary 3.8, $y_{ij}^{p^{j-i-2}} \notin \mathrm{Image}(i^*: H(U) \to H^*(I_{ij}))$. q.e.d.

The cohomology ring $H^*(U)/\sqrt{0}$ is not generated by only $v_{ij}(b)$ [6]. However we have the following.

Theorem 3.11. <u>The cohomology ring</u> $H^*(U)$ <u>is finitely generated as a</u> $k[v_{ij}(b)]$<u>-module</u>.

For the proof, we use the following lemma.

Lemma 3.12. <u>If there is a central extension</u>

$$0 \to k \xrightarrow{i} G \xrightarrow{q} G' \longrightarrow 1$$

<u>and if</u> $y^{p^s}(b)$, $0 \leq b \leq d-1$ <u>are lowest dimensional permanent elements in</u> $H^*(k)/\sqrt{0}$ <u>and</u> $y^{p^s}(b)=i^*(u_b)$, <u>then</u> $H^*(G)$ <u>is finitely generated as a</u> $k[u_b \mid 0 \leq b \leq d-1] \otimes H^*(G')$<u>-module</u>.

Proof. Since $y^{p^s}(b)$ is a permanent cycle in the spectral sequence induced by the above central extension, the E_r-term is written

$$E_r^{*,*} \cong k[y(b)^{p^s}] \otimes \sum E^{I\varepsilon}{}_r\, y(0)^{i_0} \ldots y(d-1)^{i_{d-1}} x(0)^{\varepsilon_0} \ldots x(d-1)^{\varepsilon_{d-1}}$$

where $0 \leq i_0,\ldots i_{d-1} \leq p^s-1$ and $\varepsilon_0,\ldots \varepsilon_{d-1} = 0$ or 1. By the induction each $E_r^{I\varepsilon}$ is finitely generated as a $H^*(G')$-module. q.e.d.

Proof of Theorem 3.11. Suppose $H^*(U_{n-1})$ is a finitely generated $k[v_{ij}(b) \mid i>1]$-module. Apply Lemma 3.11 to the exact sequence

$$0 \longrightarrow \langle x_{12}\rangle \rightarrow U_n/I^{12} \longrightarrow U_n/I^{11}=U_{n-1} \rightarrow 1$$

where $I^{1i}=\langle x_{1i+1},\ldots,x_{1n}\rangle$. Then $H^*(U_n/I^{12})$ is finitely generated as a $k[v_{12}(b)]\otimes k[v_{ij}(b) \mid i>1]$-module. Contining this argument, we have $H^*(U_n/I^{1n})=H(U_n)$ is finitely generated as a $k[v_{ij}(b)]$-module. q.e.d.

Conjecture 3.13. *Let V be the subalgebra of $H^*(U)$ generated by $v_{ij}(b)$. Then the inclusion V into $H^*(U)$ is an F-isomorphism, namely, for all $x\in H^*(U)$ there is s such that $x^{p^s}\in V$.*

Example 3.14. The case d=1, n=3 $p\geq 3$ (G.Lewis [6]).

$$H^*(U)/\sqrt{0} \simeq (F_p[v_{12},v_{23}]/(v_{12}{}^p v_{23}-v_{12}v_{23}{}^p)\oplus F_p\{b\})\otimes F_p[v_{13}]$$

where the product for b is given $b^2=v_{12}{}^{p-1}v_{23}{}^{p-1}$, $bv_{12}=v_{12}v_{23}{}^{p-1}$ and $bv_{23}=v_{12}{}^{p-1}v_{23}$. Therefore in this case the conjecture is right.

§4. The cohomology ring $H^*(G)$.

First we recall the theorem of Cartan-Eilenberg XII.9 [2]. Let π be a subgroup of a finite group Π such that p does not divide $[\Pi;\pi]$. Given a double coset decomposition

$$\Pi = \bigcup \pi x_i \pi .$$

The element $a\in H^*(\pi)$ is said to be stable if for each x_i,

$$a|\pi_{x_i} = x_i^*(a|\pi_{x_i^{-1}}) \qquad \text{where } \pi_{x_i}=\pi\cap x_i\pi x_i^{-1}.$$

Theorem 4.1. (Cartan-Eilenberg) *An element $a\in H^*(\pi)$ is stable if and only if $a\in \mathrm{Image}(i^*:H^*(\Pi)\rightarrow H(\pi))$.*

Let B be the Borel subgroup of G defined by uppper triangular matrices. It is immediate that

$$B = TU = \bigcup_{t\in T} UtU$$

where T is the subgroup of G of diagonal matrices. If t is the diagonal matrix with $i\times i$-entry d_i, then using (2.6) it is easily seen that

$$tx_{ij}t^{-1}=x_{ij}(d_i/d_j) \quad \text{and} \quad t^*y_{ij}(b)=(d_i/d_j)^{p^b}y_{ij}(b).$$

Lemma 4.2. *If* $\Sigma \ell_b p^b \equiv 0$ *mod* p^d-1, *then*

$$\prod_b v_{ij}(b)^{\ell_b} \in \text{Image}(i^*: H^*(B) \to H^*(U)).$$

Proof. Since $U \cap tUt^{-1} = U$, from Theorem 4.1 we have

$$H^*(U)^T = \text{Image}(i^*: H^*(B) \to H^*(U)).$$

Let $\Sigma \ell_b p^b \equiv 0 \bmod p^d-1$. Then

$$\begin{aligned} t^* \prod v_{ij}(b)^{\ell_b} &= t^* \prod (N[\ ]y_{ij}(b))^{\ell_b} \\ &= N[\ ](\prod (t^* y_{ij}(b)^{\ell_b})) \\ &= N[\ ]((d_i/d_j)^{\Sigma p^b \ell_b} \prod y_{ij}(b)^{\ell_b}) \\ &= N[\ ](\prod y_{ij}(b)^{\ell_b}) \\ &= \prod (N[\ ]y_{ij}(b)^{\ell_b}) = \prod v_{ij}(b)^{\ell_b}. \end{aligned}$$

q.e.d.

Hereafter we consider only two typical cases which satisfy Lemma 4.2, they are $v_{ij}(b)^{p^d-1}$, $\prod_b v_{ij}(b)^{p-1}$. For simplicity of notation, denote either of them by v_{ij}

It is wellknown there is a decomposition

$$G = \bigcup_{w \in W} BwB$$

where W is the Weyl group of T which is isomorphic to the symmetric group S_n. In this situation Cline-Parshall-Scott [3] and Glauberman show that $a \in H^*(B)$ is stable if

$$a|B_{w_i} = w_i^*(a|B_{w_i}) \quad \text{for all } w_i = (i,i+1) \quad \text{(transposition)}.$$

Since $H^*(B_{w_i}) \to H^*(U_{w_i})$ is injective, we have

Lemma 4.3. *An element* $a \in H(U)^T$ *is in* Image$(i^*: H^*(G) \to H^*(U))$ *if and only if* $a|U_{w_i} = w_i^*(a|U_{w_i})$ *for each* $w_i = (i,i+1)$, $1 \le i \le n-1$.

If $a \in \text{Image}(i^*: H^*(G) \to H^*(U))$, we only write $a \in H^*(G)$, identifying $i^*H(G)$ with $H^*(G)$.

Theorem 4.4. $\Delta = v_{12}v_{23}\cdots v_{n-1n} \in H^*(G)$ *and if* $p \ge n$, $\Delta \ne 0$.

Proof. Since $U_{w_i} = U - \langle x_{ii+1}\rangle$, $v_{ii+1}|U_{w_i} = 0$. This shows $\Delta|U_{w_i} = 0$. From Lemma 4.3, $\Delta \in H^*(G)$. To prove $\Delta \ne 0$, we consider the subgroup generated by

$$\langle I+F\rangle = \langle x_{12}x_{23}\cdots x_{n-1n}\rangle.$$

Then $(I+F)^p = I+F^p$. Assume $p \ge n$. Then $(I+F)^p = I$ and $y_{ii+1}|\langle I+F\rangle = y$ where $H^*(\langle I+F\rangle)/\sqrt{0} \simeq k[y(b)]$. This implies

$\Delta\langle I+F\rangle = y^{(p^d-1)(n-1)}$ (or $y^{(p-1)d(n-1)}$) $\neq 0$. q.e.d.

Lemma 4.5. *If* $k,j \neq i,\ i+1$, *then* $v_{k,j}$ *is invariant under* w^*_i.

Proof $w^*_i N[U-I_{kj} \subsetneq U](y_{kj})|U_{w_i} = N[U-w_i(I_{kj}) \subsetneq U](w_i^* y_{kj})|U_{w_i}$

It is immedite $w_i(I_{kj})=I_{kj}$ and $w_i^* y_{kj}=y_{kj}$. q.e.d.

Lemma 4.6. *If* $k<i$, *then*

$$v_{ki}|U-x_{ii+1} = N[U-I_{ki-1}-x_{ii+1} \subsetneq U-x_{ii+1}](y_{ki}).$$

Proof. $v_{ki}|U-x_{ii+1} = N[U-I_{ki-1} \subsetneq U](y_{ki})|U-x_{ii+1}$

$$= \prod_{(U-x_{ii+1})g(U-I_{ki-1})} N[U-x_{ii+1} \cap g^{-1}(U-I_{ki-1})g \subsetneq U-x_{ii+1}](i_g^* y).$$

Since $G=(U-x_{ii+1})\cdot 1\cdot(U-I_{ki-1})$, we have the lemma. q.e.d.

Lemma4.7. *If* $k<i$, *then*

$$v_{ki+1}|U-x_{ii+1} = N[U-I_{ki-1}-x_{ii+1} \subsetneq U-x_{ii+1}](N[\langle x_{ki+1}\rangle \subsetneq \langle x_{ki},x_{ki+1}\rangle](y_{ki+1}).$$

Proof. Since $G=(U-x_{ii+1})\cdot 1\cdot(U-I_{ki-1}-x_{ki})$, we have

$$v_{ki+1}|U-x_{ii+1} = N[U-I_{ki-1}-x_{ki}-x_{ii+1} \subsetneq U-x_{ii+1}](y_{ki+1})$$

$$= N[U-I_{ki-1}-x_{ii+1} \subsetneq U-x_{ii+1}]$$

(4.8) $$(N[U-I_{ki-1}-x_{ki}-x_{ii+1} \subsetneq U-I_{ki-1}-x_{ii+1}](y_{ki+1}))$$

here we use the transitive property2.1.

There are projections q_1, q_2,

$$\begin{array}{ccc} U-I_{ki-1}-x_{ki}-x_{ii+1} & \hookrightarrow & U-I_{ki-1}-x_{ii+1} \\ q_1 \downarrow & & q_2 \downarrow \\ \langle x_{ki+1}\rangle & \hookrightarrow & \langle x_{ki},x_{ki+1}\rangle. \end{array}$$

hence by naturality, (4.8) is

$$N[\langle x_{ki+1}\rangle \subsetneq \langle x_{ki},x_{ki+1}\rangle](y_{ki+1}). \quad \text{q.e.d.}$$

Lemma.4.9. *Let* $k=F_p$ (i.e.,d=1). *If* $k<i$ *then* $v_{ki}v_{ki+1}|U_{w_i}$ *is invariant under* w_i^*.

Proof. From Remark 2.12,

$$N[\langle x_{ki+1}\rangle \subsetneq \langle x_{ki},x_{ki+1}\rangle](y_{k+1}) = \prod_{\lambda\in F_p}(y_{ki+1}+\lambda y_{ki}).$$

From Lemma 4.6 and Lemma 4.7,

$$\begin{aligned} v_{ki}v_{ki+1}|U_{w_i} &= N[\ ](y_{ki})\cdot N[\ ](\textstyle\prod(y_{ki+1}+\lambda y_{ki})) \\ &= N[\ ](y_{ki}\textstyle\prod(y_{ki+1}+\lambda y_{ki})). \end{aligned}$$

This is symmetric under the transiposition (i,i+1). q.e.d.

Lemma 4.10. *Let* $k=F_p$. *If* $k>i+1$, *then* $v_{ik}v_{i+1k}|U_{w_i}$ *is invariant under* w_i^*.

Proof. $v_{ik}|U-x_{ii+1} = N[U-I_{ik-1} \subsetneq U](y_{ik})|U-x_{ii+1}$

$$= \prod_{U-x_{ii+1}gU-I_{ik-1}} N[U-x_{ii+1} \cap g^{-1}(U-I_{ik-1})g \subsetneq U-x_{ii+1}](i_g^* y_{ik})$$

$$= \prod_{x(\lambda)} N[U-x_{ii+1} \cap x_{ii+1}(\lambda)(U-I_{ik-1})x_{ii+1}(\lambda)^{-1} \subsetneq U-x_{ii+1}](y_{ik}+\lambda y_{i+1k})$$

$$= \prod_{x(\lambda)} N[\langle U-I_{ik-1}-I_{i+1k-1}, x_{i+1i+2}(\lambda)x_{ii+2},\ldots,x_{i+1k-1}(\lambda)x_{ik-1}\rangle \subsetneq U-x_{ii+1}](y_{ik}+\lambda y_{i+1k}).$$

Therefore $v_{ik}v_{i+1k}|U-x_{ii+1}$ is invariant under (i,i+1). q.e.d.

Theorem 4.11. *Let* $k=F_p$. *Then* $0 \neq \prod_{k\leq i,\ell\geq i+1} v_{k\ell} \in H^*(G)$.

Proof. From Lemma 4.5, Lemma 4.9 and Lemma 4.10, the above element is invariant under w_j^* for $j\neq i$. Since its restriction image to $U-x_{ii+1}$ is zero, it is also invariant under w_i. Moreover this element is non zero because the restriction to $\langle x_{k\ell} | k\leq i, \ell\geq i+1\rangle$ is non zero. q.e.d.

Example 4.12. The case $k=F_p$ and $n=3$, $p\geq 3$, the subalgebra of $H^*(U)/\sqrt{0}$ generated by the following (1)-(3) is F-isomorphic to $H^*(G)/\sqrt{0}$.

(1) $v_{13}+v_{12}{}^p+v_{23}{}^p$

(2) $v_{12}v_{13}$, $v_{23}v_{13}$

(3) $\Delta = v_{12}v_{23}$.

Remark 4.13. The arguments of (3.6), Theorem 3.10, Lemma 4.2 and Lemma 4.3 are extended to the case G are finite universal Chevalley groups.

Of course there are many other non zero elements in $H^*(G)$, indeed;

Theorem 4.14. (Evens [4] , Quillen [9]) *Let* A *be an elementary* p-*group of* G. *Let* $[N_G(A);A]=qh$ *where* $(p,h)=1$ *and* $q=p^s$. *Then there exists* v_A *in* H (G) *such that*

$$v_A|A' = \begin{cases} 0 & \text{if } A \not\to A' \\ e_A{}^q & \text{if } A \to A' \end{cases}$$

where $e_A = \prod_{0\neq u\in A} \beta u^\#$ *and* $\to$ *means an injection to a conjugation of* A.

However for general n and p, all elements in $k[V_{ij}]$ satifying Lemma 4.3 are only given by Theorem 4.4 and Theorem 4.11. In the next section we shall see these in the concrete c ses.

§5. The case $Gl_4(F_2)$.

The cohomology ring $H^*(GL_3(F_p);Z_{(p)})$ is known. In this section we compute $H^*(GL_4(F_2))$ and consider the relation between the cohomology theory and the invariant theory. Throughout this section we assume p=2, n=4 and d=1.
For ease of arguments we simply write subscrpts (12) (resp. (23) (34),(13),(2,4),(14)) by 1 (resp. 2,3,4,5,6), for example $y_1=y_{12}$, $v_5=v_{24}$.

Proposition 5.1. Let M be $U/\langle x_6\rangle$. Then

$$H^*(M)\simeq (F_2[y_1,y_3,k]\tilde{\oplus} F_2[y_2])\otimes F_2[v_4,v_5]/(y_1{}^2v_5+y_3{}^2v_4=k^2+y_1y_3k)$$

where $A\tilde{\oplus}B$ means $A\otimes B/(A^+\otimes B^+)$, A^+ is degree positive part.

Proof. Consider the cental extension

$$0\longrightarrow \langle x_4,x_5\rangle \rightarrow M \longrightarrow \langle x_1,x_2,x_3\rangle \longrightarrow 0.$$

The differential d_2 of induced spectral sequence is

$$d_2y_4=y_1y_2 \quad\text{and}\quad d_2y_5=y_2y_3.$$

Hence $E_3^{*,0}\simeq F_2[y_1,y_3]\tilde{\oplus} F_2[y_2]$.

The differential $d_2(y_4y_5)=ky_2$ where $k=y_4y_3+y_1y_5$,

$$\text{Ker } d_2^{*,1}= F_2[y_1,y_2,y_3]\otimes F_2[k]\in E_2^{*,1}$$

Since there is the element $v_5=N[M-\langle x_1\rangle\hookrightarrow M)(y_5)$, $y_5{}^2$ is a permanent cycle and so is $y_4{}^2$. Hence

$$E_\infty^{*,*}\simeq E_3^{*,*}=(F_2[y_1,y_3,y_4y_3+y_1y_5]\tilde{\oplus} F_2[y_2])\otimes F_2[y_4{}^2,y_5{}^2]$$

$$\hookrightarrow (F_2[y_1,y_3]\tilde{\oplus} F_2[y_2])\otimes F_2[y_4,y_5\ ].$$

If we can show

$$r_1=ky_2=0 \quad\text{and}\quad r_2=y_1{}^2v_5+y_3{}^2v_4+k^2+y_1y_3k=0$$

then we can easily prove the proposition. Let write

$$ky_2\ =\ ay_2{}^3+by_1{}^3+cy_1{}^2y_3+\dots$$

Then $0=ky_2|M-x_2\ =\ by_1{}^3+cy_1{}^2y_3+\dots$ in $H^*(M-x_2)\simeq F_2[y_1,y_3,y_4,y_5]$.

This induces b=c=0 and that $ky_2|M-x_1-x_3=0$ induces $a=0$.

The second relation is proved by using

$$v|M-x_2 = y_4^2-y_1y_4 \quad \text{and} \quad v|M-x_2 = y_5^2-y_3y_5.$$

q.e.d.

Theorem 5.2.

$$H^*(U) \simeq (F_2[y_1,y_3,v_4,v_5]/(y_1v_5+y_3v_4, y_1^2v_5+y_3^2v_4, y_1v_5+y_1y_3^2v_5+y_3v_4+y_1^2y_3v_4) \tilde{\oplus}(F_2[y_2]\otimes\Lambda(d_1,d_2))\otimes F_2[v_4,v_5])\otimes F_2[v_6]$$

except the products of y_2, d_1 and d_2, which is given by

$$d_1^2 \equiv y_2d_2 \quad , d_1 \equiv y_2v_6. \quad \text{modulo } F_2[y_1,y_3,v_4,v_5].$$

Proof. Consider the central extensions

$${}_1E \quad : \quad 0 \to \langle x_6\rangle \to U \longrightarrow M \longrightarrow 1$$

$${}_2E \quad : \quad 0 \to \langle x_6\rangle \to U-x_2 \longrightarrow M-x_2 \longrightarrow 1.$$

Comparing spectral sequences induced by the above extensions, the differential d_2 of the spectral sequence of ${}_1E$ is

$$d_2y_6=y_1y_5+y_3y_4 = k.$$

Therefore

$$E_3^{*,2n} \simeq (F_2[y_1,y_3]\tilde{\oplus} F_2[y_2])\otimes F_2[v_4,v_5]/(y_1^2v_5+y_3^2v_4)$$

$$E_3^{*,2n+1} \simeq F_2[v_4,v_5]\tilde{\otimes} F_2[y_2] .$$

Next we consider d_3 in the spectral sequence of ${}_2E$

$$d_3(y_6^2) = d_3(S_q^1y_6) = S_q^1k$$

$$= y_1(y_5^2+y_3y_5)+y_3(y_4^2+y_1y_4)+(y_1+y_3)(y_1y_5+y_3y_4)$$

Therefore in ${}_1E$

$$d_3(y_6^2) = y_1v_5+y_3v_4 .$$

The above element is non zero divisor in $E_3^{*,0}/(\text{Ideal } y_2)$.
This fact and $\text{Ker}d_3=\text{Ideal}(y_2y_6^s)$ follow

$$E_4^{*,4*} \simeq (F_2[y_1,y_3]\tilde{\oplus} F_2[y_2])\otimes F_2[v_4,v_5]/(-,-)$$

$$\oplus_{i=1}^{3} E_4^{*,4*+i} \simeq (F_2[v_4,v_5]\otimes F_2[y_2])\otimes F_2\{y_2y_6, y_2y_6^2, y_2y_6^3\}$$

Lastly we consider $d_4(y_2y_6^3)$. Since $y_1v_5+y_3v_4=0$ in $H^*(U)$, its Sq^2-image is also zero. In $H^*(U-x_2)$,

$$Sq^2(y_1v_5+y_3v_4) = S_q^2S_q^1k$$

$$= y_1y_5^4+y_1^4y_5+y_3y_4^4+y_3^4y_4$$

$$= y_1v_5^2+y_1y_3^2v_5+y_3v_4^2+y_1^2y_3v_4 \qquad (a)$$
$$+(x_1^3+x_3^3)(x_1x_5+x_3x_4).$$

Therefore the element (a) must be zero in $H^*(U)$. But

$$Sq^2Sq^1k \notin (Sq^1k,\ x_1^2v_5+x_3^2v_4).$$

Otherwise by the dimensional reason $Sq^2Sq^1k \in (Sq^1k)$ and this contradicts the fact that (Sq^1k, Sq^2Sq^1k) is regular sequence (,see Quillen [11]). Hence (a) is in Image d. Since y_6^4 is a permanent cycle, we have

$$d_4(y_2y_6^3) = (a).$$

Since $E_4^{*,0}$ is a free $F_2[v_4,v_5]$-module, $\mathrm{Ker}d_4^{*,3} \cong F[v_4,v_5]\{y_2^2y_6^3\}$. The permanent property of y_6^4 induces

$$E_5^{*,*} \simeq E_\infty^{*,*}.$$

Therefore we can prove the theorem. q.e.d.

In this case B=U. Before considering $H^*(G)$, we recall the Quillen's theorem [9].

Theorem 5.3. <u>If</u> $x|A=0$ <u>for all elementary abelian</u> p-<u>group</u> A <u>in a given finite group</u> K, <u>then</u> $x=0$ <u>in</u> $H^*(K)/\sqrt{0}$.

To seek stable elements, from 4.3, we have to check the action w_1, w_2 and w_3. The group $U_{w_1}=U-x_1$ is isomorphic to M by the map $f : U_{w_1} \to M$ so that x_2 (resp.5,3,6,4) $\mapsto x_1$ (resp.4,2,5,3).
The cohomology ring $H^*(M)$ has nonilpotent element, since so is $E_3^{*,*}$ in Proposition 5.1. From the above theorem, if we can prove

$$(5.4) \qquad (w_1^*x-x)|A = 0$$

for each maximal elmentary 2-group A in U_{w_1} then we have

$$w_1^*x = x \quad \text{in} \quad H^*(U_{w_1})$$

Lemma 5.5. <u>The maximal elementary abelian group in</u> U_{w_1} <u>are</u>

(1) $J = \langle x_3, x_5, x_6\rangle$

(2) $R = \langle x_2, x_4, x_5, x_6\rangle$

Proof. Let A be a maximal elementary abelian 2-group. Since x_5, x_6 is the center, it is contained in A. If $x_2^\alpha x_3^\beta x_4^\gamma \in A$, then

$$(x_2^\alpha x_3^\beta x_4^\gamma)^2 = x_5^{\alpha\beta} x_6^{\gamma\beta}.$$

If $\beta \neq 0$ then $\alpha=\gamma=0$ and so A=J. Otherwise A=R. q.e.d.

Let $G_1, G_2 \in GL_2(F_2)$. Then

$$\begin{bmatrix} G_1 & \\ & G_2 \end{bmatrix}\begin{bmatrix} 1 & & R & \\ & 1 & & \\ & & 1 & \\ & & & 1 \end{bmatrix}\begin{bmatrix} G_1^{-1} & \\ & G_2^{-1} \end{bmatrix} = \begin{bmatrix} 1 & & G_1RG_2^{-1} & \\ & 1 & & \\ & & 1 & \\ & & & 1 \end{bmatrix}.$$

induces

(5.6) Imge $(H^*(GL_4(F_2)) \to H^*(R)) \subset F_2[y_2,\dots,x_6]^{GL_2(F_2)|GL_2(F_2)}$

(5.7) Image$(H^*(U) \to H^*(R)) \subset F_2[y_2,\dots,y_6]^{U|U}$

where we write $F_2[-]^{G|G'}$ as the invariant ring of left G-action and right G'-action.

From the stability theorem (Lemma 4.3) and Lemma 5.5 , and (5.4), we have the following theorem.

Theorem 5.8. In $H^*(U)$,

$$H^*(G) = i_1^{*-1}(H^*(U-x_2)^{(2,3)*}) \cap i_2^{*-1}(H^*(R)^{GL_2|GL_2}) \cap i_3^{*-1}(H^*(I)^{GL_3}) \cap i_4^{*-1}(H^*(J)^{GL_3})$$

where i_s^* are the restriction maps from $H^*(U)$.

We now study each invariant.

Lemma 5.9. (Dickson [7]) The invariant $F_2[y_1,y_4,y_6]^{GL_3}$ is the subalgebra generated by

(1) $v_6+v_4^2+v_1^4$ (2) $v_6v_4+v_6v_1^2+v_4^2v_1^2$ (3) $v_6v_4v_1$

Remark 5.10. For general case, Dickson showed

$$F_p[y_{12},\dots,y_{1n}]^{GL_{n-1}(F_p)} = F_p[Q_{n-1,0},\dots,Q_{n-1,n-2}].$$

and we can easily see that (, reference Proposition 2.6 (ii)[7])

$$Q_{n-1,k} = \sum_{i_0>i_1>\dots>i_k} v_{1i_0}^{p^{n-i_0}} v_{1i_1}^{p^{n-i_1-1}} \cdots v_{1i_k}^{p^{n-i_k-k}} \text{ where } v=\tilde{v}^{p-1}.$$

Lemma 5.11. The invariant $F_2[R]^{U|U}$ is the subalgebra generated by

(1) $v_2|R=y_2$ (2) $v_4|R=\Pi y_4+\lambda y_2$, $v_5|R=\Pi y_5+\lambda y_2$

(3) $v_6|R=\Pi y_6+\lambda y_4+\mu(y_5+\lambda y_2)$

(4) $d=y_2y_6+y_4y_5$, $Sq^1d=y_2^2y_6+y_2y_6^2+y_4^2y_5+y_4y_5^2$.

Proof. Let $f\in F_2[y_2,y_4,y_5]$ be an invariant. Considering the right U_2-actin, we have $f\in F_2[(1),v_4|R,y_5]$. By the left U_2-action shows $f\in F_2[(1),(2)]$. Let $f\in F_2[y_2,y_4,y_5]\{1, y_6,y_6^2,y_6^3\}$. Then from the fact $d(y_2y_6^3)\neq 0$ in Theorem 5.2, there is no invariants such that

$y_2y_6^3+ay_6^2+\dots+d$, $y_6^3+ay_6^2+\dots+d$

Therefore we can prove $f\in F_2[(1)-(3)]\{1,d,Sq^1d,dSq^1d\}$.

Assume $f\in F_2[y_2,\dots,y_6]$ is invariant. We can write

$$f = a_mv_6^m+\dots+a_1v_6+a_0,\qquad a_i\in F_2[y_2,y_4,y_5]\{1,\dots,y_6^3\}$$

The $U_2|U_2$-action ; $y_6\mapsto y_6+y_4+(y_3+y_2)$ implies a_m must be an invariant. Therefore $f-a_mv_6^m$ is also invariant. We can inductively prove $f\in F_2[(1)-(4)]$. q.e.d.

Corollary 5.12. Image$(H^*(U)\to H^*(R)) = F_2[\ ]^{U_2|U_2}$.

Theorem 5.13. The invariant $F_2[\ ]^{GL_2(F_2)|GL_2(F_2)}$ is the subalgebra generated by

(1) $W_1 = v_6+v_4^2+v_5^2+dv_4+dv_5+dv_2^2$

(2) $W_2 = v_6v_4+v_5^2v_2^2+d_2v_5v_2$, $W_2' = v_6v_5+v_4^2v_2^2+d_2v_4v_2$

(3) $W_3 = v_6v_5v_4+v_6v_4v_2^2+v_6v_5v_2^2+(d_2+dv_2)v_6v_5v_4$

(4) $W_4 = v_6v_5v_4v_2$

(5) d ,$d_2=Sq^1d$.

Proof. A direct computation shows $v_6+v_4{}^2+dv_4$ is invariant by $(3,4)_*$. The element W_1 is expressed as

$$W_1 = (v_6+v_4^2+dv_4)+(v_5+v_2^2) +d(v_5+v_2^2)$$

so it is invariant by $(3,4)_*$. Since it is symmetric by $23\leftrightarrow 24$, it is also invariant by $(1,2)_*$. The formula W_2 is invariant by $(3,4)$ and some computations show the invariance by $(1,2)_*$. The formula W_5 is expressed as

$$W_3 = v_6v_4(v_5+v_2^2)+v_5v_2(v_6v_2+(d_2+dv_2)v_4)\ .$$

We can show by direct computation $v_6v_2+(d_2+dv_2)v_4$ is invariant by $(3,4)_*$. These show (1)-(5) are invariant.

Next we will prove all invariants expressed as a polynomial of (1)-(5). Assume f is an invariant. We give an order for each term of f by

$$y_6^{a_1}y_4^{a_2}y_5^{a_3}y_2^{a_4} < y_6^{b_1}y_4^{b_2}y_5^{b_3}y_2^{b_4}$$

if $(a_1,a_2+a_3,a_4) < (b_1,b_2+b_3,b_4)$ by the left lexicographic order. Let write

$$f = \sum d^sd_2^tv_6^{a_1}v_4^{a_2}v_5^{a_3}v_2^{a_4} +g(d,d_2,v_2,\dots,v_6).$$

where the sum runs all terms which contain the largest terms.

Here we note

(a) $a_1 \geqq a_2$, $a_3 \geqq a_4$.

For example, assume that $f = v_6^{a_1} v_5^{a_2} + ..$ and $a_1 < a_2$. Then $v_6^{a_1} v_5^{a_2}$ contains $y_6^{2a_1} y_5^{2(a_1+a_2)}$. So also it contains $y_6^{2(a_1+a_2)} y_5^{2a_1}$. This contradicts the fact that the largest term is $y_6^{4a_1} y_5^{2a_2}$. The similar argument proves (a).

Let

$$g = f - \sum d^s d_2^t \left\{ \begin{array}{ll} W_4^{a_4} W_3^{a_3-a_4} W_2^{a_2-a_3} W_1^{a_1-a_2} & \text{if } a_2 \geqq a_3 \\ W_4^{a_4} W_3^{a_2-a_4} W_2^{a_3-a_2} W_1^{a_1-a_3} & \text{if } a_2 < a_3 \end{array} \right\}$$

Then g is also invariant and its order is larger than f. Continuing this, we have the theorem. q.e.d.

Lemma 5.14. (Qillen [11]) <u>There is a ring isomorphism</u>

$$H(U-x_2) \simeq F_2[y_1, y_3, y_4, y_5]/(k, Sq^1k, Sq^2Sq^1k)$$

<u>where</u> $k = y_1y_5 + y_3y_4$. <u>Moreover</u> <u>Image</u>$i_1^* \cap H^*(U-x_2)^{(2,3)}$ <u>is the subalgebra generated by</u>

(1) v_1v_4, v_3v_5 (2) $v_1^2+v_4$, $v_3^2+v_5$ (3) v_6

(2) $d = y_1y_4 + y_3y_5$, $Sq^1d = y_1^2y_4 + y_1y_4^2 + y_3^2y_5 + y_3y_5^2$.

Proof. The group $U-x_2$ is the extra special 2-group, which is computed by Quillen (Theorem4.6 in [11]). The action induced from the trasposition (2,3) is given on the subscripts $1 \leftrightarrow 4$, $5 \leftrightarrow 3$ and $6 \leftrightarrow 6$. Hence $H^*(U-x_2)^{(2,3)}$ is generated by

$$y_1+y_4,\ y_1y_4,\ y_3+y_5,\ y_3y_5,\ v_6.$$

Since Image$i_1H^*(U)$ is generated by v_i, d_j, we can see this lemma. q.e.d.

Theorem 5.15. <u>The cohomology ring</u> $H^*(GL_4(F_2))$ <u>is isomorphic to the subalgebraof</u> $H^*(U)$ (Theorem 5.2) <u>generated by</u>

(1) $v\langle 6\rangle = W_1 + (\tilde{d}-d)v_4 + (\tilde{d}-d)v_5 + \tilde{d}v_1^2 + \tilde{d}v_3^2 + v_1^4 + v_3^4$

(2) $v\langle 4,6\rangle = W_2 + v_6v_1^2 + v_4^2v_1^2 + \tilde{d}v_4v_1^2$, $v\langle 5,6\rangle = W_2 + v_6v_3^2 + v_5^2v_3^2 + \tilde{d}v_5v_3^2$

(3) $v\langle 4,5,6\rangle = W_3 + v_6v_1^2v_3^2$

(4) $v\langle 2,4,5,6\rangle = W_4$

(5) $v\langle 2,6\rangle = \tilde{d} = d + y_1y_3$, $v\langle 2,6\rangle_2 = Sq^1\tilde{d} = d_2 + y_1^2y_3 + y_1y_3^2$

(6) $v\langle 1,4,6\rangle = v_1v_4v_6$, $v\langle 3,5,6\rangle = v_3v_5v_6$

(7) $u_s = v_3v_5(v_1^2+v_4)v_6^s$, $u_s = v_1v_4(v_3^2+v_5)v_6^s$ $0 \leqq s \leqq 3$

(8) $t_s = v_1v_3v_4v_5v_6^s$ $0 \leqq s \leqq 3$.

Proof. We can check that all elements (1)-(8) satisfy Theorem 5.8. For example,

$$v\langle 6\rangle | H^*(R) = w_1 \in H^*(R)^{GL_2 \wr GL_2}$$

$$v\langle 6\rangle | H^*(U-x_2) = v_6 + (v_4+v_1^2)^2 + (v_5+v_3^2)^2 + d(v_4+v_1^2+v_5+v_3^2) \in H^*(U-x_2)^{(2,3)}$$

$$v\langle 6\rangle | H^*(I) = v_6+v_4^2+v_1^4 \in H^*(I)^{GL_3}$$

$$v\langle 6\rangle | H^*(J) = v_6+v_5^2+v_3^4 \in H^*(J)^{GL_3} .$$

Therefore $v\langle 6\rangle \in H^*(GL_4(F_2))$.

In the theorem, $v\langle i_1,..,i_s\rangle$ is expressed the element such that

$$v\langle i_1,...,i_s\rangle | A = \begin{cases} \neq 0 & \text{for } A=\langle x_{i_1},...x_{i_s}\rangle \\ =0 & \text{for } A \subsetneq \langle \quad \rangle \end{cases}$$

and of lowest dimensional (except $v\langle 2,6\rangle_2$) in such elements.

From Theorem 5.13, the image i_2^* of (1)-(5) generates $H^*(R)^{GL_2 \wr GL_2}$ The elements (6) are in Kernel i_2^* and Image i_3^* (resp. i_4^*) of (1),(2),(6) generated

$$H^*(I)^{GL_3} \text{ (resp. } H^*(J)^{GL_3}).$$

We can prove that the element a such that $a \in \mathrm{Ker}\cdot i_2^* \cap \mathrm{Ker}\cdot i_3^* \cap \mathrm{Ker}\cdot i_4^*$ and $i_2^*(a) \in H^*(U-x_2)^{(2,3)}$, is contained in the ideal generated by (8).

Therefore all elements satisfy Theorem 5.8 are generated by (1)-(8). q.e.d.

BIBLIOGRAPHY

1. J.Aguadé, The cohomology of the GL_2 of a finite field, Arch. Math. 34 (1980) 509-516.

2. H.Cartan and S.Eilenberg, Homological algebra, Princeton Univ. Press. 1956.

3. E.Cline, B.Parshall and L.Scott, Cohomology of finite group of Lie type I, Publ. IHES (1974) 169-191.

4. L.Evens, A generalization of transfer map in the cohomology of groups, Trans. Amer. Math. Soc. 108 (1963) 54-65.

5. ______, On the Chern class of representations of finite groups, Trans.Amer. Math. Soc. 115 (1965) 180-193.

6. G.Lewis, The integral cohomology rings of groups of order p^3, Trans. Amer. Math. Soc. 132 (1968) 501-529.

7. H.Mui, Modular invariant theory and cohomology algebras of symmetric groups, J. Fac. Sci. Tokyo Univ. 22 (1976) 552-586.

8. S.Priddy, On the mod p cohomology of $U_n(F_p)$ and $GL_n(F_p)$, unpublished.

9. D.Quillen, A cohomological criterion for p-nilpotency, J. Pure and applied Algebra 1 (1971) 361-372.

10. ______, On the cohomology and K-theory of general linear group over a finite field, Ann. Math. 96 (1972) 552-582.

11. ______, The mod 2 cohomology ring of extra-special 2-groups and the Spinor group, Math. Ann. 194 (1971), 197-223.

12. M.Tezuka and N.Yagita, The mod p cohomology ring of $GL_3(F_p)$, to appear.

Michishige Tezuka
Department of mathematics
Tokyo Institute of Technology
Ohokayama, Meguroku
Tokyo, Japan

Nobuaki Yagita
Department of Mathematics
Musashi Institute of Technology
Tamazutsumi, Setagaya
Tokyo, Japan

Contemporary Mathematics
Volume 19, 1983

FILTRATIONS ON THE REPRESENTATION RING OF A FINITE GROUP

C. B. Thomas

1. MOTIVATION. Let Γ be a virtual duality group of arithmetic type, for example $SL(n,Z)$. It is reasonable to suppose that $K(B\Gamma)$ will be easier to calculate than $H^*(\Gamma,Z)$, since in certain cases it is possible to show that much of the torsion, which is known to be hard to calculate in H^*, disappears in the Atiyah-Hirzebruch spectral sequence

$$H^p(\Gamma,K^q(\text{point})) \Rightarrow K^{p+q}_{\text{compact}}(B\Gamma) .$$

Here $K^*_{\text{compact}}(B\Gamma)$ is defined by the short exact sequence

$$0 \to R^1 \lim K^{*-1}(X_r) \to K^*(B\Gamma) \to K^*_{\text{compact}}(B\Gamma) \to 0 ,$$

as X_r runs over some cofinal and suitably ordered subfamily of the finite subcomplexes of $B\Gamma$. Let us consider the example $\Gamma = SL(n,Z)$ more closely, and denote by $j_{(\ell,r)} : SL(n,Z) \to SL(n,Z/\ell^r)$ the natural projection onto a linear finite quotient group and by $Ch(\Gamma)$ the subring of the even dimensional integral cohomology generated by the Chern classes of finite dimensional representations of Γ.

THEOREM 1. For some subgroup Γ_1 of finite index in $\Gamma = SL(n,Z)$, $n \geq 3$, the Chern ring $Ch(\Gamma_1)$ is generated by elements of the form $j^!_{(\ell,r)}(x)$, $x \in Ch(\Gamma_{1,(\ell,r)})$.

SKETCH PROOF. The congruence subgroup theorem for $SL(n,Z)$, $n \geq 3$, implies that an irreducible finite dimensional representation ρ is the tensor product of the natural representation defined by inclusion in $SL(n,C)$ and some representation which vanishes on a subgroup of finite index. (I am indebted to R. Steinberg for this particular version of the theorem, presented during a lecture in Bonn, Summer 1980. For the background, see [4], particularly §16.) Hence the Chern subring will be defined by inflation from finite quotients, together with classes coming from the natural representation. However these become trivial on passing to a suitable finite covering space $B\Gamma_1$ of $B\Gamma$ [5].

Turning to K-theory we know that "up to finite indeterminacy" the universal cycles in the Atiyah-Hirzebruch spectral sequence are generated by

0271-4132/83 $1.00 + $.25 per page

Chern classes, and hence that up to some finite cokernel, $K^{0}_{\text{compact}}(B\Gamma_1)$ is determined by inflation from finite quotients. These in turn can be studied using the representation theory of finite linear groups and characteristic classes. Moreover the question of finite indeterminacy suggests that we look again at the paper of M. Atiyah [2] in which he posed the problem (in the category of finite groups) of characterising the filtration on $R(\Gamma)$ associated with the $E^{\infty}_{p,q}$-terms of the spectral sequence in purely algebraic terms. In this paper we shall summarise what we know about this problem, concentrating on two aspects: (1) those p-groups for which we can provide an algebraic characterisation and (2) an infinite class of projective special linear groups for which Atiyah's original conjecture fails.

2. PRELIMINARIES. Let $\alpha : R(\Gamma) \to K(B\Gamma)$ be the homomorphism which associates to each virtual complex representation of the finite group the corresponding integral combination of flat bundles over the classifying space $B\Gamma$. We define the (even) topological filtration on $R(\Gamma)$ by

$$R^{top}_{2k}(\Gamma) = \operatorname{Ker}(R(\Gamma) \underset{\alpha}{\to} K(B\Gamma) \underset{r}{\to} K(B\Gamma^{2k-1})) .$$

We define a related algebraic filtration by

$R^{\gamma}_{2k}(\Gamma)$ equals the subgroup generated by all monomials of the form $\gamma^{i_1}(x_{i_1})\gamma^{i_2}(x_{i_2})\ldots\gamma^{i_r}(x_{i_r})$, where $i_1 + i_2 + \ldots + i_r \geq k$ and x_{i_j} belongs to the augmentation ideal. Here

$$1 + \gamma^1(x)t + \ldots + \gamma^r(x)t^r + \ldots = \gamma_t(x) = \lambda_{t/1-t}(x) ,$$

where the coefficients of the λ-power series are given by the exterior powers of the representation x in $R(\Gamma)$. The Adams operation $\psi^r(x)$ is similarly defined to be $N_r(\lambda^1(x), \ldots \lambda^r(x))$ the rth. Newton polynomial in the exterior powers. We have the recurrence relation

$$\psi^r(x) - \lambda^1(x)\psi^{r-1}(x) + \ldots + (-1)^r r\lambda^r(x) = 0 .$$

Furthermore, see [2, Lemma 12.1], if $\rho_1, \ldots, \rho_\ell$ are the irreducible complex representations of Γ, and i is the weight of the exterior power λ^i, then $R^{\gamma}_{2k}(\Gamma)$ is generated by all monomials of total weight greater than or equal to k in the elements

$$\gamma^i(\rho_j - \varepsilon(\rho_j)) = \lambda^i(\rho_j - \varepsilon(\rho_j) + i - 1) , \ i = 1,2,\ldots,\varepsilon(\rho_j) , \ j = 1,2,\ldots,\ell .$$

In [2, Proposition 12.6] Atiyah proved that for an arbitrary finite group $R^{\gamma}_{2k} \subseteq R^{top}_{2k}$ for all values of k, with equality for $k = 0,1,2$. Although in general equality fails for higher values of k (see section 4 below), it is an attractive conjecture that it does hold for groups of prime power order.

Furthermore the two filtrations define the same topology on $R(\Gamma)$, a point to which we shall return in the last section.

In order to see the connection of the conjecture with the remarks made following Theorem 1, recall that the kth. Chern class $c_k(\rho)$ of the representation ρ is defined to be the kth. Chern class of the (virtual) flat bundle $\alpha(\rho)$ over $B\Gamma$. It is easy to see that these characteristic classes represent universal cycles in the Atiyah-Hirzebruch spectral sequence

$$H^p(B\Gamma, K^q(pt.)) \Rightarrow K^{p+q}(B\Gamma) \cong R(\Gamma)^\wedge ,$$

with completion with respect to $I = \mathrm{Ker}(\varepsilon)$, the augmentation ideal. For any group Γ such that the filtrations above are equal for all values of k , it follows that the graded ring associated to $R(\Gamma)$ by the topological filtration is generated by the images of Chern classes (and conversely).

3. PARTIAL RESULTS FOR p-GROUPS. We start with a result which shows that the filtration conjecture holds for groups of prime power order having abelian subgroups of low rank. (The p-rank of a group Γ , $rk_p(\Gamma)$, equals the number of generators of a maximal p-elementary abelian subgroup of Γ .)

THEOREM 2. (i) Suppose either that Γ is a 2-group which has a cyclic subgroup of index 2 , or

(ii) that $p \geq 5$, and Γ is a p-group with $rk_p(\Gamma) \leq 2$. Then $H^{even}(\Gamma, Z)$ is generated by Chern classes, and the filtration conjecture holds for Γ .

PROOF. (i) It is known that Γ (of order 2^{s+2}) is either dihedral, binary dihedral, semidihedral or of a type in which the relation $A^B = A^{-1+2^s}$ in the standard semidihedral presentation is replaced by $A^B = A^{1+2^s}$. The binary dihedral group has periodic cohomology $(rk_2\Gamma = 1!)$, and $H^{even}(\Gamma, Z)$ is generated by the first Chern classes of two one-dimensional representations and the second Chern class of the standard inclusion in $SU(2)$. In the remaining cases the defining extension in terms of the normal cyclic subgroup of index 2 is split, and the spectral sequence of the extension collapses [12]. If $\sigma : A \to e^{2\pi i}/2^{s+1}$ is the standard representation of the normal subgroup, $c_2(i_! \sigma) \in H^4(\Gamma; Z)$ restricts to a generator of the fibre term $E_2^{o,4}$, and the entire spectral sequence is vertically periodic. (This is an example of the class of elements used by L. Evens in [5] to prove that $H^*(\Gamma, Z)$ is a Noetherian ring.) The remaining generators of the even dimensional cohomology are first Chern classes corresponding to generators of $E_2^{2,o}$ and $E_2^{o,2}$. Even powers of the three dimensional generator ν make no contribution, since either $\nu^2 = 0$ (exceptional case) or $\nu^2 = \mu c_2(i_! \sigma)$ ((semi)dihedral case). A more thorough analysis of the spectral sequence gives the complete product structure of the integral cohomology, but is not necessary here, since we are only interested in identifying universal cycles.

(ii) The argument is formally similar to that just sketched, see [11] and [1], but with formidable computational difficulties. We start from the classification of rank two p-groups for $p \geq 5$ given by N. Blackburn, see [8, Satz 12.4]. Note that the rank one case is trivial. The groups Γ fall into three classes:

(1) not necessarily split metacyclic groups (2) groups modelled on the non-abelian group of order p^3 and exponent p, but with generating commutator of the form C^{p^s}, and (3) another extension of $Z^B/_p \times Z^C/_{p^s}$ by $Z^A/_p$ with slightly more complicated relations (A no longer centralises C). It is now a question of examining the spectral sequences of these and related extensions in order to identify a family of generators for the even dimensional cohomology.

For more general p-groups one has the following weak result, in which the conclusion is really a statement about neighbourhoods of the identity in the topology on $R(\Gamma)$.

Let Γ be a p-group of order p^n, and write $a(n) = a$ for the smallest positive integer such that $(a + 1)a \geq 2n$.

THEOREM 3. *If* n *is*
$$\left.\begin{matrix} \underline{\text{odd}},\ R^{top}_{(2k-1)n+1} \\ \\ \underline{\text{even}},\ R^{top}_{(2k-1)n+2} \end{matrix}\right\} \subseteq R^{\gamma}_{2k} \subseteq R^{top}_{2k},$$
for all values of k, *such that* $k \equiv 0 \bmod p^{n-a}$.

PROOF. We inductively construct a free action of Γ on the Cartesian product of n copies of the sphere S^{2k-1}. Given a group $\Gamma(m)$ of order p^m, $m \leq n$, we choose a central cyclic subgroup C of order p, and embed it in a maximal abelian normal subgroup A (this is where the integer $a(n)$ enters the argument, see [8, Satz 7.3]). Define a one dimensional representation of A to be faithful on the largest cyclic subgroup containing C, trivial outside, and transfer up to $\Gamma(m)$. By taking the direct sum of this representation with itself as many times as may be necessary we may define an action of $\Gamma(m)$ on the appropriate factor S^{2k-1}, which is free when restricted to the subgroup C. By the inductive hypothesis we may assume that the quotient group $\Gamma(m - 1) = \Gamma(m)/C$ acts freely on $(m - 1)S^{2k-1}$. In special cases one can improve the numerical bounds in this construction, for example, if Γ is a metacyclic group of order $p^{n_1+n_2}$ (with quotient group of order p^{n_2}) then Γ acts freely on $S^{2k-1} \times S^{2k-1}$ (with $k \equiv 0 \bmod p^{n_2}$). Denote the (2k-2)-connected orbit space by $S^{2k-1} \times \ldots \times S^{2k-1}/\Gamma = Y_k$, and note that Y_k is a finite approximation to a model for the classifying space $B\Gamma$. Indeed inspection of cells shows that we can find a model $B\Gamma$ such that

$$B\Gamma^{(2k-1)n} \supseteq Y_k \supseteq B\Gamma^{2k-1} \quad .$$

Consider now the trivial Γ-disc bundle

$$\begin{array}{c} S^{2k-1} \times \ldots \times S^{2k-1} \times D^{2k} \\ \downarrow \pi_n \\ \underbrace{S^{2k-1} \times \ldots \times S^{2k-1}}_{n-1} \end{array} \quad .$$

By the equivariant Thom isomorphism theorem [10, §3], $\tilde{K}_\Gamma(\pi_n)$ is isomorphic to $K_\Gamma(S^{2k-1} \times \ldots \times S^{2k-1})$, and the long exact sequence of the disc bundle modulo its boundary gives

$$\ldots \to K_\Gamma((n-1)S^{2k-1}) \xrightarrow{\lambda_{-1}\rho_n} K_\Gamma((n-1)S^{2k-1}) \to K_\Gamma(nS^{2k-1}) \to \ldots \quad .$$

Here $\lambda_{-1}\rho_n$ is the K-theoretic Euler class of the representation defining the action of Γ on the last sphere S^{2k-1} , and hence $K_\Gamma(nS^{2k-1})$ contains a subgroup isomorphic to some quotient of $K_\Gamma((n-1)S^{2k-1})$. Once more arguing inductively, and noting at the last step that $K_\Gamma(\text{point}) = R(\Gamma)$, we see that the image of $R(\Gamma)$ in $K_\Gamma(nS^{2k-1})$ is isomorphic to the quotient group $R(\Gamma)/\langle\lambda_{-1}(\rho_1)\ldots\lambda_{-1}(\rho_n)\rangle$. Since each class $\lambda_{-1}(\rho_i)$ is of level $2k$ in the γ-filtration, we conclude that the kernel of the constructed map $R(\Gamma) \to K(Y_k)$ is contained in $R^{\gamma}_{2k}(\Gamma)$. Doing this compatibly for pairs of integers (k_1,k_2) , which satisfy the given congruence relation, and using the fact that $\operatorname{colim}_{k\to\infty} Y_k = Y_\infty \simeq B\Gamma$, we see that the constructed map factors through $K(B\Gamma^{(2k-1)n})$, in such a way that the composition

$$R(\Gamma) \to K(B\Gamma^{(2k-1)n}) \to K(Y_k) \to K(B\Gamma^{2k-1})$$

coincides with the map whose kernel is $R^{top}_{2k}(\Gamma)$. It follows that

$R^{top}_{(2k-1)n+1}(\Gamma)$ (for n odd) or $R^{top}_{(2k-1)n+2}(\Gamma)$ (for n even) is of level $2k$ with respect to the γ-filtration.

Note that Theorem 3 provides a direct proof that the topologies defined by the two filtrations on the representation ring coincide. However as a means of studying the γ-filtration the theorem is very inefficient, as one sees from an elementary abelian group of order p^n . Here $a(n) = n$ and k is arbitrary, but we can only construct a free action on n copies of S^1 .

However by restricting to cyclic subgroups one can see in this case that the passage from $K(Y_k)$ to $K(B\Gamma^{2k-1})$ adds nothing to the kernel, see also [2, page 61]. The proof given is modelled on that for those groups with periodic cohomology, and possibly composite order, for which one can construct free linear actions on a single sphere. In this case one proves that the two filtrations agree at least on a cofinal subset of the positive integers.

4. AN INFINITE FAMILY OF COUNTEREXAMPLES. Let A_4 be the alternating group on four symbols. By calculating $R_{2k}^{top}(A_4)$ directly from the Atiyah-Hirzebruch spectral sequence, and comparing it with $R_{2k}^{\gamma}(A_4)$, E. Weiss [13] showed that

$$R_{4k}^{top}(A_4) = R_{4k}^{\gamma}(A_4)\text{ , but that}$$

$$R_{4k+2}^{top}(A_4) \supseteq R_{4k+2}^{\gamma}(A_4) \quad \text{as a subgroup of index two,} \quad k \geq 1 \ .$$

By again using the relationship between the K-theoretic classes γ^i and the ordinary Chern classes c_i we shall generalise this counterexample as follows:

THEOREM 4. *If* $\Gamma = PSL_2(F_\ell)$, $\ell \equiv \pm 3(8)$, *then* $R_6^{\gamma}(\Gamma)$ *has index two in* $R_6^{top}(\Gamma)$.

PROOF. Consider the smallest group in this sequence $A_4 \cong PSL_2(F_3)$ and examine Weiss' argument from the point of view of ordinary cohomology. With the presentation $A_4 = \{X,P,Q\colon X^3 = P^2 = Q^2 = 1 , PQ = QP , P^X = Q , Q^X = PQ\}$ we may summarise the table of simple characters as follows:

$\{1\}$	1	1	1	3
$\langle X\rangle$	1	ω	ω^2	$1+\omega+\omega^2$
$\langle P,Q\rangle$	1	1	1	$\hat{p}+\hat{q}+(\hat{p}\otimes\hat{q})$

Here we have described the restriction of each irreducible representation to representative 3- and 2-Sylow subgroups. The symbol ω denotes the representation $X \to e^{2\pi i/3}$, $\hat{p}$ the representation $\langle P,Q\rangle \to \langle -1,1\rangle$, etc. The latter notation has been chosen so that we may write $c_1(\hat{p}) = p \in H^2(Z/2 \times Z/2,Z)$, and similarly for $\hat{q}$. If ρ denotes the irreducible three dimensional representation of A_4, it follows that the total Chern class

$$c_{\cdot}(\rho)_{(2)} = 1 + (p^2 + pq + q^2) + pq(p + q) \ .$$

On the other hand

$$H^*(A_4,Z) \cong H^*(Z^X/_3,Z) \otimes H^*(Z^P/_2 \times Z^Q/_2,Z)^{inv} \ .$$

Therefore in dimension 6, the 2-torsion $H^6(A_4,Z)_{(2)}$ contains the elements

$pq(p+q)$, $p^3+q^3+p^2q$ and $p^3+q^3+pq^2$. The latter two are not Chern classes, but survive to infinity in the Atiyah-Hirzebruch spectral sequence, as one sees by comparison with $Z^P/_2 \times Z^Q/_2$ and the calculation on page 61 in [2].

We now generalise the calculation to $PSL_2(F_\ell)$, $\ell \geq 5$. The assumption that $\ell \equiv \pm 3 \bmod 8$ implies that a 2-Sylow subgroup Γ_2 is isomorphic to the Klein 4-group; Γ_2 has $\ell(\ell^2-1)/_{24}$ conjugates and the normaliser of our representative Γ_2 is isomorphic to A_4. (This is a standard group theory, see [8, §8].) Since Γ_2 is abelian the stable elements in $H^*(\Gamma_2,Z)$ coincide with those elements invariant under the action of the normaliser, that is, under the action of A_4. Hence $H^*(PSL_2(F_\ell),Z)_{(2)} \cong H^*(PSL_2(F_3),Z)_{(2)}$, and we obtain the same elements in $H^6_{(2)}$ as before. Since P,Q and PQ all belong to the same conjugacy class, the restriction of any representation σ to the subgroup $\langle P,Q\rangle$ must be symmetric in $\hat{p},\hat{q}$ and $\hat{p} \otimes \hat{q}$. Therefore

$$c_{\cdot}(\sigma)_{(2)} = [(1+p)(1+q)(1+p+q)]^n ,$$

which for dimensional reasons has only the component $pq(p+q)$ in dimension six. Restricting the spectral sequence as before we see that $R_6^{top}/_{R_6^\gamma} \cong Z/_2$.

EXERCISE. Show that the filtrations also differ when $\ell \equiv \pm 1(8)$. The points to watch here are that Γ_2 is dihedral, and one must complete the calculation sketched in Theorem 2. Furthermore since Γ_2 is no longer abelian, a more careful examination of the stable elements in cohomology is needed.

If one turns to the other classical infinite family of simple groups A_n, $n \geq 6$ (to avoid repetitions), similar calculations show that the filtrations differ for $n = 6$, $n = 7$. We have not pushed these calculations further, but it is amusing to speculate that the original conjecture fails for all finite simple groups.

5. ADMISSIBLE FILTRATIONS ON THE REPRESENTATION RING.

In his examination of the relationship between the topological and γ-filtrations [9] J. Ritter introduced the notion of an admissible filtration $R^a_{2k}(\Gamma)$, which should satisfy the properties:

(i) $R^a_{2k}(\Gamma)$ is decreasing and compatible with multiplication,

(ii) $R^a_o(\Gamma) = R(\Gamma)$, $R^a_2(\Gamma) = I(\Gamma)$, $R^a_2/_{R^a_4(\Gamma)} \cong H^2(\Gamma,Z)$

(iii) $R^a_{2k}(\Gamma) = I^k(\Gamma)$, if Γ is cyclic for all $k \geq 0$, and

(iv) $R^a_*(\Gamma)$ is compatible with the induced representation map $i_!$.

Properties (i) - (iii) are the minimal requirements for such a filtration, and from the summary given in section two above, are satisfied by both R_{2k}^{top} and R_{2k}^{γ}. The topological filtration satisfies (iv) for an arbitrary group Γ; Theorem 4 implies that for the projective special linear $PSL_2(F_\ell)$, $\ell \equiv \pm 3(8)$, the γ-filtration does not. This follows from [9, Satz A], in which it is shown that $R_{2k}^{a}(\Gamma)$ is uniquely determined by its values on subgroups of prime power order. For $\Gamma = PSL_2(F_\ell)$, $\ell \equiv \pm 3(8)$, these are all abelian, for which it is known that the γ- and topological filtrations coincide. In the positive direction Ritter proves that any admissible filtration on the category of p-groups defines a unique topology on the representation ring (op.cit., Lemma 6). It is therefore natural to pose the

PROBLEM: <u>is the γ-filtration admissible on the category of p-groups?</u>

A positive answer to the problem would however not characterise the topological filtration in algebraic terms, since there are simple examples of distinct admissible filtrations on the binary dihedral groups. However using the well-known theorem of Blichfeldt that an irreducible representation of a p-group of dimension greater than one is induced up from a one dimensional representation of a proper subgroup, together with the equation $i_! i^!(x) = x . i_! [1]$, one would be able to show that an admissible γ-filtration determines the topological filtration up to inversion of certain elements.

As a first step towards an answer to the problem above one should derive a formula for the Grothendieck classes of an induced representation $\gamma^n(i_! \rho)$, where ρ is one-dimensional. As in the case of the ordinary Chern classes one starts by embedding the subgroup K, on which ρ is defined, in the Wreath product $H \wr S_{[\Gamma:H]}$ and defining $i_! \rho$ by restriction from the larger group. The problem then is to analyse the spectral sequence for the universal example:

$$H^p(S_n, K^q(BT^n)) \Rightarrow K^{p+q}(B(T^1 \wr S_n)) \text{ , where } n = [\Gamma:K] .$$

I surmise that the universal formula will be similar to that in [7] for $c_r(i_! \rho)$, but that for p-groups, i.e. after restriction to $\Gamma \subseteq H \wr S_n$, there will be a certain amount of simplification, due to the fact that in $R(\Gamma)^{\wedge}$ we have p-adic coefficients. This type of argument, applied to the Newton polynomials ψ^r, certainly allows us to recover the known result that $i_! \psi^r = \psi^r i_!$, $(r < p-1)$, see [3]. It is perhaps interesting to note that both the argument suggested above for ψ^r and that of M. Atiyah and D. Tall exploit properties of the Galois group $G(\overline{Q}:Q)$. In any event this should be regarded as work in progress.

REFERENCES

[1] K. Alzubaidy, Thesis (University of London), 1979.

[2] M.F. Atiyah, "Characters and cohomology of finite groups", Publ. Math. de l'IHES, 9 (1961), 23-64.

[3] M.F. Atiyah and D. Tall, "Group representations, λ-rings and the J-homomorphism", Topology, 8 (1969), 253-297.

[4] H. Bass, J. Milnor & J.P. Serre, "Solution of the congruence sub-group problem for SL_n $(n \geq 3)$ and Sp_{2n} $(n \geq 2)$", Publ. Math. de l'IHES, 33 (1967), 59-137.

[5] P. Deligne & D. Sullivan, "Fibrés vectoriels complexes à groupe structural discret", C.R. Acad. Sci. Paris, Sér. A-B 281 (1975), no. 24, Ai, A1081-A1083.

[6] L. Evens, "The cohomology ring of a finite group", Trans. Amer. Math. Soc., 101 (1961), 224-239.

[7] L. Evens, "On the Chern classes of representations of finite groups", Trans. Amer. Math. Soc., 115 (1965), 180-193.

[8] B. Huppert, Endliche Gruppen I, Springer-Verlag, Heidelberg, 1967.

[9] J. Ritter, "Zur kohomologie von gruppen mit abelschen Sylowuntergruppen", Manuscripta Math., 3 (1970), 69-95.

[10] G.B. Segal,"Equivariant K-theory", Publ. Math. de l'IHES, 34 (1968), 129-151.

[11] C.B. Thomas, "Riemann-Roch formulae for group representations", Matematika, 20 (1973), 253-262.

[12] C.T.C. Wall, "On the cohomology of certain groups", Proc. Cambridge Phil. Soc., 57 (1961), 731-733.

[13] E.A. Weiss, Bonner Mathematische Schriften, 1969.

DEPARTMENT OF PURE MATHEMATICS AND MATHEMATICAL STATISTICS
UNIVERSITY OF CAMBRIDGE
16 MILL LANE
CAMBRIDGE CB2 1SB
ENGLAND

1980 Mathematics Subject Classification: 55R40,55 R50

Research partially supported by FIM-ETH, Zürich

Contemporary Mathematics
Volume 19, 1983

THE HOMOTOPY LIMIT PROBLEM

R. W. Thomason*

ABSTRACT. I describe a problem which encompasses Segal's Burnside ring conjecture, Atiyah's theorem on the K-theory of classifying spaces, my descent theorem for algebraic K-theory, Quillen's conjecture on the algebraic K-groups of an algebraically closed field in characteristic p, and results of Giffen, Karoubi, and Guin on the relation between K- and L-theory.

The general problem is that of the relation of the lax limit of a group action on a symmetric monoidal category and the homotopy limit of the group action on the associated spectrum.

I would like to thank C. Giffen, who first formulated the general homotopy limit problem and who incited my interest in it. The formulation I use here is somewhat different from Giffen's, which is better adapted to L-theory.

1. I begin by recalling some basic facts about homotopy limits of group actions. Let G be a group acting on a space X. Let EG be the free acyclic G space which is the classifying space of the category $\underline{\underline{EG}}$ below. The homotopy limit of G acting on X is the space of equivariant maps from EG to X (1.1).

$$\text{(1.1)} \qquad \operatorname*{holim}_{G} X = \mathrm{Map}_G(EG, X)$$

This construction commutes with the loop space functor, so if G acts on a spectrum X, the homotopy limit is a spectrum.

Filtering EG by skeleta induces a tower of fibrations on the homotopy limit. The resulting long exact sequences of homotopy groups assemble into an exact couple, yielding a spectral sequence

$$\text{(1.2)} \qquad E_2^{p,q} = H^p(G;\ \pi_q X) \Rightarrow \pi_{q-p}\mathrm{Map}_G(EG, X)$$

The indexing is funny, so the differential d_r has bidegree $(r, r-1)$. For X a spectrum, this is a half-plane spectral sequence. If X is a space,

AMS (MOS) classification 18G99
* partially supported by NSF

there is some trouble in that $\pi_q X$ is not abelian if $q = 1$, not a group if $q = 0$, and not defined for $q < 0$. In both cases, convergence of the spectral sequence is problematic.

If X is a K(M, 0) spectrum for M a G-module, the homotopy limit is a generalized Eilenberg-MacLane spectrum whose homotopy groups are the cohomology groups of G with coefficients in M. If X is a generalized Eilenberg-MacLane spectrum, it corresponds to a chain complex under the Dold-Kan equivalence of categories. The homotopy groups of the homotopy limit are then the hypercohomology groups of G with coefficients in the chain complex corresponding to X. In general, I interpret the homotopy limit (1.1) as the hypercohomology spectrum of G with coefficients in X. This point of view is inspired by Quillen's "homotopical algebra" which generalizes homological algebra by replacing the category of chain complexes in an abelian category with more general categories like the category of spectra. This generalization is necessary for the development of higher algebraic K-theory, and accounts for the role of topology in that subject.

Note that if G acts trivially on X, there is an isomorphism

$$\mathrm{Map}_G(EG, X) = \mathrm{Map}(BG, X) \tag{1.3}$$

The notion of homotopy limit may be defined more generally for any diagram of spaces or spectra parameterized by a small category $\underline{\underline{K}}$. The basic reference, written in terms of simplicial sets, is chapter XI of [BK]. The case of diagrams of spectra is explicitly developed in [T] §5. Justification for the statements above is found in these references.

2. Let the group G act on a small category $\underline{\underline{C}}$. Let $\underline{\underline{EG}}$ be the category whose objects are the elements of G, and with a unique morphism between any two objects. The lax limit of G acting on $\underline{\underline{C}}$ is the category of equivariant functors and natural transformations from $\underline{\underline{EG}}$ to $\underline{\underline{C}}$

$$\operatorname*{laxlim}_G \underline{\underline{C}} = \mathrm{Cat}_G(\underline{\underline{EG}}, \underline{\underline{C}}) \tag{2.1}$$

An explicit description is this: an object of the lax limit is a pair (C, ψ) where C is an object of $\underline{\underline{C}}$ and ψ is a function assigning to each $g \in G$ a morphism $\psi(g)$ in $\underline{\underline{C}}$

$$\psi(g): \ C \longrightarrow gC \tag{2.2}$$

The function ψ must satisfy normalization and cocycle identities

$\psi(1) = 1$

(2.3)
$$\begin{array}{ccc} C & \xrightarrow{\psi(g)} & gC \\ & {\scriptstyle\psi(gh)}\searrow & \downarrow{\scriptstyle g\psi(h)} \\ & & ghC \end{array}$$

$$\psi(gh) = g\psi(h) \cdot \psi(g)$$

Note $1 = \psi(1) = \psi(gg^{-1}) = g\psi(g^{-1}) \cdot \psi(g)$, and $1 = g1 = g(g^{-1}\psi(g)\ \psi(g^{-1}))$ $= \psi(g) \cdot g\psi(g^{-1})$, so each $\psi(g)$ is an isomorphism

A morphism $(C,\psi) \longrightarrow (C', \psi')$ in the lax limit is a morphism $c\colon\ C \longrightarrow C'$ in $\underline{\underline{C}}$ such that (2.4) commutes

(2.4)
$$\begin{array}{ccc} C & \xrightarrow{\psi(g)} & gC \\ {\scriptstyle c}\downarrow & & \downarrow{\scriptstyle gc} \\ C' & \xrightarrow[\psi'(g)]{} & gC' \end{array}$$

This construction produces many interesting categories as we'll see below. For now, note that if G acts trivially on $\underline{\underline{C}}$, the lax limit is the category of representations of G in $\underline{\underline{C}}$.

If $\underline{\underline{C}}$ is a symmetric monoidal category and the G action respects this structure, the lax limit inherits a symmetric monoidal structure with

(2.5) $$(C,\psi) \oplus (C',\psi') = (C \oplus C', \psi \oplus \psi')$$

The concept of lax limit is defined for any diagram of small categories parameterized by a small category $\underline{\underline{K}}$, and even for pseudo- and lax-diagrams. The construction in its explicit form is due to Street [S].

3. Let N: $\underline{\text{Cat}} \longrightarrow \underline{\Delta^{op}\text{-sets}}$ be the functor sending a category to its nerve. The geometric realization of the nerve is the usual classifying space of a category. Let the category $\underline{\underline{n}}$ have objects the integers 0, 1, 2, ..., n, with a morphism $i \to j$ if i is less than j. Then $N\underline{\underline{n}}$ is the standard n-simplex $\Delta[n]$.

An n-simplex of N $\text{Cat}_G(\underline{\underline{EG}}, \underline{\underline{C}})$ is a functor $\underline{\underline{n}} \longrightarrow \text{Cat}_G(\underline{\underline{EG}}, \underline{\underline{C}})$, which corresponds to an equivariant functor $\underline{\underline{EG}} \times \underline{\underline{n}} \longrightarrow \underline{\underline{C}}$. As N is full and faithful, this equivariant functor corresponds to an equivariant simplicial map

$$(3.1) \qquad EG \times \Delta[n] = N(\underline{\underline{EG}} \times \underline{\underline{n}}) \longrightarrow N\underline{\underline{C}}$$

This in turn corresponds to an n-simplex of the simplicial mapping space $Map_G(EG, N\underline{\underline{C}})$. Thus there is an isomorphism of simplicial sets

$$(3.2) \qquad N\ Cat_G(\underline{\underline{EG}}, \underline{\underline{C}}) \cong Map_G(EG, N\underline{\underline{C}})$$

Applying geometric realization, which is a closed functor and so is compatible with mapping space constructions, one gets a canonical map of spaces

$$(3.3) \qquad B\ Cat_G(\underline{\underline{EG}}, \underline{\underline{C}}) \cong |Map_G(EG, N\underline{\underline{C}})| \longrightarrow Map_G(EG, B\underline{\underline{C}})$$

If $\underline{\underline{C}}$ is symmetric monoidal or permutative, one gets a similar map of the associated spectra built by infinite loop space machines

$$(3.4) \qquad Spt(Cat_G(\underline{\underline{EG}}, \underline{\underline{C}})) \longrightarrow Map_G(EG, Spt\ \underline{\underline{C}})$$

These are maps from the lax limit to the homotopy limit. The homotopy limit problem asks whether the maps (3.3), (3.4) are homotopy equivalences, or rather how close they come to being homotopy equivalences.

If every morphism in $\underline{\underline{C}}$ is an isomorphism, then $N\underline{\underline{C}}$ is a Kan complex and (3.3) is a homotopy equivalence. If $\underline{\underline{C}}$ is also symmetric monoidal, (3.4) needn't be a homotopy equivalence. The map (3.3) is not a homotopy equivalence for general $\underline{\underline{C}}$. However, many examples below show that (3.4) becomes a homotopy equivalence if some appropriate modification is made. I would like a general principle to explain this phenomenon.

There are general principles that relate lax colimits with homotopy colimits, both in the case of spaces and of spectra [TH], [TF1], [TF2]. This dual problem appears much easier.

I should point out that the isomorphism (3.2) was first published by John Gray in [G]. I now turn form the general problem to a series of interesting examples.

4. Suppose that $\underline{\underline{C}}$ is a symmetric monoidal category on which G acts trivially. Then laxlim $\underline{\underline{C}}$ is the category Rep(G, $\underline{\underline{C}}$) of representations of G in $\underline{\underline{C}}$. The homotopy limit is the spectrum of maps from BG to Spt $\underline{\underline{C}}$. The homotopy limit problem asks how close the canonical map (4.1) is to being a homotopy equivalence.

$$(4.1) \qquad \mathrm{Spt}(\mathrm{Rep}(G, \underline{\underline{C}})) \longrightarrow \mathrm{Map}(BG, \mathrm{Spt}\ \underline{\underline{C}})$$

If $\underline{\underline{C}}$ is the category of finite sets and isomorphisms, with symmetric monoidal structure given by disjoint union, then Spt $\underline{\underline{C}}$ is the sphere spectrum by the Barratt-Priddy-Quillen theorem. Rep(G, $\underline{\underline{C}}$) is the category of finite G-sets and isomorphisms. For finite G, π_0 Spt(Rep(G, $\underline{\underline{C}}$)) is thus the Burnside ring A(G). One form of Segal's conjecture is that (4.1) induces an isomorphism on homotopy groups π_* after completing the homotopy groups on the left side with respect to the augmentation ideal of A(G). Carlsson says this is true at least in non-positive degrees. Note that the necessity of completing with respect to the augmentation ideal shows that (4.1) is not strictly a homotopy equivalence.

Now let $\underline{\underline{C}}$ be the category of finite dimensional vector spaces over a field k. The morphisms are linear isomorphisms, and the symmetric monoidal structure is given by direct sum. Spt $\underline{\underline{C}}$ is the spectrum K(k) whose homotopy groups are the algebraic K-groups of k. If G is finite with order invertible in k, Rep(G, $\underline{\underline{C}}$) is the category of finitely generated projective k[G] modules and isomorphisms. Thus Spt(Rep(G, $\underline{\underline{C}}$)) is the algebraic K-theory spectrum K(k[G]).

One can extend the formalism to the case where $\underline{\underline{C}}$ is a topological category. If $\underline{\underline{C}}$ is the category of finite dimensional complex vector spaces and isomorphims with the usual topology on $GL_n(\mathbb{C})$, Spt($\underline{\underline{C}}$) is the connective spectrum for topological K-theory, bu. For G finite, π_0 Spt(Rep(G, $\underline{\underline{C}}$)) is the complex representation ring R(G). Spt(Rep(G, $\underline{\underline{C}}$)) is a product of bu's indexed by a basis of R(G). The homotopy limit problem asks how close this product is to Map(BG, bu). Atiyah's theorem [A] says that if bu is replaced by the periodic spectrum BU obtained by inverting the Bott element and if R(G) is completed with respect to the augmentation ideal, then (4.1) becomes a homotopy equivalence

$$(4.2) \qquad \widehat{R(G)} \otimes_{\mathbb{Z}} BU \xrightarrow{\sim} \mathrm{Map}(BG, BU)$$

There is an analogue for Atiyah's theorem in algebraic K-theory. For simplicity, I'll restrict to the case where G is a finite ℓ-group for a prime ℓ. Instead of completing, I'll reduce all spectra mod a power of ℓ by smashing with a $\mathbb{Z}/\ell^\nu$ Moore spectrum. I get periodic spectra by inverting the Bott element in algebraic K-theory [T].

<u>Theorem 4.1</u>: Let G be a finite ℓ-group, ℓ a prime. Let k be a field of characteristic not ℓ (and which contains primitive 16th or 9th roots of unity if ℓ = 2 or 3 respectively). Then there is a homotopy equivalence of

K-theory spectra, induced by (4.1)

$$(4.3) \qquad K/\ell^\nu(k[G])[\beta^{-1}] \simeq \mathrm{Map}(BG, K/\ell^\nu(k)[\beta^{-1}])$$

If in addition $k = \mathbb{F}_q$ is a finite field, there is a homotopy equivalence of non-periodic spaces (not spectra)

$$(4.4) \qquad K/\ell^\nu(\mathbb{F}_q[G]) \simeq \mathrm{Map}(BG, K/\ell^\nu(\mathbb{F}_q))$$

Pf: If k contains all $|G|$th roots of unity, k[G] is Morita equivalent to a product of copies of k indexed by a basis of R(G). Thus Atiyah's proof in [A] generalizes to this case. If k doesn't contain enough roots of unity, the result follows by etale cohomological descent [T] from an extension of k. The result for $\mathbb{F}_q$ follows as inverting β affects only the negative K-groups of $\mathbb{F}_q$ and $\mathbb{F}_q[G]$, which is Morita equivalent to a products of various finite fields.

5. Let L'/L be a Galois extension of fields with Galois group G. Let $\underline{\underline{C}}$ be the category of finite dimensional vector spaces over L' and isomorphisms. G acts on $\underline{\underline{C}}$ via its action on L'. If V' is in $\underline{\underline{C}}$, gV' is the abelian group V' with new L' action given by pulling back the old action along $g^{-1}: L' \to L'$. The category laxlim $\underline{\underline{C}}$ is the category of semilinear representations of G; its objects are vector spaces V' together with a compatible family of isomorphisms $\psi(g)$: $V' \to gV'$. If V is a vector space over L, let (V', ψ) be defined by (5.1)

$$(5.1) \qquad V' = L' \otimes_L V \quad , \qquad \psi(g) = g \otimes 1 : L' \otimes_L V \xrightarrow{\ \cong\ } gL' \otimes_L V$$

This extends to a functor from the category of finite dimensional vector spaces over L to laxlim $\underline{\underline{C}}$. The theory of faithfully flat descent says that this functor is an equivalence of categories. See [SGA $4\frac{1}{2}$], Arcata I.4 for more details, in particular for an explanation of how this equivalence implies Hilbert's Theorem 90.

In this case, the homotopy limit problem asks whether there is an equivalence of K-theory spectra

$$(5.2) \qquad K(L) \simeq \mathrm{Map}_G(EG, K(L'))$$

If this were an equivalence, the spectral sequence (1.2) would be the spectral sequence relating the K-groups of a field extension as conjectured by Quillen and Lichtenbaum. This conjecture turns out to be false. However my cohomological descent theorem [T] §2 shows this conjecture is true after reducing mod a prime power and inverting the Bott element.

Theorem 5.1: Let L'/L be a finite Galois extension with Galois group G. Let ℓ be a prime invertible in L, and suppose L contains primitive 16th or 9th roots of unity if $\ell = 2$ or 3 respectively. Then the map (3.4) induces a homotopy equivalence

$$(5.3) \qquad K/\ell^{\nu}(L)[\beta^{-1}] \longrightarrow Map_G(EG, K/\ell^{\nu}(L')[\beta^{-1}])$$

Pf: This is proved in [T]. Specifically it results from [T] 2.20, 2.21, 2.30, (3.29), and 3.23. The cases $\ell = 2,3$ will be handled in the second edition of [T], and in [TE].

This theorem is an important step in understanding the relation between algebraic and topological K-theory. One of the other important steps also fits in the framework of the homotopy limit problem. Let k be an algebraically closed field of characteristic p. Let ϕ^q be the qth power Frobenius map on k. The infinite cyclic group $\mathbb{Z}$ acts on the category of finite vector spaces over k with a generator acting via pullback along ϕ^q. Lang's theorem [L] identifies the lax limit to the category of finite vector spaces over $\mathbb{F}_q$. If the map (3.4) were an equivalence, at least on connected components of zeroth spaces, the spectral sequence (1.2) would yield a fibration sequence of K-theory spaces

$$(5.4) \qquad BGL(\mathbb{F}_q)^+ \longrightarrow BGL(k)^+ \xrightarrow{1-\phi^q} BGL(k)^+$$

This sequence was conjectured by Quillen. It's importance is that it is equivalent to the sequence (5.5) as shown by Hiller [H]

$$(5.5) \qquad BGL(\overline{\mathbb{F}}_p)^+ \longrightarrow BGL(k)^+ \longrightarrow BGL(k)^+ \otimes \mathbb{Q}$$

This sequence and Quillen's computation of the K-groups of the algebraically closed field $\overline{\mathbb{F}}_q$ would yield a calculation of the mod ℓ^{ν} K-groups of the general algebraically closed field of characteristic p, k. If follows from [TQ] that these fibration sequences do exist for mod ℓ^{ν} K-theory after inverting the Bott element. This suffices for most applications as these also need the descent theorem 5.1 which itself requires inverting the Bott element. However, it would be nice to know if (5.4) and (5.5) are true as stated.

6. Let R be a ring with involution, and $\underline{\underline{C}}$ the category of finitely generated free R modules and isomorphisms, $\coprod GL_n(R)$. The group $G = \mathbb{Z}/2$ acts on $\underline{\underline{C}}$ by sending an isomorphism represented by a matrix M to $(\overline{M}^t)^{-1}$, the conjugate transpose inverse. The laxlimit is the category of free R-modules together with an invertible matrix M satisfying the cocycle

condition, $M \cdot (\overline{M}^t)^{-1} = 1$. Thus laxlim $\underline{\underline{C}}$ is the category of free R modules with non-singular hermitian symmetric form. The higher homotopy groups of Spt(laxlim $\underline{\underline{C}}$) are the Karoubi L-groups of R. If the map (3.4) were an equivalence, the spectral sequence (1.2) would relate $\mathbb{Z}/2$ cohomology with coefficients in $K_*(R)$ to the Karoubi L-groups. Guin [Gu] has results similar to this, and his paper led me to try Karoubi periodicity as a general method of attack on the homotopy limit problem. While this method fails to prove the Segal conjecture, it did lead to the proof of cohomological descent for algebraic K-theory, and it can be used to prove Atiyah's theorem.

Giffen has shown how to extend the general formalism to include skew-hermitian and other types of forms, and has proved general Karoubi periodicity theorems in L-theory. None of the L-theory Karoubi periodicities are as simple as the form discussed below because of the alternation between hermitian and skew-hermitian in L-theory.

7. In this section I indicate how a Karoubi periodicity theorem can solve a homotopy limit problem. Let G be a finite group acting symmetrically monoidally on $\underline{\underline{C}}$. There is the usual forgetful functor sending (C, ψ) to C

$$(7.1) \qquad \lambda^* : \operatorname*{laxlim}_G \underline{\underline{C}} \longrightarrow \underline{\underline{C}}$$

There is also a transfer functor λ_*, with $\lambda_* C = (\oplus\, gC, \psi)$ where the "sum" is taken over all $g \in G$ and $\psi(h)\colon\ \oplus\, gC \cong \oplus\, hgC$ is the obvious permutation isomorphism.

$$(7.2) \qquad \lambda_* : \underline{\underline{C}} \longrightarrow \operatorname*{laxlim}_G \underline{\underline{C}}$$

If G is the Galois group of L' over L and $\underline{\underline{C}}$ is the category of finite dimensional vector spaces over L', $\mathrm{Spt}(\lambda^*)$ and $\mathrm{Spt}(\lambda_*)$ are the usual and transfer maps respectively on the algebraic K-theory spectra.

If one works with spectra reduced mod ℓ^ν and has a sufficiently general class of homotopy limit problems to be solved, one can reduce to the case $G = \mathbb{Z}/\ell$. See [A] and [T] §2 for examples of each reduction. Henceforth, I'll assume $G = \mathbb{Z}/\ell$ for ℓ prime, and let τ be a generator of G. Define U and V by the extended homotopy fibre sequences (7.3)

$$(7.3) \qquad \begin{array}{l} \Omega\mathrm{Spt}(\underline{\underline{C}}) \xrightarrow{\partial} U(\underline{\underline{C}}) \xrightarrow{u} \mathrm{Spt}(\mathrm{laxlim}\ \underline{\underline{C}}) \xrightarrow{\lambda^*} \mathrm{Spt}(\underline{\underline{C}}) \\ \Omega\mathrm{Spt}(\mathrm{laxlim}\ \underline{\underline{C}}) \xrightarrow{\partial} V(\underline{\underline{C}}) \xrightarrow{v} \mathrm{Spt}(\underline{\underline{C}}) \xrightarrow{\lambda_*} \mathrm{Spt}(\mathrm{laxlim}\ \underline{\underline{C}}) \end{array}$$

My slogan for remembering which is which is "U for usual, V for Verlagerung". However, when this is specialized to the L-theory examples of §6, my U becomes Karoubi's V and vice-versa.

Suppose there is a Karoubi periodicity homotopy equivalence

$$(7.4) \qquad \theta : U(\underline{\underline{C}}) \xrightarrow{\sim} \Omega V(\underline{\underline{C}})$$

such that $\Omega v \cdot \theta \cdot \partial = 1-\tau$ as an endomorphism of $\Omega \mathrm{Spt}(\underline{\underline{C}})$. Then there is a horizontal tower of fibrations with fibre sequence triangles

(7.5)

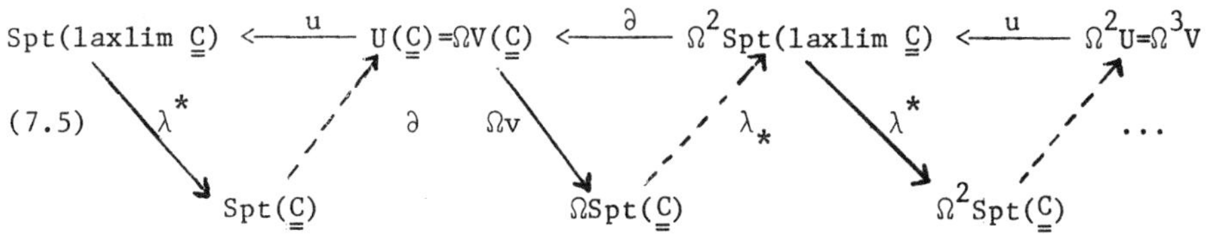

The tower extends to the right periodically. Suppose that the homotopy inverse limit of the tower is contractible, as happens if $u \cdot \theta^{-1} \cdot \partial$ is null homotopic. Then just as an Adams resolution yields the Adams spectral sequence, this tower yields a spectral sequence converging to $\pi_* \mathrm{Spt}(\mathrm{laxlim}\ \underline{\underline{C}})$. The E_1 term is a sum of copies of $\pi_* \mathrm{Spt}(\underline{\underline{C}})$, with differential d_1 induced by $\Omega v \cdot \theta \cdot \partial = 1-\tau$, and $\lambda^* \lambda_* = 1+\tau+\tau^2+\ldots+\tau^{\ell-1}$. This E_1 term is thus the canonical periodic resolution for computing the cohomology of the cyclic group $\mathbb{Z}/\ell$ with coefficients in $\pi_* \mathrm{Spt}(\underline{\underline{C}})$. Thus from the E_2 term on, the spectral sequence is

$$(7.6) \qquad E_2^{p,q} = H^p(\mathbb{Z}/\ell\,;\ \pi_q\ \mathrm{Spt}(\underline{\underline{C}})) \Longrightarrow \pi_{q-p}\ \mathrm{Spt}(\mathrm{laxlim}\ \underline{\underline{C}})$$

This looks suspiciously like the spectral sequence (1.2). In fact, if Karoubi periodicity also holds for various $\mathrm{Cat}(\prod_1^n G,\ \underline{\underline{C}})$ with twisted G-action, one can show that the map (3.4) is a homotopy equivalence. For details in the Galois case, see [T] 2.25-2.30. Similar results hold after reducing all spectra mod ℓ^ν or after inverting the Bott element β, provided Karoubi periodicity holds after doing this.

8. I'll now describe the ideas behind the proof of Karoubi periodicity for Atiyah's theorem and for my descent theorem. The technical details of the latter are quite elaborate, and the discussion below contains several omissions, oversimplifications, and outright lies. For an accurate and honest account, see [T] and [TE]. Karoubi's proof of Karoubi periodicity in [K] was the paradigm for my proof, but is technically simpler.

For Atiyah's theorem, the sequences (7.3) become

$$(8.1) \qquad \begin{array}{l} U \longrightarrow \prod_1^\ell bu \longrightarrow bu \\ V \longrightarrow bu \longrightarrow \prod_1^\ell bu \end{array}$$

Here the products are indexed by the characters of $\mathbb{Z}/\ell$, a natural basis of the representation ring $R(\mathbb{Z}/\ell)$. Thus V is a product of $\ell-1$ copies of Ωbu. Inverting the Bott element replaces bu by the periodic spectrum BU. The equivalence $BU \simeq \Omega^2 BU$ yields a Karoubi periodicity equivalence $U \simeq \Omega V$ after inverting the Bott element. It's necessary to choose the correct equivalence and to complete with respect to the augmentation ideal to get convergence in the tower (7.5).

Now let L'/L be a Galois extension of fields with Galois group $\mathbb{Z}/\ell$. For any algebra A over L, one has fibration sequences (8.2) of algebraic K-theory spectra

(8.2)
$$\begin{array}{l} U(A) \longrightarrow K(A) \xrightarrow{\lambda^*} K(A \otimes_L L') \\[1em] V(A) \longrightarrow K(A \otimes_L L') \xrightarrow{\lambda_*} K(A) \end{array}$$

Consider (8.3)

$$\begin{array}{ccccc} \Omega K(L) & \xrightarrow{\lambda^*} & \Omega K(L') & \xrightarrow{\partial} & U(L) \\ & \swarrow & \Big\downarrow 1-\tau & \swarrow & \\ \Omega V(L) & \xrightarrow{v} & \Omega K(L') & \xrightarrow{\lambda_*} & \Omega K(L) \end{array}$$

The composition of $1-\tau$ with either λ^* or λ_* is null homotopic. Choice of null homotopies provide lifts $U(L) \longrightarrow \Omega K(L')$ and $\Omega K(L') \longrightarrow \Omega V(L)$. There is a choice of compatible null homotopies, so the Toda bracket $\langle \lambda_*, 1-\tau, \lambda^* \rangle$ vanishes and there is a $\theta : U(L) \longrightarrow \Omega V(L)$ compatible with the lifts. The compatibilities are non-trivial, so the map θ is non-trivial. To convert this mush into mathematics, one deploys a maze of fibre sequences to express the Toda bracket on the primary homotopy level. In fact, one uses diagram (8.4), in which all columns are fibre sequences

(8.4)
$$\begin{array}{ccccc} U(A) & & \xrightarrow{\theta(A)} & & \Omega V(A) \\ \Big\downarrow & & & & \Big\downarrow \\ K(A) & \xrightarrow{\cup x} & D(A) & \longrightarrow & D(A) \\ \Big\downarrow \lambda^* & & \Big\downarrow \lambda^* & & \Big\downarrow d \\ K(A \otimes L') & \xrightarrow{\cup x} & D(A \otimes L') & \longrightarrow & V(A \otimes L') \\ & & & & \Big\downarrow \lambda_* \\ & & & & V(A) \end{array}$$

Here the cup product map is induced by an $x \in \pi_0 D(L)$. $\pi_0 V(L')$ is the augmentation ideal of the group ring $\mathbb{Z}[\mathbb{Z}/\ell]$, and x is defined so dx is $1-\tau$. The existence of the lift x expresses the compatibilities mentioned above, as is evident by the construction in [TE].

This $\theta(L)$ satisfies the condition $\Omega v \cdot \theta \cdot \partial = 1-\tau^{-1}$. If it were a homotopy equivalence, all would be well. One calculates low dimensional homotopy groups of $U(L)$ and $\Omega V(L)$ from (8.2).

$$\pi_{-2}\, U(L) = 0 \qquad \pi_{-2}\, \Omega V(L) = \mathbb{Z}/\ell$$

$$\pi_{-1}\, U(L) = 0 \qquad \pi_{-1}\, \Omega V(L) = \operatorname{coker}(\lambda_*\colon\ L'^* \to L^*)$$

(8.5) $$\pi_0\ U(L) = L'^*/L^*$$

$$0 \longrightarrow \operatorname{coker} \lambda_* K_2 \longrightarrow \pi_0 \Omega V(L) \longrightarrow \ker(\lambda_*\colon\ L'^* \to L^*) \longrightarrow 0$$

Note $\pi_{-2}\ \theta$ is not an isomorphism, so something must be done. On the other hand, Hilbert's Theorem 90 says that $1-\tau^{-1}$ or θ induces an isomorphism of L'^*/L^* on the kernal of the norm map, ker λ_*. Hilbert's Theorem 90 is part of the faithfully flat descent theory discussed in §5. Hilbert's Theorem 90 is also the basis of Kummer theory, which says that if L is of characteristic not ℓ and contains a primitive ℓth root of unity ζ, then $L' = L(\alpha)$ for α an ℓth root of $a \in L$. One can chose α so that $\tau\alpha/\alpha = \zeta$. I use this information to show $\theta[\beta^{-1}]$ is an equivalence.

Let $y \in \pi_0\ U(L)$ be the image of $\alpha \in K_1(L') = L'^*$ under the boundary map. Then $\Omega v\theta(y) = \Omega v \theta \partial(\alpha) = \alpha/\tau^{-1}\alpha = \zeta$. Suppose L' contains a primitive ℓ^2 root of unity γ, then $\zeta = \gamma^\ell$, so ζ is 0 mod ℓ in $K_1(L')$. Consider the fibre sequence of spectra reduced mod ℓ

(8.6) $$F/\ell(L) \longrightarrow U/\ell(L) \xrightarrow{\ \Omega v\theta\ } \Omega K/\ell(L')$$

I've shown that the reduction of y in $\pi_0 U/\ell(L)$ dies under $\pi_0 \Omega v\theta$, so it lifts to a t in $\pi_0 F/\ell(L)$. If θ were an equivalence, $F/\ell(L)$ would be $\Omega^2 K/\ell(L)$, and (8.6) would be a shift of a sequence in (8.2). The element t would be a basis of $\Omega^2 K/\ell(L)$ as a module over $K/\ell(L)$. One does have (8.7), inducing the map ϕ on fibres.

(8.7) $$\begin{array}{ccc} F/\ell(L) & \overset{\phi}{\dashrightarrow} & \Omega^2 K/\ell(L) \\ \downarrow & & \downarrow \\ U/\ell(L) & \xrightarrow{\ \theta\ } & \Omega V/\ell(L) \\ {\scriptstyle \Omega v\theta}\downarrow & & \downarrow \\ \Omega K/\ell(L') & \xrightarrow{\ 1\ } & \Omega K/\ell(L') \end{array}$$

Everything in sight is a module spectrum over the ring spectrum $K/\ell(L)$, and all maps are module maps. Thus the element t determines a cup product

map $\cup t:$ $K/\ell(L) \longrightarrow F/\ell(L)$. The Bockstein lemma reveals that
$\phi \cdot \cup t:$ $K/\ell(L) \longrightarrow \Omega^2 K/\ell(L)$ is cup product with the Bott element β .
This is because β Booksteins to $\zeta = \Omega v \theta(y)$ in $K_1(L')$, and it's the
divisibility of this element by ℓ that allows t to exist.

Consider the map Ξ induced by the shifted vertical fibre sequences on
the left half of (8.8). Here Ω^{-1} is a delooping functor on spectra.

(8.8)

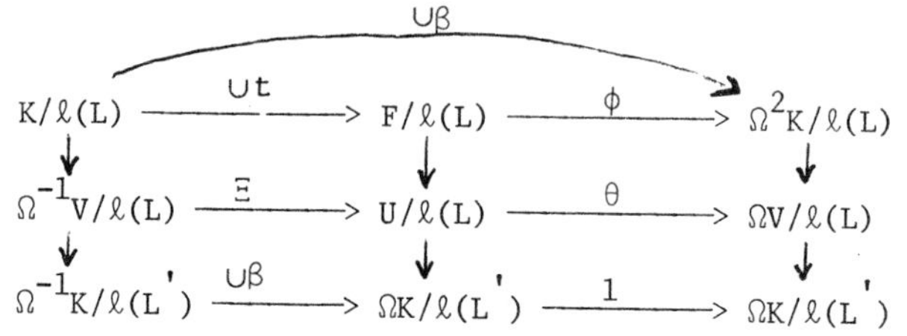

If one inverts β so that $\cup\beta$ is a homotopy equivalence, the 5-lemma shows
that $\theta \equiv [\beta^{-1}]$ is a homotopy equivalence.

There is also a commutative diagram (8.9), which shows Ξ θ $[\beta^{-1}]$ is a
homotopy equivalence.

$$
\begin{array}{ccc}
\Omega K/\ell(L)[\beta^{-1}] & \xrightarrow{\cup\beta} & \Omega^3 K/\ell(L)[\beta^{-1}] \\
\downarrow & & \downarrow \\
\Omega K/\ell(L')[\beta^{-1}] & \xrightarrow{\cup\beta} & \Omega^3 K/\ell(L')[\beta^{-1}] \\
\partial\downarrow & & \downarrow \\
U/\ell(L)[\beta^{-1}] & \xrightarrow{\Xi\theta} & \Omega^2 U/\ell(L)[\beta^{-1}]
\end{array}
$$

(8.9)

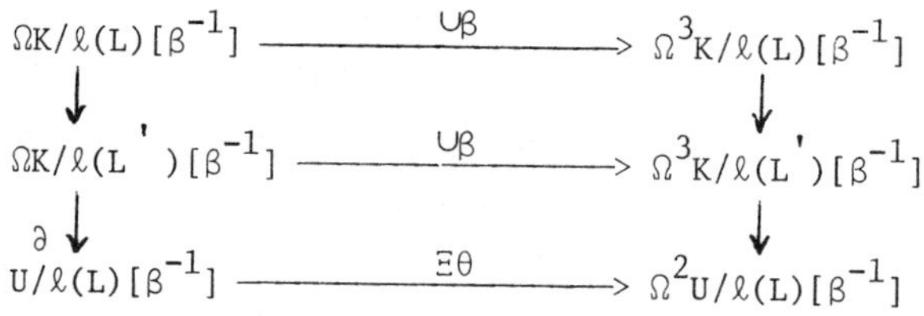

Thus $\theta:$ $U/\ell(L)[\beta^{-1}] \longrightarrow \Omega V/\ell(L)[\beta^{-1}]$ is the required Karoubi periodicity
equivalence.

The diagram (8.9) results from the commutative diagram (8.10), which
requires some work to verify.

(8.10)

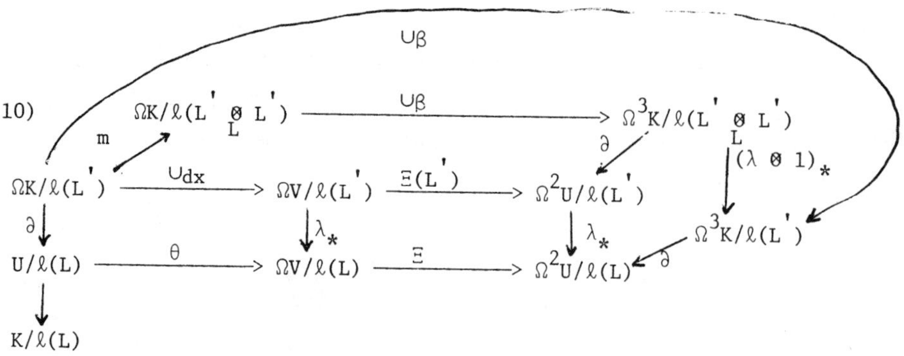

I hope this makes the idea of the proof clear. To get a real proof one
just patches all the holes.

Bibliography

[A] M.F. Atiyah, "Characters and cohomology of finite groups", Publ. Math. IHES 9(1961) 23-64

[BK] A.K. Bousfield, D.M. Kan, Homotopy Limits, Completions, and Localizations, Springer Lecture Notes in Math. 304 (1972)

[G] J.W. Gray, "Closed categories, lax limits, and homotopy limits", J. Pure Applied Alg., 19 (1980), 127-158

[Gu] D. Guin, "Une suite spectral en K-theorie hermitienne. Invariants des formes quadratiques", C.R. Acad. Sc. Paris, t. 289 (1978), 727-729

[H] H.L. Hiller, "λ-rings and algebraic K-theory", J. Pure Applied Alg., 20 (1981), 241-266

[K] M. Karoubi, "Le theoreme fondamental de la K-theorie hermitienne", Annals of Math., 112 (1980), 259-282

[L] S. Lang, "Algebraic groups over finite fields", Amer. J. Math. 78 (1956), 555-563

[S] R. Street, "Two constructions on lax functors", Cahier de Top. Geom. Diff., XIII (1972), 217-264

[SGA $4\frac{1}{2}$] P. Deligne et al. SGA $4\frac{1}{2}$, Springer Lecture Notes in Math. 569 (1977)

[T] R.W. Thomason, "Algebraic K-theory and etale cohomology", preprint (1980)

[TE] R.W. Thomason, "Erratum to 'Algebraic K-theory and etale cohomology'", xerographed scrawl (1982)
[TE] fixes an error in the proof of descent for infinite Galois extensions given descent for finite extensions, and indicates how to modify [T] to work for $\ell = 2$ or 3.

[TF1] R.W. Thomason, "First quadrant spectral sequences in algebraic K-theory", Algebraic Topology: Aarhus 1978, Springer Lecture Notes in Math. 763 (1979), 332-355

[TF2] R.W. Thomason, "First quadrant spectral sequences in algebraic K-theory via homotopy colimits", to appear, Comm. in Algebra.

[TH] R.W. Thomason, "Homotopy colimits in the category of small categories", Math. Proc. Camb. Phil. Soc. 85 (1979), 91-109

[TQ] R.W. Thomason, "The Lichtenbaum-Quillen conjecture for $K/\ell_*[\beta^{-1}]$", to appear, proceedings of the 1981 conference on algebraic topology at Univ. of Western Ontario, AMS Contemporary Mathematics

Mathematics Dept.
M.I.T.
Cambridge, MA 02139

Contemporary Mathematics
Volume 19, 1983

A PRIMER ON THE DICKSON INVARIANTS

Clarence Wilkerson

The ring of invariant forms of the full linear group $GL(V)$ of a finite dimensional vector space over the finite field F_q was computed early in the 20th century by L. E. Dickson [5], and was found to be a graded polynomial algebra on certain generators $\{c_{n,i}\}$. This ring of invariants, for $q = p$, has found use in algebraic topology in work of Milgram-Mann [9], Singer [16, 17], Adams-Wilkerson [1], Rector [13], Lam [6], Mui [11], and Smith-Switzer [18]. The aim of this exposition is to give a simple proof of the structure of the ring of invariants, and to compute the action of the Steenrod algebra on the generators of the invariants. The methods used are implicit in Adams-Wilkerson.

Dickson's viewpoint was to vastly generalize the defining equation of $\mathbb{F}_q$, $X^q = X$. This equation was replaced by the fundamental equation

$$f_n(X) = \prod_{v \in V} (X-v) = X^{q^n} + \sum_{i=0}^{n-1} (-1)^{n-i} c_{n,i} X^{q^i} .$$

The polynomial $f_n(X)$ has the property that its roots from a $\mathbb{F}_q$-vector space. The sparse nature of the coefficients almost immediately gives the structure of the ring of invariants, and the Steenrod algebra action (for q=p) can be easily determined from Dickson's fundamental equation. Other authors (Milgram-Mann, Singer, Smith-Switzer) have determined the action by other means, but it seems instructive to give a direct proof from Dickson's original viewpoint.

I. The Fundamental Equation and the Ring of Invariants

Let K be a field containing the $\mathbb{F}_q$-space V of dimension n. Here $q = p$ for p a rational integer prime. We first prove the equivalence of the two forms of the fundamental equation.

<u>Proposition 1.1</u>: If $f_n(X)$ is a monic separable polynomial in $K[X]$ such that its roots are the elements of V, then

0271-4132/83 $1.00 + $.25 per page

$$f_n(X) = X^{q^n} + \sum_{i=0}^{n-1} (-1)^{n-i} c_{n,i} X^{q^i}$$

for $c_{n,i} \in K$.

Note: The choice of signs in $f_n(X)$ is best explained by the proof.

Proof: $f_n(X) = \prod_{v \in V} (X-v)$. Choose a basis for V over F_q , $\langle x_1, \ldots, x_n \rangle$, and define V_{n-1} to be the subspace spanned by the first $n-1$ basis elements. The determinant

$$\Delta_n(X) = \mathrm{DET} \begin{bmatrix} x_1 & \cdots & X \\ x_1^q & \cdots & X^q \\ \cdot & \cdots & \cdot \\ \cdot & \cdots & \cdot \\ \cdot & \cdots & \cdot \\ x_1^{q^n} & \cdots & X^{q^n} \end{bmatrix}$$

is seen by column operations to have among its roots the elements of V . Hence $\Delta_n(X) = \Delta_{n-1}(x_n) f_n(X)$. It remains to verify that the constant $\Delta_{n-1}(x_n)$ is nonzero. For $n = 1, \Delta_1(X) = x_1 f_1(x) \neq 0$, so the induction can be started. If the statement is true for vector spaces of dimension less than n , then

$$\Delta_{n-1}(X) = \Delta_{n-2}(x_{n-1}) f_{n-1}(X) \neq 0 .$$

If V_{n-1} is used to define $\Delta_{n-1}(X)$, then since x_n is not in V_{n-1} , it is not a root and hence $\Delta_{n-1}(x_n) \neq 0$. It is also easily seen by this induction that $\Delta_{n-1}(x_n) = \prod_{u \in V^*} u$, where V^* is the subset of V which are non-zero and for which the last non-zero coordinate is 1 .

Now in the application to Dickson's theorem, we take K as the field of fractions of the symmetric algebra $S(V)$ over F_q . By a choice of basis, this can be identified with a polynomial ring on n-variables. It is convenient to grade $S(V)$ and K with V having grading 2 (resp. 1 for $p = 2$). Then $GL(V)$ acts on $S(V)$, preserving the grading.

Theorem 1.2: (Dickson [5])

$$S(V)^{GL(V)} = F_q[c_{n,n-1}, \ldots, c_{n,0}]$$

where $f_n(X) = \prod_{v \in V} (X-v) = X^{q^n} + \sum_{i=0}^{n-1} (-1)^{n-i} c_{n,i} X^{q^i}$.

The $\{c_{n,i}\}$ have degrees $\{2(q^n - q^i)\}$, respectively $\{q^n - q^i\}$ for $p = 2$. The $\{c_{n,i}\}$ are the unique, up to scalar multiple, non-zero invariant elements in these degrees.

Proof: Clearly $\{c_{n,i}\} \subset S(V)^{GL(V)}$, and hence the algebra R^* generated by the $\{c_{n,i}\}$ is invariant. But $S(V)$ is integral over R^*, so the transcendence degree of R^* over F_q is also n. There are exactly n of the $c_{n,i}$, so the elements of $\{c_{n,i}\}$ are algebraically independent. Thus

$$R^* = F_q[c_{n,0},\ldots,c_{n,n-1}] \subset S(V)^{GL(V)} .$$

Let L^* be the graded field of fractions of R^*, and K^* that of $S(V)$. Since K^* is the splitting field for $f_n(X)$ over L^*, the extension is Galois, with Galois group W. Then $GL(V) \subset W$, since L^* is $GL(V)$ invariant, and the action of $GL(V)$ on K^* is faithful. But $WV \subset V$, since the Galois group permutes the roots of $f(X)$. The action of W on K^* is L^* linear, and in particular, it is F_q linear. Since the action on V determines the action on K^*, $GL(V)$ is a subgroup of W and W is a subgroup of $GL(V)$. That is, $W = GL(V)$. Hence $K^{*GL(V)} = L^*$. Since R^* is a polynomial algebra, it is integrally closed. $S(V)^{GL(V)}$ is integral over R^*, and hence

$$F_q[c_{n,0},\ldots,c_{n,n-1}] = R^* = S(V)^{GL(V)} .$$

Finally, we provide some explicit formulae for the invariants.

Proposition 1.3: Let $B = \langle x_1,\ldots,x_n\rangle$ be an ordered basis for V over F_q.

a) If A_B is the $(n+1) \times n$ matrix with entries $\{x_j^{q^i} : 0 \le i \le n,\ 1 \le j \le n\}$, and $A_B(i)$ is this matrix with the i-th row deleted, then $c_{n,i} = \det(A_B(i))/\Delta_{n-1}(x_n)$.

b) Let V_{n-1} be span $\langle x_1,\ldots,x_{n-1}\rangle$, and $\{c_{n-1,i}\}$ be the Dickson generators for the invariants of $GL(V_{n-1})$. Then $c_{n,i} = c_{n-1,i-1}{}^q - c_{n-1,i-1}f_{n-1}(x_n)^{q-1}$, where $f_{n-1}(x_n) = \prod_{v \in V_{n-1}} (x_n - v)$ as in Prop. 1.1.

Proof: a) This follows immediately from $\Delta_n(X) = \Delta_{n-1}(x_n)f_n(X)$ and expansion by minors of $\Delta_n(X)$.

b)

$$f_n(X) = \prod_{v\in V}(X-v) = \prod_{a\in \mathbb{F}_q}\ \prod_{v\in V_{n-1}} (X-ax-v)$$

$$= \prod_{a\in \mathbb{F}_q} f_{n-1}(X-ax_n) = \prod_a (f_{n-1}(X) - af_{n-1}(x_n))$$

$$= f_{n-1}(X)^q - f_{n-1}(X)f_{n-1}(x_n)^{q-1} .$$

Hence by comparing coefficients, we have

$$c_{n,i} = c_{n-1,i-1}{}^q - c_{n-1,i}f_{n-1}(x_n)^{q-1} .$$

Corollary 1.4: If $\varphi : V \to U$ is surjective, then

$$\varphi^* : S(V)^{GL(V)} \to S(U)$$

has image exactly

$$(S(U)^{GL(U)})^{q^{\dim y - \dim y}} .$$

[Rector,13]

II. The Steenrod Algebra Action

We now restrict our attention to $q = p$. Then $S(V)$ has an unique action of the Steenrod algebra A_p compatible with the unstable axiom and the Cartan formula. Namely,

$$\beta v = 0$$

$$P^0 v = v$$

$$P^1 v = v^p$$

$$P^{k+1} v = 0 \quad \text{for} \quad v \in V \quad \text{and} \quad k \geq 1 .$$

The action of $GL(V)$ commutes with this A_p-action, so the invariants inherit an unstable A_p-action. This section gives explicit formulae for the action on the generators $\{c_{n,i}\}$ that are useful in applications. In this section, the indeterminate X is treated as a 2-dimensional (1 for p=2) element with an unstable A_p-action on the algebra it generates.

Proposition 2.1: For $f(X) = \prod_{v\in V}(X-v) = X^{q^n} + \sum_{i=0}^{n-1} (-1)^{n-i}c_{n,i}X^{q^i}$, $f(X)$ divides $P^k f(X)$ in $\mathbb{F}_q[c_{n,0},\ldots,c_{n,n-1}][X]$. Hence

$$P^k f(X) = -P^{k-p^{n-1}} c_{n,n-1} f(X) \quad \text{for} \quad k \neq p^n \quad \text{or} \quad 0$$

$$P^{p^n} f(X) = f(X)^p .$$

Here $P^j \equiv 0$ if $j < 0$, $c_{n,n} = 1$, and $c_{n,j} = 0$ if $j < 0$.

Example: The "total" Steenrod class for $p = 2$ is

$$Sq_T(f(X)) =$$

$$f(X)[f(X)T^{2^n} + c_{n,n-1}{}^2T^{2^{n-1}} + c_{n,0}T^{2^{n-1}-1} + c_{n,1}T^{2^{n-1}-2} + \ldots + c_{n,n-2}T^{2^{n-2}} + 1] .$$

The Steenrod algebra has Milnor's [10] primitives $\{P^{\Delta_i}\}$. These were denoted by $\{Q^i\}$ in Adams-Wilkerson [1].

Proposition 2.2: Let P^{Δ_i} be the primitive of dimension $2(p^i-1)$ (respectively $2^i - 1$ for $p = 2$) in the Milnor basis of A_p dual to ξ_i. Then

$$P^{\Delta_i} v = v^{p^i}$$

$$P^{\Delta_i} f(X) = 0 \quad \text{for} \quad i < n$$

$$P^{\Delta_n} f(X) = (-1)^n c_{n,0} f(X)$$

$$P^{\Delta_i} f(X) = P^{p^{i-1}} P^{\Delta_{i-1}} f(X), \text{ for } i > n .$$

Given Propositions 2.1 and 2.2, one can quickly read off a recursion relation for the Steenrod algebra action.

Corollary 2.3:

a) $P^k c_{n,i} = P^{k-p^{i-1}} c_{n,i-1} + (-1)^{n-i-1}(P^{k-p^{n-1}} c_{n,n-1})c_{n,i}$, if $k \neq 0$ or p^n

b) $P^{\Delta_j} c_{n,i} = 0$ if $i \neq j$, $0 \le j < n$

$$P^{\Delta_j} c_{n,j} = (-1)^{j+1} c_{n,0} \quad \text{if} \quad 0 \le j < n$$

$$P^{\Delta_n} c_{n,i} = (-1)^n c_{n,0} c_{n,j}$$

$$P^{\Delta_j} c_{n,i} = P^{p^{j-1}} P^{\Delta_{j-1}} c_{n,i} \quad \text{for} \quad j > n .$$

Corollary 2.4:

a) $P^{p^k} c_{n,i} = 0$ if $p^k < p^{n-1}$ and $k \neq i-1$.

b) $P^{p^{i-1}} c_{n,i} = c_{n,i-1}$ for $i > 0$.

c) $P^{p^{n-1}} c_{n,i} = (-1)^{n-i-1} c_{n,i} c_{n,n-1}$

Proof of Proposition 2.1:

$$P^k f(X) = P^k(\prod_{v\in V}(X-v)) = \sum_I P^1(X-v_{i_1}) \cdots P^1(X-v_{i_k}) \prod_{V-\{v_1\}}(X-v)$$

$$= f(X) \sum_I (X^{p-1} - v_{i_1}^{p-1}) \cdots (X^{p-1} - v_{i_k}^{p-1}) \text{ by the Cartan formula.}$$

Hence $f(X)$ divides $P^k f(X)$ in $S(V)[X]$, and hence in $S(V)^{GL(V)}[X]$.

Also from the Cartan formula, if x_2 is 2-dimensional (respectively 1-dimensional for $p = 2$) ,

$$P^j x^{p^i} = 0 \quad \text{if } j \neq p^i$$

$$P^{p^i} x^{p^i} = x^{p^{i+1}}$$

Thus

$$P^k f(X) = \sum_{i=0}^{n-1} (-1)^{n-i} P^{k-p^{n-1}} c_{n,i} X^{p^{i+1}} + \sum_{i=0}^{n-1} (-1)^{n-i} (P^k c_{n,i}) X^{p^i}$$

for $k \neq 0$ or p^n . This has degree at most p^n in X so

$$P^k f(X) = -f(X) P^{k-p^{n-1}} c_{n,n-1} \quad \text{for } k \neq 0 \text{ or } p^n .$$

Hence, comparing coefficients we have

$$P^k c_{n,i} = P^{k-p^{i-1}} c_{n,i-1} + (-1)^{n-i-1} c_{n,i} P^{k-p^{n-1}} c_{n,n-1}$$

where P^j and $c_{n,j}$ are interpreted as 0 for $j < 0$.

Proof of Prop. 2.2: The P^{Δ_i} act as derivations, since they are primitive in A_p . Furthermore, $P^{\Delta_i} x_2 = x_2^{p^i}$. Hence $P^{\Delta_j} f(X) = f(X) \sum_{v\in V} (X^{p^j-1} - v^{p^j-1})$ in $S(V)[X]$. Thus $f(X)$ divides $P^{\Delta_j} f(X)$ in $S(V)^{GL(V)}[X]$ also. By computation, using the derivation property,

$$P^{\Delta_j} f(X) = \sum_{i=0}^{n-1} (-1)^{n-i} P^{\Delta_j} c_{n,i}\, X^{p^i} + (-1)^n c_{n,0} X^{p^j} .$$

For $j < n$, this has degree less than p^n in X, and hence is zero. For $j = n$, $P^{\Delta_n} f(X) = (-1)^n c_{n,0} f(X)$. Comparing coefficients, we obtain

$$P^{\Delta_j} c_{n,i} = 0 \quad \text{for } i \neq j, \text{ and } j < n$$

$$P^{\Delta_i} c_{n,i} = (-1)^{i+1} c_{n,0}$$

and

$$P^{\Delta_n} c_{n,i} = (-1)^n c_{n,0} c_{n,i} .$$

In general, $P^{\Delta_j} = P^{p^{j-1}} P^{\Delta_{j-1}} - P^{\Delta_{j-1}} P^{p^{j-1}}$, while on an unstable class of dimension less than $2p^{j-1}$, the latter term on the RHS is zero. Hence the last assertion is formal. The usual re-indexing changes for $p = 2$ are required.

Proof of the Corollaries:

Corollary 2.3 has been proved in the course of the proof of Proposition 2.1 and 2.2.

a) $k \neq i-1$ and $k < n-1$, then $m_j = p^k - p^{i-1} \dots - p^{i-j} \neq 0$ for $1 \le j \le i$. Hence $P^{p^k} c_{n,i} = P^{m_i} c_{n,0}$ from Corollary 2.3(a), since $m_i < p^{n-1}$.

b) This follows directly from Corollary 2.3(a).

c) $P^{p^{n-1}} c_{n,i} = (-1)^{n-i-1} c_{n,i}\, c_{n,n-1} + P^{p^{n-1} - p^{i-1}} c_{n,i-1}$. As in a) this last term is zero, since $s = p^{n-1} - p^{i-1} \dots -1$ is less than p^{n-1}.

III. Other Examples of Polynomial Invariants in Characteristic p

If W is a generalized reflection group over the complex numbers such that the reflection representation can be defined over $\mathbb{F}_q$, where q is prime to the order of W, then one can show (e.g. Chevalley [2], Shephard-Todd [15], or Serre [14]) that the invariant subalgebra is again polynomial.

In this section, examples with the order of W not prime to p are discussed. The example of the Weyl group of the compact Lie group F_4 for $p = 3$ (Toda [20]), shows that the reflection condition is not sufffficient to ensure polynomial invariants.

In the analysis of the Dickson invariants of Section I, there were three main steps:

1) Make a good guess for a set of n algebra generators for the invariants!

2) Show that $S(V)$ is integral over the subalgebra R generated by the elements guessed in 1). If so, R must be polynomial.

3) Show that the degree of the underlying field extension from R to $S(V)$ is the order of the group W. Then the field of fractions of R must be the fixed subfield of W, and since R is integrally closed, and $S(V)$ is integral over R,

$$R = S(V)^W$$

In addition to the case $W = GL(V)$, three other cases are amenable to this strategy:

a) W the symmetric group on some basis of V

b) W the special linear group $SL(V)$

c) W one of certain p-subgroups of $GL(V)$, including the case of the upper triangular matrices with diagonal 1's .

Theorem 3.1:

a) $S(V)^{\Sigma_n} = \mathbb{F}_q[\sigma_1,\ldots,\sigma_n]$ where Σ_n is the set of permutation matrices with respect to some fixed basis of V, and σ_i is the i-th elementary symmetric polynomial in the elements of this basis.

b) $S(V)^{SL(V)} = \mathbb{F}_q[u,c_{n,1},\ldots,c_{n,n-1}]$ where $\{c_{n,i}\}$ are the Dickson invariants, and u is the product of a non-zero element v_L chosen from each line L in V.

c) If W is a p-group contained in $GL(V)$ such that there is an ordered basis $\{x_i\}$ of V with $Wx_i - x_i$ contained in the span of $\langle x_1,\ldots,x_{i-1}\rangle$ and $\prod_i \mathrm{card}(Wx_i) = |W|$, then $S(V)^W$ is a polynomial algebra on generators

$$\{y_i \mid y = \prod_{\mathrm{orbit}\, W(x_i)} z\} .$$

A useful lemma for computing the field extension degree in the graded case comes from Adams-Wilkerson [1, corr.].

Lemma 3.2: If $R = \mathbb{F}_q[z_1,\dots,z_n]$ is contained in $S(V)$ such that $S(V)$ is integral over R, then the degree of the underlying field extension is $\Pi(\deg(z_i)/2)$, respectively $\Pi \deg(z_i)$ for $p=2$.

Of course, the counting of degrees of generators will work in the $GL(V)$ case also, but it is unnecessary there.

Proof of Theorem 3.1:

a) This is left to the reader. It could be viewed as an easy non-inductive proof of the Fundamental Theorem of Symmetric Functions.

b) $c_{n,0} = u^{q-1}$, and hence $S(V)$ is integral over R since it is integral over a subalgebra of R. The product of the degrees of $\{c_{n,i}, i > 0\}$ and that of u is the order of $SL(V)$, so Theorem 3.1(b) is established if u is invariant. Obviously, $wu = \varphi(w)u$ for some map $\varphi : GL(V) \to \mathbb{F}_q^*$, the units, since the lines in V are permuted by the action of $GL(V)$, up to scalar multiples, φ is a homomorphism, and factors through $GL(V)_{ab}$. Hence $\varphi = (\det)^a$. In particular $\varphi|SL(V) \equiv 1$, and u is invariant.

c) The degrees of the $\{y_i\}$ are correct by fiat, and the $\{y_i\}$ are obviously invariant. It remains only to show that $S(V)$ is integral over $R = \mathbb{F}_q[y_1,\dots,y_n]$. It is enough to show that $\{x_1,\dots,x_n\}$ are integral over R. Of course, $x_1 \in R$. Write W_i for the stablizer of x_i in W, and set

$$y_i = \prod_{W/W_i} wx_i = \Pi(x_i - (x_i - wx_i)) .$$

Then x_i is integral over $\mathbb{F}_q[x_1,\dots,x_{i-1},y_i]$. By the inductive hypothesis, $\{x_1,\dots,x_{i-1}\}$ are all integral over R, so x_i is integral over R. Note that this case includes the results of Mui [11] for the upper triangular groups (but does not treat his case of exterior generators).

Remarks:

1) The Steenrod operations on $S(V)^{SL(V)}$ are easily derived from the formulae of Section 2, using the fact that $u^{q-1} = -c_{n,0}$.

2) The regular representation of the cyclic group of order $p, p > 2$ gives an example of a p-group with invariants not a polynomial algebra.

IV. Dickson Invariants, the Dyer-Lashof Algebra, and the Lambda Algebra

I want to sketch an application of the Dickson invariants to a description of the Dyer-Lashof algebra and the lambda algebra. This description arises in the work of W. Singer [17] on the lambda algebra, but some parts of the description of the Dyer-Lashof algebra were known to Milgram, Madsen, and Priddy.

One essential ingredient is Milgram's observation in [Quillen, 12] that the characteristic classes of the regular representation of the $\mathbb{F}_p$-vector space V are exactly the Dickson invariants:

Proposition 4.1: (a) Let $\rho_n : V \to O(2^n)$ be the regular representation of the mod 2 vector space V. Then the Stiefel-Whitney classes of ρ_n are

$$w_{2^n-2^i} = c_{n,i}(H^1(V,F_2)) \quad \text{and } 0 \text{ otherwise.}$$

b) If $\rho_n : V \to U(p^n)$ is the regular representation of V, then the Chern classes of ρ_n are

$$c_{p^n-p^i} = \pm\, c_{n,i}(H^2(V)) \quad \text{and } 0 \text{ otherwise.}$$

This follows easily from the decomposition of the regular representation into a sum of one-dimensional representations, together with the identification of the Dickson invariants as elementary symmetric functions.

In the following, p will be 2 unless otherwise noted. Details about the structure of the Dyer-Lashof algebra and its action can be found in Madsen [8] or the book of Cohen-Lada-May [3]. I want to only summarize the work needed to give the connection with the rings of invariants. I thank F. Cohen and W. Dwyer for providing me with the background material on the Dyer-Lashof algebra.

Now the infinite loop space QS^0 can be constructed from the spaces $B\Sigma_n$. In any case, there are maps $\{B\Sigma_n \to (QS^0)_n\}$. The Dyer-Lashof algebra R acts on the homology of QS^0. From the structure of R, it follows that the sub-coalgebra $R[k]$ generated by monomials of length k is closed under the action of the Steenrod algebra defined by the Nishida relations. Madsen [8] calculated the linear dual of $R[k]$ and found it to be a rank k graded polynomial algebra on certain generators $\xi_{n,i}$ with an unstable action of the Steenrod algebra. It is easy to observe that $R[k]$ is isomorphic as an algebra over the Steenrod algebra to the Dickson invariants, and that the $\{c_{n,i}\}$ are just a re-indexing of Madsen's generators.

In fact, the duals can be identified by evaluating $R[k]$ on the homology class [1] in QS^0 and seeing that this is the image of the homology of

$$BV_k \to B\Sigma_{2^k} \to (QS^0)_{2^k} .$$

This follows from the fact

$$\text{image } H^*(B\Sigma_{2^k}) = (H^*(BV_k))^{GL(V)} ,$$

since the normalizer of V in Σ_{2^k} contains $GL(V)$.

Madsen also describes explicitly a "coproduct"

$$\tau_{k+j,k,j} : R[k+j]^* \to R[k]^* \otimes R[j]^* .$$

It is possible to give a description of this also in terms of invariant theory: it is just the "first-order approximation" to the natural inclusion

$$i_{k,j} : \text{invariants}(k+j) \to \text{invariants}(k) \otimes \text{invariants}(j) ,$$

i.e., restriction from $GL(V_{k+j})$ invariants to $GL(U_k) \times GL(W_j)$ invariants. One easily computes this on the $\{c_{n,i}\}$ by using properties of the polynomial $f_n(X)$ of Section I., truncating the W-invariant terms at height 1, and extending multipicatively. The formulae produced by this procedure agree with those of Madsen, so this has mnemonic value at the least. Presumably, the procedure could be justified by computing the deviation between the multiplication in R of elements x and y followed by evaluation on [1], and the composition product $x[1]y[1]$.

Thus far, this has just reproduced descriptions known in part to the experts. But this does shed some light on W. Singer's description of the dual to the lambda algebra. Curtis [4], for example, observed that the Dyer-Lashof algebra is a quotient DGA of the lambda algebra: namely the admissable monomials of negative excess generate the kernel. To describe the dual of Λ then, one can describe first the dual of R as the product of the $R[k]^*$ for all k , and then seek to adjoin enough generators to get up to Λ . Singer in effect shows that

$$\{c_{n,0}{}^{-s} c^I\}$$

can be adjoined to $R[n]^*$ in accordance with an excess rule on (I,s) so as to form $\Lambda[n]^*$.

Finally, Singer observes that the induced action of the Steenrod algebra on $\Lambda[n]^*$ is linear for the differential. On the other hand, Wellington [21] forces a formal action of the Steenrod algebra on the dual of $\Lambda[n]$ by the Nishida relations. This action is not linear over the differential, but there are interesting formulae relating it to the differential. Thus far, no applications have been suggested for either action, so the proper interpretation of these actions is still uncertain.

For p odd, some approximation of the above should be true. It is not true exactly, since the duals of the component coalgebras of the Dyer-Lashof algebra fail to be the entire ring of invariants

$$H^*(BV,F_p)^{GL(V)} .$$

However, if $c_{n,0}$ is inverted, this seems to be true. More details might appear in joint work with F. Cohen.

This work was supported by Wayne State University, the National Science Foundation, the Alfred P. Sloan Foundation, and the Institute for Advanced Studies of Hebrew University of Jerusalem.

References

1. Adams, J.F., Wilkerson, C.W., "Finite H-spaces and algebras over the Steenrod algebra", Annals of Math. 111(1980), 95-143, and "Corrections", Annals of Math. (1981).

2. Chevalley, C., Invariants of finite groups generated by reflections, Amer. J. Math. 77(1955), 778-782.

3. Cohen, F.R., Lada, T., and May, J.P., The Homology of Iterated Loop Spaces, Lecture Notes in Math., Vol. 533, Springer-Verlag, New York, 1976.

4. Curtis, E.B., The Dyer-Lashof algebra and the lambda-algebra, Illinois Journal of Math. 18(1975), 231-246.

5. Dickson, L.E., A fundamental system of invariants of the general modular linear group with a solution of the form problem. Trans. A.M.S. 12(1911), 75-98.

6. Lam, S.P., Cambridge thesis, 1982.

7. Li, Hu Hsiang, W.M. Singer, Resolutions of modules over the Steenrod Algebra and the classical theory of invariants, preprint 1982.

8. Madsen, Ib, On the action of the Dyer-Lashof algebra in H*(G), Pac. J. Math., 60(1975) 1, 235-275.

9. Mann, B.M., Milgram, R.J., "On the Chern classes of the regular representation of some finite groups", to appear in Proc. of the Edinburgh Math. Soc.

10. Milnor, J. The Steenrod algebra and its dual, Annals of Math. v. 67 no. 1, (1958), 150-171.

11. Mui, H., Modular invariant theory and the cohomology algebras of symmetric spaces, Journ. Fac. Sci. U. of Tokyo 22(1975), 319-369.

12. Quillen, D., The Mod 2 Cohomology Rings of Extra-special 2-groups and the Spinor Groups, Math. Ann. 194, 197-212(1971).

13. Rector, D.L., Noetherian Cohomology Rings and Finite Loop Spaces with Torsion, preprint, 1982.

14. Serre, J.-P., Groupes finis d'automorphismes d'anneaux locaux reguliers, Colloque d'Algebra Paris, ENSJF (1967), exp. 8, 11pp.

15. Shephard, G.C., Todd, J.A., Finite unitary reflections generated by reflections, Canadian Math. Journ. 6(1954), 274-304.

16. Singer, W.M., A new chain complex for the homology of the Steenrod algebra, Proc. Cambridge Phil. Soc. 90(1981), 279-292.

17. Singer, W.M., Invariant theory and the lambda algebra, to appear in Trans. A.M.S.

18. Smith, L., Switzer, R., On the non-realizability of rings of invariants, preprint, 1982.

19. Smith, L., Switzer, R., Variations on a theme of Adams and Wilkerson, preprint, 1982.

20. Toda, H., Cohomology mod 3 of the classifying space of the exceptional group F_4 , J. Math. Kyoto Univ., 13(1973), 97-115.

21. Wellington, R.J., The unstable Adams spectral sequence for free iterated loop spaces, Memoirs of the A.M.S. no. 258, v. 36, March, 1982.

DEPARTMENT OF MATHEMATICS
WAYNE STATE UNIVERSITY
DETROIT, MI 48202

Contemporary Mathematics
Volume 19, 1983

CURRENT PROBLEMS IN HOMOTOPY THEORY AND RELATED TOPICS

The problems below were presented at a session of the Northwestern University Homotopy Theory Conference. They are grouped under the name of the person who presented them. We have restricted our editorial efforts to supplying an occasional reference or update. Our thanks go to Bruce Williams for his assistance in assembling this list, and for the able job he did as organizer of the session itself.

John Jones

1. The Kervaire invariant problem. Does there exist an element $\theta_k \in \pi^s_{2^{k+1}-2}$ with Kervaire invariant one?

The Strong Kervaire invariant problem. Does there exist an element $\theta_k \in \pi^s_{2^{k+1}-2}$ of order 2, with Kervaire invariant one?

2. Browder brought up the problem of studying the Kervaire invariant in the cobordism theories corresponding to the groups $\pi^s_* \mathbb{R}P^\infty$, $\pi^s_* \mathbf{C}P^\infty$, $\pi^s_* \mathbf{H}P^\infty$, in the proceedings of the Stanford conference in 1976. This raises the following homotopy theoretic problem: Is $h_k^2 x_1 \in \mathrm{Ext}_A(H^* \mathbb{R}P^\infty, \mathbf{Z}/2)$ an infinite cycle in the Adams spectral sequence for $\mathbb{R}P^\infty$? Here $x_1 \in H_1 \mathbb{R}P^\infty$ is the non-zero class.

Ralph Cohen and Mark Mahowald have shown that $h_k^2 y_2 \in \mathrm{Ext}_A(H^* \mathbf{C}P^\infty, \mathbf{Z}/2)$ is an infinite cycle ($y_2 \in H_2\, \mathbf{C}P^\infty$ is the non-zero class). This corresponds to an immersion of an oriented manifold of dimension $2^{k+1}-2$ in $\mathbb{R}^{2^{k+1}}$ with non-zero Kervaire invariant. Can one explicitly construct this immersion? (This problem was also suggested by Ralph Cohen.)

3. (with Ralph Cohen) Is $h_1 h_k^2$ an infinite cycle in the Adams spectral sequence for S^0? This is related to the existence of $(2^{j+1}-2)$-dimensional manifolds (not necessarily orientable) immersed in $\mathbb{R}^{2^{j+1}-1}$ with Kervaire invariant 1.

0271-4132/83 $1.00 + $.25 per page

4. Another version of the Kervaire invariant problem is to ask if there is a stable map $f : S^{2^{k+1}-2} \to \mathbb{R}P^{\infty}$ detected by the functional cohomology operation $Sq_f^{2^k}$. Can one construct a stable map $f : S^{2^{k+1}-2} \to \mathbb{R}P_{\ell}^{\infty}$, for some ℓ, detected by $Sq_f^{2^k}$? Here $\mathbb{R}P_{\ell}^{\infty} = \mathbb{R}P^{\infty}/\mathbb{R}P^{\ell-1}$.

5. (Mahowald) Let $\beta_k \in \pi_{s(k)}^{s}$ be the k-th non-trivial generator of the image of J. What is the minimum integer $\ell \in \mathbb{Z}$ such that β_k factors stably as

$$\begin{array}{ccc} S^{2^{k+1}-2} & \longrightarrow & \mathbb{R}P_{\ell}^{2^{k+1}-2-s(k)} \\ & \beta_k \searrow \quad \swarrow \text{ collapse} & \\ & S^{2^{k+1}-2-s(k)} & \end{array}$$

(negative ℓ is allowed by James periodicity). This integer ℓ is > -1 ; $\ell = 0$ implies the existence of θ_k .

<u>Conjecture</u>: $\ell = 0$ if and only if θ_k exists. In any case if ℓ is the minimum such integer, the obstruction to a factorization with ℓ replaced by $\ell-1$ defines a coset of homotopy classes θ_k' ; if θ_k' is in the $2^{k+1}-2$ stem then θ_k' is made up of elements with Kervaire invariant one. This creates an infinite family of homotopy classes and translates the Kervaire invariant problem into the question, what is the stem of θ_k' ? There are other similar ways of constructing infinite families of homotopy classes which, if they are in the $2^{k+1}-2$ stem, have Kervaire invariant one.

6. If θ_k exists, does it desuspend to the $2^{k+1}-1-s(k)$ sphere? It is known that it can desuspend no further (see article in these proceedings).

7. Does the existence of θ_k imply the existence of θ_{k-1} ?

All these problems look hard, but one feels that some steps toward solutions would represent significant progress. A more accessible problem whose solution would be interesting is this:

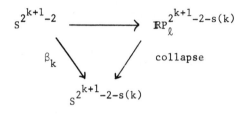

8. Compute the Kervaire invariants of Lie groups with their various framings, and also hypersurfaces with their various framings.

Mark Mahowald

1. Detailed calculations suggest that the following is true.

(a) $\pi_{j+n}(S^n) \to \pi_j^s(S^0)$ is onto if $n \geq 2^i$ and $j < 2^{i+1}-2$.

(b) $\pi_{2^{i+1}-2+n}(S^n) \to \pi^s_{2^{i+1}-2}(S^0)$ has cokernel $\mathbf{Z}/2$ generated by the Kervaire invariant class if $n < 2^{i+1}-\phi(i)$ where $\phi(4a+b) = 8a+2^b$, $0 \leq b \leq 3$, and 0 if $n \geq 2^{i+1}-\phi(i)$.

2. There is some evidence for the following which could be used to prove 1.

$$\mathrm{Ext}_{A_i}^{s,t}(\tilde{H}^*(P^{2^i}),\mathbf{Z}/2) \cong \mathrm{Ext}_A^{s+1,t+1}(\mathbf{Z}/2,\mathbf{Z}/2) \quad \text{for} \quad 0 < t-s < 2^{i+1}-2.$$

Here A_i is the subalgebra of the Steenrod algebra A generated by $\{Sq^n : n < 2^{i+1}\}$. Note that there is no obvious map giving the isomorphism.

3. Let $M(2)$ be the $\mathbf{Z}/4$ Moore space. There is a map $\Sigma^8 M(2) \to M(2)$ representing v_2^4. Let $M(2,4)$ be the cofiber. There is a choice of $M(2,4)$ which is a ring spectrum. There is a self map $\Sigma^{48}M(2,4) \to M(2,4)$ representing v_2^8 (unpublished, but see Davis and Mahowald, Amer. J. Math. 103(1981)). Let $M(2,4,8)$ be the cofiber.

(a) Show $M(2,4,8)$ can be chosen to be a ring spectrum.

(b) Construct the rest of the ring spectra $M(2,4,8,\ldots,2^i)$ with self map representing $v_i^{2^{i+1}}$ defining $M(2,4,\ldots,2^{i+1})$ as the cofiber. A map represents $v_i^{2^{i+1}}$ if it is detected in BP by $v_i^{2^{i+1}}$.

(c) As an easier first step construct these things in Ext_A using $v_i^{2^{i+1}}$ as a map of filtration $2^{i+1}(-\varepsilon)$.

4. Call a 2-local spectrum E with unit $\eta : S^0 \to E$ <u>atomic</u> if (i) η generates $\pi_0(E)$ over $\mathbf{Z}_{(2)}$ and (ii) any map $f : E \to E$ such that $f\eta = \eta$ is a homotopy equivalence. Consider the transfer map $\mathbb{R}P^\infty = P \to S^0$. This gives

a map $\overline{\lambda} : E \wedge P \to E$. Using the Snaith splitting, we have a map $s : QS^0 \to \Omega^\infty(P \wedge E)$. Let $A = \ker \overline{\lambda}_* \cap \operatorname{im} s_* \subseteq \pi_*(P \wedge E)$.

Conjecture. If E is atomic, A finitely generated over $\mathbf{Z}_{(2)}$, and $\overline{\lambda}_*$ onto in positive dimensions, then E is $S^0_{(2)}$, $K(\mathbf{Z}_{(2)})$, or $K(\mathbf{Z}/2^i)$ for some i.

Don Davis

1. Conjecture. All nondesuspensions of stunted complex projective spaces are detected by the Adams operations in K-theory; i.e. if $b(X)$ is the dimension of the bottom cell of a maximal desuspension of X, then

$$b(\mathbb{C}P_n^{n+k}) = \max\left\{ j - \nu_p\binom{n}{j} : p \text{ prime}, \ j < \left[\frac{k}{p-1}\right]\right\}$$

or zero if this maximum is negative.

Evidence: True when $k = 2$. True when $b = 0$. Adams operations in BP do no better.

2. Let bo be the spectrum representing real connective K-theory. Give an interpretation of the equation

$$P_{-\infty}^{\infty} \wedge bo \simeq \bigvee_{i \in \mathbf{Z}} \Sigma^{4i-1} K\hat{\mathbf{Z}}_2 .$$

Evidence: (a) $\lim\limits_{\overleftarrow{N}} \pi_i(P_{-N} \wedge bo) = \hat{\mathbf{Z}}_2 \quad i \equiv 3(r)$

$= 0 \quad$ otherwise.

(b) If $M = \text{cofiber } (\lambda : P_1 \to S^0)$ then $M \quad bo \simeq \bigvee_{i \geq 0} \Sigma^{4i} K\mathbf{Z}_{(2)}$, and this can be used to give maps $P_{-4N+1} \wedge bo \to S^{-4N} \wedge bo$ whose cofibers are $\bigvee_{i \geq -N} \Sigma^{4i} K\mathbf{Z}_{(2)}$.

(This would be useful in connection with the obstruction theory studied by Davis, Gitler, Iberkleid, and Mahowald in Bol. Soc. Mat. Mex. 24(1979).)

Ralph Cohen

1. Immersions. Stiefel-Whitney classes suggest that in dimensions $n \not\equiv 4k+1$ orientable n-manifolds immerse in $\mathbb{R}^{2n-\alpha(n)-1}$. When is this true?

Remarks. (a) It's known to be false if n is a power of 2 or if $n = 2^r+2^s$ with $r > s > 1$. ($\mathbb{C}P[n/2]$'s are counterexamples.) (b) It's true if $n = 2^n+2$ (Mahowald and Peterson). (c) Weaker question: when is it true up to cobordism? (It's true up to cobordism if $n = 4k+2$. - M. deLemos.)

2. Find a description of the ideal of relations among characteristic classes (of various sorts) of n-manifolds in terms of monomials in these classes.

3. Problems related to Mahowald's η_j family.

We call an element in $\pi^S_{2^j}(S^0)_{(2)}$ "η_j" if it is represented by h_1h_j $\mathrm{Ext}^2_A(\mathbb{Z}_2,\mathbb{Z}_2)$ in the Adams spectral sequence. Mahowald proved that such elements exist (Topology 16(1977)).

It's known that $2\eta_j$ is represented by $h_1^2h_{j-1}^2$, which for $j > 4$ is non-zero.

Questions. (a) Is there a family $\{\eta_j : j \geq 1\}$ with the property that $2\eta_j = \eta_{j-1}^2$?

(b) Is there an η_j such that $4\eta_j = 0$?

Remark. The odd primary analogues to the above conjectures are all known to be true.

F. P. Peterson

Relate the structure of $H^*(BO)$ as a free right A-module to the algebra structure. To be more precise, there is a coalgebra map $\theta : N \to H^*(MO)$ which gives an A-basis; i.e. θ can be extended to a coalgebra, left A-module isomorphism $\bar{\theta} : A \otimes N \to H^*(MO)$. Apply the Thom isomorphism to get $\theta' : N \otimes A \to H^*(BO)$. This is an isomorphism of coalgebras and right A-modules.

Problem. Find the algebra structure on $N \otimes A$ so that θ' is a map of algebras.

It might be easier to work in homology. Let $\hat{\theta} = (\theta')^* : H_*(BO) \to N_* \otimes A_*$; $\hat{\theta}$ is a map of polynomial algebras. Thus we need only find the coproduct on each generator. Brown and Peterson (1974 Northwestern Conference Proceedings, Mex. Math. Soc.) proved that there exists θ such that $1 \otimes \xi_k$ is primitive for each k. This problem is related to Ralph Cohen's problem 2 in this list.

Stewart Priddy

1. Compute $H^*(BGL_n(\mathbb{F}_p);\mathbb{F}_p)$. Is this cohomology detected by elementary abelian p-groups? One is led to conjecture e.g. that $\tilde{H}^n(BGL_{2n}(\mathbb{F}_2);\mathbb{F}_2) \neq 0$ and that this is the first non-zero group (true for $n = 1,2,3$).

2. (with Haynes Miller) Does there exist an infinite loop fibration (at 2) $BU \xrightarrow{f} G/O \to IBO$, where f comes from the stable Adams conjecture and IBO is the fiber of the unit map $QS^0 \to BO \times \mathbf{Z}$. See On G and the stable Adams conjecture, Geometric Applications of Homotopy Theory II, Springer Lecture Notes in Math. 658, 1977. Tørnehave has shown that this is correct as transfer spaces, and (using the Segal conjecture for D_8) that there is essentially a unique such transfer-commuting fibration.

Fred Cohen

1. Barratt's conjecture. If the suspension order of the identity of $\Sigma^2 B$ is p^k, then $p^{k+1}\pi_*\Sigma^2 B = 0$.

2. Moore's conjecture. Let X be a simply-connected finite complex. Consider ${}_p\pi_*X$, the p-primary component of π_*X. Then for each prime p there exists a k such that $p^k\,{}_p\pi_*X = 0$ if and only if $\pi_*X \otimes \mathbb{Q}$ is a finite dimensional vector space.

3. Find an analogue of the Kahn-Priddy theorem for S^{2n+1}, $n < \infty$, which gives the Kahn-Priddy theorem in the limit. For instance, the natural map $\Sigma^{2n+1}\mathbb{R}P^{2n} \to S^{2n+1}$ might be onto the 2-torsion. It can't split, but the odd primary analogue might.

4. Suppose that Y is a simply-connected complex of finite type (not necessarily finite). Conjecture: If $Y \wedge S^j$ has an exponent at p bounded by p^k for all $j \geq 1$, then Y has an exponent at p (which is perhaps bounded by p^{k+1}).

A related problem concerns the order of compositions for double suspensions. Assume that $\alpha : B \quad S^2 \to X$ has order p^k in $[B \wedge S^2, X]$ and let $f : A \wedge S^2 \to B \wedge S^2$ be any map. Notice that the composition αf has finite order if A is a finite complex.

5. Show that αf has order at most p^{k+1} in $[A \wedge S^2, X]$.

6. Let $2 : \Omega X \to \Omega X$ denote the H-space squaring map and let $(\Omega X)\{2\}$ denote the homotopy theoretic fibre of 2. Let $S^3\langle 3\rangle$ denote the 3-connective cover of S^3. Is $\Omega^2 S^3\langle 3\rangle$ a 2-local retract of $(\Omega^2 S^5)\{2\}$? It suffices to exhibit a map $\hat{\eta} : \Omega^2 S^5 \to \Omega S^3$ which induces an epimorphism on π_3 and which has order two in $[\Omega^2 S^5, \Omega S^3]$. Notice that $\Omega^2\bar{\eta} : \Omega^2 S^5 \to \Omega^2 BS^3$, where $\bar{\eta}$ is a choice of generator of $\pi_5 BS^3 \cong \mathbf{Z}/2$, has order 4.

Desuspension questions are intimately related to the above problems.

7. Since $S^k \wedge (\mathbb{R}P^{n+k-1}/\mathbb{R}P^{n-1})$ is a stable retract of $\Omega^n S^{n+k}$, one might ask to find the smallest integer L such that $S^{k+L} \wedge (\mathbb{R}P^{n+k-1}/\mathbb{R}P^{n-1})$ is a retract of $\Sigma^L \Omega^n S^{n+k}$. We conjecture that L equals the embedding dimension of $\mathbb{R}P^{n-1} \times \mathbb{R}^{n+1}$.

8. Localize at an odd prime p. Is the p-adic part of $\Omega^{2n} S^{2n+1}$, D_p, a retract of $\Omega^{2n} S^{2n+1}$ after suspending $(2n+1)$ times? $(2n)$ times?

9. Localize at an odd prime. What is the structure of $\Sigma^2\Omega^2\Sigma^2 X$?

J. Neisendorfer

1. The natural filtration of U by $U(n)$ leads to a spectral sequence with $E^1_{p,q} = \pi_{p+q}(S^{2p-1})$. Bott's computation of $\pi_k(U)$ indicates the presence of many differentials. Describe these differentials without computing E^1.

2. Localize at an odd prime p. There exists a map $\theta : \Omega^2 S^3\langle 3\rangle \to \Omega^{2p} S^3\langle 3\rangle$ where $S^3\langle 3\rangle$ is the 3-connected cover of S^3 and $\theta_*(\alpha_i) = \alpha_{i+1}$ for the homotopy classes of Adams and Toda. Compute the homotopy groups of the fibre of θ.

3. Do there exist loop spaces such that the Lie algebra of its mod-p homotopy groups, $p > 3$ (with modifications in case $p = 3$), is isomorphic to a free Lie algebra on at least two generators?

John Harper

Incompressibility problems.

Def. A map $f : X \to Y$ is incompressible if there is no factorization up to homotopy through a finite complex,

$$\begin{array}{ccc} X & \xrightarrow{\quad f \quad} & Y \\ & \searrow \quad \nearrow & \\ & K & \end{array}$$

Steve Weingram gave this definition (Ann. of Math. 93(1971)) and proved:

Theorem. If G is a finitely generated abelian group, then any essential map $f : \Omega S^{2n+1} \to K(G,2n)$ is incompressible.

Questions. Are all essential maps

(a) $f : \Omega S^{n+1} \to K(Z/2,n) \qquad n \neq 2^i-1$

(b) $g : \Omega^2 S^{2n+1} \to K(Z/p,2n-1) \qquad n \neq 1$, p prime

incompressible?

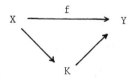

(c) In general $f : X \to Y$ induces $\hat{f} : JX \to SP^{\infty}Y$, where J is the James construction and SP^{∞} is the infinite symmetric product. What can be said about incompressibility of $\hat{f}$?

Remarks. S^1, S^3, S^7, E_8, Spin (2^i) provide compressions for the exceptional cases in (a). The Cohen-Moore-Neisendorfer map $\pi : \Omega^2 S^{m+1} \to S^{m-1}$ provides a compression for (b) if Z/p is replaced by Z/p^k, $k \geq 2$ and $p \mid \deg g$. Examples of this sort are also known for $p = 2$.

C. A. McGibbon

1. Conjecture. Let X be a 1-connected finite CW-complex. If $\tilde{H}_*(X; Z/p) \neq 0$, then there is p-torsion in the kernel of the double suspension

$$E^2 : \pi_* X \to \pi_* \Omega^2 \Sigma^2 X .$$

2. Problem: Localize all spaces at an odd prime p. Let B be a classifying space for S^{2n-1} where n divides $p-1$. It is known [Topology 20(1981), pp.109-118] that the infinite suspension $E^{\infty} : B \to Q(B)$ sends all torsion in $\pi_* B$ to zero. What, if anything, does this imply about the image of $\pi_* S^{2n-1}$ in $\pi_*^s S^0$?

Discussion: In the paper just cited, it was conjectured that for $n \leq p-1$, $E^{\infty} \pi_* S^{2n-1}$ ImJ, mod decomposables. This optimism was based on the observation that K-theory faithfully detects the stable orders of the attaching maps of B and on known results about the spheres of origin for the β and γ families.

Unfortunately, this conjecture is false. The first known counterexample was pointed out by Oka and is due to Toda, [Proc. Japan Acad. 43(1967), pp.839-842]. It is an element ϕ of order p^2, of stable dimension $2(p^2+p)(p-1)-3$, which desuspends to S^7 for $p > 3$.

2. Question. Let $BT^n = \mathbb{C}P^{\infty} \times \dots \times \mathbb{C}P^{\infty}$ (n-factors, $n \geq 2$), and let $Q(X) = \lim\limits_{\to} \Omega^n \Sigma^n X$. Is $Q(BT^n) \simeq Z_n \times F_n$, where Z_n is a space whose integral

homology groups and homotopy groups are torsion free, and F_n is a space with finite homotopy groups?

Jim Stasheff

Without using Quillen's work, find a natural d.g. Lie algebra $L(X)$ for any (simply connected) space X such that $H(L(X))$ is naturally equivalent to $\pi_*(\Omega X) \otimes \mathbb{Q}$ as Lie algebras and/or $\mathscr{C}(L(X))$ is naturally equivalent to $C_*(X)$ as d.g. coalgebras. Here $\mathscr{C}$ denotes the d.g. Koszul construction of Quillen, Ann. of Math. 90(1969).

Peter May

1. Compute the ordinary $RO(G)$-graded cohomology of a point. That is, compute $H_G^*(pt; A)$ where A is the Burnside ring coefficient system and $*$ runs through the real representation ring. This theory was introduced by Lewis, May, and McClure and also Waner (for G finite) and is the equivariant analog of ordinary integral cohomology.

2. Generalize McClure's computation of $K_*(QX; \mathbb{Z}_p)$ to a computation of $K_*(\Omega^n\Sigma^n X; \mathbb{Z}_p)$ for $n < \infty$.

3. Prove the odd primary Whitehead conjecture concerning the symmetric powers of the sphere spectrum, generalizing N. J. Kuhn's solution in the case $p = 2$.

4. Understand in some conceptual way for which G-spectra k_G the natural homomorphism $\pi_*(k_G) \to k_G^{-*}(EG)$ induces an isomorphism upon completion with respect to the Burnside ring. This holds for periodic equivariant K-theory and for equivariant cohomotopy, by Carlsson's validation of the Segal conjecture. It fails for ordinary cohomology with Burnside ring coefficients.

5. Prove that BP is an H_∞ ring spectrum; McClure has reduced this to an explicit computation in $MU_*(\mathbf{CP}^\infty)$. Try to compute $BP_*(QX; \mathbb{Z}_p)$ or $K(n)_*(QX)$, where $K(n)$ is the nth Morava K-theory.

6. Find a useful combinatorial model for $(\Omega^n\Sigma^n X)_{(p)}$. The existing global model, for $\Omega^n\Sigma^n X$, suffers from the defect that attempts to use it to prove intrinsically local phenomena by use of combinatorial maps are doomed to failure.

7. Generalize Triantafillou's algebraization of equivariant rational homotopy theory from finite to compact Lie groups.

8. Determine algebraically the rational equivariant stable category (as constructed for compact Lie groups by Lewis and May).

Doug Ravenel

1. Generalize the results of Mahowald's "The image of J in the EHP sequence" (Ann. of Math. 116(1982)) to odd primes. They should be easier to state and prove here than in the case $p = 2$.

2. Let $V(n)$ be Toda's complex with $H^*(V(n); \mathbb{Z}_p)) = E(Q_0,Q_1,\ldots,Q_n)$. Toda showed $V(2)$ exists for $p = 5$. Show that $V(3)$ does or does not exist. Calculations show there is a nontrivial obstruction group. A positive solution would imply that the elements γ_t exist in π_*^s and that $\beta_1^{19} \neq 0$ for $p = 5$.

3. Spectra X, Y are Bousfield equivalent $(X \sim Y)$ if $X \wedge E = pt \Longleftrightarrow Y \wedge E = pt$ for all spectra E (Bousfield, Comm. Math. Helv. 54(1979)).

(a) Let $MU\{2n\}$ denote the Thom spectrum associated with the (2n-1)-connected cover of BU (e.g. $MU\{2\} = MU$, $MU\{4\} = MSU$). Show $MU\{2n\} \sim MU$.

(b) Let $MO\{n\}$ be similarly defined, e.g. $MO\{4\} = MSpin$. Show $MO\{n\}_{(2)} \sim BP\langle m-1\rangle$ where m is the number of nontrivial homotopy groups of BO in dimensions less than n. Here $BP\langle m-1\rangle$ is as in Johnson and Wilson, Topology 12(1973).

(c) Say $X > Y$ if $X \quad E = pt \Rightarrow Y \wedge E = pt$ for any spectrum E. Let $MU\langle k\rangle$ and $MO\langle k\rangle$ denote the Thom spectra associated with the k-fold Whitney sum of the universal bundle. Show $MO\langle 2k\rangle \sim MU\langle k\rangle$ and $MU\langle j\rangle > MU\langle k\rangle$ if $k \mid j$.

4. With notation as in 3, let $L(n) = Sp^{p^n}(S^0)/Sp^{p^{n-1}}(S^0)$ and let $K(n)$ denote Morava K-theory at the prime p, $L(0) = S_{(p)} = S_{(p)}$, $K(0) = S\mathbb{Q}$. Show $L(n) \sim K(n) \vee L(n+1)$.

Haynes Miller

1. (with Stewart Priddy) Construct a discrete model for MU or allied spectra using formal group laws. For example, the automorphism group of a formal group law F acts on any reasonable category of "formal representations" of F. Perhaps the associated spectrum represents an oriented theory with formal group F. This bears on the nilpotence conjectures advertised by Ravenel, as well as May's problem 5 in this list.

2. Carry out the program of Adams-Wilkerson [Ann. of Math. 111(1980)] in BP. If it can be done, this should overcome the separability restrictions in their work, and eliminate many more polynomial algebras as possible cohomology algebras.

3. (with Mark Mahowald) The arithmetic properties of a prime number p seem not to enter into the description of the p-component of homotopy groups of spheres. We make the following conjecture, which we state only for stable homotopy and which we intentionally leave rather vague.

Conjecture. For each $n \geq 1$ there is a number $N(n)$ such that for all primes $p \geq N(n)$, the p-component of $\pi_*^s(S^0)$ can be described in dimensions up to p^n by expressions polynomial in p.

By this we mean, for example, that the algebra generators occur in collections of cardinality which is a polynomial in p, with dimensions depending polynomially on p, and the relations have coefficients which are polynomial in p. All higher-order structure appears to exhibit this behavior as well. Thus, one may perhaps speak of "homotopy-theory at the infinite prime."

C. H. Giffen

1. Exhibit specific quaternionic line bundles η_q over BSp(1) with characteristic classes

$$q^2(\text{gen}) \in H^4 BSp(1)$$

$q = 3, 5, 7, \ldots$

(These exist, thanks to Sullivan.)

2. If G is a finite group and

$$H_1G \oplus H_2G \oplus H_3G = 0$$

then $G = \{e\}$.

Ian Hambleton

Homology of 2-groups.

1. (with Taylor) Given $k \geq 0$, describe the smallest class of finite 2-groups, $\mathcal{G}_k$, such that

$$\bigoplus_{\substack{\tau \leq \pi \\ \tau \in \mathcal{G}_k}} H_k(\tau; \mathbb{Z}/2) \to H_k(\pi; \mathbb{Z}/2) \text{ is onto for all finite 2-groups } \pi.$$

Surgery on Closed Manifolds.

References: (a) Cappell and Shaneson, A counterexample on the oozing problem for closed manifolds, Algebraic Topology: Aarhus, 1978, Springer Lecture Notes in Math. 763.

(b) Hambleton, Projective surgery obstructions on closed manifolds, preprint.

(c) Taylor and Williams, Surgery Spaces: Formulae and structure, Algebraic Topology, Waterloo, 1978, Springer Lecture Notes in Math. 741.

(d) Taylor and Williams, Surgery on closed manifolds, preprint.

If we localize at (2), then the surgery obstruction map σ_* factors as follows (see (c) and (d)):

$$\begin{array}{ccc}
[M^n, G/TOP] & \xrightarrow{\quad\sigma_*\quad} & L_n^h(\mathbf{Z}\pi, w) \\
\Big\downarrow c & & \Big\downarrow + \\
\bigoplus_i (H_{n-4i}(M, \mathbf{Z}_{(2)}) \oplus H_{n-4i-2}(M, \mathbf{Z}/2)) & \xrightarrow{\sum_i (\mathscr{I}_{n-4i} + \theta_{n-4i-2})} & \bigoplus_i L_{n-4i}^h(\mathbf{Z}\pi, w)
\end{array}$$

Henceforth, we shall assume π is finite.

2. Does $\theta_k = 0$ for $k > 3$?

<u>Conjecture</u>: (a) Given k, there exists π such that $\theta_k \neq 0$.

(b) (with Taylor and Williams) Given π, there exists $N = N(\pi)$ such that $\theta_k = 0$ for $k > N(\pi)$.

(Problem 1 was motivated by Problem 2.)

3. <u>Conjecture</u>: If $\theta_k(x) \neq 0$, then there exist 1- or 2-dimensional classes $\alpha_1, \ldots, \alpha_t$, in $H^*(\pi, \mathbf{Z}/2)$ such that

$$x \cap (\alpha_1 \cup \ldots \cup \alpha_t) = 1 .$$

<u>J. P. E. Hodgson</u>

Let K be a simply-connected finite CW complex which admits a Poincaré Complex n-thickening whose boundary is a sphere. Thus there is a map $f : S^{n-1} \to K$ whose mapping cone satisfies Poincaré duality. When can one find a map $g : S^{n+1} \to SK$ whose mapping cone satisfies Poincaré duality, so that SK admits a Poincaré Complex thickening whose boundary is a sphere?

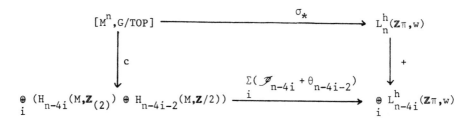

Ib Madsen

1. If G is a cyclic group of odd order then $\Omega_*^G(X) \to KO_*^G(X) \otimes \mathbb{Z}[1/2]$ is onto and

$$\Omega_*^G(X) \otimes_{\Omega_*^G(pt)} KO_*^G(pt) \otimes \mathbb{Z}[1/2] \simeq KO_*^G(X) \otimes \mathbb{Z}[1/2] .$$

(i) What happens at 2 ?

(ii) What happens for odd order groups in general?

(iii) Even order groups?

2. What is the homotopy type at the prime 2 of F/PL in the equivariant setting?

3. G odd order group. Does stable G-transversality hold in the locally linear PL category?

Reinhard Schultz

1. Manifolds with no group actions. Is there a closed simply connected manifold with no effective finite group actions?

Remarks. (a) There are many examples of closed nonsimply connected manifolds (Conner-Raymond, Bloomberg, Schultz). From these one can construct examples of compact bounded simply connected manifolds.

(b) The work of Petrie contains substantial information relevant to the case of homotopy complex projective spaces.

(c) Weaker question: For smooth manifolds, can one find an example with no smooth actions?

(d) Another variant (P. Löffler). Given M^n closed and simply connected, is M rationally homotopy equivalent to a manifold with a circle action? (Löffler and M. Raußen have studied this problem for M sufficiently highly connected).

2. Mann's Conjecture. Let Σ^n be an exotic n-sphere, and suppose that T^r acts smoothly and effectively on Σ. Then $r \leq (n + 3)/4$.

Remarks. (a) Such an estimate is essentially best possible if Σ^n bounds a parallelizable manifold.

(b) One can ask a similar question for actions of $(\mathbb{Z}/p)^r$, assuming $p\Sigma^n = 0$ in Θ_n.

3. Existence of circle actions. If M^n is a manifold admitting $\mathbf{Z}/p$ actions for all primes p that are sufficiently large, does M^n admit a topological circle action?

Remarks. (a) The connected sum of T^n with an exotic sphere is a smooth counterexample.

(b) One can ask a similar question regarding $T^r \times (\mathbf{Z}/p)$ and T^{r+1} actions.

4. Universal coverings of aspherical manifolds. Let M be a closed aspherical manifold with one-ended universal covering. Is the fundamental group at infinity necessarily trivial or uncountable?

Remark. Work of M. Davis shows that the fundamental group need not be trivial.

5. Aspherical manifolds associated to reflection groups (examples of M. Davis). Let M be such a manifold, and let π denote its fundamental group. Under what circumstances can a homomorphism $G \to \mathrm{Out}(\pi)$ [G finite] be realized by a (continuous or smooth) group action on M^n ? For manifolds of the form $\Gamma H/K$ (K maximal compact, Γ a suitable discrete subgroup), results of Conner, Raymond, and K. B. Lee suggest that some obvious algebraic necessary conditions may always be sufficient.

6. Converses to the P. A. Smith Theorem. The work of Lowell Jones and subsequent work by other mathematicians has given a good overall understanding

of the fixed point sets for smooth or PL Z/p actions on disks or spheres (p prime). Roughly speaking, the necessary conditions imposed by Smith theory are also the principal sufficient conditions. Here are some similar questions in slightly more general settings:

(a) Topological actions. The local version of Smith theory implies that the fixed point set of a topological Z/p action on an n-disk or an n-sphere must be a Z/p homology manifold in an appropriate sense. What further conditions suffice for a space to be the fixed point set of a topological Z/p action on an n-disk or an n-sphere?

Remark. Let L be a (possibly infinite) set of primes not containing p. One can find a fixed point set whose Cech cohomology contains q-torsion if and only if q belongs to L (see Bredon, in Proc. Conf. on Transf. Gps. (New Orleans, 1967), page 58, Thm. (3.4)).

(b) Actions of larger groups. For the sake of simplicity, let $G = Z/p^2$. If G acts nontrivially but not semifreely, then the fixed point set of $H = Z/p < G$ is G-invariant and inherits a $G/H = Z/p$ action. Which smooth Z/p homology disks or spheres with Z/p actions can be equivariantly realized as $FIX(Z/p = H)$ for some smooth G-action on a disk or a sphere?

Remarks. Similar questions may be asked for Z/p^r actions and the fixed point set of Z/p^{r-s}. For semifree actions on disks, the work of Jones et al should apply directly, but the nonsemifree case is much less transparent.

(c) Realizing nonunitary homology spheres as fixed point sets. (Proposed verbally by R. Strong) Let p be an odd prime, and let H be a closed smooth Z/p homology sphere. If H is the fixed point set of a smooth Z/p action on S^n, then H inherits a unitary structure. Can H be realized as the fixed point set of a smooth action on some Z/p homology sphere even if H admits no unitary structure?

Remark. The ambient manifold cannot admit a unitary structure either for such a choice of H, and its homology must contain 2-torsion. So the strongest result possible would be that H is the fixed point set of a smooth Z/p action on a simply connected Z[1/2]-homology sphere.

Sylvan Cappell

Topological Equivalence of Matrices (Reference: Cappell and Shaneson).

1. Conjecture: In dim. ≤ 8, topological equivalence implies linear equivalence for orthogonal matrices.

Evidence: True for $n \leq 5$ ($n \leq 4$ is easy, but $n = 5$ is hard).
False for $n = 9$.
True for many classes of matrices in dim. ≤ 8.

2. Question: What does the topological classification of non-diagonalizable matrices (with eigenvalues of modulus 1) look like?

Frank Quinn

1. Sullivan's Conjecture: The following fixed point conjecture is a generalization of Sullivan's conjecture that $[RP^{\infty}, K] = *$, for K a finite CW complex.

Suppose $\mathbb{Z}/p$ acts on a complex X, which is homotopy equivalent to a finite complex. Then $\mathbb{Z}/p$ acts on the mapping space $\mathrm{Map}(E\,\mathbb{Z}/p, X)$ by conjugation.

Conjecture: This action is $\mathbb{Z}_{(p)}$ dominate.
Roughly, an action on K is $\mathbb{Z}_{(p)}$ dominate if $H_*(K^{\mathbb{Z}/p}; \mathbb{Z}_{(p)})$ is finitely generated and the chain complex $C_*(K, K^{\mathbb{Z}/p}; \mathbb{Z}_{(p)})$ is $\mathbb{Z}_{(p)}[\mathbb{Z}/p]$ equivalent to a finitely generated projective $\mathbb{Z}_{(p)}[\mathbb{Z}/p]$ complex. See Quinn, Finite nilpotent group actions on finite complexes, Geometric Applications of Homotopy Theory, I, Springer Lecture Notes in Math. 657, 1977.

Sullivan's original question concerned $\mathbf{Z}/2$ actions and was part of an effort to recover a real algebraic variety as the fixed points of the involution on the étale homotopy type of the complex variety. The conjecture that $[RP^{\infty},K] = *$ is π_0 of the fixed point conjecture in the case $\mathbf{Z}/2$ acts trivially on K.

(H. Miller has proven that $\pi_* \operatorname{Map}_*(B\mathbf{Z}/p,X) = 0$ for a class of connected spaces including spheres.)

2. Surgery Theory: The objective is to find a variant of the L groups which permits systematic computation.

Define: $L_n^0(\mathbf{Z}\pi)$ to be the usual L_n^p-group (forms on projective $\mathbf{Z}\pi$-modules), and L_n^{-i} to be the inverse limit of $L_{n+i}^0(\mathbf{Z}[\pi \times \mathbf{Z}^i])$ under homomorphisms induced by $\mathbf{Z}^i \to \mathbf{Z}^i$ represented by diagonal matrices with positive entries.

Define: $L_n^{-\infty}(\mathbf{Z}\pi) = \lim\limits_{i\to\infty} L_n^{-i}(\mathbf{Z}\pi)$, under maps induces by

$$\times S^1 : L_{n+i}^0(\mathbf{Z}[\pi \times \mathbf{Z}^i]) \to L_{n+i+1}^0(\mathbf{Z}[\pi \times \mathbf{Z}^{i+1}]).$$

Problem: Compute $L_*^{-\infty}(\mathbf{Z})$.

Facts: 1. There is a Rothenberg type sequence

$$\cdots \to L_n^{-i}(\mathbf{Z}\pi) \to L_n^{-i-1}(\mathbf{Z}\pi) \to H^n(\mathbf{Z}/2;K_{-i}(\mathbf{Z}\pi)) \to \cdots$$

2. There is an exact product formula

$$L_n^{-\infty}(\mathbf{Z}[\pi \times \mathbf{Z}]) \to L_n^{-\infty}(\mathbf{Z}[\pi]) \oplus L_{n-1}^{-\infty}(\mathbf{Z}[\pi])$$

3. For finite π, since $K_{-i} = 0$ for $-i < -1$ (Carter), these stabilize:

$$L_n^{-i}(\mathbf{Z}[\pi]) = L_n^{-\infty}(\mathbf{Z}) \quad \text{for} \quad -i < -1.$$

Probably Facts:

4. $K_{-i}(\mathbf{Z}\pi) = 0$ for sufficiently large i, if π is poly- (cyclic or finite). This follows from Farrell-Hsiang and Quinn, Ends of Maps II, Inv. math. 68(1982). Thus L_*^{-i} also stabilizes for these groups.

5. $L_*^{-\infty}(\mathbf{Z}\pi)$ for π poly- (cyclic or finite) can be computed in terms of $L_*^{-\infty}$ of finite subgroups of π (using methods of Farrell-Hsiang, and Yamasahi, whose thesis does this mod 2-torsion for $L_*^s(\mathbf{Z}\pi)$).

Jack Morava

Problem: To understand Adams' operations as "complex multiplications." Specifically, is there a nice p-adic topological K-theory for valuated, perhaps not locally compact, topological p-fields [cf. J. P. Serre, Bull. Math. Soc. France 89(1961)] such as the cyclotomic closure of $\mathbf{Q}_p$?

One hopes for an answer that looks like $K_{\mathbf{C}}^*(-) \otimes W(\text{residue field})$, the Galois group

$$\mathrm{Gal}(\mathbf{Q}_p^{ab}/\mathbf{Q}_p) = \hat{\mathbf{Z}} \times \hat{\mathbf{Z}}_p^*$$

acting through its second factor by the p-adic Adams operations on $K_{\mathbf{C}}^*(-) \otimes \hat{\mathbf{Z}}_p$. For some further speculations, see [J. Morava, Asterisque 63(1979)].

ABCDEFGHIJ-AMS-89876543